nullity T	The nullity of the linear transformation T, §3.3
O	The zero matrix, §2.1
$\mathbf{0}$	The zero vector, §1.1, §1.2, §8.1
P	The vector space of polynomials, §8.1
P_n	The vector space of polynomials of degree $\leq n$, §8.2
$\overline{PQ}$	The directed line segment from the point P to the point Q, §1.1
$\overrightarrow{PQ}$	The vector representation with head at the point Q and tail at the point P, §1.1
rank A	The rank of the matrix A, §3.3
rank T	The rank of the linear transformation T, §3.3
R	The set of real numbers, §1.2
R^n	The set of n-tuples of real numbers, §1.2
$R(T)$	The range of the linear transformation T, §2.5
Span of S	The span of the set S, §1.3, §8.2
$S^\perp$	The orthogonal complement of the set S, §7.1.
T, U	Linear transformations, §2.4
T_0	The zero transformation, §2.4
tr A	The trace of the matrix A, §2.1
$[T]_S$	The matrix representation of the linear transformation T relative to the ordered basis S, §4.2, §8.3
$[T]_S^{S'}$	The matrix representation of the linear transformation T relative to the ordered bases S and S', §8.3
$\mathbf{x}, \mathbf{y}, \mathbf{z}$	Vectors in R^n, §1.1, §1.2
$[\mathbf{x}]_S$	The coordinate vector of the vector $\mathbf{x}$ relative to the ordered basis S, §4.1, §8.3
$\|\mathbf{x}\|$	The norm of the vector $\mathbf{x}$, §1.6, §8.4
$\mathbf{x} \cdot \mathbf{y}$	The dot product of the vectors $\mathbf{x}$ and $\mathbf{y}$, §1.6
$\mathbf{x} \times \mathbf{y}$	The cross product of the vectors $\mathbf{x}$ and $\mathbf{y}$, §1.6
$(\mathbf{x}, \mathbf{y})$	The inner product of the vectors $\mathbf{x}$ and $\mathbf{y}$, §8.4
V, W	Vector spaces or subspaces, §1.3, §8.1, §8.2

Introduction
to Linear Algebra
with Applications

Introduction to Linear Algebra with Applications

Stephen H. Friedberg
Arnold J. Insel

Illinois State University

Prentice-Hall, Englewood Cliffs, N.J.

Library of Congress Cataloging in Publication Data

Friedberg, Stephen H.
  Introduction to linear algebra with applications.

  Includes index.
  1. Algebras, Linear. I. Insel, Arnold J.
II. Title.
QA184.F74 1986   512'.5   85-19126
ISBN 0-13-485988-X

Editorial/production supervision and
    interior design: Fay Ahuja
Cover design: Lundgren Graphics, Ltd.
Cover painting: ''Horizon Structuré'' by Yvaral
Manufacturing buyer: John B. Hall

Printed in the United States of America

10 9 8 7 6 5 4 3 2 1

ISBN 0-13-485988-X 01

Prentice-Hall International (UK) Limited, *London*
Prentice-Hall of Australia Pty. Limited, *Sydney*
Prentice-Hall Canada Inc., *Toronto*
Prentice-Hall Hispanoamericana, S.A., *Mexico*
Prentice-Hall of India Private Limited, *New Delhi*
Prentice-Hall of Japan, Inc., *Tokyo*
Prentice-Hall of Southeast Asia Pte. Ltd., *Singapore*
Editora Prentice-Hall do Brasil, Ltda., *Rio de Janeiro*
Whitehall Books Limited, *Wellington, New Zealand*

To my wife, Ruth Ann,
and our children, Rachel, Jessica, and Jeremy

S.H.F.

To my wife, Barbara,
and our children, Tom and Sara

A.J.I.

Contents

9 Numerical Methods _____ 371

10 Introduction to Linear Programming _____ 409

References _____ 426

Appendix A _____ 428

Appendix B _____ 432

Selected Answers for the Exercises _____ 443

Index _____ 453

Preface

In recent years the demand for skills in linear algebra has increased at a rapid pace. In addition to engineering, physics, and economics majors, enrollment in linear algebra classes now includes those majoring in computer science, operations research, psychology, and biology. Of course, the use of linear algebra in multivariate statistics has made the subject a natural requirement for those students wishing to pursue a career in other quantitative areas. The presence of so diverse an audience as well as the importance of applications to mathematics majors has influenced our decision to include a wide variety of significant applications. Rather than postpone these applications to the end of the text, we have made an effort to introduce them as the necessary background is developed.

The material is aimed at the sophomore–junior student. There is no use of calculus until the introduction of function spaces in Chapter 8. The core topics include: the vector space properties and Euclidean n-space, systems of linear equations, matrices, linear transformations, determinants, and eigenvalues and eigenvectors. In addition, orthogonal diagonalization, abstract vector spaces, numerical methods for solving systems and for finding eigenvalues and eigenvectors, and linear programming are covered.

Approach: Through experience, we have discovered that students without a background in abstract mathematics find abstract vector spaces very difficult to comprehend. We have therefore decided to introduce the notion of vector spaces and their properties through the familiar Euclidean n-spaces. What we believe is *unique* about this text is the *spiral* approach that is employed. Instead of introducing all the material about systems, followed by all the material on vector spaces, followed by all the material on linear transformations and matrices, we have made an effort to ease the student gently into the elementary properties of all these topics before the advanced properties are introduced.

Not only are the concepts easier to assimilate, but the student sees the interplay between the various structures early in the development. For example, after linear combinations are introduced, systems are presented as a means of discovering if a given vector is a linear combination of other vectirs. Matrices are then needed as a convenient notation for using Gaussian elimination. Once matrices are introduced, their arithmetic is motivated by an example of a Markov process. Matrix multiplication provides the motivation for the definition of a linear transformation and its matrix representation. Now it makes sense to talk about the subspaces associated with a linear transformation, namely, the null space and range. All of this is done *before* the more difficult concepts of linear independence, basis, and dimension are introduced.

We are firm believers in the use of **geometry** to motivate as well as clarify many of the topics of linear algebra. For example, the application of the elementary properties of vector arithmetic are used to show that the diagonals of a rhombus bisect one another. The determination of the null space, range, eigenspaces, and other characteristics of the geometric transformations—rotations, projections, reflections, and shear transformations—permeate the exercises and examples.

The **microcomputer** as a tool is ever present in this text. Throughout the exercise sets, problems preceded by (*) are to be done on a microcomputer. A diskette, which includes a number of useful programs for working these problems (see Appendix B), is provided to adopters of the text. In addition, the idea of operation counts (for example, in the solution to systems or in the computation of determinants) is used as a measure of computational efficiency.

Chapter 1 introduces the elementary properties of vector operations; norm and dot product in Euclidean 2-, 3-, and n-space; linear combinations and subspaces; systems of linear equations and Gaussian elimination; and matrices as a tool for manipulating systems. Included are examples illustrating the power of linear algebra to establish results in geometry.

Chapter 2 introduces the elementary properties of matrix arithmetic. Five significant applications of matrices are given to illustrate their power. Left-multiplication by a matrix introduces the concept of a linear transformation. The null space and range of a linear transformation are given as examples of subspaces and as tools for studying additional properties of systems.

Chapter 3 moves the reader into the more sophisticated concepts of linear independence, basis, rank, and dimension. These ideas are united with the earlier concepts of system, null space, and range to provide deeper insights. For example, the dimension theorem is proved and used to establish information about solution spaces. The properties of matrix inverses are established. The construction of the inverse of a matrix is accomplished with the introduction of elementary matrices. The results are then applied to the Leontief closed and open economic models. Finally, additional theoretical results about systems are proved.

Chapter 4 introduces change of coordinate vectors and various matrix representations of a linear transformation. This material forms the necessary background for diagonalization.

Chapter 5 begins with the definition of the determinant of a 2×2 matrix and its properties. These properties are then extended to determinants of $n \times n$ matrices. The basic approach is to establish the properties for elementary matrices and then use the fact, established earlier, that every invertible matrix is a product of elementary matrices. The chapter is concluded with Cramer's rule, the classical adjoint, and an application to cryptography.

Chapter 6 introduces perhaps the most important concept in linear algebra—*diagonalization*. The basic results concerning eigenvalues, eigenvectors, and necessary and sufficient conditions for the diagonalization of a matrix or linear transformation are established. The results are then applied to solving difference equations, examining the long-term behavior of Markov chains, and solving systems of differential equations.

Chapter 7 is concerned with the properties of orthogonal sets. The Gram–Schmidt process is used to prove that every subspace of R^n has an orthonormal basis. Diagonalization of a symmetric matrix by an orthogonal matrix is used to transform a quadratic expression into standard form. Orthogonal projections are carefully developed and applied to derive the least-squares formula. Finally, rotation matrices are applied to computer graphics.

Chapter 8 utilizes most of the previous material to develop abstract vector spaces. The emphasis, however, is on function spaces. The differential operator is given as a special case of a linear transformation on an infinite-dimensional vector space. The chapter concludes with the elementary properties of inner product spaces.

Chapter 9 provides a somewhat more extensive treatment of numerical methods than most texts at this level. For solving systems we describe the direct methods—pivoting, the *LU*- and Cholesky decompositions, and the iterative methods—the Jacobi and the Gauss–Seidel methods. For estimating eigenvalues, Gerschgorin's theorem is proved. Finally, for estimating eigenvalues and eigenvectors, the power and inverse power methods as well as the deflation method are developed and illustrated. Exercises are given which use the programs listed in Appendix B and are included on the diskette. The sections of this chapter are independent and thus may be covered in any order or omitted.

Chapter 10 introduces linear programming. The development is divided into two parts. The first part introduces some terminology and the graphical method. The second part describes the simplex method.

Complex numbers: On the advice of several reviewers, we decided to include examples where complex numbers play an important role, especially when eigenvalues are considered. For example, harmonic motion is illustrated as a case when complex eigenvalues are particularly important. Also, Gerschgorin's theorem takes on a different geometrical interpretation if complex eigenvalues are considered. Appendix A is included to establish the elementary properties of complex numbers. No interruption of the flow of the material will occur if complex numbers are entirely deleted from the presentation.

Notation: The results of problems preceded by a dagger (†) are used in subsequent sections.

Dependencies within the material: All the applications are independent of one another and of the rest of the material. In Section 5.3, the classical adjoint is used only in the application to cryptography. The following flowchart indicates the other dependencies within the material.

Numbering: We have numbered our theorems, lemmas, and corollaries by chapter

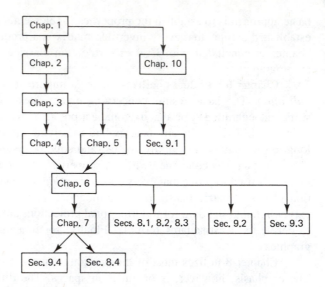

and section. For example, Theorem 3.2.1 is the first theorem in the second section of Chapter 3. Tables and figures are each numbered similarly. Examples are numbered sequentially within each section.

We would like to express our thanks to the following people who carefully reviewed our manuscript and made many helpful suggestions: Carl C. Cowen, Mathematics Department, Purdue University; Vincent Giambalvo, Mathematics Department, University of Connecticut; Terry L. Herdman, Mathematics Department, Virginia Polytechnic Institute and State University; Kenneth Kalmanson, Computer Science Department, Montclair State College; Robert H. Lohman, Mathematical Sciences Department, Clemson University.

In addition, special thanks go to the staff of Prentice-Hall, especially Fay Ahuja, production editor, and Robert Sickles, mathematics editor. We are also grateful to Ken Hirschel for proofreading the galleys.

Finally, we are indebted to our colleagues and students for their insightful comments and their encouragement.

SHF
AJI

Introduction to Linear Algebra with Applications

1

Introduction to the Vector Space Properties of R^n and Systems of Linear Equations

In this chapter we introduce the *vector space* properties of Euclidean *n*-space. Although the notion of *vector* was introduced in the nineteenth century primarily by the Irish mathematician W. R. Hamilton, its usefulness in real-world applications, particularly in physics, was not recognized until the twentieth century. More recently, the important properties of vectors have been exploited in such areas as the social and biological sciences, as well as in statistics.

We begin this chapter with the most important Euclidean space, namely, the plane. It is in this space that we can take advantage of our ability to visualize many of the geometric properties of vectors. With this introduction in place, it will be easier to understand the properties of vectors in Euclidean *n*-space.

1.1 VECTORS IN R^2

Certain physical quantities, such as length, area, speed, and mass, can be described by a single number or magnitude. However, others, such as velocity and force, require both a magnitude and a direction.

Example 1   The *velocity* of an object is described by giving its speed and direction. The *speed*, a nonnegative number, is the magnitude of the velocity. Suppose that **x** denotes the velocity of an object traveling 20 miles per hour in a northeast direction. Geometrically, **x** can be represented as an arrow, that is, a directed line segment of length 20 which points northeast (see Figure 1.1.1).

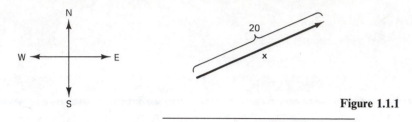

Figure 1.1.1

Example 2   Let **f** be the force exerted by the weight of a 200-pound man. The magnitude of this force is 200 and the direction is toward the center of the earth. Geometrically, this force can be represented as an arrow of length 200 which points from the man to the earth's center (see Figure 1.1.2).

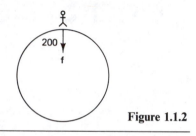

Figure 1.1.2

Any quantity determined by both a magnitude and a direction is called a ***vector***. The velocity and force described in Examples 1 and 2, respectively, are each examples of vectors. Acceleration is another example of a vector. Two vectors are considered to be ***equal*** if they have the same direction and magnitude.

A vector may be represented geometrically as an arrow or directed line segment. The line segment is pointing in the direction of the vector, and its length is the magnitude of the vector. The tail and head of this arrow are called the ***initial*** and ***terminal points***, respectively. The arrow whose initial point is A and whose terminal point is B will be denoted by $\overline{AB}$.

Referring to Figure 1.1.3, we see that $\overline{AB}$ and $\overline{CD}$ have the same magnitude and direction. Therefore, they represent the same vector. The fact that different directed line segments may represent the same vector is very useful. For example, the vector that represents the velocity of a car traveling northeast at 30 miles per hour on a particular road also represents the velocity of that car if it were traveling at the same speed and direction on another road. The arrow $\overline{PQ}$ is pointing in the same direction

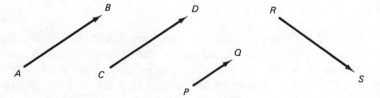

Figure 1.1.3

as $\overline{AB}$ but it is shorter and so it does not represent the same vector as $\overline{AB}$. The arrow $\overline{RS}$ has the same length as $\overline{AB}$ but it is pointing in a different direction. Therefore, it does not represent the same vector as $\overline{AB}$.

Two directed line segments such as $\overline{AB}$ and $\overline{CD}$ in Figure 1.1.3 which determine the same vector, that is, which have the same direction and are of equal length, are called *equivalent*. The vector they both represent is denoted by either $\overrightarrow{AB}$ or $\overrightarrow{CD}$. We may write

$$\overrightarrow{AB} = \overrightarrow{CD}$$

Vectors in the Plane

It is useful to consider vectors in the context of a rectangular coordinate system. Consider the familiar coordinate system of the xy-plane. We identify a point in the plane with its coordinates. $O = (0, 0)$ is called the *origin* of the system. Consider a vector $\overrightarrow{PQ}$ in the xy-plane. Suppose that $P = (a, b)$ and $Q = (c, d)$ (see Figure 1.1.4). Setting $C = (c - a, d - b)$, $A = (c - a, 0)$, and $R = (c, b)$, we see that the (right) triangles OCA and PQR have legs of equal length and so are congruent. Therefore, the vectors $\overrightarrow{OC}$ and $\overrightarrow{PQ}$ have the same magnitude. Since $\overrightarrow{OA}$ and $\overrightarrow{PR}$ have the same direction, it follows that $\overrightarrow{OC}$ and $\overrightarrow{PQ}$ also have the same direction and hence are equal. Summarizing, we have:

For any points $P = (a, b)$ and $Q = (c, d)$ in the plane, the vector $\overrightarrow{PQ}$ can be identified with the vector $\overrightarrow{OC}$, where $C = (c - a, d - b)$.

We call $c - a$ the *first component* (or *x-component*) and $d - b$ the *second component* (or *y-component*) of the vector $\overrightarrow{PQ}$. In this manner, any vector $\mathbf{x}$ in the xy-plane can be associated with a unique ordered pair of real numbers, namely, its components. In our notation above, if $\mathbf{x} = \overrightarrow{PQ}$, we may also write $\mathbf{x} = (c - a, b - d)$. Conversely, given any point $D = (p, q)$ in the xy-plane, the vector $\mathbf{y} = \overrightarrow{OD}$ has components p and q and we may write $\mathbf{y} = (p, q)$. Thus, there is a one-to-one correspondence between vectors in the plane and ordered pairs. We also denote the set of all vectors in the plane by R^2.

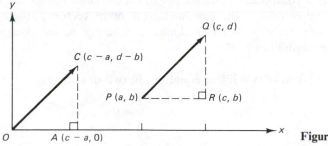

Figure 1.1.4

Some authors use a different notation for points than for vectors. Usually, the context in which an ordered pair is used will make it clear whether we mean a point or a vector.

From this definition of components, we see that two vectors are equal if and only if their corresponding components are equal.

Example 3   Let $A = (2, 3)$, $B = (5, 2)$, $C = (3, 8)$, and $D = (6, 7)$. We wish to show that $\overrightarrow{AB} = \overrightarrow{CD}$. Clearly, the first component of $\overrightarrow{AB}$ is $5 - 2 = 3$, and the second component is $2 - 3 = -1$. Similarly, the first and second components of $\overrightarrow{CD}$ are 3 and -1, respectively. Therefore, $\overrightarrow{AB} = \overrightarrow{CD}$.

Consider again the velocity vector **x** of Example 1. If the positive direction of the y-axis points north and the initial point of the directed line segment representing **x** is placed at the origin, it can be shown using trigonometry that the terminal point of **x** is the ordered pair $(10\sqrt{2}, 10\sqrt{2})$. Thus, these coordinates are the components of the vector **x** [see Figure 1.1.5(a)], that is, $\mathbf{x} = (10\sqrt{2}, 10\sqrt{2})$.

The components of the force **f** of Example 2 depend on the orientation of the coordinate system. If the coordinate system is oriented so that the man is at the origin and the negative direction of the y-axis points toward the earth's center, the first component of **f** is 0 and the second is -200 [see Figure 1.1.5(b)]. So $\mathbf{f} = (0, -200)$.

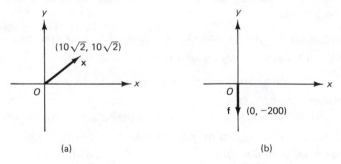

(a)   (b)   **Figure 1.1.5**

Vectors have uses other than to represent physical quantities such as velocity and force. For example, a vector might be constructed for each person in a group in which the first component gives the height (in feet) and the second component gives the weight (in pounds). In this case, we have as many vectors as there are people in the group. In the study of Markov chains in Chapter 6, the components represent particular probabilities.

Example 4   Consider a system of two linear equations in two unknowns:

$$2x + y = 0$$
$$3x - 2y = 7$$

We shall learn how to solve such systems later in this chapter. It is easy to verify that $x = 1$ and $y = -2$ is a solution to this system. Rather than expressing this solution by two equations, we can make use of one (vector) equation using the vector $\mathbf{y} = (x, y) = (1, -2)$. This notation will be of considerable use in Chapter 2.

Vectors would be of little interest if they were used merely to describe properties, as in the examples above. They are useful because there is an arithmetic defined on them. This arithmetic allows us to make certain calculations in order to draw new conclusions about vector properties.

Suppose that a train is traveling at a speed of 60 miles per hour while a passenger is walking toward the front of the train at 2 miles per hour. Then the velocity of the passenger relative to the ground is $60 + 2 = 62$. Similarly, if the passenger is walking toward the back of the train at 2 miles per hour, her velocity relative to the train is -2 miles per hour, and her velocity relative to the ground is $60 + (-2) = 58$ miles per hour. It is clear that the operation of addition can be used to combine velocities.

Now we consider a comparable situation in the xy-plane. Suppose that a pilot aims an airplane in a particular direction and that the airplane moves at a certain "airspeed," that is, the speed relative to the surrounding air. Then the airplane has a velocity $\mathbf{x}$ relative to the air around it. Now suppose that at the same time the air or wind is moving relative to the ground with a given speed and direction. Suppose that the motion of the wind has a velocity $\mathbf{y}$. We combine the velocity of the airplane relative to the air and the velocity of the wind relative to the ground to determine the velocity of the airplane relative to the ground. Call this combined velocity $\mathbf{z}$. Let us see how to compute $\mathbf{z}$. Imposing a coordinate system whose positive x-axis points east and whose positive y-axis points north, suppose that $\mathbf{x} = (a, b)$ and $\mathbf{y} = (c, d)$. Then a and c represent the horizontal (east-west) components of $\mathbf{x}$ and $\mathbf{y}$, respectively. As with the example of the passenger on the train, it follows from physics that these components can be added to yield the horizontal component of the combined velocity, namely, $a + c$. Similarly, $b + d$ is the vertical (north-south) component of the combined velocity. Therefore, $\mathbf{z} = (a + c, b + d)$.

The operation of combining two vectors by adding the corresponding components is called *vector addition*.

Definition   Let $\mathbf{x} = (a, b)$ and $\mathbf{y} = (c, d)$ be two vectors in R^2. We define the *sum* of $\mathbf{x}$ and $\mathbf{y}$, denoted by $\mathbf{x} + \mathbf{y}$, to be the vector in R^2 defined by

$$\mathbf{x} + \mathbf{y} = (a + c, b + d)$$

For example, if $\mathbf{x} = (1, -2)$ and $\mathbf{y} = (2, 4)$, then $\mathbf{x} + \mathbf{y} = (3, 2)$.

There is also a geometric interpretation of vector addition which is quite useful. Using facts about congruent (right) triangles (see Figure 1.1.6), it follows that if the sides of a parallelogram are determined by the vectors $\mathbf{x}$ and $\mathbf{y}$, the sum $\mathbf{z} = \mathbf{x} + \mathbf{y}$ is given by the diagonal of the parallelogram. This result is called the *parallelogram law of vector addition*.

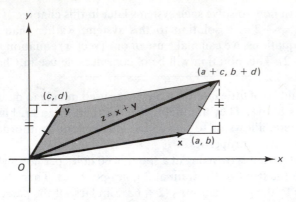

Figure 1.1.6

It is interesting to consider the parallelogram law in the context of our airplane example. Recall that the velocity (relative to the air) of the airplane is given by **x**, the wind velocity is given by **y**, and the resulting velocity (relative to the ground) of the airplane is given by **z**. If **x** and **y** are in approximately the same direction, then the airplane's speed (relative to the ground), which is given by the magnitude of **z**, is increased [see Figure 1.1.7(a)]. On the other hand, if **x** and **y** are approximately in opposite directions, then the magnitude of **z** is decreased [see Figure 1.1.7(b)]. Of course, this agrees with our intuition about the effect of wind on the velocity of an object.

The same physical interpretation that we have given to the sum of velocity vectors may also be applied to the sum of vectors that represent forces [see Exercise 2(b) and (d)].

Vector addition may be easily extended to sums of more than two vectors. For example, if **x**, **y**, and **u** are three vectors, we may define their sum **z** as $\mathbf{z} = (\mathbf{x} + \mathbf{y}) + \mathbf{u}$. It is easy to see that **z** may also be determined by $\mathbf{x} + (\mathbf{y} + \mathbf{u})$. For example, if $\mathbf{x} = (a, b)$, $\mathbf{y} = (c, d)$, and $\mathbf{u} = (e, f)$, then

$$\mathbf{z} = (\mathbf{x} + \mathbf{y}) + \mathbf{u}$$
$$= (a + c, b + d) + (e, f)$$
$$= ((a + c) + e, (b + d) + f)$$
$$= (a + (c + e), b + (d + f))$$
$$= (a, b) + (c + e, d + f)$$
$$= \mathbf{x} + (\mathbf{y} + \mathbf{u})$$

The result above is the ***associative law for vector addition***. It allows us to omit parentheses and write $\mathbf{z} = \mathbf{x} + \mathbf{y} + \mathbf{u}$. The short proof of this fact illustrates a technique which will be employed again. Namely, the associative law of addition of real numbers has been used to establish the associative law of vector addition. The fact

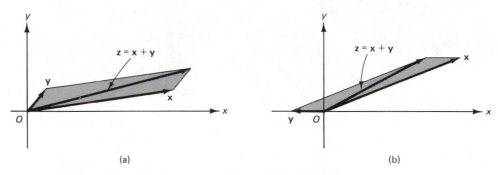

Figure 1.1.7

that many of the arithmetic properties of real numbers extend to vectors will be established later.

If we reconsider the parallelogram law for constructing $z = x + y$, it is easy to see that z may be determined by placing the tail of y at the head of x [see Figure 1.1.8(a)] and then draw z as indicated. This method is called ***tail-to-head addition***. It provides an easy view of vector addition of more than two vectors. Figure 1.1.8(b) also provides a geometric verification of the associative law of vector addition.

It is often desirable to compare vectors of different magnitudes but pointing in the same or opposite directions. For example, comparing vectors $x = (2, 1)$ and $y = (4, 2)$, we see (see Figure 1.1.9) that x and y have the same direction but have

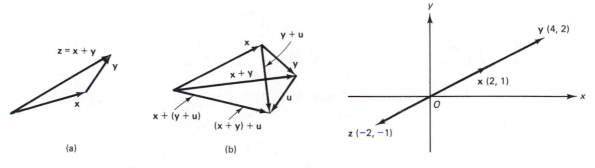

Figure 1.1.8   Figure 1.1.9

different magnitudes. In fact, the magnitude of y is twice that of x. Furthermore, the vector $z = (-2, -1)$ has the opposite direction of x. We may describe these relationships more easily by making use of the following definition.

Definition   Let $x = (a, b)$ be a vector in R^2. Let t be any real number. We define the ***product*** of t and x, denoted tx, by

$$tx = (ta, tb)$$

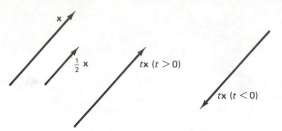

Figure 1.1.10

For example, if $t = 3$ and $\mathbf{x} = (2, 0)$, then $t\mathbf{x} = 3(2, 0) = (6, 0)$.

Because real numbers are often called *scalars*, we refer to this operation between scalars and vectors as *scalar multiplication*. It should be noted that in many texts, the word *scalar* refers to either a complex or real number. Most of the theory we develop will hold for complex numbers as well as real numbers.

Let $\mathbf{x}$ and $\mathbf{y}$ be vectors in R^2. If $\mathbf{y} = t\mathbf{x}$ for some scalar t, we say that $\mathbf{y}$ is *parallel* to $\mathbf{x}$. If $t > 0$, we say that $\mathbf{y}$ is pointing in the *same direction* as $\mathbf{x}$. If $t < 0$, we say that $\mathbf{y}$ is pointing in the *opposite direction* of $\mathbf{x}$. We may visualize these relationships as in Figure 1.1.10.

The fact that the vector $t\mathbf{x}$ is parallel to the vector $\mathbf{x}$ and $|t|$ times as long follows from theorems in geometry concerning similar triangles.

If $t = 0$, then $\mathbf{y} = t\mathbf{x} = (0, 0)$. This vector is called the *zero vector* and is denoted by $\mathbf{0}$. The zero vector, which has no direction, is by definition parallel to every vector.

We may use both of the vector operations to introduce two new terms. Given vectors $\mathbf{x}$ and $\mathbf{y}$, we define the *inverse* of $\mathbf{x}$, denoted by $-\mathbf{x}$, as $(-1)\mathbf{x}$ and the *difference* of $\mathbf{x}$ and $\mathbf{y}$, denoted by $\mathbf{x} - \mathbf{y}$, as $\mathbf{x} + (-\mathbf{y})$.

So if $\mathbf{x} = (2, 3)$ and $\mathbf{y} = (5, -8)$, we have that $-\mathbf{x} = (-2, -3)$ and $\mathbf{x} - \mathbf{y} = (-3, 11)$. In general, if $\mathbf{x} = (a, b)$ and $\mathbf{y} = (c, d)$, it is easy to see that $-\mathbf{y} = (-c, -d)$ and $\mathbf{x} - \mathbf{y} = (a - c, b - d)$.

To see the vector difference $\mathbf{x} - \mathbf{y}$ geometrically, let $\mathbf{x} = \overrightarrow{PQ}$ and $\mathbf{y} = \overrightarrow{PR}$, where $P = (a, b)$, $Q = (c, d)$, and $R = (e, f)$. Then

$$\begin{aligned}
\mathbf{x} - \mathbf{y} &= \overrightarrow{PQ} - \overrightarrow{PR} \\
&= (c - a, d - b) - (e - a, f - b) \\
&= (c - e, d - f) \\
&= \overrightarrow{RQ}
\end{aligned}$$

As a consequence, if $\mathbf{x}$ and $\mathbf{y}$ are vectors represented by arrows with the same initial point, $\mathbf{x} - \mathbf{y}$ is the vector represented by the arrow connecting the head of $\mathbf{y}$ to the head of $\mathbf{x}$ (see Figure 1.1.11).

Although these geometric representations of the vector operations are useful when dealing with one, two, or three vectors, the situation becomes too complex when

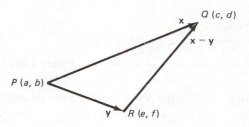

Figure 1.1.11

there are many vectors involved. Just as we showed that the associative law of vector addition follows from the similar property for scalars, we will see in Theorem 1.1.1 that most of the familiar arithmetic properties of scalars also hold for vectors.

Theorem 1.1.1   *Let* **x**, **y**, *and* **z** *be vectors in* R^2, *and let* s *and* t *be any scalars. Then*

(a) $\mathbf{x} + \mathbf{y} = \mathbf{y} + \mathbf{x}$ *(commutative law of vector addition)*

(b) $(\mathbf{x} + \mathbf{y}) + \mathbf{z} = \mathbf{x} + (\mathbf{y} + \mathbf{z})$ *(associative law of vector addition)*

(c) $\mathbf{x} + \mathbf{0} = \mathbf{x}$ **(0** *is the identity)*

(d) $\mathbf{x} + (-\mathbf{x}) = \mathbf{0}$ $(-\mathbf{x}$ *is the inverse of* **x***)*

(e) $1\mathbf{x} = \mathbf{x}$

(f) $(st)\mathbf{x} = s(t\mathbf{x})$

(g) $s(\mathbf{x} + \mathbf{y}) = s\mathbf{x} + s\mathbf{y}$ *(scalar multiplication distributes over vector addition)*

(h) $(s + t)\mathbf{x} = s\mathbf{x} + t\mathbf{x}$

Proof

We already have proved part (b). We will prove part (f) and leave the other parts as exercises.

Let $\mathbf{x} = (a, b)$. Then

$$(st)\mathbf{x} = (st)(a, b)$$

$$= (sta, stb)$$

$$= s(ta, tb)$$

$$= s(t(a, b))$$

$$= s(t\mathbf{x}) \qquad \blacksquare$$

Geometric Application

We will use the arithmetic properties of vector addition and scalar multiplication to prove:

> *The line segment joining the midpoints of two sides of a triangle is parallel to and one-half of the length of the third side.*

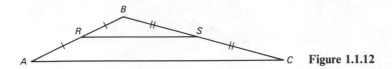

Figure 1.1.12

Consider a triangle ABC in the plane. Let R be the midpoint of the segment $\overline{AB}$, and let S be the midpoint of the segment $\overline{BC}$ (see Figure 1.1.12). Notice that because R is the midpoint of $\overline{AB}$, it follows that $\overrightarrow{RB} = \frac{1}{2}\overrightarrow{AB}$. Similarly, $\overrightarrow{BS} = \frac{1}{2}\overrightarrow{BC}$.

Using tail-to-head addition, we have

$$\overrightarrow{RS} = \overrightarrow{RB} + \overrightarrow{BS}$$
$$= \tfrac{1}{2}\overrightarrow{AB} + \tfrac{1}{2}\overrightarrow{BC}$$
$$= \tfrac{1}{2}(\overrightarrow{AB} + \overrightarrow{BC}) \qquad \text{by Theorem 1.1.1(g)}$$
$$= \tfrac{1}{2}\overrightarrow{AC}$$

Thus, $\overrightarrow{RS}$ is parallel to and one-half the length of $\overrightarrow{AC}$.

Exercises

1. For points A, B, C, and D, determine whether or not $\overrightarrow{AB}$ equals $\overrightarrow{CD}$. In each case sketch the vectors.
   (a) $A = (1, 2)$, $B = (3, 4)$, $C = (1, 5)$, and $D = (3, 7)$.
   (b) $A = (1, -1)$, $B = (2, 1)$, $C = (4, 2)$, and $D = (5, 1)$.
   (c) $A = (-3, 2)$, $B = (2, 1)$, $C = (0, 0)$, and $D = (5, -1)$.
   (d) $A = (2, 1)$, $B = (1, 4)$, $C = (1, -3)$, and $D = (0, 0)$.
   (e) $A = (-2, 3)$, $B = (2, 3)$, $C = (0, 1)$, and $D = (0, 0)$.

2. For each of the following descriptions of a vector $\mathbf{x}$, determine the components of $\mathbf{x}$. In each case assume an xy-coordinate system with the positive direction of the x-axis pointing east and the positive direction of the y-axis pointing to the north.
   (a) The velocity of a car heading northwest at a speed of 50 miles per hour.
   (b) The force of a rocket thruster propelling a toy car southward. The force has a magnitude of 2 pounds.

   (c) The velocity of a boat relative to the shore of a river. The river is flowing southward at 4 miles per hour. The boat is pointing to the east and is moving at a speed of 3 miles per hour relative to the water.
   (d) A combination of two separate forces acting on an airplane. The first force, due to the engine, is pushing the plane to the west with a magnitude of 2000 pounds. The second force, due to a wind, is pushing the plane to the north with a magnitude of 200 pounds.
   (e) The vector determined by the directed line segment with initial point $(1, 3)$ and terminal point $(-2, 5)$.

3. Let $\mathbf{x}$ be a vector in the plane with magnitude r, and such that $\mathbf{x}$ makes an angle of t radians with the positive x-axis. Show that $\mathbf{x} = (r \cos t, r \sin t)$. [*Hint*: Consider $\mathbf{x} = (x, y)$ in polar coordinates.]

4. A pilot steers his airplane northeastward while maintaining an airspeed of 300 miles

per hour. A wind from the west is blowing eastward at 50 miles per hour.

(a) Find a vector that describes the velocity of the airplane relative to the ground.

(b) What is the speed of the airplane relative to the ground?

5. Prove parts (a), (c), (d), (e), (g), and (h) of Theorem 1.1.1.

6. Suppose that the line segments connecting $(0, 0)$ to $(1, 2)$ and $(0, 0)$ to $(3, 1)$ are adjacent sides of a parallelogram. Find the fourth vertex of the parallelogram.

7. Suppose that the line segments connecting $(2, 3)$ to $(5, 7)$ and $(2, 3)$ to $(4, 4)$ are adjacent sides of a parallelogram. Find the fourth vertex of the parallelogram.

8. (a) Let $\mathbf{x}$ be a vector in R^2 and let t be a positive scalar. Show that the magnitude of $t\mathbf{x}$ is the product of t and the magnitude of $\mathbf{x}$.

(b) How can the conclusion of part (a) be modified if t is negative?

9. Suppose that we are given points (a, b) and (c, d) in the xy-plane. Show that the midpoint of the line segment connecting the points is $((a + c)/2, (b + d)/2)$.

10. Show that for the additive identity $\mathbf{0}$ of R^2 and for any scalar t, we have that $t\mathbf{0} = \mathbf{0}$.

11. Show that for any vector $\mathbf{x}$ in R^2, we have that $0\mathbf{x} = \mathbf{0}$.

12. Let $P = (1, -1)$ be a point and let $\mathbf{x} = (2, 3)$ be a vector in R^2. Find the coordinates of the point Q in R^2 such that $\overrightarrow{PQ} = \mathbf{x}$.

13. Show that if the midpoints of adjacent sides of a quadrilateral are joined, the resulting figure is a parallelogram. (*Hint*: Use Exercise 9.)

14. Show that the diagonals of a parallelogram bisect each other.

1.2 VECTORS IN R^3 AND R^n

We begin this section with a study of the geometry of vectors in Euclidean 3-space. Then we introduce the concept of a vector in n-space. The more interesting properties of vectors are postponed until additional background in linear algebra is developed.

The description of vectors in Euclidean 3-space shares many of the similarities with the description of vectors in the xy-plane. Geometrically, vectors are again represented by arrows or directed line segments and are defined to be *equal* if they have the same direction and magnitude.

The only immediate differences are found when these vectors are studied in the context of a coordinate system. Instead of representing a vector by two components, we need three components, for example, $\mathbf{x} = (a, b, c)$ (see Figure 1.2.1).

As in the xy-plane, we may associate each vector with a unique ordered triple of scalars, namely, its components. That is, if $P = (a, b, c)$ and $Q = (a', b', c')$, then $\overrightarrow{PQ} = (a' - a, b' - b, c' - c)$. This correspondence allows us to identify the vectors and points in Euclidean 3-space.

Vector addition may be seen geometrically as in the parallelogram law, head-to-tail addition, or by component addition (see Figure 1.2.2).

For example, if $\mathbf{x} = (a, b, c)$ and $\mathbf{y} = (a', b', c')$, then

$$\mathbf{x} + \mathbf{y} = (a + a', b + b', c + c')$$

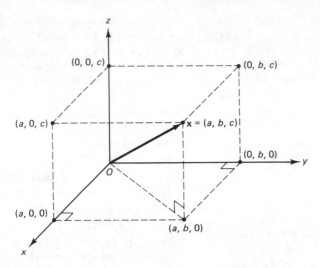

Figure 1.2.1

Similarly, for any scalar t and vector $\mathbf{x}$, the **product** of t and $\mathbf{x}$, denoted by $t\mathbf{x}$, is defined as

$$t\mathbf{x} = (ta, tb, tc)$$

The geometric representation of the scalar product is also the same as the one in the xy-plane].

If we define the **zero vector**, denoted $\mathbf{0}$, as $(0, 0, 0)$ and the **inverse** of $\mathbf{x}$ as $-\mathbf{x} = (-1)\mathbf{x}$, then all of the arithmetic properties stated in Theorem 1.1.1 for vectors in R^2 hold for vectors in R^3. The proofs are virtually identical.

Example 1   Suppose that we have a group of 20 people. We are interested in their ages, heights, and weights. To each person we may associate a vector in R^3, where the first component represents their age in years, the second component represents their height

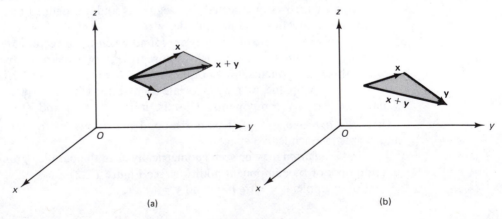

(a)   (b)

Figure 1.2.2

in inches, and the third component represents their weight in pounds. For example, with person 1 we might associate the vector $\mathbf{x}_1 = (13, 60, 85)$. Suppose that we associate the vector $\mathbf{x}_i = (a_i, b_i, c_i)$ with person i. To find the group averages for these three measurements, we compute

$$\frac{1}{20}(\mathbf{x}_1 + \mathbf{x}_2 + \cdots + \mathbf{x}_{20})$$

$$= \frac{1}{20}[(a_1, b_1, c_1) + \cdots + (a_{20}, b_{20}, c_{20})]$$

$$= \frac{1}{20}(a_1 + \cdots + a_{20}, b_1 + \cdots + b_{20}, c_1 + \cdots + c_{20})$$

$$= \left(\frac{a_1 + \cdots + a_{20}}{20}, \frac{b_1 + \cdots + b_{20}}{20}, \frac{c_1 + \cdots + c_{20}}{20}\right)$$

$$= (\text{average age, average height, average weight})$$

Thus, to find the average of each of the three measurements, we take the "average of the vectors," denoted by $\bar{\mathbf{x}}$ (read "x bar"), that is,

$$\bar{\mathbf{x}} = \frac{1}{20}(\mathbf{x}_1 + \mathbf{x}_2 + \cdots + \mathbf{x}_{20})$$

The computation above has a geometric interpretation. If we plot all the measurements as points in 3-space, we obtain a *scatter plot* of the information (see Figure 1.2.3). The vector $\bar{\mathbf{x}}$ represents a *center of mass* of the data. In statistical applications, the representation of data by vectors and their average by $\bar{\mathbf{x}}$ has considerable advantages.

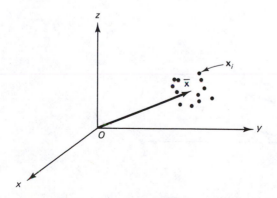

Figure 1.2.3

Lines in 3-Space

Unlike lines in the xy-plane which may be represented by one scalar equation, such as $y = 2x + 3$, lines in 3-space cannot be represented so simply. On the other hand, there are many similarities between the representations of lines in the xy-plane and 3-space which may be seen through a vector approach.

We begin with the following characterization of a line:

A line is uniquely determined by a point on the line and a (nonzero) vector parallel to the line.

Let $\mathbf{y} = (a, b, c)$ be parallel to a line L. $\mathbf{y}$ is called the ***direction vector*** of L. Let $P_0 = (x_0, y_0, z_0)$ be a point on L (see Figure 1.2.4).

It is clear that a point $P = (x, y, z)$ lies on L if and only if $\overrightarrow{P_0P}$ is parallel to $\mathbf{y}$. That is, P lies on L if and only if there exists a scalar t such that

$$\overrightarrow{P_0P} = t\mathbf{y}$$

or, in terms of components,

$$(x - x_0, y - y_0, z - z_0) = (ta, tb, tc)$$

If we identify corresponding components, we obtain the *three* scalar equations

$$x = x_0 + ta$$

$$y = y_0 + tb$$

$$z = z_0 + tc$$

These equations are called the ***parametric equations*** of the line L, and t is called a ***parameter***. As t varies through all real numbers, we obtain all the points on the line.

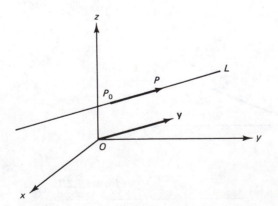

Figure 1.2.4

Example 2   Find the parametric equations of L if $P_0 = (1, -2, 4)$ and $\mathbf{y} = (2, -4, 6)$. Do the points $(-3, 6, -8)$ and $(3, -6, 4)$ lie on L?

In the notation above, we have $x_0 = 1$, $y_0 = -2$, $z_0 = 4$, $a = 2$, $b = -4$, and $c = 6$. Thus, the parametric equations for L are

$$
\begin{aligned}
x &= 1 + 2t \\
y &= -2 - 4t \\
z &= 4 + 6t
\end{aligned}
$$

To see if the point $(-3, 6, -8)$ lies on L, we must determine if there is a value of t such that

$$
\begin{aligned}
-3 &= 1 + 2t \\
6 &= -2 - 4t \\
-8 &= 4 + 6t
\end{aligned}
$$

Each equation implies that $t = -2$. Thus, the point lies on L.

To see if the point $(3, -6, 4)$ lies on L, we must again determine if there is a value of t such that

$$
\begin{aligned}
3 &= 1 + 2t \\
-6 &= -2 - 4t \\
4 &= 4 + 6t
\end{aligned}
$$

The first two equations imply that $t = 1$ but the third equation implies that $t = 0$. Thus, there is no value of t that satisfies all three equations, so the point $(3, -6, 4)$ does not lie on L.

Another view of a vector equation of a line L is to use the form (see Figure 1.2.5)

$$
\overrightarrow{OP} = \overrightarrow{OP_0} + t\mathbf{y}
$$

This equation resembles the scalar equation $y = mx + b$, which represents a line in the xy-plane. However, the vector equation has the advantage that the variable t may be used to represent a physical quantity such as time. In this context, the equation not only indicates the location of, say, an object traveling along a line (at constant speed), but also the time that the object is at a particular point. For example, in the parametric equations of Example 2, if t measures the time in seconds after the object was at the point P_0, we would know that the object reached the point $(3, -6, 10)$ after 1 second and was at the point $(7, -14, 22)$ after 3 seconds.

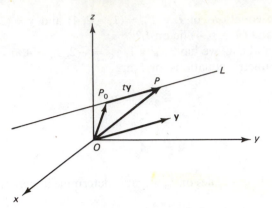

Figure 1.2.5

Vectors in Euclidean *n*-Space

As we have seen, ordered pairs and triples have uses other than designating vectors and points in the *xy*-plane and 3-space. We have seen that an ordered triple is a convenient notation for representing the age, height, and weight of an individual. Suppose that we also want to record the person's pulse rate (in beats per second). We would then need an ordered *4-tuple*, say, (13, 60, 85, 78). Of course, there may be 10 or 100 measurements for each person that we might wish to record. To accommodate such descriptions, mathematicians use ordered *n*-tuples.

Definition   An ***ordered n-tuple*** is a sequence of *n* real numbers or scalars. An ordered *n*-tuple **x** is denoted by

$$\mathbf{x} = (x_1, x_2, \ldots, x_n)$$

The scalar x_i is called the ***ith coordinate*** or ***ith component*** of **x**. Two ordered *n*-tuples **x** = $(x_1, x_2, \ldots, x_n)$ and **y** = $(y_1, y_2, \ldots, y_n)$ are ***equal*** if their corresponding coordinates are equal, that is, if $x_i = y_i$ for all *i*. The set of all ordered *n*-tuples is denoted by R^n.

The ordered 4-tuple (2, 0, 2, −3) has first coordinate 2 and fourth coordinate −3.

Since we have identified ordered pairs and triples with vectors, we frequently call any ordered *n*-tuple a ***vector***. Of course, we now may no longer represent the vectors with more than three coordinates geometrically.

Instead of writing R^1, we simply write *R*. The elements of the set *R* are scalars, which may be considered as a special case of vectors with one coordinate.

We now consider additional examples in which ordered *n*-tuples are used to describe properties other than magnitude and direction.

Example 3   Suppose that we are measuring the age distribution of a colony of 95 animals. Assume that 25 members of the colony are of age less than 1, 40 are of age 1, and 30 are of age 2. One could summarize this distribution of ages by means of the vector $\mathbf{x} = (25, 40, 30)$. More generally, suppose that all the animals in the colony are less than n years of age, for some positive integer n. Let the number of animals of age less than 1 be x_1, and in general, the number of animals of age i be x_{i+1}. We could then express the distribution of ages by the vector $\mathbf{x} = (x_1, x_2, \ldots, x_n)$.

Example 4   Consider the result of tossing two coins on a table. There are three possible outcomes of such a toss: both coins can land heads up, one can land heads up and the other heads down, or both can land heads down. By the use of elementary probability theory, it can be shown that the probabilities of these outcomes are 0.25, 0.5, and 0.25, respectively. This distribution of probabilities can be summarized by the vector $\mathbf{p} = (0.25, 0.5, 0.25)$. More generally, suppose that there are n distinct outcomes of an event, and suppose that the probability of the ith outcome is p_i. Then the ordered n-tuple $\mathbf{p} = (p_1, p_2, \ldots, p_n)$ characterizes this distribution of probabilities. The vector $\mathbf{p}$ is called a *probability vector*. Probability vectors will be studied in Chapters 2 and 6.

It is easy to see how to extend the operations of vector addition and scalar multiplication to vectors in R^n.

Definition   Let $\mathbf{x} = (x_1, x_2, \ldots, x_n)$ and $\mathbf{y} = (y_1, y_2, \ldots, y_n)$ be vectors in R^n and let t be any scalar. We define the *vector sum*, denoted $\mathbf{x} + \mathbf{y}$, and the *product* of t and $\mathbf{x}$, denoted $t\mathbf{x}$, by

$$\mathbf{x} + \mathbf{y} = (x_1 + y_1, x_2 + y_2, \ldots, x_n + y_n)$$

and

$$t\mathbf{x} = (tx_1, tx_2, \ldots, tx_n)$$

We also extend the definition of the *zero vector* of R^n to be the vector $\mathbf{0} = (0, 0, \ldots, 0)$ and the definition of the *inverse* $-\mathbf{x} = (-1)\mathbf{x}$ of a vector $\mathbf{x}$. The notation $\mathbf{x} - \mathbf{y}$ will denote the vector $\mathbf{x} + (-\mathbf{y})$.

Example 5   Let $\mathbf{x} = (2, 1, -3, 0)$ and $\mathbf{y} = (3, 4, 6, 2)$. Then $\mathbf{x} + \mathbf{y} = (5, 5, 3, 2)$ and $\mathbf{x} - \mathbf{y} = (-1, -3, -9, -2) = -(1, 3, 9, 2)$.

Because the arithmetic properties of the vector operations for vectors in R^2 depend on the corresponding properties of scalars, it is not surprising that the same properties hold for the vectors in R^n. We list them below in Theorem 1.2.1.

Theorem 1.2.1 *Let* **x**, **y**, *and* **z** *be vectors in* R^n, *and let s and t be any scalars. Then*

    (a) $\mathbf{x} + \mathbf{y} = \mathbf{y} + \mathbf{x}$ *(commutative law of vector addition)*

    (b) $(\mathbf{x} + \mathbf{y}) + \mathbf{z} = \mathbf{x} + (\mathbf{y} + \mathbf{z})$ *(associative law of vector addition)*

    (c) $\mathbf{x} + \mathbf{0} = \mathbf{x}$ *(0 is the identity)*

    (d) $\mathbf{x} + (-\mathbf{x}) = \mathbf{0}$ *(−**x** is the inverse of **x**)*

    (e) $1\mathbf{x} = \mathbf{x}$

    (f) $(st)\mathbf{x} = s(t\mathbf{x})$

    (g) $s(\mathbf{x} + \mathbf{y}) = s\mathbf{x} + s\mathbf{y}$ *(scalar multiplication*

                                      distributes over vector addition)

Because of part (b) of the theorem, the expression $\mathbf{x} + \mathbf{y} + \mathbf{z}$ can be written unambiguously without parentheses.

Proof

We shall prove only parts (d) and (f). The rest will be left as exercises.

For the proofs below we let $\mathbf{x} = (x_1, x_2, \ldots, x_n)$ and $\mathbf{y} = (y_1, y_2, \ldots, y_n)$.

(d)
$$\mathbf{x} + (-\mathbf{x}) = (x_1, x_2, \ldots, x_n) + (-x_1, -x_2, \ldots, -x_n)$$
$$= (x_1 - x_1, x_2 - x_2, \ldots, x_n - x_n)$$
$$= (0, 0, \ldots, 0)$$
$$= \mathbf{0}$$

(f) Let s and t be scalars.

$$(st)\mathbf{x} = (st)(x_1, x_2, \ldots, x_n)$$
$$= ((st)x_1, (st)x_2, \ldots, (st)x_n)$$
$$= (s(tx_1), s(tx_2), \ldots, s(tx_n))$$
$$= s(tx_1, tx_2, \ldots, tx_n)$$
$$= s(t\mathbf{x}) \qquad\blacksquare$$

Example 6   Suppose that 15 people in a class each take 10 tests. For each person a vector in R^{10} may be used to represent the person's scores; that is, the ith coordinate represents the person's score on the ith test. Let the 15 vectors associated with the class be denoted by $\mathbf{x}_1, \ldots, \mathbf{x}_{15}$. Using the same notation as in Example 1, we have that the vector

$$\bar{\mathbf{x}} = \frac{1}{15}(\mathbf{x}_1 + \cdots + \mathbf{x}_{15})$$

may be interpreted in the following way: The ith coordinate of $\bar{\mathbf{x}}$ represents the class average on the ith test.

Exercises

1. Determine whether or not $\overrightarrow{AB}$ equals $\overrightarrow{CD}$. In each case, sketch the vectors.
   (a) $A = (1, 2, 3)$, $B = (0, 1, -1)$, $C = (2, 3, 5)$, and $D = (1, 2, 1)$.
   (b) $A = (0, 0, 0)$, $B = (1, -3, -2)$, $C = (-1, 3, 2)$, and $D = (0, 0, 0)$.
   (c) $A = (-1, 0, 1)$, $B = (2, 1, 3)$, $C = (2, 3, 5)$, and $D = (3, 4, 7)$.

2. For the point P and the vector $\mathbf{y}$, find the point Q such that $\overrightarrow{PQ} = \mathbf{y}$.
   (a) $P = (2, 3, 4)$ and $\mathbf{y} = (1, 2, 0)$.
   (b) $P = (1, -4, 3)$ and $\mathbf{y} = (-4, 4, 2)$.

3. For the point $P_0 = (x_0, y_0, z_0)$ and the vector $\mathbf{y} = (a, b, c)$, find the parametric equations of the line containing P_0 and parallel to $\mathbf{y}$.
   (a) $P_0 = (2, -1, 3)$ and $\mathbf{y} = (1, 3, 1)$.
   (b) $P_0 = (-1, 0, 3)$ and $\mathbf{y} = (2, -1, 1)$.

4. For the points $P = (x_0, y_0, z_0)$ and $Q = (x_1, y_1, z_1)$, find the parametric equations of the line containing P and Q.
   (a) $P = (2, 3, 5)$ and $Q = (4, 5, 1)$.
   (b) $P = (-1, 2, -1)$ and $Q = (1, 0, 3)$.

5. An object located at a point $P = (2, 4, 3)$ begins moving along a line with a constant velocity described by the vector $\mathbf{y} = (-1, 2, 3)$. Find its location after two units of time.

6. In each of the following situations, choose an appropriate ordered n-tuple to describe the data.
   (a) A grocer has a stock of 100 small eggs, 150 medium eggs, 300 large eggs, and 100 extra-large eggs.
   (b) A dog breeder has 100 dalmatians, 200 terriers, and 50 poodles.
   (c) On a three-problem quiz, a student earned 10 points on the first problem, 15 points on the second problem, and 20 points on the third problem.

7. Suppose that a grocer has a stock of 50 small eggs, 200 medium eggs, 100 large eggs, and no extra-large eggs. He merges this stock with the stock described in Exercise 6(a). Describe how the arithmetic of vectors can be applied to compute the distribution of the merged stock.

8. An airplane is flying with a ground speed of 300 miles per hour at an angle of 30° east of due north (see Figure 1.2.6). In addition, the airplane is climbing at the rate of 10 miles per hour. Determine the vector in R^3 that represents the velocity of the airplane.

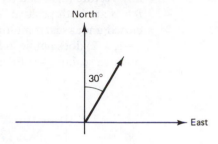

Figure 1.2.6

9. Suppose that the line segments connecting $(1, -1, 3)$ to $(0, 1, 2)$ and $(1, -1, 3)$ to $(2, 2, 5)$ are adjacent sides of a parallelogram. Find the fourth vertex of the parallelogram.

10. Show that for the additive identity $\mathbf{0}$ in R^n and any scalar t, we have that $t\mathbf{0} = \mathbf{0}$.

11. Show that for any vector $\mathbf{x}$ in R^n, we have that $0\mathbf{x} = \mathbf{0}$.

12. Prove parts (a), (b), (c), (e), (g), and (h) of Theorem 1.2.1.

13. Find the probability vector of outcomes for the event of throwing a pair of dice.

1.3 SUBSPACES AND LINEAR COMBINATIONS

The arithmetic vector operations defined in Sections 1.1 and 1.2 can be used to distinguish certain important subsets of R^n, called "subspaces." These subspaces will be used to represent geometric figures such as lines and planes as well as solutions to systems of linear equations.

Definition   A subset S of R^n is ***closed under (vector) addition*** if for any vectors $\mathbf{x}$ and $\mathbf{y}$ in S, the sum $\mathbf{x} + \mathbf{y}$ is also in S.

A subset S of R^n is ***closed under scalar multiplication*** if for any scalar t in R and any vector $\mathbf{x}$ in S, the product $t\mathbf{x}$ lies in S.

It should be noted that if a subset S of R^n is closed under addition, then for any finite collection of vectors $\mathbf{x}_1, \mathbf{x}_2, \ldots, \mathbf{x}_k$ in S, the sum

$$\mathbf{x}_1 + \mathbf{x}_2 + \cdots + \mathbf{x}_k$$

is also in S.

Example 1   Let S be the first quadrant of the xy-plane. Recall that an ordered pair lies in S if each of its coordinates is positive. We will show that S is closed under addition. Consider any vectors (a, b) and (c, d) in S. Then each of a, b, c, and d is positive. So $a + c$ and $b + d$ are both positive. We conclude that $(a, b) + (c, d)$ lies in S. Notice that S is *not* closed under scalar multiplication. For example, although $(1, 2)$ lies in S, $(-1)(1, 2) = (-1, -2)$ does not lie in S. It is true, however, that S is closed under multiplication by positive scalars [see Figure 1.3.1(a)].

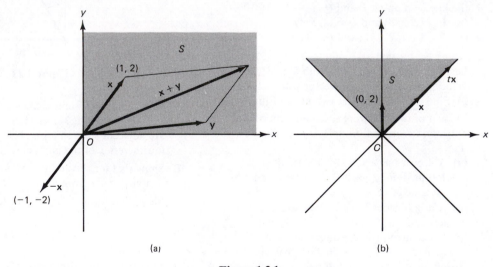

(a)   (b)

Figure 1.3.1

Example 2   Let $S = \{(x, y) : x^2 = y^2\}$. Show that S is closed under scalar multiplication.

Suppose that (a, b) lies in S and that t is any scalar. Because $a^2 = b^2$, we have $t^2 a^2 = t^2 b^2$. Therefore, $(ta)^2 = (tb)^2$. So $(ta, tb) = t(a, b)$ lies in S. However, notice that S is not closed under addition. For example, although $(1, 1)$ and $(-1, 1)$ both lie in S, their sum $(0, 2)$ does not lie in S [see Figure 1.3.1(b)].

Definition   A nonempty subset V of R^n is called a **subspace** of R^n if V is closed under both addition and scalar multiplication.

Notice that the subsets of R^n given in Examples 1 and 2 are not subspaces of R^n. The empty subset of R^n *vacuously* satisfies both closure conditions but is not a subspace since it contains no vectors. Clearly, R^n is a subspace of itself. It is also obvious that the subset $\{0\}$ consisting of the single vector $\mathbf{0}$ is a subspace of R^n. This set is nonempty, and it is easy to verify that it is closed under the two operations. It is called the **zero subspace** of R^n.

Example 3   Let V be the subset of R^3 defined by

$$V = \{(x, y, z) : x + y = 2z\}$$

In common language, V consists of all those vectors in R^3 that have the property that the sum of the first two components equals twice the third component. We shall verify that V is a subspace of R^3. First, notice that V is not empty. For example, $\mathbf{0} = (0, 0, 0)$ lies in V. Next, observe that V is closed under addition. For if (a, b, c) and (a', b', c') lie in V, then we must have that

$$a + b = 2c \qquad \text{and} \qquad a' + b' = 2c'$$

Thus,

$$(a + a') + (b + b') = 2c + 2c' = 2(c + c')$$

We conclude that $(a, b, c) + (a', b', c') = (a + a', b + b', c + c')$ lies in V.

Next, observe that V is closed under scalar multiplication. For if (a, b, c) lies in V and t is any scalar, we have that

$$a + b = 2c$$

and hence

$$ta + tb = 2tc$$

We conclude that $t(a, b, c) = (ta, tb, tc)$ lies in V. Thus, V is a subspace of R^3.

We shall show in Section 1.6 that the subset of Example 3 is a plane in 3-space which contains the origin. In fact, it can be shown that any plane in R^3 which passes through the origin is a subspace of R^3. Example 4 will demonstrate that any line in the xy-plane which passes through the origin is also a subspace.

Example 4   Any line in R^2 that passes through the origin is a subspace of R^2. Let L be such a line. Either L is the y-axis or L has a slope. The verification that the y-axis is a subspace of R^2 is left as an exercise. So suppose that L is not the y-axis. Then the equation of L is of the form $y = mx$ for some real number m, the slope of L. Clearly, L is not empty. For example, $\mathbf{0} = (0, 0)$ lies in L. Now suppose that (a, b) and (c, d) lie in L. Then $b = ma$ and $d = mc$. Therefore, $b + d = ma + mc = m(a + c)$. We conclude that $(a + c, b + d) = (a, b) + (c, d)$ lies in L. Thus, L is closed under vector addition. Finally, if (a, b) is in L and t is any scalar, then since $b = ma$, we have $tb = t(ma) = m(ta)$. Thus, $(ta, tb) = t(a, b)$ lies in L. We conclude that L is a subspace of R^2.

We shall see that subspaces satisfy many of the arithmetic properties of R^n. For example, the next theorem asserts that all the subspaces of R^n contain the zero vector as well as the inverses of all its elements.

Theorem 1.3.1   *Let V be a subspace of R^n. Then:*
   (a) *The zero vector $\mathbf{0}$ lies in V.*
   (b) *For any vector $\mathbf{x}$ in V, the inverse $-\mathbf{x}$ lies in V.*

Proof
(a) Since V is not empty, it contains a vector $\mathbf{x}$. Because V is closed under scalar multiplication, we have that $0\mathbf{x} = \mathbf{0}$ lies in V.
(b) Let $\mathbf{x}$ be any vector in V. Since $-\mathbf{x} = (-1)\mathbf{x}$ and V is closed under scalar multiplication, we have that $-\mathbf{x}$ is contained in V.   ∎

Let us apply Theorem 1.3.1 to combine subspaces of R^n to form new subspaces of R^n. Given two sets S and S', recall that the *intersection* of S and S', denoted by $S \cap S'$, is the set consisting of all vectors contained in both S and S'. More generally, the intersection of several sets,

$$S_1 \cap S_2 \cap \cdots \cap S_k$$

is the set consisting of the vectors common to all the sets.

Theorem 1.3.2   *Let V_1 and V_2 be subspaces of R^n. Then the intersection $V_1 \cap V_2$ is also a subspace of R^n.*

Proof
We must verify that $V_1 \cap V_2$ satisfies three properties: It is not empty, it is closed under addition, and it is closed under scalar multiplication.
   Since V_1 and V_2 are each subspaces of R^n, they each contain $\mathbf{0}$. Therefore, $\mathbf{0}$ lies in the intersection, so we conclude that $V_1 \cap V_2$ is not empty.
   Next we show that $V_1 \cap V_2$ is closed under vector addition. Suppose that $\mathbf{x}$ and $\mathbf{y}$ lie in the intersection. Then $\mathbf{x}$ and $\mathbf{y}$ both lie in V_1, a subspace of R^n. Hence, $\mathbf{x} + \mathbf{y}$ lies in V_1. Similarly, $\mathbf{x} + \mathbf{y}$ lies in V_2. We conclude that $\mathbf{x} + \mathbf{y}$ lies in the intersection of the two spaces. Therefore, the intersection is closed under addition.
   The proof that $V_1 \cap V_2$ is closed under scalar multiplication is left as an exercise.   ∎

Corollary 1.3.3

Let $V_1, V_2, \ldots, V_k$ be subspaces of R^n. Then the intersection

$$V_1 \cap V_2 \cap \cdots \cap V_k$$

is also a subspace of R^n.

The proof is left as an exercise.

Notice that the union of two subspaces is not necessarily a subspace. For example, suppose that V_1 is the subspace of all vectors lying along the x-axis in R^2, and that V_2 is the subspace of all vectors lying along the y-axis in R^2. Then $(1, 0)$ lies in V_1 and $(0, 1)$ lies in V_2, but $(1, 0) + (0, 1) = (1, 1)$ does not lie in the union (see Figure 1.3.2).

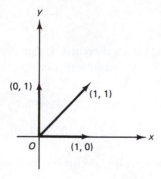

Figure 1.3.2

Example 5

Recall the subspace $V = \{(x, y, z) : x + y = 2z\}$ of R^3 defined in Example 3. Let $W = \{(x, y, z) : x + y + z = 0\}$. As in Example 3 it can be shown that W is also a subspace of R^3. So by Theorem 1.3.2 we conclude that $V \cap W$ is a subspace of R^3. This subspace is actually the set of solutions to the system of equations

$$x + y - 2z = 0$$
$$x + y + z = 0$$

We shall learn how to find all solutions to systems such as this in Section 1.4.

Some insight into the structure of subspaces of R^n can be gained by studying ways of constructing a subspace from a finite set of vectors. Given such a set, one can use the operations of vector addition and scalar multiplication to combine vectors of the set in various ways, thus obtaining new vectors. As we shall see, the collection of all such vectors determines a subspace of R^n. We begin by considering the process of combining vectors of a finite set.

Definition   Let $S = \{\mathbf{x}_1, \mathbf{x}_2, \ldots, \mathbf{x}_k\}$ be a finite (nonempty) subset of R^n. A vector $\mathbf{x}$ in R^n is called a ***linear combination*** of the vectors in S if there exist scalars $t_1, t_2, \ldots, t_k$ (not necessarily distinct) such that

$$\mathbf{x} = t_1\mathbf{x}_1 + t_2\mathbf{x}_2 + \cdots + t_k\mathbf{x}_k$$

We shall often refer to the t_i's as the ***coefficients*** of the linear combination.

As with sums of numbers it is often useful to represent the sum of several vectors

$$t_1\mathbf{x}_1 + t_2\mathbf{x}_2 + \cdots + t_k\mathbf{x}_k$$

with the sigma notation

$$\sum_{i=1}^{k} t_i\mathbf{x}_i$$

To help the reader become accustomed to this notation, we shall use it together with the ordinary notation to designate sums in this section.

Example 6   Let

$$S = \{(1, 0, 0), (0, 1, 0), (1, 1, 0)\}$$

The vector $\mathbf{x} = (2, 1, 0)$ is a linear combination of the vectors in S since

$$(2, 1, 0) = 2(1, 0, 0) + 1(0, 1, 0) + 0(1, 1, 0)$$

Note that $\mathbf{x}$ could have been represented as another linear combination:

$$(2, 1, 0) = 1(1, 0, 0) + 0(0, 1, 0) + 1(1, 1, 0)$$

The vector $(1, 2, 3)$ is *not* a linear combination of the vectors in S. For if it were, we would have

$$(1, 2, 3) = x(1, 0, 0) + y(0, 1, 0) + z(1, 1, 0)$$

for some scalars x, y, and z. Adding the terms on the right-hand side of the foregoing equation, we have

$$(1, 2, 3) = (x + z, y + z, 0)$$

Equating the third coordinates of both sides, we have that $3 = 0$—an impossibility! We conclude that $(1, 2, 3)$ is not a linear combination of the vectors in S.

Example 7   Show that the vector $(1, 0)$ is a linear combination of the vectors $(1, -3)$ and $(-1, 4)$. To do this we must find scalars x and y such that

$$(1, 0) = x(1, -3) + y(-1, 4)$$
$$= (x - y, -3x + 4y)$$

Equating corresponding coordinates, we obtain

$$x - \ y = 1$$
$$-3x + 4y = 0$$

This is a system of two linear equations in two unknowns. Fortunately, we have already acquired the skills to solve such a system. Multiplying both sides of the first equation by 3 and adding the resulting equation to the second equation, we have

$$3x - 3y = 3$$
$$y = 3$$

Substituting $y = 3$ into the first equation, we find that $x = 4$. Thus, we have

$$(1, 0) = 4(1, -3) + 3(-1, 4)$$

In general, when attempting to determine whether or not a given vector is a linear combination of a finite set of vectors, one encounters systems of linear equations in several unknowns. The general problem of solving such systems will be discussed in Section 1.4.

There are certain vectors in R^n which are often useful when considering linear combinations. For each $i = 1, 2, \ldots, n$, let $\mathbf{e}_i$ denote the ordered n-tuple which has a 1 in its ith position and zeros elsewhere. That is,

$$\mathbf{e}_1 = (1, 0, \ldots, 0),\ \mathbf{e}_2 = (0, 1, 0, \ldots, 0), \ldots, \mathbf{e}_n = (0, 0, \ldots, 1)$$

These vectors are called the ***standard vectors*** of R^n. They are useful because any vector in R^n can be represented easily as a linear combination of standard vectors. For if $\mathbf{x} = (a_1, a_2, \ldots, a_n)$ is a vector in R^n, it is easy to see that

$$\mathbf{x} = a_1\mathbf{e}_1 + a_2\mathbf{e}_2 + \cdots + a_n\mathbf{e}_n$$

$$= \sum_{i=1}^{n} a_i\mathbf{e}_i$$

For example,

$$(1, 0, 4) = 1\mathbf{e}_1 + 0\mathbf{e}_2 + 4\mathbf{e}_3$$

Because all the vectors in R^n may be represented as linear combinations of the set of standard vectors, it is of interest to ask if other subspaces of R^n may be described similarly as linear combinations of some finite subset of vectors.

Definition   The collection of all linear combinations of vectors in a finite subset S of R^n is called the ***span*** of S. If W is the span of S, we shall say that S ***spans*** W.

We have just shown that the span of the set of standard vectors in R^n is all of R^n. It is easy to see that the span of the set $\{\mathbf{x}\}$ consisting only of the vector $\mathbf{x}$ is the set of all scalar multiples of $\mathbf{x}$. In R^2 or R^3 if $\mathbf{x} \neq \mathbf{0}$, this set may be represented geometrically as a line through the origin.

Example 8   Describe geometrically the set V which is the span of $\{(1, 0, 0), (0, 1, 0)\}$ of R^3. An arbitrary vector $\mathbf{x}$ in V is of the form

$$\mathbf{x} = s(1, 0, 0) + t(0, 1, 0)$$
$$= (s, t, 0)$$

for some scalars s and t. It is clear that $\mathbf{x}$ lies in V if and only if the third coordinate of $\mathbf{x}$ is 0. Thus, V is the xy-plane in R^3.

We shall use the parallelogram law for vector addition to show that the span of any two nonparallel vectors $\mathbf{x}$ and $\mathbf{y}$ in the xy-plane is all of R^2. For this purpose, let $\mathbf{z}$ be any vector in R^2. Determine multiples of the vectors $\mathbf{x}$ and $\mathbf{y}$ as in Figure 1.3.3 so that $\mathbf{z}$ is the diagonal of the parallelogram whose sides are determined by $s\mathbf{x}$ and $t\mathbf{y}$. Then $\mathbf{z} = s\mathbf{x} + t\mathbf{y}$.

Notice that in all the examples above, the span of a set is a subspace. The next theorem asserts that the span of a subset S is the *smallest* subspace of R^n that contains S.

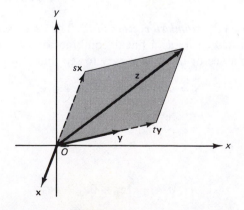

Figure 1.3.3

Theorem 1.3.4 *Let S be any nonempty finite subset of R^n. Then:*

(a) *S is a subset of the span of S.*

(b) *The span of S is a subspace of R^n.*

(c) *If S is contained in a subspace V of R^n, then the span of S is also contained in V.*

Proof

(a) Suppose that $S = \{x_1, x_2, \ldots, x_k\}$ is a subset of R^n. Then, for any i,

$$x_i = 0x_1 + 0x_2 + \cdots + 1x_i + \cdots + 0x_k$$

Thus, x_i is a linear combination of the vectors in S, so x_i lies in the span of S. We conclude that S is a subset of the span of S.

(b) Clearly, the span of S is a nonempty subset of R^n since it contains S. We must verify that the span of S is closed under addition and scalar multiplication. Consider any two vectors x and y in the span of S. Since x and y are linear combinations of vectors of S, there are scalars $s_1, s_2, \ldots, s_k, t_1, t_2, \ldots, t_k$, such that

$$x = s_1 x_1 + s_2 x_2 + \cdots + s_k x_k$$

$$= \sum_{i=1}^{k} s_i x_i$$

and

$$y = t_1 x_1 + t_2 x_2 + \cdots + t_k x_k$$

$$= \sum_{i=1}^{k} t_i x_i$$

It follows that

$$x + y = (s_1 + t_1)x_1 + (s_2 + t_2)x_2 + \cdots + (s_k + t_k)x_k$$

$$= \sum_{i=1}^{k} (s_i + t_i)x_i$$

So $x + y$ is a linear combination of vectors of S; that is, $x + y$ lies in the span of S. We conclude that the span of S is closed under addition. The proof that the span of S is closed under scalar multiplication is left as an exercise.

(c) Suppose that V is a subspace of R^n which contains S. We wish to show that V contains the span of S; that is, we wish to show that any linear combination of vectors of S lies in V. Let x lie in the span of S. Then

$$x = s_1 x_1 + s_2 x_2 + \cdots + s_k x_k$$

$$= \sum_{i=1}^{k} s_i x_i$$

Since each x_i lies in V and since V is closed under scalar multiplication, we have that $s_i x_i$ lies in V for each i. Using the fact that V is closed under addition, we have that x lies in V. Thus, any vector in the span of S lies in V. Therefore, the span of S is a subspace of V. ∎

      Intuitively, part (c) tells us that the span of S is the smallest subspace which contains S. In light of this perspective, we extend the definition of "span" to include the span of the empty subset of R^n. Since the smallest subspace of R^n is the zero subspace, we define the *span of the empty set* to be the zero subspace.

Example 9   Let V be the subspace $\{(x, y, z): z = 2x\}$, and let $S = \{(1, 1, 2), (1, -1, 2)\}$. Show that the span of S equals V.

    Clearly, S is contained in V. So, by part (c) of Theorem 1.3.4, the span of S is contained in V. Now there are two ways of showing that V is the span of S.

    Geometric approach: Because the vectors of S are not parallel, we know that the span of S is a plane. Since V is also a plane, it follows that the span of S equals V.

    Algebraic approach: Let x be an arbitrary vector in V. We must show that x is a linear combination of the vectors in S. For this purpose, let $x = (x, y, z) = (x, y, 2x)$. We must find scalars a and b such that

$$(x, y, 2x) = a(1, 1, 2) + b(1, -1, 2)$$

or

$$x = a + b$$
$$y = a - b$$
$$2x = 2a + 2b$$

Because the third equation is a multiple of the first equation, we need only solve for a and b in the first two equations. The solution is

$$a = \frac{x + y}{2} \quad \text{and} \quad b = \frac{x - y}{2}$$

Exercises

1. For each of the following sets, test for closure under vector addition and scalar multiplication.

   (a) $\{(x, y): xy = 0\}$
   (b) $\{(x, y, z): x + y = z\}$
   (c) $\{(x, y, z): x > y \text{ and } z < 0\}$
   (d) $\{(x, y, z): xy = z^2\}$
   (e) $\{(x, y, z): (xy)^2 \geq 0\}$
   (f) $\{(x, y): x + y > 0\}$

2. Determine which of the following subsets of R^n are subspaces. Justify your conclusion.

   (a) $\{(x, y): x^2 + y^2 = 1\}$

(b) $\{(x, y, z) : 2x - y + z = 0\}$

(c) $\{(x, y, z) : x = yz\}$

(d) $\{(p, q) : p \text{ and } q \text{ are integers}\}$

(e) The set of all nonnegative numbers

(f) $\{(x, y, z) : x - 2y = 3z\}$

(g) $\{(x, y, z, w) : x + y + z + w = 1\}$

(h) $\{(x, y, z) : x \geq y \geq z\}$

3. Complete the proof of Theorem 1.3.2 by showing that the intersection of two subspaces of R^n is closed under scalar multiplication.

4. Determine whether or not the given vector **x** is a linear combination of the vectors in the set S.

(a) $\mathbf{x} = (1, 1)$;
   $S = \{(1, 0), (0, 1)\}$

(b) $\mathbf{x} = (1, -1)$;
   $S = \{(1, 1)\}$

(c) $\mathbf{x} = (1, 1, 2)$;
   $S = \{(1, 0, 1), (1, 0, -1)\}$

(d) $\mathbf{x} = 3$;
   $S = \{1\}$

(e) $\mathbf{x} = (1, 1)$;
   $S = \{(1, 0), (0, -1), (0, 0)\}$

(f) $\mathbf{x} = (-1, 11)$;
   $S = \{(1, 3), (2, -1)\}$

(g) $\mathbf{x} = (0, 5, 2)$;
   $S = \{(1, 2, 1), (1, 0, 1), (-1, 1, 1)\}$

5. For each of the following, determine whether or not the subset S of R^n spans the subspace V of R^n. Justify your conclusion.

(a) $S = \{(1, 2), (2, 3)\}$,
   $V = R^2$

(b) $S = \{(1, 1, 1), (3, 3, 3)\}$,
   $V = \{(x, y, z) : x = y = z\}$

(c) $S = \{(1, 2, -3)\}$,
   $V = \{(x, y, z) : x + y + z = 0\}$

(d) $S = \{(1, 2, 1), (1, 1, 1)\}$,
   $V = \{(x, y, z) : x = z\}$

(e) $S = \{(1, 2, 3, 4), (1, -1, 5, 2)\}$,
   $V = \{(x, y, z, w) : 2w = x + 2y + z\}$

6. Verify that each of the sets V in Exercise 5 is a subspace.

7. Complete the proof of Theorem 1.3.4 by showing that the span of a nonempty set is closed under the operation of scalar multiplication.

8. Let **x** be a vector in R^2.
   (a) Describe span of $\{\mathbf{x}\}$ geometrically if $\mathbf{x} \neq \mathbf{0}$.
   (b) What is span of $\{\mathbf{x}\}$ if $\mathbf{x} = 0$?

9. Show that L, the set of all points on the y-axis in the plane (see Example 4), is a subspace of R^2.

10. Show that a line in R^3 is a subspace of R^3 if and only if the line passes through the origin.

11. Let S be a nonempty subset of R^n. Prove that S is a subspace of R^n if and only if for any vectors **x** and **y** in S and for any scalar c, we have that $\mathbf{x} + c\mathbf{y}$ is in S.

12. Prove Corollary 1.3.3.

13. Let $V = \{(0, x, y) : x \text{ and } y \text{ are any scalars}\}$.
   (a) Prove that V is a subspace of R^3.
   (b) Find a set of two vectors whose span is V. Justify your answer.

14. For any vectors **x** and **y** of R^n, show that the span of $(\mathbf{x}, \mathbf{y})$ is the same as the span of $\{\mathbf{x} + \mathbf{y}, \mathbf{x} - \mathbf{y}\}$.

1.4 INTRODUCTION TO SYSTEMS OF LINEAR EQUATIONS

Thus far we have encountered only very small systems of equations, say, two equations in two or three unknowns. These systems arose in the context of determining if a given vector is a linear combination of other vectors. In this section we address the general problem of finding all solutions to a system of linear equations. As we shall discover, such a system can have a single solution, no solutions, or infinitely many solutions.

Definition   A *system of m linear equations in n unknowns*, $x_1, x_2, \ldots, x_n$ is a collection of m equations of the form

$$a_{11}x_1 + a_{12}x_2 + \cdots + a_{1n}x_n = b_1$$
$$a_{21}x_1 + a_{22}x_2 + \cdots + a_{2n}x_n = b_2$$
$$\vdots$$
$$a_{m1}x_1 + a_{m2}x_2 + \cdots + a_{mn}x_n = b_m$$

where a_{ij} and b_i are scalars for $1 \le i \le m$ and $1 \le j \le n$. The scalars a_{ij} are called the *coefficients* of the system and the scalars b_i are called the *constants* of the system.

The system of equations given in Example 5 of Section 1.3 is an example of a system of two linear equations in the three unknowns x, y, and z.

A vector $\mathbf{y} = (c_1, c_2, \ldots, c_n)$ in R^n is called a *solution* to a system of m equations in the n unknowns $x_1, x_2, \ldots, x_n$ if the substitution of c_i for each x_i into each of the m equations yields a valid equation. It is easy to show that $(-2, 2, 0)$ is a solution to the system mentioned above. The set of all solutions to a system of linear equations is called the *solution set* of the system.

We now consider examples of how systems of linear equations can arise.

Example 1   Two flasks each contain a solution of alcohol and sugar in water. The amount of alcohol and sugar in each flask is given as a percentage of the solution by weight according to Table 1.4.1. Is it possible to mix a certain amount of liquid from each flask to obtain a mixture that contains 4 grams of alcohol and 2.6 grams of sugar?

TABLE 1.4.1

	Alcohol	Sugar
Flask 1	10	5
Flask 2	5	4

Let x_1 be the weight of the liquid taken from the first flask and let x_2 be the weight of the liquid from the second flask so that the resulting mixture satisfies these requirements. Then taken

$$0.1x_1 + 0.05x_2 = 4$$
$$0.05x_1 + 0.04x_2 = 2.6$$

Notice that $\mathbf{y} = (20, 40)$ is a solution to this system.

Example 2   A nutritious breakfast drink can be made by mixing whole egg, milk, and orange juice in a blender. The food energy and protein for the ingredients are given in Table 1.4.2. How much of each ingredient should be blended to produce a drink with 540 calories of energy and 25 grams of protein?

TABLE 1.4.2

	Food Energy (Calories)	Protein (Grams)
1 egg	80	6
1 cup milk	160	9
1 cup orange juice	110	2

Let x be the number of eggs, y the amount of milk in cups, and z the amount of orange juice in cups. Then

$$80x + 160y + 110z = 540$$

$$6x + 9y + 2z = 25$$

Unlike the system of equations of Example 1, this system consists of more unknowns than equations. As we shall see, such a system ordinarily has infinitely many solutions. One solution to this system is $\mathbf{y} = (2, 1, 2)$; another is $\mathbf{z} = (0.325, 2.25, 1.4)$.

The Geometry of Systems

One way to gain insight into the nature of the solution set of a system of linear equations is to regard it as a geometrical object in R^n. Given a system of m equations in n unknowns, a vector $\mathbf{y}$ in R^n is a solution to the system if and only if it is a solution to each individual equation. Therefore, the solution set of the system is the intersection of the solution sets of each of the individual equations.

Consider, for example, the case of two unknowns. Since the graph of a single linear equation in two unknowns is a straight line in the xy-plane, this line is the solution set of the equation. Thus, the solution set of a system of two equations in two unknowns is the intersection of two lines. From our knowledge about lines in the xy-plane, we know that one of three cases is possible: the lines intersect at a single point, the lines do not intersect at all (they are parallel), or the two lines intersect to form one line (the lines coincide).

Example 3   The system

$$x - y = 1$$

$$x + y = 2$$

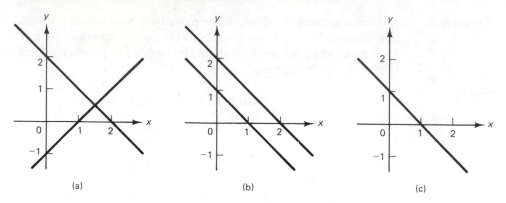

Figure 1.4.1

consists of equations whose graphs are lines which intersect at a single point. Thus, the system has a unique solution [see Figure 1.4.1(a)].

The system

$$x + y = 1$$
$$x + y = 2$$

consists of equations whose graphs are parallel lines that do not intersect. This system has no solution [see Figure 1.4.1(b)].

The system

$$x + \ y = 1$$
$$2x + 2y = 2$$

consists of equations whose graphs are the same line. Therefore, the system has infinitely many solutions, namely, all points on one line [see Figure 1.4.1(c)].

In Section 1.6 we shall show that the solution set of one linear equation in three unknowns is a plane in R^3. The solution set of a system of several equations in three unknowns is therefore the intersection of several planes. When two planes intersect, they form another plane, a line, or the empty set.

Solutions to Systems of Linear Equations

Two systems of linear equations are called *equivalent* if they have the same solution set. We now develop a technique for transforming a system of linear equations into an equivalent system whose solutions are easy to find. The idea is to systematically eliminate unknowns from as many equations as possible. Before we state any results about equivalent systems, we will look at a specific example.

Consider the following system of three linear equations in three unknowns.

$$2x_1 + x_2 + 3x_3 = 15$$
$$-2x_1 + 2x_2 - 2x_3 = -14$$
$$-4x_1 + 4x_2 + x_3 = -8$$

We shall eliminate x_1 from the second equation by adding the first equation to the second equation. We obtain the new system

$$2x_1 + x_2 + 3x_3 = 15$$
$$3x_2 + x_3 = 1$$
$$-4x_1 + 4x_2 + x_3 = -8$$

Similarly, to eliminate x_1 from the third equation, we add 2 times the first equation to the third equation. We obtain

$$2x_1 + x_2 + 3x_3 = 15$$
$$3x_2 + x_3 = 1$$
$$6x_2 + 7x_3 = 22$$

To simplify the system further, we add -2 times the second equation to the third equation. This yields

$$2x_1 + x_2 + 3x_3 = 15$$
$$3x_2 + x_3 = 1$$
$$5x_3 = 20$$

Notice that each successive equation contains at least one fewer unknowns than the preceding equation. This final system is very easy to solve. The method used for this purpose is called ***back substitution***. In applying this method to the system above, we first solve the third equation to obtain $x_3 = 4$. We insert this result into the second equation and obtain

$$3x_2 + 4 = 1$$

or $x_2 = -1$. Now we substitute both of these results into the first equation and arrive at

$$2x_1 + (-1) + 3(4) = 15$$

or $x_1 = 2$. So a solution vector for the last system is $\mathbf{y} = (2, -1, 4)$. If these values are substituted into the original system, it is easy to see that $\mathbf{y}$ is also a solution of this system. As the next theorem will show, this is no coincidence. At each step our system was modified by adding a multiple of one equation to another. It is this type of operation that produces an equivalent system. We state this fact as a theorem. The proof is omitted.

Theorem 1.4.1 *If a multiple of one equation of a system of linear equations is added to another equation in the system, then the new system is equivalent to the original one.*

This particular method of transforming a system of linear equations into an equivalent one is an example of an elementary operation.

Definition An ***elementary operation*** on a system of linear equations is an operation that results in a new system of linear equations in one of three ways:

1. Interchange two equations of the original system.
2. Multiply the terms of one of the equations of the original system by a nonzero constant.
3. Add a multiple of one of the equations of the original system to another equation of the original system.

An elementary operation is of ***type 1, 2, or 3*** depending on whether it is obtained by operation 1, 2, or 3.

We managed to solve the previous system by applying only type 3 operations. The other two types of elementary operations also transform systems of linear equations into equivalent ones. We state this as a theorem and leave the proof as an exercise.

Theorem 1.4.2 *Any elementary operation transforms a system of linear equations into an equivalent system.*

The first system we solved has a unique solution. We will now consider a system with infinitely many solutions.

Example 4 Find all solutions to the system of linear equations

$$x_1 + x_2 + x_3 = 2$$
$$x_1 + 3x_2 - x_3 = 8$$

As in the Example on page 33, we eliminate x_1 from the second equation by means of a type 3 operation, obtaining

$$x_1 + x_2 + x_3 = 2$$
$$2x_2 - 2x_3 = 6$$

Dividing the second equation of the new system by 2 (a type 2 operation), we obtain

$$x_1 + x_2 + x_3 = 2$$
$$x_2 - x_3 = 3$$

We cannot eliminate more unknowns as before. Notice that if we assign an arbitrary value to x_3, we can solve for x_2 and x_1 using back substitution. Rather than committing ourselves to a particular value for x_3, we assign the *parameter s* to x_3, that is, $x_3 = s$. Solving for x_2, we have

$$x_2 = 3 + x_3$$
$$= 3 + s$$

Solving for x_1, we have

$$x_1 = 2 - x_2 - x_3$$
$$= 2 - (3 + s) - s$$
$$= -1 - 2s$$

We conclude that an arbitrary solution to this system has the form

$$\mathbf{y} = (x_1, x_2, x_3)$$
$$= (-1 - 2s, 3 + s, s)$$
$$= (-1, 3, 0) + (-2s, s, s)$$
$$= (-1, 3, 0) + s(-2, 1, 1)$$

Geometrically, the solution set of this system is a line in R^3 (see Section 1.2).

In the next section we will see a notationally simpler approach to finding solutions of systems of linear equations.

Homogeneous Systems of Linear Equations

One of the most important cases of a system of linear equations is one in which the solution set is a subspace of R^n. From earlier examples we saw that this occurred when the constants of the system were zeros. We begin with an example.

Example 5   Solve the following system of linear equations.

$$x_1 + 4x_2 + x_3 = 0$$
$$x_1 - 2x_2 + 3x_3 = 0$$
$$3x_1 + 6x_2 + 5x_3 = 0$$

By adding the appropriate multiples of the first equation to the second and third equations (two type 3 operations), we obtain

$$x_1 + 4x_2 + x_3 = 0$$
$$- 6x_2 + 2x_3 = 0$$
$$- 6x_2 + 2x_3 = 0$$

By performing another type 3 operation, we have

$$x_1 + 4x_2 + x_3 = 0$$
$$- 6x_2 + 2x_3 = 0$$
$$0 = 0$$

Because the last equation provides no information, we need only consider the system consisting of the first two equations. If we assign the parameter s to x_3 and then use back substitution, we have

$$x_2 = \frac{1}{3} x_3$$

$$= \frac{1}{3} s$$

and

$$x_1 = -4x_2 - x_3$$

$$= -\frac{4}{3} s - s$$

$$= -\frac{7}{3} s$$

We conclude that an arbitrary solution has the form

$$\mathbf{y} = (x_1, x_2, x_3)$$

$$= \left(-\frac{7}{3}s, \frac{1}{3}s, s\right)$$

$$= s\left(\frac{-7}{3}, \frac{1}{3}, 1\right)$$

Thus, the solution set consists of all scalar multiples of the vector $(-\frac{7}{3}, \frac{1}{3}, 1)$, that is, the span of $\{(-\frac{7}{3}, \frac{1}{3}, 1)\}$. Hence, the solution set is a subspace spanned by one vector. Geometrically, the solution set may be represented by a line through the origin.

Definition   A system of linear equations is called a ***homogeneous system of linear equations*** if all the constants are zero. The solution set of a homogeneous system of linear equations is called the ***solution space*** of the system.

Using this definition, we may conclude that the solution space of the homogeneous system in Example 5 is the span of $\{(-\frac{7}{3}, \frac{1}{3}, 1)\}$. In Corollary 2.5.2 we shall show that a solution space is always a subspace.

Example 6   Solve the following homogeneous system of one linear equation in three unknowns.

$$x_1 - x_2 + 3x_3 = 0$$

Because there is only one equation, we need not be concerned with the elementary row operations. Proceeding with back substitution, we assign the parameters s and t to x_2 and x_3, respectively. We obtain

$$x_1 = x_2 - 3x_3$$

$$= s - 3t$$

Thus, an arbitrary solution to this system may be represented as

$$\mathbf{y} = (x_1, x_2, x_3)$$

$$= (s - 3t, s, t)$$

$$= (s, s, 0) + (-3t, 0, t)$$

$$= s(1, 1, 0) + t(-3, 0, 1)$$

So the solution space consists of all linear combinations of the two vectors above, that is, the span of $\{(1, 1, 0), (-3, 0, 1)\}$. This means that the solution space is a subspace of R^3. Because the vectors are not parallel, the solution space may be represented as a plane through the origin.

In Section 1.5 we will construct a general procedure for solving systems of linear equations.

Exercises

1. Determine whether or not each of the following systems of equations is linear. Justify your answer.

   (a) $\dfrac{1}{x} + \dfrac{2}{y} = 3$

   $x - 3y = 7$

   (b) $x + 3y - z = \frac{1}{3}$

   $x - 4y \quad\;\; = 2$

   (c) $yx + zy - xz = 2$

   $3x + 5y + z^2 = 4$

   (d) $2 - 3x + y = z$

   $x - 3 \; + z = 2$

2. Find all solutions to each of the given systems of linear equations.

   (a) $x - y + 2z = 3$

   $3x + 2y + z = 4$

   $2x - 3y + 3z = 1$

   (b) $2x + y = 4$

   (c) $x + 3y - z = 4$

   $2x + 5y + z = 6$

   (d) $3x + 2y - z = 1$

   $x + y + z = 3$

   $x \quad\;\; - 3z = -5$

   (e) $3x + 2y + z - w = 0$

   $x + 2y + z + w = 0$

   (f) $2x + 3y - z + w = 4$

   (g) $x + 2y \quad\;\; - w = 2$

   $2x + 5y - 3z \quad\;\; = 3$

   $x + 3y - 5z + 3w = 5$

   (h) $x + 2y + z - w = 3$

   $2x + 5y - z \quad\;\; = 10$

   $x + y + 4z - 3w = -1$

3. Liquid A is a solution of 10 grams of salt and 20 grams of sugar per liter of water. Liquid B is a solution of 15 grams of salt and 8 grams of sugar per liter of water. What volumes of the liquids should be combined to produce a solution containing exactly 7 grams of salt and $\frac{20}{3}$ grams of sugar?

4. Find three numbers whose sum is 10, and such that the sum of the first and second numbers is 4 while the sum of the first and third numbers is 8.

5. Companies A, B, and C each produce variety packs of nuts consisting of walnuts, filberts, and almonds. The amount of each kind of nut in pounds is described according to Table 1.4.3. How many variety packs should be purchased from each company to obtain a total of 11 pounds of walnuts, 8 pounds of filberts, and 6 pounds of almonds?

TABLE 1.4.3

Company	Walnuts	Filberts	Almonds
A	3	2	1
B	1	1	1
C	2	1	1

6. Referring to Exercise 5, describe the different ways of purchasing variety packs from companies A, B, and C to obtain 11 pounds of walnuts and 8 pounds of filberts, ignoring the amount of almonds obtained.

7. Prove that a type 1 or type 2 elementary operation transforms a system of linear equations into an equivalent system.

8. Prove that a type 3 elementary operation transforms a system of linear equations into an equivalent system.

9. Prove that the result of *dividing* the terms of one of the equations of a system of linear equations by a nonzero constant is a type 2 elementary operation.

10. Prove that the result of *subtracting* a multiple of one the equations of a system of linear equations from another equation of the system is a type 3 elementary operation.

1.5 MATRICES AND GAUSSIAN ELIMINATION

In this section we introduce one of the most important tools of linear algebra: the "matrix." Although our purpose for defining matrices in this section is as an aid to finding solutions to systems of linear equations, we will see matrix applications throughout the remainder of this text.

A system of linear equations is completely determined by the coefficients and the constants of the system. For example, the system

$$2x_1 + x_2 - x_3 = 5$$
$$x_1 + 2x_3 = 1$$

is completely determined by the array

$$
\begin{array}{cccc}
2 & 1 & -1 & 5 \\
1 & 0 & 2 & 1
\end{array}
$$

The first three entries of each row are the coefficients of the unknowns of the corresponding equation, and the last entry of the row is the constant of the corresponding equation. In solving a system of linear equations, we can more conveniently manipulate with the rows of such an array than with the actual equations. This motivates the following definition.

Definition   A rectangular array of scalars is called a ***matrix***. The scalars are called the ***entries*** of the matrix. If the matrix has m rows and n columns, we say that it is an ***m by n matrix***, or that it has ***size m by n***, written $m \times n$. If $m = n$, we say that the matrix is ***square*** and that the matrix is of ***order n***.

We usually use capital letters at the beginning of the alphabet to denote particular matrices. Customarily, the array is surrounded by brackets. For example, if

$$
A = \begin{bmatrix} 1 & 2 & 3 \\ 1 & -1 & 2 \end{bmatrix} \quad
B = \begin{bmatrix} 0 & 1 \\ 1 & 2 \end{bmatrix} \quad
C = \begin{bmatrix} -1 & 0 \\ 2 & 1 \\ 4 & 3 \end{bmatrix}
$$

then A is a 2×3 matrix, B is a 2×2 matrix, and C is a 3×2 matrix. For any $m \times n$ matrix A, we let A_{ij}, called the *ijth entry* of A, denote the entry in the ith row and jth column of A. For example, with respect to the matrices A, B, and C above,

$$A_{11} = 1, \quad A_{23} = 2, \quad B_{12} = 1, \quad C_{31} = 4$$

We now relate matrices to systems of equations.

Definition   Given the system of m linear equations in n unknowns

$$a_{11}x_1 + a_{12}x_2 + \cdots + a_{1n}x_n = b_1$$
$$a_{21}x_1 + a_{22}x_2 + \cdots + a_{2n}x_n = b_2$$
$$\vdots$$
$$a_{m1}x_1 + a_{m2}x_2 + \cdots + a_{mn}x_n = b_m$$

the $m \times n$ matrix

$$\begin{bmatrix} a_{11} & a_{12} & \cdots & a_{1n} \\ a_{21} & a_{22} & \cdots & a_{2n} \\ & & \vdots & \\ a_{m1} & a_{m2} & \cdots & a_{mn} \end{bmatrix}$$

is called the *coefficient matrix* of the system. The $m \times (n + 1)$ matrix

$$\begin{bmatrix} a_{11} & a_{12} & \cdots & a_{1n} & b_1 \\ a_{21} & a_{22} & \cdots & a_{2n} & b_2 \\ & & \vdots & & \\ a_{m1} & a_{m2} & \cdots & a_{mn} & b_m \end{bmatrix}$$

is called the *augmented matrix* of the system. Notice that the first n columns of the augmented matrix consist of the columns of the coefficient matrix of the system.

For example, the system

$$2x_1 + x_2 - x_3 = 4$$
$$x_1 \qquad + x_3 = 7$$

has coefficient matrix

$$\begin{bmatrix} 2 & 1 & -1 \\ 1 & 0 & 1 \end{bmatrix}$$

and augmented matrix

$$\begin{bmatrix} 2 & 1 & -1 & 4 \\ 1 & 0 & 1 & 7 \end{bmatrix}$$

Notice the entry 0 in the 2, 2 position of this matrix. If an unknown is absent in an equation, it has coefficient equal to zero.

Certain manipulations of the rows of the augmented matrix of a system of linear equations produce the effect of performing elementary operations on the system. Since any row of an $m \times n$ matrix has n entries, it can be regarded as an ordered n-tuple. With this in mind, we reformulate the concept of an elementary operation on a system of linear equations in the context of rows of a matrix.

Definition   An *elementary row operation* on a matrix is an operation that results in a new matrix in one of three ways:

1. Interchange two rows of the matrix.
2. Multiply a row of the matrix by a nonzero scalar.
3. Add a multiple of one of the rows of the matrix to another row of the matrix.

An elementary row operation is of *type 1, 2, or 3* depending on whether it is obtained by operation 1, 2, or 3.

_____

For example, consider the matrix

$$A = \begin{bmatrix} 1 & 3 & -1 & 0 \\ 1 & 2 & 5 & 4 \\ 2 & -1 & 0 & 3 \end{bmatrix}$$

Adding -2 times the first row of A to the third row of A, we obtain the matrix

$$B = \begin{bmatrix} 1 & 3 & -1 & 0 \\ 1 & 2 & 5 & 4 \\ 0 & -7 & 2 & 3 \end{bmatrix}$$

Notice that A is transformed into B by means of an elementary row operation of type 3. As with elementary operations on linear equations, we can describe this operation as subtracting 2 times the first row of A from the third row of A. In general, a type 3 elementary row operation on a matrix can be described in terms of subtraction (see Exercise 4).

If we compare elementary operations on a system of linear equations to elementary row operations on a matrix, it is obvious that the effect of performing an elementary row operation on the augmented matrix of a system of linear equations is identical to that of performing the comparable elementary operation on the system. Consequently, we have the following reformulation of Theorem 1.4.2.

Theorem 1.5.1 *An elementary row operation transforms the augmented matrix of a system of linear equations into the augmented matrix of an equivalent system of linear equations.*

This theorem will allow us to replace a system by an equivalent system which is in a very convenient form to solve. This form is called "triangular" and it is defined below.

Definition A matrix is in **triangular form** if it satisfies the following two conditions:

1. Any row consisting only of zeros is below any row that contains at least one nonzero entry.
2. The first (reading from left to right) nonzero entry of any row is to the left of the first nonzero entry of any lower row.

The following three matrices are in triangular form.

$$\begin{bmatrix} 5 & 2 & 3 & 4 \\ 0 & 0 & 7 & 6 \\ 0 & 0 & 0 & 1 \\ 0 & 0 & 0 & 0 \end{bmatrix} \qquad \begin{bmatrix} 0 & -1 & 2 & -1 & 0 \\ 0 & 0 & 3 & 0 & 0 \\ 0 & 0 & 0 & 0 & -6 \end{bmatrix} \qquad \begin{bmatrix} 1 & 0 & 0 \\ 0 & 2 & 0 \\ 0 & 0 & 2 \end{bmatrix}$$

Notice that because of condition 2, any entry of a matrix in triangular form which lies in the same column but below the first nonzero entry of a row must be equal to zero.

The following two matrices are *not* in triangular form.

$$A = \begin{bmatrix} 2 & 0 & 0 & 0 \\ 3 & 0 & 0 & 0 \\ 0 & 0 & 0 & 0 \end{bmatrix} \qquad B = \begin{bmatrix} 0 & 0 & 0 & 0 \\ 0 & 2 & 0 & 0 \\ 0 & 0 & 4 & 0 \end{bmatrix}$$

The matrix A is not in triangular form because the first nonzero entry of the first row of A is not to the left of the first nonzero entry of the second row of A. The matrix B is not in triangular form because its top row consists only of zeros.

Some authors add the condition:

3. The first nonzero entry of each row is 1.

A matrix that satisfies all three conditions is said to be in **row echelon form**. To transform a matrix that is in triangular form to one in row echelon form requires several divisions. If a calculator or a computer is used, these divisions will often introduce unnecessary round-off errors. Because our main objective in the use of these forms is to achieve accurate solutions to systems of linear equations, we prefer the triangular form.

We shall apply Theorem 1.5.1 to solve the following system of linear equations.

$$2x_1 + 4x_2 \qquad - 8x_4 = -10$$
$$x_1 + 2x_2 + x_3 - x_4 = \quad -2$$
$$x_3 + 3x_4 = \qquad 3$$

The augmented matrix for this system is

$$\begin{bmatrix} 2 & 4 & 0 & -8 & -10 \\ 1 & 2 & 1 & -1 & -2 \\ 0 & 0 & 1 & 3 & 3 \end{bmatrix}$$

In a parallel approach to the method in the preceding section, we will construct new matrices which are the augmented matrices of equivalent systems by systematically eliminating unknowns. However, now we will apply the elementary row operations. Interchanging the first and second rows (a type 1 operation) transforms the matrix into

$$\begin{bmatrix} 1 & 2 & 1 & -1 & -2 \\ 2 & 4 & 0 & -8 & -10 \\ 0 & 0 & 1 & 3 & 3 \end{bmatrix}$$

Adding -2 times the first row to the second row (a type 3 elementary row operation) transforms the matrix into

$$\begin{bmatrix} 1 & 2 & 1 & -1 & -2 \\ 0 & 0 & -2 & -6 & -6 \\ 0 & 0 & 1 & 3 & 3 \end{bmatrix}$$

Multiplying the second row by $-\frac{1}{2}$ (a type 2 operation), we obtain

$$\begin{bmatrix} 1 & 2 & 1 & -1 & -2 \\ 0 & 0 & 1 & 3 & 3 \\ 0 & 0 & 1 & 3 & 3 \end{bmatrix}$$

Finally, adding -1 times the second row to the third row (a type 3 operation), we obtain

$$\begin{bmatrix} 1 & 2 & 1 & -1 & -2 \\ 0 & 0 & 1 & 3 & 3 \\ 0 & 0 & 0 & 0 & 0 \end{bmatrix}$$

This matrix is the augmented matrix of the equivalent system of linear equations

$$x_1 + 2x_2 + x_3 - x_4 = -2$$
$$x_3 + 3x_4 = 3$$

Notice that since the last row of the matrix above consists of all zeros, it yields the equation $0 = 0$, which can be disregarded. Also, observe that whereas in the preceding section we performed elementary operations on the equations in a system to eliminate unknowns, now we apply row operations on the augmented matrix to produce a matrix in triangular form.

An unknown of a system of linear equations whose augmented matrix is in triangular form is called an **_initial unknown_** if it appears as the first unknown in one of the equations. In the system above, x_1 and x_3 are the initial unknowns. Observe that if the unknowns which are not initial are assigned arbitrary values, the system can be solved using back substitution. For this reason, we assign parameters to these noninitial unknowns, that is,

$$x_2 = s \quad \text{and} \quad x_4 = t$$

We can now solve for the initial unknowns by back substitution:

$$x_3 = 3 - 3x_4$$
$$= 3 - 3t$$

and

$$x_1 = -2 - 2x_2 - x_3 + x_4$$
$$= -2 - 2s - (3 - 3t) + t$$
$$= -5 - 2s + 4t$$

In vector form the general solution to the system is

$$\mathbf{y} = (x_1, x_2, x_3, x_4)$$
$$= (-5 - 2s + 4t, s, 3 - 3t, t)$$
$$= (-5, 0, 3, 0) + s(-2, 1, 0, 0) + t(4, 0, -3, 1)$$

In order for the technique above to work in general, we need to know that every matrix can be transformed by elementary row operations to a triangular form. This result seems reasonable and may be proved by mathematical induction. We state the result as a theorem without proof.

Theorem 1.5.2 *Any matrix may be transformed into an upper triangular matrix by a finite number of elementary row operations.*

This theorem suggests a general technique for solving any system of linear equations. The process described below is called *Gaussian elimination*, after Carl Friedrich Gauss, a famous nineteenth-century German mathematician. We list the steps of the procedure.

Gaussian Elimination

1. Construct the augmented matrix for the system.
2. By means of elementary row operations, transform this augmented matrix into a matrix in triangular form.
3. The matrix in triangular form is the augmented matrix of a system of linear equations. Determine this system.
4. After assigning parametric values to the noninitial unknowns, apply the method of back substitution to solve the system in 3.

Example 1   By means of the method of Gaussian elimination, find all solutions to the system of linear equations

$$
\begin{aligned}
2x_1 - 3x_2 - \ x_3 + 2x_4 + \ 3x_5 &= \ \ 4 \\
4x_1 - 4x_2 - \ x_3 + 4x_4 + 11x_5 &= \ \ 4 \\
2x_1 - 5x_2 - 2x_3 + 2x_4 - \ \ x_5 &= \ \ 9 \\
2x_2 + \ x_3 \qquad\quad + \ 4x_5 &= -5
\end{aligned}
$$

The augmented matrix for the system is

$$
\begin{bmatrix}
2 & -3 & -1 & 2 & 3 & 4 \\
4 & -4 & -1 & 4 & 11 & 4 \\
2 & -5 & -2 & 2 & -1 & 9 \\
0 & 2 & 1 & 0 & 4 & -5
\end{bmatrix}
$$

We must transform this matrix into a matrix in triangular form. By adding the appropriate multiples of the first row to the other rows (three type 3 operations), we obtain

$$
\begin{bmatrix}
2 & -3 & -1 & 2 & 3 & 4 \\
0 & 2 & 1 & 0 & 5 & -4 \\
0 & -2 & -1 & 0 & -4 & 5 \\
0 & 2 & 1 & 0 & 4 & -5
\end{bmatrix}
$$

By applying two more type 3 operations, we have

$$\begin{bmatrix} 2 & -3 & -1 & 2 & 3 & 4 \\ 0 & 2 & 1 & 0 & 5 & -4 \\ 0 & 0 & 0 & 0 & 1 & 1 \\ 0 & 0 & 0 & 0 & -1 & -1 \end{bmatrix}$$

Finally, if we add row 3 to row 4, we have

$$\begin{bmatrix} 2 & -3 & -1 & 2 & 3 & 4 \\ 0 & 2 & 1 & 0 & 5 & -4 \\ 0 & 0 & 0 & 0 & 1 & 1 \\ 0 & 0 & 0 & 0 & 0 & 0 \end{bmatrix}$$

This matrix is in triangular form. It is the augmented matrix for a system of linear equations equivalent to the original one. In writing the system associated with this matrix we disregard the last row which consists of all zeros. Thus, we have

$$2x_1 - 3x_2 - x_3 + 2x_4 + 3x_5 = 4$$
$$2x_2 + x_3 \qquad + 5x_5 = -4$$
$$x_5 = 1$$

We now solve this system using the technique of back substitution. The initial unknowns are x_1, x_2, and x_5. We set $x_3 = s$ and $x_4 = t$. Clearly, $x_5 = 1$. Using the second equation to solve for x_2, we have

$$x_2 = -2 - \frac{1}{2}x_3 - \frac{5}{2}x_5$$

$$= -2 - \frac{1}{2}s - \frac{5}{2}(1)$$

$$= -\frac{9}{2} - \frac{1}{2}s$$

Using the first equation to solve for x_1, we have

$$x_1 = 2 + \frac{3}{2}x_2 + \frac{1}{2}x_3 - x_4 - \frac{3}{2}x_5$$

$$= 2 + \frac{3}{2}\left(-\frac{9}{2} - \frac{1}{2}s\right) + \frac{1}{2}s - t - \frac{3}{2}(1)$$

$$= -\frac{25}{4} - \frac{1}{4}s - t$$

In vector form the general solution to the system is

$$(x_1, x_2, x_3, x_4, x_5) = \left(-\frac{25}{4} - \frac{1}{4}s - t, \ -\frac{9}{2} - \frac{1}{2}s, \ s, \ t, \ 1\right)$$

$$= \left(-\frac{25}{4}, \ -\frac{9}{2}, \ 0, 0, 1\right)$$

$$+ s\left(-\frac{1}{4}, \ -\frac{1}{2}, \ 1, 0, 0\right) + t(-1, 0, 0, 1, 0)$$

Example 2   By means of the method of Gaussian elimination, find all solutions to the system of linear equations

$$2x_1 + 4x_2 - 4x_3 + 4x_4 + \ x_5 = 7$$
$$x_1 + \ x_2 - 3x_3 + 2x_4 + \ x_5 = 2$$
$$x_1 + 2x_2 - 2x_3 + 2x_4 + 3x_5 = 6$$
$$x_1 + 2x_2 - 2x_3 + 2x_4 - 2x_5 = 1$$

We shall apply the elementary row operations to the augmented matrix to produce a matrix in triangular form. Below we have indicated which operations are used. We leave it to the reader to fill in the details.

$$\begin{bmatrix} 2 & 4 & -4 & 4 & 1 & 7 \\ 1 & 1 & -3 & 2 & 1 & 2 \\ 1 & 2 & -2 & 2 & 3 & 6 \\ 1 & 2 & -2 & 2 & -2 & 1 \end{bmatrix} \xrightarrow[\text{type 1}]{\text{one}} \begin{bmatrix} 1 & 1 & -3 & 2 & 1 & 2 \\ 2 & 4 & -4 & 4 & 1 & 7 \\ 1 & 2 & -2 & 2 & 3 & 6 \\ 1 & 2 & -2 & 2 & -2 & 1 \end{bmatrix} \xrightarrow[\text{type 3}]{\text{three}}$$

$$\begin{bmatrix} 1 & 1 & -3 & 2 & 1 & 2 \\ 0 & 2 & 2 & 0 & -1 & 3 \\ 0 & 1 & 1 & 0 & 2 & 4 \\ 0 & 1 & 1 & 0 & -3 & -1 \end{bmatrix} \xrightarrow[\text{type 1}]{\text{one}} \begin{bmatrix} 1 & 1 & -3 & 2 & 1 & 2 \\ 0 & 1 & 1 & 0 & -3 & -1 \\ 0 & 1 & 1 & 0 & 2 & 4 \\ 0 & 2 & 2 & 0 & -1 & 3 \end{bmatrix} \xrightarrow[\text{type 3}]{\text{two}}$$

$$\begin{bmatrix} 1 & 1 & -3 & 2 & 1 & 2 \\ 0 & 1 & 1 & 0 & -3 & -1 \\ 0 & 0 & 0 & 0 & 5 & 5 \\ 0 & 0 & 0 & 0 & 5 & 5 \end{bmatrix} \xrightarrow[\text{type 3}]{\text{one}} \begin{bmatrix} 1 & 1 & -3 & 2 & 1 & 2 \\ 0 & 1 & 1 & 0 & -3 & -1 \\ 0 & 0 & 0 & 0 & 5 & 5 \\ 0 & 0 & 0 & 0 & 0 & 0 \end{bmatrix}$$

The last matrix is in triangular form. It is the augmented matrix for a system of linear equations equivalent to the one given. In writing the system associated with this matrix, we disregard the last row, which consists of all zeros. We have

$$
\begin{aligned}
x_1 + x_2 - 3x_3 + 2x_4 + \ x_5 &= \ \ 2 \\
x_2 + \ x_3 \qquad\quad - 3x_5 &= -1 \\
5x_5 &= \ \ 5
\end{aligned}
$$

We now solve this system using back substitution. The initial unknowns are x_1, x_2, and x_5. Set $x_3 = s$ and $x_4 = t$. Clearly, $x_5 = 1$. Using the second equation to solve for x_2, we have

$$
\begin{aligned}
x_2 &= -1 - x_3 + 3x_5 \\
&= -1 - s + 3 \\
&= 2 - s
\end{aligned}
$$

Using the first equation to solve for x_1, we have

$$
\begin{aligned}
x_1 &= 2 - x_2 + 3x_3 - 2x_4 - x_5 \\
&= 2 - (2 - s) + 3s - 2t - 1 \\
&= -1 + 4s - 2t
\end{aligned}
$$

In vector form the general solution to the system is

$$
\begin{aligned}
(x_1, x_2, x_3, x_4, x_5) &= (-1 + 4s - 2t, 2 - s, s, t, 1) \\
&= (-1, 2, 0, 0, 1) + s(4, -1, 1, 0, 0) + t(-2, 0, 0, 1, 0)
\end{aligned}
$$

Inconsistent Systems

As we have seen earlier, there are systems of linear equations with no solutions. Such systems are said to be *inconsistent* or *overdetermined*. Otherwise, the system is said to be *consistent*. We usually cannot recognize an inconsistent system immediately. However, the inconsistency of such a system becomes apparent once the augmented matrix of the system is transformed into a matrix in triangular form. Such a matrix is the augmented matrix of a system of linear equations equivalent to the given one, so it, too, will be inconsistent. This leads us to the following question: What can be said about an inconsistent system of linear equations whose augmented matrix is in triangular form?

The answer is that the last nontrivial equation of the system contains no unknowns with nonzero coefficients. For if it did, it would have an initial unknown that could be solved for and substituted into the preceding equation. This process could be continued by back substitution, producing a solution to the system. In terms of the associated augmented matrix, this means that the last nontrivial row must be of the form $(0, 0, \ldots, 0, c)$, where $c \neq 0$. We may restate this result as:

Theorem 1.5.3   *A system of linear equations is inconsistent if and only if it is equivalent to a triangular system whose augmented matrix has a row of the form $(0, 0, \ldots, 0, c)$, where $c \neq 0$.*

Example 3   Find all solutions to the system

$$x_1 + 3x_2 + 2x_3 = 2$$
$$x_1 + 2x_2 + x_3 = 1$$
$$x_1 + 5x_2 + 4x_3 = 5$$

The augmented matrix of this system

$$\begin{bmatrix} 1 & 3 & 2 & 2 \\ 1 & 2 & 1 & 1 \\ 1 & 5 & 4 & 5 \end{bmatrix}$$

can be transformed to the matrix in triangular form (we omit the details):

$$\begin{bmatrix} 1 & 3 & 2 & 2 \\ 0 & 1 & 1 & 1 \\ 0 & 0 & 0 & 1 \end{bmatrix}$$

Because of the form of the last row of the matrix above, we may conclude that the original system is inconsistent.

Systems of Linear Equations and Subspaces

We now apply the techniques of solving systems of linear equations to determine if a given vector belongs to a subspace.

Suppose that W is the subspace spanned by the set

$$S = \{(1, 2, 1, 1), (0, 1, -1, 1), (1, 0, 2, 3), (1, -1, 2, 6)\}$$

Consider the vectors $\mathbf{y}$ and $\mathbf{z}$ given by

$$\mathbf{y} = (1, 5, -1, 0) \quad \text{and} \quad \mathbf{z} = (11, -3, -4, -1)$$

Determine whether or not $\mathbf{y}$ and $\mathbf{z}$ are contained in W. This is equivalent to checking if $\mathbf{y}$ and $\mathbf{z}$ are linear combinations of the vectors of S.

First, consider the vector **y**. The vector **y** is a linear combination of the vectors of S if and only if there exist scalars x_1, x_2, x_3, and x_4 such that

$$(1, 5, -1, 0) = x_1(1, 2, 1, 1) + x_2(0, 1, -1, 1) + x_3(1, 0, 2, 3) + x_4(1, -1, 2, 6)$$

Adding the vectors on the right-hand side of the equation above and equating corresponding components we obtain the following system of equations:

$$
\begin{aligned}
x_1 \quad\ \ + x_3 + x_4 &= \ \ 1 \\
2x_1 + x_2 \qquad\ - x_4 &= \ \ 5 \\
x_1 - x_2 + 2x_3 + 2x_4 &= -1 \\
x_1 + x_2 + 3x_3 + 6x_4 &= \ \ 0
\end{aligned}
$$

Solving this system, we obtain

$$
\begin{aligned}
x_1 &= \ \ 2 + s \\
x_2 &= \ \ 1 - s \\
x_3 &= -1 - 2s \\
x_4 &= \quad s
\end{aligned}
$$

Therefore, there are infinitely many solutions. In particular, if $s = 0$, the values $x_1 = 2$, $x_2 = 1$, $x_3 = -1$, and $x_4 = 0$ can be used to represent **y** as a linear combination of vectors in S:

$$\mathbf{y} = (1, 5, -1, 0) = 2(1, 2, 1, 1) + (0, 1, -1, 1) - (1, 0, 2, 3)$$

We now consider whether or not the vector **z** is in W. The vector **z** is a linear combination of the vectors in S if and only if there exist scalars x_1, x_2, x_3, and x_4 such that

$$(11, -3, -4, -1) = x_1(1, 2, 1, 1) + x_2(0, 1, -1, 1) + x_3(1, 0, 2, 3) + x_4(1, -1, 2, 6)$$

Adding the vectors on the right-hand side of the equation above and equating corresponding coordinates, we obtain the system of equations

$$
\begin{aligned}
x_1 \quad\ \ + x_3 \ + x_4 &= \ \ 11 \\
2x_1 + x_2 \qquad\ - x_4 &= -3 \\
x_1 - x_2 + 2x_3 + 2x_4 &= -4 \\
x_1 + x_2 + 3x_3 + 6x_4 &= -1
\end{aligned}
$$

The augmented matrix for this system is

$$\begin{bmatrix} 1 & 0 & 1 & 1 & 11 \\ 2 & 1 & 0 & -1 & -3 \\ 1 & -1 & 2 & 2 & -4 \\ 1 & 1 & 3 & 6 & -1 \end{bmatrix}$$

Reducing this matrix to triangular form, we obtain

$$\begin{bmatrix} 1 & 0 & 1 & 1 & 11 \\ 0 & 1 & -2 & -3 & -25 \\ 0 & 0 & 1 & 2 & 40 \\ 0 & 0 & 0 & 0 & 1 \end{bmatrix}$$

Because the last row of the matrix is $(0, 0, 0, 0, 1)$, we know that the system is inconsistent; that is, there are no solutions. Thus, the vector $\mathbf{z}$ is not a linear combination of the vectors of S. So $\mathbf{z}$ is not a vector in W.

Numerical Considerations

Although we shall return to Gaussian elimination as well as to other methods for solving systems of linear equations in Chapter 9, it is important to discuss what we have done thus far from the point of view of both numerical efficiency and accuracy.

In many applications it is necessary to solve systems consisting of hundreds of equations in hundreds of unknowns. As the computations described in this section would be virtually impossible to perform by hand, a computer must be employed. In this situation the time a method takes, the necessary storage inside the computer, and the numerical accuracy in terms of round-off errors must all be considered.

To obtain a rough approximation for the length of time it will take for a computer to carry out Gaussian elimination, we shall perform an operations count. As computer time is approximately proportional to the number of operations it must perform, we shall have a measure with which to compare various methods. For the sake of simplicity, we assume that we have a system of n equations in n unknowns, and we only count multiplications (we consider divisions as multiplications).

To eliminate x_1 from the ith equation we must form $-a_{i1}/a_{11}$ and multiply it by the entries of row 1 (except that we do not multiply a_{11} by this term since we know that the result is equal to 1) and then add this row to row i. This procedure requires $n + 1$ multiplications (as well as $n + 1$ additions). As we must repeat this for the remaining $n - 1$ rows to eliminate x_1 from the second equation on, we shall have performed $(n + 1)(n - 1)$ multiplications. To eliminate x_2 from the third equation on, we must perform $n(n - 2)$ multiplications. Thus, to obtain a system in triangular form, we are required to perform

$$\sum_{k=3}^{n+1} k(k - 2) = \frac{n(n - 1)(2n + 5)}{6}$$

$$= \frac{n^3}{3} + \frac{n^2}{2} - \frac{5n}{6}$$

multiplications (see Exercise 10).

For large values of n we may neglect terms in the expression above which contain powers of n less than 3. So for large systems it takes approximately $n^3/3$ multiplications to reduce the system to triangular form. It is left as an exercise to show that the number of multiplications to back-solve the system is $n(n + 1)/2$. We can see from these computations that back substitution takes only a small fraction of the time.

Another popular method for solving systems is to use the ith equation to eliminate the unknown x_i from the first $i - 1$ equations as well as the last $n - i$ equations, creating a *diagonal* system. This method is called *Gauss–Jordan elimination* and it is usually far more costly to perform (see Exercise 12).

As a rule, more operations cause more round-off errors. Since a calculator or computer can only hold a finite number of decimal places for each entry, the errors may become magnified as the number of operations increase. In Section 9.1 we discuss *pivoting* strategies and iterative methods to reduce these errors.

Although an analysis of the storage of these computations in a computer for a large system properly belongs in a course in numerical methods, we should mention that for a large class of problems, the coefficient matrices have very few nonzero entries (such matrices are called *sparse*). In such cases these "empty" locations may be used to hold various computed quantities. Similar techniques may be used for matrices exhibiting some form of symmetry.

Exercises

1. For each matrix:
   (i) Determine whether or not the matrix is in triangular form.
   (ii) Determine whether or not the matrix is in row echelon form.

(a) $\begin{bmatrix} 2 & 1 & 0 \\ 0 & 0 & 1 \\ 0 & 0 & 0 \end{bmatrix}$ (b) $\begin{bmatrix} 1 & 2 & 3 & 4 \\ 0 & 1 & 0 & 0 \\ 0 & 0 & 1 & 2 \end{bmatrix}$

(c) $\begin{bmatrix} 0 & 0 & 0 & 1 \\ 1 & 0 & 0 & 1 \\ 0 & 1 & 0 & 0 \end{bmatrix}$ (d) $\begin{bmatrix} 3 & 2 & 1 & 0 \\ 0 & 0 & 0 & 0 \\ 0 & 1 & 2 & 1 \end{bmatrix}$

(e) $\begin{bmatrix} 0 & 0 & 1 \\ 0 & 1 & 0 \\ 1 & 0 & 0 \end{bmatrix}$ (f) $[1 \quad 2 \quad 0 \quad 0]$

(g) $[0 \quad 2 \quad 1 \quad 0]$ (h) $\begin{bmatrix} 1 & 0 \\ 2 & 1 \\ 0 & 0 \end{bmatrix}$

(i) $\begin{bmatrix} 2 \\ 0 \\ 0 \end{bmatrix}$

2. Use elementary row operations to transform the given matrix into a matrix in triangular form.

(a) $\begin{bmatrix} 1 & 3 & 1 \\ 0 & 0 & 0 \\ 2 & 0 & 1 \end{bmatrix}$

(b) $\begin{bmatrix} 1 & 2 & 1 & -1 \\ 3 & 1 & 2 & 0 \\ 1 & -8 & -1 & 5 \end{bmatrix}$

(c) $\begin{bmatrix} 1 & 2 \\ 1 & 3 \\ 2 & 1 \\ 0 & 1 \end{bmatrix}$

(d) $\begin{bmatrix} 3 & 1 & 0 & 1 \\ 2 & -1 & 0 & 3 \\ 0 & 0 & 1 & 1 \\ 1 & 2 & 1 & -1 \end{bmatrix}$

(e) $\begin{bmatrix} 1 & 2 & 1 & 2 \\ 2 & 4 & 5 & 1 \\ 3 & 6 & 3 & 2 \end{bmatrix}$

3. Use the method of Gaussian elimination to solve each of the systems of linear equations.

(a) $2x + y + 2z = 5$
$x + 2y + 3z = 5$
$3x + 3z = 3$

(b) $3x + y + z = 5$
$x + 2y - z = 7$
$x - 3y - 3z = 9$

(c) $2x - y + 2z = 3$
$4x + y - z = 2$
$6x + 3y - 4z = 4$

(d) $3x + 2y - z = 4$

(e) $x + y + z + 3w = 6$
$2x - 5y + 2z - w = -2$

(f) $2x_1 - x_2 + 3x_3 + x_4 + 2x_5 = 11$
$3x_1 + 2x_2 - 2x_4 + x_5 = -2$
$x_1 + 2x_2 - x_3 + 2x_4 - x_5 = 0$

4. Show that the procedure of *subtracting* a multiple of one of the rows of a matrix from another row of the same matrix is equivalent to an elementary row operation of type 3 on the matrix.

5. For each system of linear equations, answer the following questions:
(i) For what value(s) of c is the system inconsistent?

(ii) For what value(s) of c does the system have a *unique* solution?

(a) $x + y + z = 1$
$x - y + z = 2$
$x + z = c$

(b) $x + cy = 3$
$x - y = 3$

6. Determine whether each of the following systems of linear equations is consistent or inconsistent.

(a) $x + y = 3$
$x - y = 2$

(b) $x + 2y = 1$
$2x + y = -1$
$x - 4y = -5$

(c) $x + 2y = 3$
$2x + y = 2$
$x + 5y = 1$

(d) $x + 2y + z = -1$
$2x + 4y + 2z = -2$

(e) $x + 2y + 3z = 1$
$2x + 5y + 5z = 3$
$x + 3y + 2z = 4$

(f) $2x + z - w = 1$
$x + 2y + 3z + w = 1$
$2x + y - z + w = 2$

(g) $2x + 3y + z - w = 2$
$x + 2y + 3z = -2$
$x + y - 2z - w = 4$

(h) $3x + 2y + z + 2w = 1$
$x + 2y - z + w = 0$
$2x + y + 2z + w = 3$
$2x + 3y - 2z + 2w = 2$

7. Solve the system of linear equations given on page 50.

8. For each of the following parts, determine whether or not the vector $\mathbf{y}$ is a linear combination of the vectors of the set S.

(a) $\mathbf{y} = (1, 3)$;
$S = \{(2, 1), (1, -1)\}$

(b) $\mathbf{y} = (2, 3)$;
$S = \{(1, 1), (2, 2)\}$

(c) $\mathbf{y} = (6, 3, -1)$;
$S = \{(1, 2, 1), (1, -1, 2), (2, 1, 3)\}$

(d) $\mathbf{y} = (1, 8, -1)$;
$S = \{(1, 2, 1), (1, -1, 2), (2, 1, 3)\}$

(e) $\mathbf{y} = (1, 1, 1, -2)$;
$S = \{(1, 2, 1, -1), (2, 1, 2, 3), (1, -1, 1, 0)\}$

(f) $\mathbf{y} = (2, 1, 2, 2)$;
$S = \{(1, 2, 1, -1), (2, 1, 2, 3), (1, -1, 1, 0)\}$

9. Refer to Exercise 5 of Section 1.4.
   (a) Associate with each company a vector in R^3 which characterizes the company's variety pack.
   (b) Explain how Exercise 5 of Section 1.4 can be reformulated in the context of linear combinations of vectors in part (a).

10. Use the formulas

$$\sum_{k=1}^{n} k = \frac{n(n+1)}{2}$$

and

$$\sum_{k=1}^{n} k^2 = \frac{n(n+1)(2n+1)}{6}$$

to verify the formula for the number of multiplications required to reduce a system of n linear equations in n unknowns to triangular form.

11. Verify that there are $n(n+1)/2$ multiplications required to back-solve a triangular system of n linear equations in n unknowns.

12. Determine the number of multiplications to solve a system of n linear equations in n unknowns using Gauss–Jordan elimination.

13. Determine the number of additions (or subtractions) to solve a system of n linear equations in n unknowns using Gaussian elimination.

*14. Use the program ROW REDUCTION to reduce the augmented matrix in Exercise 6(g) to triangular form.

*15. Use the program ROW REDUCTION to reduce the following matrix to a triangular matrix all of whose entries are integers.

$$\begin{bmatrix} 2 & 7 & 6 & 2 \\ 1 & 4.5 & 3.5 & 2.5 \\ -2 & -3 & -1 & 7 \\ 0.4 & 1 & 1 & 0.2 \end{bmatrix}$$

*16. Use the program LINSYST to solve the systems in parts (f), (g), and (h) of Exercise 6.

*17. Use the program LINSYST in parts (d), (e), and (f) of Exercise 8 to determine if the vector y is a linear combination of the vectors in S.

*18. Use the program LINSYST to solve the linear system

$$x + \frac{1}{2}y + \frac{1}{3}z + \frac{1}{4}w = 1$$

$$\frac{1}{2}x + \frac{1}{3}y + \frac{1}{4}z + \frac{1}{5}w = 2$$

$$\frac{1}{3}x + \frac{1}{4}y + \frac{1}{5}z + \frac{1}{6}w = 3$$

$$\frac{1}{4}x + \frac{1}{5}y + \frac{1}{6}z + \frac{1}{7}w = 4$$

Compare your result with the *exact* solution obtained by manually applying the method of Gaussian elimination.

* This problem should be solved by using one of the programs noted in Appendix B.

1.6 NORMS AND DOT PRODUCTS

In this final section of our introduction to R^n it seems fitting to return to our original notion of "vector," that is, a quantity determined by a magnitude and a direction. Thus far, we have studied the arithmetic properties of vector addition, the algebraic structure of subspaces, and the use of systems of linear equations to determine if individual vectors belonged to particular subspaces. We will use many of these previously developed concepts to study both the magnitude and direction of vectors in R^n.

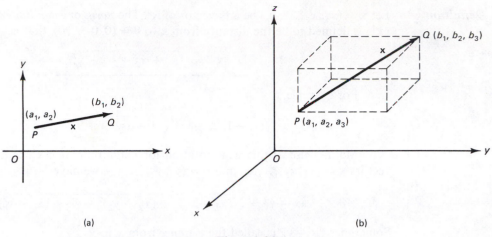

Figure 1.6.1

Recall that the magnitude of a vector in R^2 or R^3 is the length of a directed line segment which represents the vector, that is, the distance between the initial and terminal points of a directed line segment which represents the vector (see Figure 1.6.1).

With the aid of the Pythagorean theorem, we see that the magnitude of the vector **x**, or the distance from P to Q, is given by

$$\sqrt{(a_1 - b_1)^2 + (a_2 - b_2)^2}$$

in R^2, and by

$$\sqrt{(a_1 - b_1)^2 + (a_2 - b_2)^2 + (a_3 - b_3)^2}$$

in R^3.

Because of the formulas above, it is easy to extend the notion of distance to points in R^n for $n > 3$.

Definition   Let $P = (a_1, a_2, \ldots, a_n)$ and $Q = (b_1, b_2, \ldots, b_n)$ be points in R^n. We define the ***distance*** between P and Q by the expression

$$\sqrt{(a_1 - b_1)^2 + (a_2 - b_2)^2 + \cdots + (a_n - b_n)^2}$$

For example, the distance between $(1, -1, 2, 1)$ and $(2, 0, 1, -1)$ is

$$\sqrt{(1 - 2)^2 + (-1 - 0)^2 + (2 - 1)^2 + [1 - (-1)]^2} = \sqrt{7}$$

With this definition of distance one can extend the definition of magnitude to a vector in R^n for $n > 3$.

Definition   Let $\mathbf{x} = (x_1, x_2, \ldots, x_n)$ be any vector of R^n. The *norm* or *magnitude* of $\mathbf{x}$, denoted by $\|\mathbf{x}\|$, is defined to be the distance from $\mathbf{x}$ to $\mathbf{0} = (0, 0, \ldots, 0)$, that is,

$$\|\mathbf{x}\| = \sqrt{x_1^2 + x_2^2 + \cdots + x_n^2}$$

For example,

$$\|(1, -1, 2, 3)\| = \sqrt{1 + 1 + 4 + 9} = \sqrt{15}$$

Making use of this new notation for magnitude, it is clear that for any two vectors $\mathbf{x} = (x_1, x_2, \ldots, x_n)$ and $\mathbf{y} = (y_1, y_2, \ldots, y_n)$, we have

$$\|\mathbf{x} - \mathbf{y}\| = \sqrt{(x_1 - y_1)^2 + (x_2 - y_2)^2 + \cdots + (x_n - y_n)^2}$$

Sometimes $\|\mathbf{x} - \mathbf{y}\|$ is called the *distance* from $\mathbf{x}$ to $\mathbf{y}$.

We are now ready to explore the more complicated concept of "direction." We shall be concerned with the relationship between the directions of two vectors. In R^2 and R^3 vectors are represented by directed line segments. If two such segments which represent vectors $\mathbf{x}$ and $\mathbf{y}$ share the same initial point, we define the *angle* between the vectors $\mathbf{x}$ and $\mathbf{y}$ as the unique angle θ, $0 \leq \theta \leq \pi = 180°$, between the line segments (see Figure 1.6.2).

There is an operation on vectors called the "dot product" which is very useful in determining the angle between vectors.

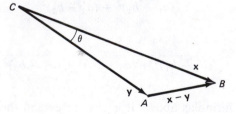

Figure 1.6.2

Definition   Let $\mathbf{x} = (x_1, x_2, \ldots, x_n)$ and $\mathbf{y} = (y_1, y_2, \ldots, y_n)$ be two vectors in R^n. The *dot product* of $\mathbf{x}$ and $\mathbf{y}$, denoted by $\mathbf{x} \cdot \mathbf{y}$, is defined to be the scalar given by

$$\mathbf{x} \cdot \mathbf{y} = x_1 y_1 + x_2 y_2 + \cdots + x_n y_n$$

Notice that unlike the sum of two vectors or the product of a scalar and a vector where the result is a vector, the dot product of two vectors is always a scalar. In addition, the dot product makes sense only when both vectors have the same number of components. Some authors reserve the name "dot product" when dealing with vectors in R^2 and R^3 but use other names, such as "scalar product" or "Euclidean inner product," for vectors with more than three components.

For example, suppose that $\mathbf{x} = (1, -1, 2)$ and $\mathbf{y} = (0, 2, 5)$. Then

$$\mathbf{x} \cdot \mathbf{y} = (1)(0) + (-1)(2) + (2)(5) = 8$$

There are certain elementary relationships involving the dot product, the norm of a vector, and the vector operations. For example, the dot product combines with the vector sum according to a "distributive law." These relationships are very convenient when doing calculations involving the dot product. We list these properties in Theorem 1.6.1.

Theorem 1.6.1   Let $\mathbf{x}$, $\mathbf{y}$, and $\mathbf{z}$ be vectors in R^n. Let t be any scalar. Then
  (a) $\mathbf{x} \cdot \mathbf{x} = \|\mathbf{x}\|^2$
  (b) $\mathbf{x} \cdot \mathbf{y} = \mathbf{y} \cdot \mathbf{x}$
  (c) $\mathbf{x} \cdot (\mathbf{y} + \mathbf{z}) = \mathbf{x} \cdot \mathbf{y} + \mathbf{x} \cdot \mathbf{z}$
  (d) $(t\mathbf{x}) \cdot \mathbf{y} = t(\mathbf{x} \cdot \mathbf{y}) = \mathbf{x} \cdot (t\mathbf{y})$
  (e) $\|t\mathbf{x}\| = |t| \|\mathbf{x}\|$

Proof

We prove parts (b) and (e), leaving the task of establishing parts (a), (c), and (d) as exercises.

(b) Let $\mathbf{x} = (x_1, x_2, \ldots, x_n)$ and $\mathbf{y} = (y_1, y_2, \ldots, y_n)$. Then

$$\mathbf{x} \cdot \mathbf{y} = x_1 y_1 + x_2 y_2 + \cdots + x_n y_n$$
$$= y_1 x_1 + y_2 x_2 + \cdots + y_n x_n$$
$$= \mathbf{y} \cdot \mathbf{x}$$

(e) First notice that for the scalar t, $|t|$ denotes the ordinary absolute value of t, and that $|t| = \sqrt{t^2}$. So

$$\|t\mathbf{x}\|^2 = (t\mathbf{x}) \cdot (t\mathbf{x}) \qquad \text{by (a)}$$
$$= t^2 (\mathbf{x} \cdot \mathbf{x}) \qquad \text{by (d)}$$
$$= t^2 \|\mathbf{x}\|^2 \qquad \text{by (a)}$$

Taking the square root of both sides of the equation, we conclude that

$$\|t\mathbf{x}\| = |t| \|\mathbf{x}\| \qquad\qquad \blacksquare$$

Part (c) of Theorem 1.6.1 is called the "left distributive law of dot product over vector addition." It should be noted that parts (b) and (c) immediately yield the alternative distributive law, the "right distributive law of dot product over vector addition":

$$(\mathbf{x} + \mathbf{y}) \cdot \mathbf{z} = \mathbf{x} \cdot \mathbf{z} + \mathbf{y} \cdot \mathbf{z}$$

for any vectors $\mathbf{x}$, $\mathbf{y}$, and $\mathbf{z}$ in R^n. The proof is left as an exercise.

We may apply these distributive laws to obtain formulas that are analogous to the formulas for scalars. For example,

$$(\mathbf{x} + \mathbf{y}) \cdot (\mathbf{x} + \mathbf{y}) = (\mathbf{x} + \mathbf{y}) \cdot \mathbf{x} + (\mathbf{x} + \mathbf{y}) \cdot \mathbf{y}$$
$$= \mathbf{x} \cdot \mathbf{x} + \mathbf{y} \cdot \mathbf{x} + \mathbf{x} \cdot \mathbf{y} + \mathbf{y} \cdot \mathbf{y}$$
$$= \mathbf{x} \cdot \mathbf{x} + 2(\mathbf{x} \cdot \mathbf{y}) + \mathbf{y} \cdot \mathbf{y}$$

Similarly,

$$(\mathbf{x} - \mathbf{y}) \cdot (\mathbf{x} - \mathbf{y}) = \mathbf{x} \cdot \mathbf{x} - 2(\mathbf{x} \cdot \mathbf{y}) + \mathbf{y} \cdot \mathbf{y} \tag{1}$$

We are now in a position to investigate the relationship between the dot product of two vectors and the angle between the two vectors. Let $\mathbf{x}$ and $\mathbf{y}$ be vectors in R^2 or in R^3. Then $\mathbf{x}$ and $\mathbf{y}$ can be represented by directed line segments sharing an initial point C (see Figure 1.6.2).

The *law of cosines* applied to triangle ABC produces

$$\|\mathbf{x} - \mathbf{y}\|^2 = \|\mathbf{x}\|^2 + \|\mathbf{y}\|^2 - 2\|\mathbf{x}\| \|\mathbf{y}\| \cos \theta$$

So, by Theorem 1.6.1(a), we have

$$(\mathbf{x} - \mathbf{y}) \cdot (\mathbf{x} - \mathbf{y}) = \mathbf{x} \cdot \mathbf{x} + \mathbf{y} \cdot \mathbf{y} - 2\|\mathbf{x}\| \|\mathbf{y}\| \cos \theta$$

Now, by equation (1), we have

$$\mathbf{x} \cdot \mathbf{x} - 2\mathbf{x} \cdot \mathbf{y} + \mathbf{y} \cdot \mathbf{y} = \mathbf{x} \cdot \mathbf{x} + \mathbf{y} \cdot \mathbf{y} - 2\|\mathbf{x}\| \|\mathbf{y}\| \cos \theta$$

Canceling $\mathbf{x} \cdot \mathbf{x}$ and $\mathbf{y} \cdot \mathbf{y}$ from each side of the equation above and then dividing both sides by -2, we arrive at

$$\mathbf{x} \cdot \mathbf{y} = \|\mathbf{x}\| \|\mathbf{y}\| \cos \theta$$

We summarize this result in the next theorem.

Theorem 1.6.2 *Let $\mathbf{x}$ and $\mathbf{y}$ be vectors in R^2 or in R^3. Let θ be the angle between $\mathbf{x}$ and $\mathbf{y}$. Then*

$$\mathbf{x} \cdot \mathbf{y} = \|\mathbf{x}\| \|\mathbf{y}\| \cos \theta$$

Example 1 For vectors $\mathbf{x} = (1, 3)$ and $\mathbf{y} = (2, 1)$, we may use Theorem 1.6.2 to determine the angle between $\mathbf{x}$ and $\mathbf{y}$. Applying the formula above, we have

$$2 + 3 = \sqrt{10} \sqrt{5} \cos \theta$$

Thus,

$$\cos \theta = \frac{1}{\sqrt{2}}$$

We conclude that θ is 45° or $\pi/4$ radians.

Of particular importance is the case when nonzero vectors $\mathbf{x}$ and $\mathbf{y}$ are "perpendicular." This occurs exactly when the angle θ between $\mathbf{x}$ and $\mathbf{y}$ is $\pi/2$, that is, $\cos \theta = 0$. In this case we must have

$$\mathbf{x} \cdot \mathbf{y} = \|\mathbf{x}\| \, \|\mathbf{y}\| \, \cos \frac{\pi}{2} = 0$$

Conversely, if $\mathbf{x} \cdot \mathbf{y} = 0$, then $\cos \theta = 0$, that is, $\theta = \pi/2$. This result suggests the following definition.

Definition Two vectors $\mathbf{x}$ and $\mathbf{y}$ in R^n are called *orthogonal* (*perpendicular*) if $\mathbf{x} \cdot \mathbf{y} = 0$.

Notice that the zero vector is by definition orthogonal to every vector. If $\mathbf{x}$ and $\mathbf{y}$ are nonzero vectors in R^2 or R^3, $\mathbf{x}$ is orthogonal to $\mathbf{y}$ if and only if $\mathbf{x}$ is perpendicular to $\mathbf{y}$; that is, the angle between $\mathbf{x}$ and $\mathbf{y}$ is 90°.

Example 2 Find the set S of vectors in R^4 which are orthogonal to both $(1, 2, 1, 1)$ and $(2, 3, 1, 2)$.
A vector (x_1, x_2, x_3, x_4) is in S if and only if

$$(x_1, x_2, x_3, x_4) \cdot (1, 2, 1, 1) = 0 \qquad \text{and} \qquad (x_1, x_2, x_3, x_4) \cdot (2, 3, 1, 2) = 0$$

That is, (x_1, x_2, x_3, x_4) must be a solution to the system

$$x_1 + 2x_2 + x_3 + x_4 = 0$$
$$2x_1 + 3x_2 + x_3 + 2x_4 = 0$$

Using Gaussian elimination, we obtain that

$$(x_1, x_2, x_3, x_4) = s(1, -1, 1, 0) + t(-1, 0, 0, 1)$$

for parameters s and t. So S is the subspace spanned by the vectors $(1, -1, 1, 0)$ and $(-1, 0, 0, 1)$.

Geometric Application

The diagonals of a parallelogram are perpendicular if and only if the parallelogram is a rhombus.

Recall that a *rhombus* is a parallelogram all of whose sides are of equal length. Consider a rhombus in R^2 or in R^3. Suppose that the sides of the rhombus are determined by the vectors $\mathbf{x}$ and $\mathbf{y}$ (see Figure 1.6.3).

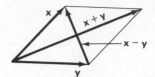

Figure 1.6.3

By definition, we have $\|\mathbf{x}\| = \|\mathbf{y}\|$. By the geometric interpretations of vector addition and subtraction, it follows that the diagonals of the rhombus are given by $\mathbf{x} + \mathbf{y}$ and $\mathbf{x} - \mathbf{y}$. To show that the diagonals are perpendicular, we will show that the dot product is zero:

$$
\begin{aligned}
(\mathbf{x} + \mathbf{y}) \cdot (\mathbf{x} - \mathbf{y}) &= \mathbf{x} \cdot (\mathbf{x} - \mathbf{y}) + \mathbf{y} \cdot (\mathbf{x} - \mathbf{y}) \\
&= \mathbf{x} \cdot \mathbf{x} - \mathbf{x} \cdot \mathbf{y} + \mathbf{y} \cdot \mathbf{x} - \mathbf{y} \cdot \mathbf{y} \\
&= \mathbf{x} \cdot \mathbf{x} - \mathbf{y} \cdot \mathbf{y} \\
&= \|\mathbf{x}\|^2 - \|\mathbf{y}\|^2 \\
&= 0
\end{aligned}
$$

Thus, the two diagonals are perpendicular. By tracing backward through the proof, we can easily show that the converse is also true.

Orthogonal Projections

We have already encountered the formula for the distance between two points. To find the distance between other types of geometric objects, for example, a point and a line, we will need the concept of an "orthogonal projection" of a vector on a line. Specifically, let $\mathbf{y}$ be a vector in R^2 (or R^3) and let L denote a line with a direction vector $\mathbf{x}$. Assume that the geometric representations of $\mathbf{x}$ and $\mathbf{y}$ have a common initial point. We begin by dropping a perpendicular from the head of $\mathbf{y}$ to the line L [see Figure 1.6.4(a)]. Notice that the vector $\overrightarrow{AB}$ is parallel to $\mathbf{x}$, so there exists a scalar c such that $\overrightarrow{AB} = c\mathbf{x}$. To find the scalar c, we use the fact that $\mathbf{x}$ and $\mathbf{y} - c\mathbf{x}$ are perpendicular and set the dot product equal to 0.

$$\mathbf{x} \cdot (\mathbf{y} - c\mathbf{x}) = 0$$

$$\mathbf{x} \cdot \mathbf{y} - c\|\mathbf{x}\|^2 = 0 \qquad \text{(Theorem 1.6.1)}$$

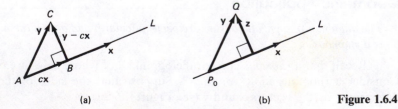

(a)   (b)   **Figure 1.6.4**

So, $c = (\mathbf{x} \cdot \mathbf{y})/\|\mathbf{x}\|^2$. We define the ***orthogonal projection of y on L*** by the formula

$$\frac{\mathbf{x} \cdot \mathbf{y}}{\|\mathbf{x}\|^2} \mathbf{x}$$

As an application of this concept, we will find the distance from a point Q to a line L which has the direction vector $\mathbf{x}$ [see Figure 1.6.4(b)]. Choose any point P_0 on L. We want the length of the vector $\mathbf{z}$ which is the vector difference of $\mathbf{y} = \overrightarrow{P_0 Q}$ and the orthogonal projection of $\mathbf{y}$ on L. That is,

$$\|\mathbf{z}\| = \left\| \mathbf{y} - \frac{\mathbf{x} \cdot \mathbf{y}}{\|\mathbf{x}\|^2} \mathbf{x} \right\|$$

As an example, suppose that $Q = (3, 7)$ and L is given by $y = 2x - 5$ in R^2. We may choose the point $P_0 = (0, -5)$ on L and the vector $\mathbf{x} = (1, 2)$ as a direction vector of L. So $\mathbf{y} = (3, 12)$ and $\mathbf{x} \cdot \mathbf{y} = 27$. Thus, the distance from Q to L is

$$\left\| (3, 12) - \frac{27}{5} (1, 2) \right\| = \frac{6}{\sqrt{5}}$$

Planes in R^3

In Section 1.2 we used vector operations to determine the parametric equations of lines in R^2 and R^3. Now we will use vector operations to determine the equation of a plane in R^3.

We will appeal to the fact that a plane is uniquely determined by a point on the plane and a vector, called a ***normal*** to the plane, which is perpendicular to the plane. Specifically, let $P_0 = (x_0, y_0, z_0)$ be a point in a plane and let $\mathbf{n} = (a, b, c)$ be a normal to the plane (see Figure 1.6.5).

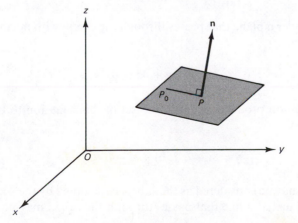

Figure 1.6.5

Clearly, a point $P = (x, y, z)$ lies in the plane if and only if $\mathbf{n}$ is orthogonal to $\overrightarrow{P_0 P}$. That is,

$$0 = \mathbf{n} \cdot \overrightarrow{P_0 P} = (a, b, c) \cdot (x - x_0, y - y_0, z - z_0)$$

or

$$a(x - x_0) + b(y - y_0) + c(z - z_0) = 0$$

For example, the equation of the plane with normal $\mathbf{n} = (1, 2, 4)$ and containing the point $P_0 = (1, -3, 2)$ is

$$(x - 1) + 2(y + 3) + 4(z - 2) = 0$$

Collecting the constants, we can rewrite this equation as

$$x + 2y + 4z = 3$$

Notice that unlike the representation of a line in R^3, a plane requires only *one* scalar equation. In fact, a plane may be viewed as the solution set to one (nontrivial) equation in three unknowns. If we solve the equation above by assigning the parameters s and t to y and z, respectively, we obtain

$$x = 3 - 2s - 4t$$

So a vector $\mathbf{y}$ is a solution to the equation if and only if it can be written in the form

$$\mathbf{y} = (3 - 2s - 4t, s, t)$$
$$= (3, 0, 0) + s(-2, 1, 0) + t(-4, 0, 1)$$

If instead we consider a plane that passes through the origin with normal $\mathbf{n} = (1, 2, 4)$, we obtain the plane

$$x + 2y + 4z = 0$$

A similar computation produces the solution set of this equation to be the set of all vectors of the form

$$\mathbf{y} = s(-2, 1, 0) + t(-4, 0, 1)$$

That is, the plane may be considered as the subspace spanned by the vectors $(-2, 1, 0)$ and $(-4, 0, 1)$. So just as a line that passes through the origin may be represented as the subspace spanned by a (nonzero) vector, we have that a plane which passes through the origin may be represented as the subspace spanned by two (nonparallel) vectors. When we study the concept of "dimension" in Chapter 3, we will return to these ideas.

It is frequently more convenient to deal with normal vectors which have length equal to 1.

Definition   A vector $\mathbf{u}$ is called a ***unit vector*** if $\|\mathbf{u}\| = 1$.

Given any nonzero vector $\mathbf{x}$, it is useful to consider a unit vector pointing in the same direction as $\mathbf{x}$. Such a vector is easily computed. Let

$$\mathbf{u} = \frac{1}{\|\mathbf{x}\|}\,\mathbf{x}$$

By Theorem 1.6.1(e),

$$\|\mathbf{u}\| = \left\|\frac{1}{\|\mathbf{x}\|}\,\mathbf{x}\right\| = \frac{1}{\|\mathbf{x}\|}\,\|\mathbf{x}\| = 1$$

Thus, $\mathbf{u}$ is a unit vector. Since $\mathbf{u}$ is a positive multiple of $\mathbf{x}$, it follows that $\mathbf{u}$ has the same direction as $\mathbf{x}$. For example, if $\mathbf{x} = (1, 2, -2)$, the unit vector $\mathbf{u}$ in the same direction as $\mathbf{x}$ is

$$\mathbf{u} = \frac{1}{\|\mathbf{x}\|}\,\mathbf{x} = \frac{1}{3}\,(1,\,2,\,-2) = \left(\frac{1}{2},\frac{2}{3},\,-\frac{2}{3}\right)$$

The Cauchy–Schwarz and the Triangle Inequalities

In Theorem 1.6.2 we stated that for vectors $\mathbf{x}$ and $\mathbf{y}$ in R^2 or R^3, we have

$$\mathbf{x}\cdot\mathbf{y} = \|\mathbf{x}\|\,\|\mathbf{y}\|\,\cos\theta$$

where θ is the angle between $\mathbf{x}$ and $\mathbf{y}$. Since $|\cos\theta| \le 1$ for any angle θ, we must have that

$$|\mathbf{x}\cdot\mathbf{y}| \le \|\mathbf{x}\|\,\|\mathbf{y}\|$$

This inequality is called the "Cauchy–Schwarz inequality." We have established it for R^2 and R^3 by making use of Theorem 1.6.2. However, the inequality is also valid in R^n for any positive integer n.

Theorem 1.6.3 (*Cauchy–Schwarz Inequality*)   *For any vectors* **x** *and* **y** *in* R^n, *we have*

$$|\mathbf{x} \cdot \mathbf{y}| \le \|\mathbf{x}\| \, \|\mathbf{y}\|$$

Proof

If either **x** or **y** is zero, the result is obvious since both sides of the inequality have value zero. So we may assume that both **x** and **y** are nonzero vectors.

First suppose that both **x** and **y** are unit vectors; that is, $\|\mathbf{x}\| = \|\mathbf{y}\| = 1$. Because $\|\mathbf{x} - \mathbf{y}\|^2$ is nonnegative, we have

$$
\begin{aligned}
0 &\le \|\mathbf{x} - \mathbf{y}\|^2 \\
&= (\mathbf{x} - \mathbf{y}) \cdot (\mathbf{x} - \mathbf{y}) \\
&= \mathbf{x} \cdot \mathbf{x} - 2\mathbf{x} \cdot \mathbf{y} + \mathbf{y} \cdot \mathbf{y} \\
&= 1 - 2\mathbf{x} \cdot \mathbf{y} + 1
\end{aligned}
$$

Thus, we have $2\mathbf{x} \cdot \mathbf{y} \le 2$ or $\mathbf{x} \cdot \mathbf{y} \le 1$.

Now suppose that **x** and **y** are any nonzero vectors. Then $(1/\|\mathbf{x}\|)\mathbf{x}$ and $(1/\|\mathbf{y}\|)\mathbf{y}$ are both unit vectors, and therefore, by the inequality above,

$$\frac{1}{\|\mathbf{x}\|} \mathbf{x} \cdot \frac{1}{\|\mathbf{y}\|} \mathbf{y} \le 1$$

Multiplying both sides of this inequality by $\|\mathbf{x}\| \, \|\mathbf{y}\|$, we obtain

$$\mathbf{x} \cdot \mathbf{y} \le \|\mathbf{x}\| \, \|\mathbf{y}\|$$

By replacing **x** with $-\mathbf{x}$ in the equation above, we have

$$
\begin{aligned}
-(\mathbf{x} \cdot \mathbf{y}) &= (-\mathbf{x}) \cdot \mathbf{y} \\
&\le \|-\mathbf{x}\| \, \|\mathbf{y}\| \\
&= \|\mathbf{x}\| \, \|\mathbf{y}\|
\end{aligned}
$$

or

$$-\|\mathbf{x}\| \, \|\mathbf{y}\| \le \mathbf{x} \cdot \mathbf{y}$$

Therefore,

$$|\mathbf{x} \cdot \mathbf{y}| \le \|\mathbf{x}\| \, \|\mathbf{y}\| \qquad \blacksquare$$

Example 3   We apply the Cauchy-Schwarz inequality to show that for any real numbers $x_1, x_2, \ldots, x_n$ (not necessarily distinct), we have

$$|x_1| + |x_2| + \cdots + |x_n| \le \sqrt{n} \sqrt{x_1^2 + x_2^2 + \cdots + x_n^2}$$

Set $\mathbf{x} = (x_1, x_2, \ldots, x_n)$ and $\mathbf{y} = (y_1, y_2, \ldots, y_n)$, where

$$y_i = \begin{cases} 1 & \text{if } x_i \geq 0 \\ -1 & \text{if } x_i < 0 \end{cases}$$

Notice that for each i, $y_i x_i = |x_i|$. If we apply the Cauchy–Schwarz inequality to $\mathbf{x}$ and $\mathbf{y}$ and observe that $\|\mathbf{y}\| = \sqrt{n}$, the result is immediate.

The Cauchy–Schwarz inequality is useful in establishing another inequality, called the "triangle inequality."

Corollary 1.6.4

(**Triangle Inequality**) *For any vectors* $\mathbf{x}$ *and* $\mathbf{y}$ *in* R^n, *we have*

$$\|\mathbf{x} + \mathbf{y}\| \leq \|\mathbf{x}\| + \|\mathbf{y}\|$$

Before establishing its validity, let us discuss the geometric significance of the triangle inequality in R^2 and R^3. Consider the triangle ABC in Figure 1.6.6. If $\mathbf{x} = \overrightarrow{AB}$ and $\mathbf{y} = \overrightarrow{BC}$, then $\mathbf{x} + \mathbf{y} = \overrightarrow{AC}$. The triangle inequality asserts the well-known

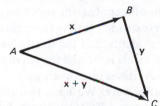

Figure 1.6.6

geometric fact that the length of any side of a triangle cannot exceed the sum of the lengths of the other two sides. We are now ready to proceed with the proof of the triangle inequality.

Proof

Let $\mathbf{x}$ and $\mathbf{y}$ be vectors in R^n. Then

$$\begin{aligned}
\|\mathbf{x} + \mathbf{y}\|^2 &= (\mathbf{x} + \mathbf{y}) \cdot (\mathbf{x} + \mathbf{y}) \\
&= \mathbf{x} \cdot \mathbf{x} + 2\mathbf{x} \cdot \mathbf{y} + \mathbf{y} \cdot \mathbf{y} \\
&\leq \mathbf{x} \cdot \mathbf{x} + 2|\mathbf{x} \cdot \mathbf{y}| + \mathbf{y} \cdot \mathbf{y} \\
&\leq \mathbf{x} \cdot \mathbf{x} + 2\|\mathbf{x}\| \|\mathbf{y}\| + \mathbf{y} \cdot \mathbf{y} \qquad \text{(Theorem 1.6.3)} \\
&= \|\mathbf{x}\|^2 + 2\|\mathbf{x}\| \|\mathbf{y}\| + \|\mathbf{y}\|^2 \\
&= (\|\mathbf{x}\| + \|\mathbf{y}\|)^2
\end{aligned}$$

Summarizing, we have

$$\|\mathbf{x} + \mathbf{y}\|^2 \leq (\|\mathbf{x}\| + \|\mathbf{y}\|)^2$$

Since norms are nonnegative numbers, the triangle inequality follows by taking the square root of both sides of the inequality above.   ∎

The case of equality is considered in the exercises.

Application to Averaging Class Size

A private school for the training of computer programmers advertised that its average class size is 20. After receiving complaints of false advertising, an investigator for the Office of Consumer Affairs obtained a list of the 60 students enrolled in the school. He polled each student and obtained the student's class size. He added these numbers and divided the result by 60 (the number of students) to obtain the figure of 27.6—significantly above the advertised average class size of 20. So he proceeded to file an official complaint against the school. However, the complaint was withdrawn by his supervisor after doing some investigating of her own.

Using the same enrollment list, the supervisor polled all 60 students. She found out that the students were divided among three classes. The first class had 25 students, the second class had 3 students, and the third class had 32 students. Notice that the sum of these three enrollments is 60. She then divided 60 by 3 to obtain a class average of 20, confirming the advertised class average.

Something is strange here. We have two different ways of computing a class average, and we arrive at two different answers. Let us consider these two processes. Suppose that m individuals are divided up into n mutually exclusive classes of sizes $x_1, x_2, \ldots, x_n$. The average of the class sizes is given by

$$\bar{x} = \frac{1}{n} \sum_{i=1}^{n} x_i = \frac{m}{n}$$

This is the method used by the supervisor.

Now consider the method used by the investigator. Suppose that each person is polled on the size of his or her class, the results are added, and the total is divided by the number of people polled. Let x^* denote this result. To compute x^*, observe that each person in the ith class will respond that his or her class has size x_i. Since x_i people in this class are polled, the sum of these responses is $x_i x_i = x_i^2$. Since this is done for each class, the total of the responses is given by

$$\sum_{i=1}^{n} x_i^2$$

Since m people are polled, this sum is divided by m to obtain the "average":

$$x^* = \frac{1}{m} \sum_{i=1}^{n} x_i^2$$

Now we investigate how $\bar{x}$ and x^* are related. Consider the two vectors $\mathbf{x} = (1/n)(1, 1, \ldots, 1)$ and $\mathbf{y} = (x_1, x_2, \ldots, x_n)$ in R^n. Then

$$\mathbf{x} \cdot \mathbf{y} = \frac{1}{n} \sum_{i=1}^{n} x_i = \frac{m}{n} = \bar{x}$$

$$\mathbf{x} \cdot \mathbf{x} = \frac{1}{n} \quad \text{and} \quad \mathbf{y} \cdot \mathbf{y} = \sum_{i=1}^{n} x_i^2$$

So

$$\bar{x} = \frac{m}{n} = \frac{n}{m}\left(\frac{m}{n}\right)^2$$

$$= \frac{n}{m}(\bar{x})^2$$

$$= \frac{n}{m}(\mathbf{x} \cdot \mathbf{y})^2$$

$$\leq \frac{n}{m}\|\mathbf{x}\|^2\|\mathbf{y}\|^2$$

$$= \frac{n}{m}\frac{1}{n}\sum_{i=1}^{n} x_i^2$$

$$= x^*$$

Thus, we have that $\bar{x} \leq x^*$.

By Exercise 5 this inequality is strict unless $\mathbf{x}$ and $\mathbf{y}$ are multiples of each other. Since the components of $\mathbf{x}$ are equal, we have that $\bar{x} < x^*$ unless each class has the same number of people.

Exercises

1. Compute the distance from P to Q.
   (a) $P = (1, 2);$ $\quad$ $Q = (3, 7)$
   (b) $P = (1, -1);$ $\quad$ $Q = (2, 1)$
   (c) $P = (1, 2, 1, -1);$ $\quad$ $Q = (2, 3, 2, 0)$
   (d) $P = (1, 3, 1);$ $\quad$ $Q = (-1, 4, 2)$

2. Compute the norms of each of the following vectors.
   (a) $(1, 2)$ $\quad\quad\quad$ (b) $(3, 4)$

   (c) $(1, -1, 3)$ $\quad\quad$ (d) $(2, 1, 0)$
   (e) $(1, 2, -1)$ $\quad\quad$ (f) $(1, 0, 1, 3)$
   (g) $(1, -1, -2, 1)$ $\quad$ (h) $(2, 3, 1, 1, 1)$

3. For each of the following parts, compute the dot product of $\mathbf{x}$ and $\mathbf{y}$ and find the cosine of the angle between $\mathbf{x}$ and $\mathbf{y}$.
   (a) $\mathbf{x} = (1, -1, 3);$ $\quad$ $\mathbf{y} = (2, 1, 0)$
   (b) $\mathbf{x} = (1, 3);$ $\quad\quad$ $\mathbf{y} = (-4, 1)$

(c) $x = (3, 4)$; $y = (5, -12)$
(d) $x = (1, 2, 2)$; $y = (-1, 1, 1)$
(e) $x = (-1, 2, 0, 3)$; $y = (1, 2, 1, 1)$

4. For the vectors x and y compute $\|x\|$, $\|y\|$, $\|x + y\|$, and $|x \cdot y|$. Use these results to verify the Cauchy–Schwarz and the triangle inequalities.
   (a) $x = (3, 1)$; $y = (-1, 3)$
   (b) $x = (1, -1, 2)$; $y = (2, 0, 1)$
   (c) $x = (1, -2, 1)$; $y = (-2, 4, -2)$
   (d) $x = (1, -2, 1)$; $y = (2, -4, 2)$

5. Under what circumstances does $|x \cdot y| = \|x\| \|y\|$? State and prove a result that determines when the equality holds. [*Hint*: Look at Exercise 4(c).]

6. Under what circumstances does $\|x + y\| = \|x\| + \|y\|$? State and prove a result that determines when the equality holds. (*Hint*: Do Exercise 5 first.)

7. Under what circumstances does $\|x + y\|^2 = \|x\|^2 + \|y\|^2$? State and prove a result that determines when the equality holds.

8. Use the triangle inequality to prove

$$| \|x\| - \|y\| | \leq \|x - y\|$$

9. Show that for any fixed vector z in R^n, $\{x : x \cdot z = 0\}$ is a subspace of R^n.

10. Let $S = \{x : x \cdot (1, -3, 2) = 0\}$. Find a finite set that spans S.

11. Prove that a plane in R^3 is a subspace of R^3 if and only if the plane contains the origin. (*Hint*: Use Exercise 9.)

12. Prove parts (a), (c), and (d) of Theorem 1.6.1.

13. Justify the right distributive law of dot product over vector addition:

$$(x + y) \cdot z = x \cdot z + y \cdot z$$

14. Let L be the line containing point $P_0 = (1, 3, -1)$ with direction vector $x = (2, 2, -1)$. Find the distance from the point $Q = (4, 3, -1)$ to L. (*Hint*: First construct the appropriate orthogonal projection.)

15. Find the equation of the plane with normal $n = (-1, 3, 1)$ that contains $P_0 = (2, 1, 3)$.

16. Let P_0 be a point that lies in a given plane in R^3. Prove that there exist vectors x and y such that for any point P in R^3, P lies in the plane if and only if

$$\overrightarrow{OP} = \overrightarrow{OP_0} + sx + ty$$

for some scalars s and t.

17. It is well known that the diagonal of a square makes an angle of $45°$ with one of the adjacent sides. Show that the angle formed by the diagonal of a cube and one of its adjacent sides has cosine equal to $1/\sqrt{3}$. [*Hint*: Choose the cube with vertices $(0, 0, 0)$, $(1, 0, 0)$, $(0, 1, 0)$, $(0, 0, 1)$, $(1, 1, 0)$, $(1, 0, 1)$, $(0, 1, 1)$, and $(1, 1, 1)$. Then the line segment connecting $(0, 0, 0)$ to $(1, 1, 1)$ is a diagonal, and the line segment connecting $(0, 0, 0)$ to $(1, 0, 0)$ is an adjacent side.]

18. For vectors $x = (a, b, c)$ and $y = (p, q, r)$, define the **cross product** of x and y by the equation

$$x \times y = (br - cq, cp - ar, aq - bp)$$

   (a) Show that the cross product of x and y is orthogonal to both x and y.
   (b) Compute the cross product of x and y for the following examples. In each case, verify that $x \times y$ is orthogonal to both x and y.
      (i) $x = (1, 0, 1)$; $y = (1, 1, 0)$
      (ii) $x = (1, -1, 1)$; $y = (0, 1, 2)$
      (iii) $x = (1, 0, 0)$; $y = (0, 1, 0)$

19. Suppose that x and y are unit vectors which are orthogonal to each other. Show that for any scalars s and t,

$$\|sx + ty\|^2 = s^2 + t^2$$

*20. In each of the following cases, use the program LINSYST to find a vector y that is orthogonal to the vectors in the subset S.
   (a) $S = \{(1, 3, 1), (2, -2, 4)\}$
   (b) $S = \{(3, 2, 2, 1), (5, 3, 2, -1)\}$
   (c) $S = \{(2, 3, 3, 1), (3, 3, 2, 2), (5, 6, -3, 4)\}$

* This problem should be solved by using one of the programs noted in Appendix B.

Key Words

Matrices and Linear Transformations

Many of the basic properties of subspaces of Euclidean n-space R^n were introduced in Chapter 1. One particular use of subspaces was related to the study of solution spaces of homogeneous systems of linear equations. In a similar context, matrices were defined and used to develop a practical method of computing solutions to systems of linear equations. In this chapter we continue to develop properties of matrices. We shall learn that matrices are closely related to functions that preserve the operations of vector addition and scalar multiplication in R^n. This relationship provides the key to a deeper understanding of the role and use of matrices in solving problems.

2.1 MATRIX ARITHMETIC

As we shall see, matrices have many uses beyond providing a convenient tool for solving systems of linear equations. In fact, they may be applied to areas as diverse as sociology, business, communication theory, and physics. We shall use some of these applications to motivate the definitions of arithmetic operations for matrices.

Recall that for an $m \times n$ matrix A, the ijth entry of A is denoted by A_{ij}. Two $m \times n$ matrices A and B are said to be *equal*, written $A = B$, if their corresponding entries are equal; that is, $A_{ij} = B_{ij}$ for all i and j.

The $m \times n$ matrix all of whose entries are zeros is called the $m \times n$ *zero matrix*, and it is denoted by 0. For example, the 2×3 zero matrix is given by

$$0 = \begin{bmatrix} 0 & 0 & 0 \\ 0 & 0 & 0 \end{bmatrix}$$

We use the next example as a motivation for the definition of "matrix addition."

Example 1   Suppose that a company owns two bookstores each of which sells magazines and paperback books. We may represent the sales (in hundreds of dollars) of these two stores for the months of July and August by the following matrices:

$$
\begin{array}{ccc}
 & \text{July} & \text{August} \\
\text{Store} & 1 \quad 2 & 1 \quad 2 \\
\text{Magazines} & \begin{bmatrix} 15 & 20 \\ 45 & 64 \end{bmatrix} & \begin{bmatrix} 18 & 31 \\ 52 & 68 \end{bmatrix} \\
\text{Books} & &
\end{array}
$$

For example, store 1 sold \$1500 worth of magazines during the month of July. These matrices are called ***inventory matrices***.

It is now an easy matter to combine the two months' sales into one matrix which represents the total sales for both months.

$$
\begin{array}{ccc}
\text{Store} & 1 & 2 \\
\text{Magazines} & \begin{bmatrix} 33 & 51 \\ 97 & 132 \end{bmatrix} \\
\text{Books} & &
\end{array}
$$

This operation of combining two matrices by adding corresponding entries is called "matrix addition."

Definition   Let A and B be $m \times n$ matrices. Define the ***sum*** of A and B, denoted by $A + B$, to be the $m \times n$ matrix whose ijth entry is $A_{ij} + B_{ij}$; that is, for all i, j we have

$$(A + B)_{ij} = A_{ij} + B_{ij}$$

Note: To form the sum of two matrices, the matrices must be of the same size. Suppose that A and B are defined as follows:

$$A = \begin{bmatrix} 3 & 4 & -2 \\ 5 & 2 & 6 \end{bmatrix} \quad \text{and} \quad B = \begin{bmatrix} 2 & -3 & 9 \\ 4 & 5 & 7 \end{bmatrix}$$

Then

$$A + B = \begin{bmatrix} 5 & 1 & 7 \\ 9 & 7 & 13 \end{bmatrix}$$

Example 2   In Example 1 suppose that we want to represent by a matrix the situation in which the July sales were doubled for both stores. It is easy to see that this may be accomplished by simply multiplying each of the entries of the first matrix by 2, obtaining the matrix

$$\begin{bmatrix} 30 & 40 \\ 90 & 128 \end{bmatrix}$$

Example 2 introduces our second operation for matrices, "scalar multiplication."

Definition   For an $m \times n$ matrix A and scalar c, we define the **product** of c and A, denoted cA, to be the $m \times n$ matrix whose ijth entry is cA_{ij}; that is, for all i, j we have

$$(cA)_{ij} = cA_{ij}$$

We shall denote $(-1)B$ by $-B$ and define the **difference** of A and B, denoted $A - B$, as $A + (-B)$.

Example 3   Let A, B, and C be given as

$$A = \begin{bmatrix} 2 & -3 & 3 \\ 6 & 5 & 1 \end{bmatrix} \qquad B = \begin{bmatrix} -4 & 0 & 7 \\ 8 & -3 & 3 \end{bmatrix} \qquad C = \begin{bmatrix} 5 & 4 & -1 \\ 4 & 1 & 2 \end{bmatrix}$$

Then

$$(A - B) + 2C = \begin{bmatrix} 6 & -3 & -4 \\ -2 & 8 & -2 \end{bmatrix} + \begin{bmatrix} 10 & 8 & -2 \\ 8 & 2 & 4 \end{bmatrix}$$

$$= \begin{bmatrix} 16 & 5 & -6 \\ 6 & 10 & 2 \end{bmatrix}$$

Most of the laws of arithmetic that are valid for scalars are also valid for matrices.

Theorem 2.1.1   *Let A, B, and C be $m \times n$ matrices and let c be a scalar. Then:*
   (a) $A + B = B + A$ (*commutative law of addition*)
   (b) $(A + B) + C = A + (B + C)$ (*associative law of addition*)
   (c) $c(A + B) = cA + cB$ (*scalar multiplication distributes over addition*)
   (d) $A + 0 = A = 0 + A$, *where 0 denotes the $m \times n$ zero matrix*

We shall always assume that the size of the zero matrix is compatible with the context in which it appears. In this case, 0 must be an $m \times n$ matrix.

Proof

We prove only part (c). The other parts are proved similarly and are left for the exercises.

We show that both sides of the equation in part (c) have the same entries. For any i and j we have

$$(c(A + B))_{ij} = c(A + B)_{ij}$$
$$= c(A_{ij} + B_{ij})$$
$$= cA_{ij} + cB_{ij}$$
$$= (cA)_{ij} + (cB)_{ij}$$
$$= (cA + cB)_{ij} \qquad \blacksquare$$

Notice that the associative law allows us to omit parentheses when writing expressions such as $A + B + C$.

There is another important operation called *matrix multiplication*. Before considering this operation, we must introduce some terminology and conventions that are used throughout the book.

In Chapter 1 we represented a vector in R^n horizontally. We may view such a representation as a $1 \times n$ matrix. However, it is frequently convenient to represent such a vector vertically, and view it as an $n \times 1$ matrix. When viewed horizontally, the representation is called a ***row vector***. When viewed vertically, the representation is called a ***column vector***. For example, if $\mathbf{x} = (1, 2, 3)$ in R^3, we may write $\mathbf{x}$ as the row vector $(1, 2, 3)$, or as the column vector

$$\mathbf{x} = \begin{bmatrix} 1 \\ 2 \\ 3 \end{bmatrix}$$

It is important to distinguish between the vector and its representations. Although the two representations are distinct, they designate the same vector. If the representations of two vectors in R^n are distinct, their dot product is still defined. For example, if

$$\mathbf{x} = (2, -1, 3) \qquad \text{and} \qquad \mathbf{y} = \begin{bmatrix} 1 \\ 5 \\ 6 \end{bmatrix}$$

then

$$\mathbf{x} \cdot \mathbf{y} = (2)(1) + (-1)(5) + (3)(6)$$
$$= 15$$

With these ideas in mind, we can now consider how row and column vectors are used in matrices. For an $m \times n$ matrix A, each of the m rows of A is a $1 \times n$ row vector and, as such, is a vector in R^n. Similarly, each of the n columns of A is an $m \times 1$ column vector and may be considered as a vector in R^m. For example, the 2×3 matrix

$$A = \begin{bmatrix} 1 & 2 & 3 \\ 4 & 5 & 6 \end{bmatrix}$$

has row vectors $(1, 2, 3)$ and $(4, 5, 6)$, while the column vectors of A are

$$\begin{bmatrix} 1 \\ 4 \end{bmatrix} \quad \begin{bmatrix} 2 \\ 5 \end{bmatrix} \quad \begin{bmatrix} 3 \\ 6 \end{bmatrix}$$

It is convenient to have a notation to designate the row and column vectors of a matrix. For an $m \times n$ matrix A, let $_iA$ denote the ith row of A, and let ^jA denote the jth column of A. For the matrix A above we have

$$_1A = (1, 2, 3) \qquad _2A = (4, 5, 6)$$

and

$$^1A = \begin{bmatrix} 1 \\ 4 \end{bmatrix} \qquad ^2A = \begin{bmatrix} 2 \\ 5 \end{bmatrix} \qquad ^3A = \begin{bmatrix} 3 \\ 6 \end{bmatrix}$$

The following example provides a motivation for the definition of matrix multiplication. This particular example will be referred to in this and subsequent sections.

Example 4   A sociologist is interested in studying the population changes in a metropolitan area as people move between the city and suburbs. From empirical evidence she has discovered that in any given year, 15% of those living in the city will move to the suburbs and 3% of those living in the suburbs will move to the city. For simplicity, we assume that the metropolitan population remains stable. This information may be represented as in the following matrix:

$$\begin{array}{cc} & \text{From} \\ & \begin{array}{cc} \text{City} & \text{Suburbs} \end{array} \\ A = \text{To} \begin{array}{c} \text{City} \\ \text{Suburbs} \end{array} & \begin{bmatrix} 0.85 & 0.03 \\ 0.15 & 0.97 \end{bmatrix} \end{array}$$

Notice that the entries of A are nonnegative and that the entries of each column sum to 1. Such a matrix is called a **transition** or **stochastic matrix**. Suppose that there are now 50 thousand people living in the city and 75 thousand people living in the

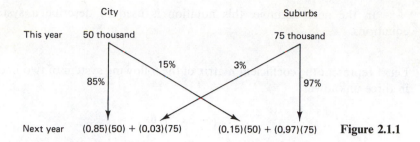

Figure 2.1.1

suburbs. The sociologist would like to know how many people will be living in each of the two areas next year. Figure 2.1.1 describes the changes of population from one year to the next. It follows that the number of people who will be living in the city next year is $(0.85)(50) + (0.03)(75) = 44.75$ thousand, and the number of people who will be living in the suburbs is $(0.15)(50) + (0.97)(75) = 80.25$ thousand.

If we let **p** denote the column vector

$$\mathbf{p} = \begin{bmatrix} 50 \\ 75 \end{bmatrix}$$

we see that the new city population may be computed as the dot product of the first row of A and **p**, that is, $_1A \cdot \mathbf{p}$. Similarly, the new suburb population may be computed as the dot product of the second row, $_2A$ and **p**.

This particular application, as well as many similar ones, provides us with a motivation for the following definition.

Definition   Let A be an $m \times n$ matrix with rows $_1A, \, _2A, \ldots, \, _mA$ and let **x** be an $n \times 1$ column vector. Define the ***matrix product*** $A\mathbf{x}$ to be the $m \times 1$ column vector whose ith entry is the dot product of the ith row of A with **x**, that is, $_iA \cdot \mathbf{x}$. The operation associated with this process is called ***matrix multiplication***.

For this definition to make sense, the number of columns of A must equal the number of rows of **x**.

We may express the results obtained in Example 4 using matrix multiplication as

$$A\mathbf{p} = \begin{bmatrix} 0.85 & 0.03 \\ 0.15 & 0.97 \end{bmatrix} \begin{bmatrix} 50 \\ 75 \end{bmatrix}$$

$$= \begin{bmatrix} (0.85)(50) + (0.03)(75) \\ (0.15)(50) + (0.97)(75) \end{bmatrix}$$

$$= \begin{bmatrix} 44.75 \\ 80.25 \end{bmatrix}$$

In the next example, this notation is used to describe a system of linear equations.

Example 5   Let A represent the coefficient matrix of the following system of two linear equations in three unknowns.

$$2x_1 - 3x_2 + 6x_3 = 7$$
$$5x_1 + 8x_2 - 2x_3 = 11$$

If we let

$$\mathbf{x} = \begin{bmatrix} x_1 \\ x_2 \\ x_3 \end{bmatrix} \quad \text{and} \quad \mathbf{b} = \begin{bmatrix} 7 \\ 11 \end{bmatrix}$$

then by the definition above, the system of linear equations reduces to the single matrix equation $A\mathbf{x} = \mathbf{b}$.

$$\begin{bmatrix} 2 & -3 & 6 \\ 5 & 8 & -2 \end{bmatrix} \begin{bmatrix} x_1 \\ x_2 \\ x_3 \end{bmatrix} = \begin{bmatrix} 7 \\ 11 \end{bmatrix}$$

After we learn more about matrices, we shall be able to exploit this method of representing a system of linear equations to our benefit. We shall see how this is done in Chapter 3.

To assure your understanding of this operation, we compute two additional matrix products:

$$\begin{bmatrix} 1 & 2 & 3 \\ -1 & 1 & 0 \end{bmatrix} \begin{bmatrix} 2 \\ 1 \\ -1 \end{bmatrix} = \begin{bmatrix} (1)(2) + (2)(1) + (3)(-1) \\ (-1)(2) + (1)(1) + (0)(-1) \end{bmatrix}$$

$$= \begin{bmatrix} 1 \\ -1 \end{bmatrix}$$

and

$$\begin{bmatrix} 2 & 3 \\ 1 & 1 \\ 0 & -4 \end{bmatrix} \begin{bmatrix} 3 \\ -4 \end{bmatrix} = \begin{bmatrix} -6 \\ -1 \\ 16 \end{bmatrix}$$

Another important application of matrix multiplication occurs in analytic geometry in the study of rotations. Using methods of trigonometry, it can be shown that if (x, y) represents the components of a vector in R^2, and if (x', y') represents the result of rotating (x, y) by an angle θ, then we have the equations

$$x' = x \cos \theta - y \sin \theta$$

$$y' = x \sin \theta + y \cos \theta$$

These equations can be expressed in terms of matrix multiplication. We define the *rotation* (by θ) *matrix* as

$$A_\theta = \begin{bmatrix} \cos \theta & -\sin \theta \\ \sin \theta & \cos \theta \end{bmatrix}$$

It follows from the equations above that for any vector $\mathbf{x}$ in R^2, the matrix product $A_\theta \mathbf{x}$, where $\mathbf{x}$ is viewed as a column vector, is the rotation (counterclockwise) of $\mathbf{x}$ by θ as shown in Figure 2.1.2. For example, if $\mathbf{x} = (1, 2)$, we have

$$\begin{bmatrix} \cos \theta & -\sin \theta \\ \sin \theta & \cos \theta \end{bmatrix} \begin{bmatrix} 1 \\ 2 \end{bmatrix} = \begin{bmatrix} \cos \theta - 2 \sin \theta \\ \sin \theta + 2 \cos \theta \end{bmatrix}$$

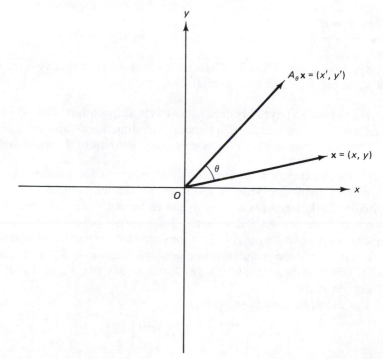

Figure 2.1.2

For a justification of this property, see Exercise 18. Examples of other geometric transformations, such as projections and reflections, are considered in the exercises. We shall make use of the rotation matrix in later work.

For what follows we represent the standard vectors $\mathbf{e}_1, \mathbf{e}_2, \ldots, \mathbf{e}_n$ of R^n as column vectors.

Theorem 2.1.2 *Let A be an $m \times n$ matrix, and let $\mathbf{x}$ and $\mathbf{y}$ be $n \times 1$ column vectors. Then:*

    (a) $A\mathbf{e}_j = {}^j A$ *for $j = 1, \ldots, n$.*

    (b) $A(c\mathbf{x}) = c(A\mathbf{x})$ *for any scalar c.*

    (c) $A(\mathbf{x} + \mathbf{y}) = A\mathbf{x} + A\mathbf{y}$.

Proof

We shall prove parts (a) and (c) and leave part (b) as an exercise.

(a) For any j, we have that $A\mathbf{e}_j$ and ${}^j A$ are each $m \times 1$ column vectors. For any i, the ith entry of $A\mathbf{e}_j$ is ${}_i A \cdot \mathbf{e}_j = A_{ij}$. Similarly, the ith entry of ${}^j A$ is A_{ij}. Therefore, $A\mathbf{e}_j = {}^j A$.

(c) For any i, we have that the ith entry of $A(\mathbf{x} + \mathbf{y})$ is the dot product

$$ {}_i A \cdot (\mathbf{x} + \mathbf{y}) = {}_i A \cdot \mathbf{x} + {}_i A \cdot \mathbf{y} \qquad \text{(Theorem 1.6.1(c))} $$

which is the ith entry of $A\mathbf{x} + A\mathbf{y}$. Since all the entries of $A(\mathbf{x} + \mathbf{y})$ and $A\mathbf{x} + A\mathbf{y}$ are equal, the matrices must be equal. ∎

Corollary 2.1.3

Let A and B be $m \times n$ matrices, and suppose that $A\mathbf{x} = B\mathbf{x}$ for all $n \times 1$ column vectors $\mathbf{x}$. Then $A = B$.

Proof

For any i, take $\mathbf{x} = \mathbf{e}_i$. Then, by Theorem 2.1.2(a), we have that ${}^i A = {}^i B$. Thus, A and B have equal columns and so are equal. ∎

Up to now we have only defined matrix multiplication between an $m \times n$ matrix and an $n \times 1$ column vector. What about extending this definition? The next example will illustrate the advantage of a more general definition of matrix multiplication.

Example 6

In reference to Example 4, recall that if the population distribution for the city and the suburb in a particular year is given by the (column) vector $\mathbf{p}$, the population distribution in the next year is given by the vector $A\mathbf{p}$, where A is the transition matrix. Suppose that we now let P denote the 2×3 matrix whose columns ${}^1 P$, ${}^2 P$, and ${}^3 P$ represent the population distributions for each of three particular years. To determine the effect of A on these population vectors, we compute $A({}^1 P)$, $A({}^2 P)$, and $A({}^3 P)$. It is useful to combine these results into one 2×3 matrix, $(A({}^1 P), A({}^2 P), A({}^3 P))$, which we shall denote as AP.

For example, if we have vectors

$$ {}^1 P = \begin{bmatrix} 50 \\ 75 \end{bmatrix} \qquad {}^2 P = \begin{bmatrix} 40 \\ 60 \end{bmatrix} \qquad {}^3 P = \begin{bmatrix} 30 \\ 70 \end{bmatrix} $$

then

$$A(^1P) = \begin{bmatrix} 44.75 \\ 80.25 \end{bmatrix} \quad A(^2P) = \begin{bmatrix} 35.80 \\ 64.20 \end{bmatrix} \quad A(^3P) = \begin{bmatrix} 27.60 \\ 72.40 \end{bmatrix}$$

So

$$AP = \begin{bmatrix} 0.85 & 0.03 \\ 0.15 & 0.97 \end{bmatrix} \begin{bmatrix} 50 & 40 & 30 \\ 75 & 60 & 70 \end{bmatrix}$$

$$= \begin{bmatrix} 0.85(50) + 0.03(75) & 0.85(40) + 0.03(60) & 0.85(30) + 0.03(70) \\ 0.15(50) + 0.97(75) & 0.15(40) + 0.97(60) & 0.15(30) + 0.97(70) \end{bmatrix}$$

$$= \begin{bmatrix} 44.75 & 35.80 & 27.60 \\ 80.25 & 64.20 & 72.40 \end{bmatrix}$$

or, using the notation of dot products, we have

$$AP = \begin{bmatrix} {}_1A \cdot {}^1P & {}_1A \cdot {}^2P & {}_1A \cdot {}^3P \\ {}_2A \cdot {}^1P & {}_2A \cdot {}^2P & {}_2A \cdot {}^3P \end{bmatrix}$$

Thus, the ijth entry of AP is the dot product of the ith row of A and the jth column of P.

Definition   Let A be an $m \times n$ matrix and let B be an $n \times p$ matrix. We define the **matrix product** of A and B to be the $m \times p$ matrix AB whose ijth entry is the dot product of the ith row of A and the jth column of B. Using dot product notation, we have

$$(AB)_{ij} = {}_iA \cdot {}^jB$$

$$= (A_{i1}, A_{i2}, \ldots, A_{im}) \begin{bmatrix} B_{1j} \\ B_{2j} \\ \vdots \\ B_{mj} \end{bmatrix}$$

$$= A_{i1}B_{1j} + A_{i2}B_{2j} + \cdots + A_{im}B_{mj}$$

$$= \sum_{k=1}^{n} A_{ik}B_{kj}$$

Notice that when the sizes of A and B are written side by side in the same order as the product, that is, $(m \times n)(n \times p)$, the inner dimensions must be equal, and the outer dimensions give the size of the product AB. Or, symbolically, $(m \times n) \times (n \times p) = (m \times p)$.

For example, suppose that A and B are defined as

$$A = \begin{bmatrix} 4 & 0 \\ 2 & 3 \\ 4 & -1 \end{bmatrix} \quad \text{and} \quad B = \begin{bmatrix} 4 & 5 & 1 & 7 \\ 6 & 3 & 0 & 2 \end{bmatrix}$$

Then

$$AB = \begin{bmatrix} 16 & 20 & 4 & 28 \\ 26 & 19 & 2 & 20 \\ 10 & 17 & 4 & 26 \end{bmatrix}$$

Notice the relationship between the various sizes: $(3 \times 2) \times (2 \times 4) = (3 \times 4)$. The product BA is not defined.

Although, as will be shown in the next section, most of the laws of arithmetic for scalars are valid for matrices, matrix multiplication is *not* commutative, that is, AB does not necessarily equal BA even when both products are defined. To see this, let

$$A = \begin{bmatrix} 1 & 2 \\ 0 & 1 \end{bmatrix} \quad \text{and} \quad B = \begin{bmatrix} 2 & 1 \\ 3 & 4 \end{bmatrix}$$

Then

$$AB = \begin{bmatrix} 8 & 9 \\ 3 & 4 \end{bmatrix} \quad \text{but} \quad BA = \begin{bmatrix} 2 & 5 \\ 3 & 10 \end{bmatrix}$$

Just as we may define positive integer powers of scalars, powers of square matrices may also be defined. If A is a square matrix, define

$$A^2 = AA \qquad A^3 = (AA)A$$

and, in general, $A^k = A^{k-1}A$ for $k \geq 2$. For example, if

$$A = \begin{bmatrix} 1 & -2 \\ 2 & -3 \end{bmatrix}$$

then

$$A^2 = \begin{bmatrix} -3 & 4 \\ -4 & 5 \end{bmatrix} \quad \text{and} \quad A^3 = \begin{bmatrix} 5 & -6 \\ 6 & -7 \end{bmatrix}$$

Definition   Let A be an $m \times n$ matrix. Define the **transpose** of A, denoted by A^t, to be the $n \times m$ matrix satisfying $(A^t)_{ij} = A_{ji}$ for $1 \leq i \leq n$, $1 \leq j \leq m$.

For example,

$$\begin{bmatrix} 2 & 1 \\ 1 & 0 \\ 4 & 3 \end{bmatrix}^{t} = \begin{bmatrix} 2 & 1 & 4 \\ 1 & 0 & 3 \end{bmatrix} \quad \text{and} \quad \begin{bmatrix} 1 & 5 \\ 5 & 3 \end{bmatrix}^{t} = \begin{bmatrix} 1 & 5 \\ 5 & 3 \end{bmatrix}$$

From this example it is clear that the rows of A^t are the columns of A, and vice versa.

Notice that the last matrix in the example above has the property that it is equal to its transpose. Such a matrix is called **symmetric**. It follows that a square matrix A is symmetric if and only if $A_{ij} = A_{ji}$ for all i and j. As we shall see in Section 7.2, symmetric matrices arise frequently in applications.

There are a number of interesting interactions between transposition and matrix operations. One of these is indicated in Theorem 2.1.4. Others can be found in the exercises.

Theorem 2.1.4   *Let A be an $m \times n$ matrix, and let B be an $n \times p$ matrix. Then*

$$(AB)^t = B^t A^t$$

Proof
The *ij*th entry of $(AB)^t$ is

$$(AB)^t_{ij} = (AB)_{ji}$$

$$= \sum_{k=1}^{n} A_{jk} B_{ki}$$

$$= \sum_{k=1}^{n} (A^t)_{kj} (B^t)_{ik}$$

$$= \sum_{k=1}^{n} (B^t)_{ik} (A^t)_{kj}$$

$$= (B^t A^t)_{ij}$$

Therefore, $(AB)^t = B^t A^t$.   ■

Application to Treaty Relationships

The study of certain relationships between objects may become very complicated as soon as the number of objects begins to increase even beyond three or four. In particular, suppose that there are five countries some of which maintain treaties with each other. To organize these treaty relationships, we shall use a 5×5 matrix A defined as follows. For $i \neq j$, we have

$$A_{ij} = \begin{cases} 1 & \text{if country } i \text{ and } j \text{ have a treaty between each other} \\ 0 & \text{otherwise} \end{cases}$$

We define $A_{ii} = 1$ for $i = 1, \ldots, 5$. So A is a matrix all of whose entries consist of 0's and 1's. Such matrices are called ***adjacency matrices***, and they are an object of study in their own right.

For purposes of illustration, suppose that

$$A = \begin{bmatrix} 1 & 0 & 1 & 1 & 0 \\ 0 & 1 & 0 & 1 & 1 \\ 1 & 0 & 1 & 0 & 1 \\ 1 & 1 & 0 & 1 & 0 \\ 0 & 1 & 1 & 0 & 1 \end{bmatrix}$$

Notice that A has the property that $A = A^t$; that is, A is symmetric. The symmetry arises here because the underlying treaty relationship is symmetric. This is true of many of the relationships that occur in nature.

At this point we have not used any of the arithmetic of matrices. We shall examine the significance of the matrix A^2. For example,

$$(A^2)_{14} = A_{11}A_{14} + A_{12}A_{24} + A_{13}A_{34} + A_{14}A_{44} + A_{15}A_{54}$$

A typical term on the right-hand side of the equation has the form $A_{1k}A_{k4}$. This term is 1 if and only if both factors are 1, that is, if and only if country 1 has a treaty with another country which has a treaty with country 4. Thus, the right side gives the number of countries that "link" country 1 and country 4 (including countries 1 and 4 themselves, if they have a treaty with each other). To see the rest of the entries, we compute

$$A^2 = \begin{bmatrix} 3 & 1 & 2 & 2 & 1 \\ 1 & 3 & 1 & 2 & 2 \\ 2 & 1 & 3 & 1 & 2 \\ 2 & 2 & 1 & 3 & 1 \\ 1 & 2 & 2 & 1 & 3 \end{bmatrix}$$

We see that $(A^2)_{14} = 2$, so that two countries link countries 1 and 4. Since $A_{14} = 1$, countries 1 and 4 have a treaty between each other. Thus, there is no third country that has a treaty with both countries 1 and 4. Other deductions are left for the exercises.

By looking at other powers of A, additional information may be obtained. For example, it can be shown that for such a matrix A, if $A + A^2 + \cdots + A^{n-1}$ has a zero entry, then there is a group of countries such that no member of this group has a treaty with any country outside the group.

As a final comment, it should be pointed out that most of the reasoning used in this example may be applied to nonsymmetric relationships.

Application to Scheduling

Suppose that the administration at a small college with m students wants to plan the times for its n courses. The goal of such planning is to avoid scheduling popular courses at the same time. To minimize the number of time conflicts, the students are surveyed. Each student is asked which courses he or she would like to take in the following semester. The results of this survey may be put into matrix form. Define the $m \times n$ matrix A as follows:

$$A_{ij} = \begin{cases} 1 & \text{if student } i \text{ wants to take course } j \\ 0 & \text{otherwise} \end{cases}$$

It is interesting that the matrix product $A^t A$ will provide important information regarding the scheduling of course times. We begin with an interpretation of the elements of this matrix. For example,

$$(A^t A)_{12} = \sum_{k=1}^{m} (A^t)_{1k} A_{k2}$$

$$= \sum_{k=1}^{m} A_{k1} A_{k2}$$

Now, $A_{k1} A_{k2} = 1$ if and only if $A_{k1} = 1$ and $A_{k2} = 1$, that is, student k wants to take both course 1 and course 2. So $(A^t A)_{12}$ represents the total number of students who want to take both course 1 and 2. In general,

$$(A^t A)_{ij} = \text{total number of students who want to take}$$
$$\text{both course } i \text{ and course } j$$

Example 7   Suppose that we have a sample of 10 students and five courses. Table 2.1.1 lists the results of the survey. By letting A represent the 10×5 array under the course numbers, we compute

$$A^t A = \begin{bmatrix} 4 & 1 & 2 & 0 & 2 \\ 1 & 3 & 1 & 1 & 1 \\ 2 & 1 & 5 & 1 & 5 \\ 0 & 1 & 1 & 3 & 1 \\ 2 & 1 & 5 & 1 & 5 \end{bmatrix}$$

TABLE 2.1.1

		Course Number			
Student	1	2	3	4	5
1	1	0	1	0	1
2	0	0	1	1	1
3	1	0	0	0	0
4	0	1	1	0	1
5	0	0	0	0	0
6	1	1	0	0	0
7	0	0	1	0	1
8	0	1	0	1	0
9	1	0	1	0	1
10	0	0	0	1	0

From this matrix we see that there are five students who want course 3 and course 5. All other pairs of courses are wanted by at most two students. In fact, if courses 1, 3, and 5 were offered at different times, there would be very few time conflicts.

Notice that the interpretation of the entries of A^tA given above allows us to conclude that the entry $(A^tA)_{ii}$ represents the number of students who desire class i. For example, we see that four students prefer course 1, three students desire course 2, and so on. Thus, the *trace* (see Exercise 15) of A^tA would equal the total enrollment (counting some students more than once) of all of the courses if they were offered at different times.

Finally, notice that the matrix A^tA is symmetric (see Exercise 14). So for large matrices it would not be efficient to compute the i, j entries for $i > j$ since $(A^tA)_{ij} = (A^tA)_{ji}$.

Exercises

Exercises 1–5 will use the following matrices:

$$A = \begin{bmatrix} 1 & -2 \\ 3 & 4 \end{bmatrix} \quad B = \begin{bmatrix} 7 & 4 \\ 1 & 2 \end{bmatrix}$$

$$C = \begin{bmatrix} 3 & 8 & 1 \\ 2 & 0 & 4 \end{bmatrix}$$

$$\mathbf{x} = \begin{bmatrix} 2 \\ 3 \end{bmatrix} \quad \mathbf{y} = \begin{bmatrix} 1 \\ 3 \\ -5 \end{bmatrix} \quad \mathbf{z} = (7, -1)$$

1. Compute each of the following or give a reason why the expression is not defined.
   (a) $A + B$ (b) $A - 2B$
   (c) $A + C$ (d) $B\mathbf{x}$
   (e) $B\mathbf{y}$ (f) $C\mathbf{y}$
   (g) $\mathbf{xz}$ (h) $A\mathbf{z}^t$

2. Compute each of the following or give a reason why the expression is not defined.
   (a) AC (b) AB
   (c) BA (d) BC

(e) CB (f) CB^t

(g) A^2 (h) A^3

3. Verify that $(A + B)^t = A^t + B^t$ for the matrices above.

4. Verify that $(AB)C = A(BC)$ for the matrices above.

5. Verify that $(AB)^t = B^t A^t$ for the matrices above.

6. Prove parts (a), (b), and (d) of Theorem 2.1.1.

7. In a metropolitan area suppose that there are 40 thousand people living in the city and 30 thousand people living in the suburbs. Use the transition matrix A in Example 4 to determine:

   (a) The number of people living in the city and suburbs after 1 year.

   (b) The number of people living in the city and suburbs after 2 years.

8. In each part below determine the rotation matrix A_θ and then compute the vector obtained by rotating (counterclockwise) by the angle θ.

   (a) $(1, 1)$ by $90°$

   (b) $(1, -2)$ by $45°$

   (c) $(3, 4)$ by $360°$

9. Define A by

$$A = \begin{bmatrix} -1 & 0 \\ 0 & 1 \end{bmatrix}$$

   (a) Show that for any column vector **x**, the vector $A\mathbf{x}$ is the *reflection* of **x** about the y-axis (see Figure 2.1.3).

   (b) How should A be defined so that $A\mathbf{x}$ is the reflection of **x** about the x-axis?

10. Define A by

$$A = \begin{bmatrix} 1 & 0 \\ 0 & 0 \end{bmatrix}$$

   (a) Show that for any column vector **x**, the vector $A\mathbf{x}$ is the *projection* of **x** onto the x-axis (see Figure 2.1.4).

   (b) How should A be defined so that $A\mathbf{x}$ represents the projection of A on the y-axis?

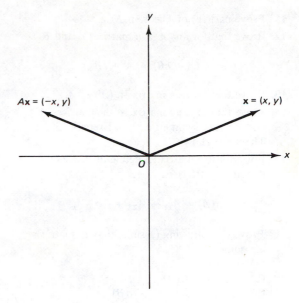

Figure 2.1.3

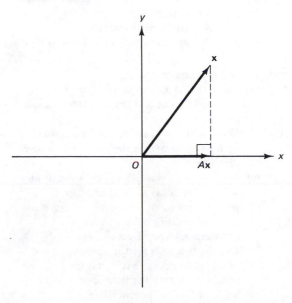

Figure 2.1.4

11. Prove part (b) of Theorem 2.1.2.

12. Prove that for any $m \times n$ matrices A and B

$$(A + cB)^t = A^t + cB^t$$

13. Prove that for any matrix A, $(A^t)^t = A$.

14. Prove that for any matrix A, the matrices A^tA and AA^t are symmetric. (Use Exercise 13 and Theorem 2.1.4.)

15. For an $n \times n$ matrix A, define the **trace** of A, denoted by $\text{tr}(A)$, as

$$\text{tr}(A) = A_{11} + A_{22} + \cdots + A_{nn}$$

Prove the following (assume that all matrices are square):
(a) $\text{tr}(0) = 0$
(b) $\text{tr}(cA) = c \, \text{tr}(A)$
(c) $\text{tr}(A + B) = \text{tr}(A) + \text{tr}(B)$
(d) $\text{tr}(AB) = \text{tr}(BA)$
(e) $\text{tr}(A^t) = \text{tr}(A)$

16. Using the definition of "trace" in Exercise 15, find matrices A and B such that $\text{tr}(AB) \neq \text{tr}(A) \, \text{tr}(B)$.

17. Suppose that A and B are symmetric matrices each of order n. Use Exercise 12 and Theorem 2.1.4 to prove that:
(a) $A + B$ is a symmetric matrix.
(b) AB is a symmetric matrix if and only if $AB = BA$.

18. For the rotation matrix A_θ:
(a) Show that $A_\theta \mathbf{e}_1$ can be obtained from $\mathbf{e}_1$ by a rotation of θ in the counterclockwise direction.
(b) Show that $A_\theta \mathbf{e}_2$ can be obtained from $\mathbf{e}_2$ by a rotation of θ in the counterclockwise direction.
(c) Let $\mathbf{x} = (a, b) = a\mathbf{e}_1 + b\mathbf{e}_2$ be a vector in R^2. Use parts (a) and (b) and the parallelogram law to show that $A_\theta \mathbf{x}$ is the result of rotating $\mathbf{x}$ by θ in the counterclockwise direction.
(d) Use the formulas for the sine and cosine of a sum to prove that $A_\theta \mathbf{x}$ is the vector obtained by rotating the vector $\mathbf{x}$ by an angle of θ in the counterclockwise direction. [*Hint*: Using formulas to transform polar to rectangular coordinates, represent $\mathbf{x}$ as $r(\cos \theta, \sin \theta)$ for some scalar r and angle θ.]

19. Let A be an $m \times n$ matrix and let B be an $n \times p$ matrix. If ^kA represents the kth column of A, and $_kB$ represents the kth row of B, prove that

$$AB = \sum_{k=1}^{n} (^kA)(_kB)$$

Problems 20 and 21 are based on the ideas developed in the application on treaty relationships.

20. Use the adjacency matrix A in the application to answer the following.
(a) Which pairs of countries have treaties with one another?
(b) How many countries link country 1 with country 3?
(c) Give an interpretation of $(A^3)_{14}$.

21. Suppose that there is a group of four people and an associated 4×4 adjacency matrix A defined by

$$A_{ij} = \begin{cases} 1 & \text{if person } i \text{ likes person } j \\ 0 & \text{otherwise} \end{cases}$$

Assume for technical reasons that $A_{ii} = 0$ for $i = 1, 2, 3, 4$. We say that persons i and j are *friends* if they like one another; that is, $A_{ij} = A_{ji} = 1$. Let A be given by

$$A = \begin{bmatrix} 0 & 1 & 0 & 1 \\ 1 & 0 & 1 & 0 \\ 0 & 1 & 0 & 1 \\ 1 & 1 & 1 & 0 \end{bmatrix}$$

(a) List all pairs of friends.
(b) Give an interpretation to the entries of A^2.

Define the 4×4 matrix B by

$$B_{ij} = \begin{cases} 1 & \text{if } i \text{ and } j \text{ are friends} \\ 0 & \text{otherwise} \end{cases}$$

(c) Write down the matrix B. Is it a symmetric matrix?

A *clique* is a set of three or more people each of whom is friendly with all the others in the set.

(d) Show that person i belongs to a clique if and only if $(B^3)_{ii} > 0$.

*(e) Use the program MATRIX to count the cliques associated with B.

Exercises 22 and 23 refer to the application on scheduling.

22. Table 2.1.2 indicates student preferences for the given courses.

(a) Give the pair(s) of courses that are desired by the largest number of students.
(b) Give the pair(s) of courses that are desired by the least number of students.
(c) Use the diagonal entries of the appropriate matrix to determine the number of students who prefer each course.

TABLE 2.1.2

Student	Course Number				
	1	2	3	4	5
1	1	0	0	0	1
2	0	0	1	1	0
3	1	0	1	0	1
4	0	1	0	0	0
5	1	0	0	0	1
6	0	1	0	0	1
7	1	0	1	0	1
8	0	1	1	0	1
9	1	0	0	1	1
10	0	0	1	1	0

23. Let A be defined as in the application to scheduling.

(a) Justify the following interpretation:

$$(AA^t)_{ij} = \text{number of classes that are desired by both students } i \text{ and } j$$

(b) Show that $(AA^t)_{12} = 2$ and $(AA^t)_{91} = 3$.
(c) Use the survey in the application to show that parts (a) and (b) are consistent.
(d) Give an interpretation of the diagonal entries of AA^t.

*24. For the matrices

$$A = \begin{bmatrix} 1 & -1 & 4 & 3 \\ 2 & 1 & 0 & -2 \\ 1 & 3 & 1 & 5 \\ 2 & 1 & -4 & 3 \end{bmatrix} \qquad B = \begin{bmatrix} 1 \\ -4 \\ 5 \\ 7 \end{bmatrix}$$

$$C = \begin{bmatrix} 2 & -1 \\ 2 & 4 \\ 3 & 6 \\ 0 & -8 \end{bmatrix}$$

$$D = \begin{bmatrix} 1 & 2 & 6 & 3 \\ 0 & 2 & 1 & -5 \end{bmatrix}$$

$$E = \begin{bmatrix} 1 & 2 & 3 & 4 \\ 0 & 1 & 5 & -2 \end{bmatrix}$$

use the program MATRIX to compute:

(a) $D(AC)$ (b) $(DA)C$
(c) $(DA)^t$ (d) $(D + E)^t$
(e) $(D + E)A$ (f) A^tC
(g) C^tA (h) AB
(i) A^{10} (j) $(CD)^{11}$

* This problem should be solved by using one of the programs noted in Appendix B.

2.2 ARITHMETIC PROPERTIES OF MATRIX MULTIPLICATION

Matrix addition and multiplication were introduced in the preceding section. There we saw that matrix addition obeys many of the arithmetic rules of scalars, but matrix multiplication is not commutative even in the case that each of the products is defined.

In this section we explore the many similarities that matrix multiplication has with scalar multiplication.

Theorem 2.1.1 shows us that the zero matrix has some of the properties of the zero scalar. It is natural to ask whether there exists a matrix that can play a role similar to the scalar 1 with respect to matrix multiplication. That is, is it possible to define a matrix I such that $AI = A = IA$ for every matrix A? For both of these products to be defined, it is necessary that both A and I be square matrices.

Definition   For any positive integer n, we define the *identity matrix of order n*, denoted I_n, by

$$(I_n)_{ij} = \begin{cases} 1 & \text{if } i = j \\ 0 & \text{if } i \neq j \end{cases}$$

We shall often omit the subscript n and simply write I if the size of I is clear from the context. For example, we have the following identity matrices:

$$I_1 = (1) \qquad I_2 = \begin{bmatrix} 1 & 0 \\ 0 & 1 \end{bmatrix} \qquad I_3 = \begin{bmatrix} 1 & 0 & 0 \\ 0 & 1 & 0 \\ 0 & 0 & 1 \end{bmatrix}$$

The following matrix product illustrates that I has the desired property.

$$\begin{bmatrix} 3 & 4 & -1 \\ 6 & 2 & 5 \\ 8 & -9 & 0 \end{bmatrix} \begin{bmatrix} 1 & 0 & 0 \\ 0 & 1 & 0 \\ 0 & 0 & 1 \end{bmatrix} = \begin{bmatrix} 3 & 4 & -1 \\ 6 & 2 & 5 \\ 8 & -9 & 0 \end{bmatrix}$$

The reader can verify that the same result occurs when the order of multiplication is reversed.

If A is a square matrix of order n, we call the entries $A_{11}, A_{22}, \ldots, A_{nn}$ the (main) *diagonal* of A. We call A a *diagonal matrix* if every entry of A that does not lie on the diagonal is 0. Both the zero and the identity matrices are diagonal matrices. In Chapter 6 we shall see many important applications of diagonal matrices. The next example illustrates one particularly important property of such matrices.

Example 1   If A is defined by

$$A = \begin{bmatrix} 2 & 0 & 0 \\ 0 & 1 & 0 \\ 0 & 0 & 3 \end{bmatrix}$$

then

$$A^2 = \begin{bmatrix} 2 & 0 & 0 \\ 0 & 1 & 0 \\ 0 & 0 & 3 \end{bmatrix} \begin{bmatrix} 2 & 0 & 0 \\ 0 & 1 & 0 \\ 0 & 0 & 3 \end{bmatrix}$$

$$= \begin{bmatrix} 4 & 0 & 0 \\ 0 & 1 & 0 \\ 0 & 0 & 9 \end{bmatrix}$$

In fact, it is easy to verify that

$$A^n = \begin{bmatrix} 2^n & 0 & 0 \\ 0 & 1 & 0 \\ 0 & 0 & 3^n \end{bmatrix}$$

for any positive integer n.

This example illustrates the fact that powers of diagonal matrices are also diagonal matrices. It is also the case that diagonal matrices of the same size commute. The proofs of these and other results concerning diagonal matrices are left for the exercises.

We are now ready to state the main theorem of this section. This theorem will allow us to treat matrices very much like scalars, with the zero and identity matrices playing the roles of 0 and 1, respectively.

Theorem 2.2.1 *Let A, B, and C be matrices for which the operations below are defined and let c be any scalar. Then:*

    (a) $A0 = 0A = 0$
    (b) $AI = A = IA$
    (c) $c(AB) = (cA)B = A(cB)$
    (d) $A(B + C) = AB + AC$ *(multiplication distributes*
    (e) $(B + C)A = BA + CA$ *over addition)*
    (f) $A(BC) = (AB)C$ *(associative law of multiplication)*

Proof

We shall prove the first half of part (b) and part (d). A direct proof of part (f) is sketched in the exercises. A more "elegant" proof of this part is provided in the next section. The proofs of the rest of the parts are left as exercises.

(b) Assume that A is a square matrix of order n. The ijth entry of AI is

$$(AI)_{ij} = \sum_{k=1}^{n} A_{ik} I_{kj}$$

$$= A_{ij}$$

since I_{kj} is 0 for $k \neq j$. The second half of part (b) is proved similarly.

(d) The *ij*th entry of $A(B + C)$ is

$$(A(B + C))_{ij} = \sum_{k=1}^{n} A_{ik}(B + C)_{kj}$$

$$= \sum_{k=1}^{n} A_{ik}(B_{kj} + C_{kj})$$

$$= \sum_{k=1}^{n} (A_{ik}B_{kj} + A_{ik}C_{kj})$$

$$= \sum_{k=1}^{n} A_{ik}B_{kj} + \sum_{k=1}^{n} A_{ik}C_{kj}$$

$$= (AB)_{ij} + (AC)_{ij}$$

$$= (AB + AC)_{ij}$$

which is the *ij*th entry of $AB + AC$.   ∎

In the preceding section we defined the rotation matrix A_θ as

$$A_\theta = \begin{bmatrix} \cos \theta & -\sin \theta \\ \sin \theta & \cos \theta \end{bmatrix}$$

Recall that for any nonzero vector $\mathbf{x}$, $A_\theta \mathbf{x}$ represents the vector obtained by rotating $\mathbf{x}$ counterclockwise by an angle θ. Similarly, $A_\phi(A_\theta \mathbf{x})$ represents the vector obtained by rotating $A_\theta \mathbf{x}$ counterclockwise by an angle ϕ, which is geometrically equivalent to rotating the vector $\mathbf{x}$ by $\theta + \phi$ (see Figure 2.2.1). That is, $A_\phi(A_\theta \mathbf{x}) = A_{\theta + \phi} \mathbf{x}$. However, by the associative law of matrix multiplication, we have that $A_\phi(A_\theta \mathbf{x}) = (A_\phi A_\theta)\mathbf{x}$. Thus,

$$(A_\phi A_\theta)\mathbf{x} = A_{\theta + \phi} \mathbf{x}$$

for any vector $\mathbf{x}$, so

$$A_\phi A_\theta = A_{\theta + \phi}$$

by Corollary 2.1.3. In particular, $A_\theta A_\phi = A_\phi A_\theta$; that is, the matrices commute. Notice the use of geometry to obtain these results about matrices.

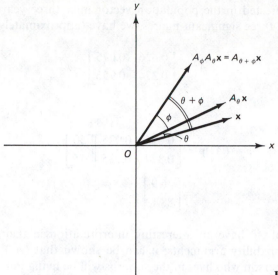

Figure 2.2.1

Substituting the appropriate rotation matrix in the equation above, we obtain

$$\begin{bmatrix} \cos \phi \cos \theta - \sin \phi \sin \theta & -\cos \phi \sin \theta - \sin \phi \cos \theta \\ \sin \phi \cos \theta + \cos \phi \sin \theta & -\sin \phi \sin \theta + \cos \phi \cos \theta \end{bmatrix}$$

$$= \begin{bmatrix} \cos(\theta + \phi) & -\sin(\theta + \phi) \\ \sin(\theta + \phi) & \cos(\theta + \phi) \end{bmatrix}$$

Now, by equating corresponding entries, we produce the familiar trigonometric identities of the sine and cosine of a sum of angles.

Application to Markov Processes

An interesting use of the associative property of matrix multiplication arises in the determination of population vectors as in Example 4 of Section 2.1. Recall that if $\mathbf{p}$ is the population vector for a given year, then $A\mathbf{p}$ represents the population vector for the following year. After two years, the population vector is given by $A(A\mathbf{p})$, which by the associative law is $(A^2)\mathbf{p}$. By applying the associative law repeatedly, it is easy to see that after n years, the population vector is given by $(A^n)\mathbf{p}$.

In our example, the transition matrix A and the population vector $\mathbf{p}$ was given by

$$A = \begin{bmatrix} 0.85 & 0.03 \\ 0.15 & 0.97 \end{bmatrix} \quad \text{and} \quad \mathbf{p} = \begin{bmatrix} 50 \\ 75 \end{bmatrix}$$

If we are interested in the population vector after three years, we could first compute A^3. Using three significant figures, we have (approximately)

$$A^3 = \begin{bmatrix} 0.626 & 0.075 \\ 0.374 & 0.925 \end{bmatrix}$$

so

$$(A^3)\mathbf{p} = \begin{bmatrix} 0.626 & 0.075 \\ 0.374 & 0.925 \end{bmatrix} \begin{bmatrix} 50 \\ 75 \end{bmatrix}$$

$$= \begin{bmatrix} 36.9 \\ 88.1 \end{bmatrix}$$

The entries of A^3 have an interesting interpretation in their own right. For example, using probability arguments, it can be shown that $(A^3)_{12}$ represents the probability that a person who lives in the suburbs will be living in the city in 3 years. In other words, approximately 7.5% of the people now living in the suburbs will be living in the city in 3 years.

To discover the long-term proportions of people living in the city and suburbs, larger and larger powers of A would have to be computed. Direct computation would be quite tedious. Later, we shall introduce "eigenvectors" to make this computation practical.

This example represents a special case of a *Markov process*, in which there are a finite number of *states* (city and suburb) and fixed probabilities of moving from one state to another state. We will return to this example in Chapter 6.

Block Multiplication

To facilitate matrix multiplication, it is often useful to regard a matrix as an array of *blocks* of smaller matrices. These blocks are formed by drawing horizontal and vertical lines through the matrix. Naturally, there are many ways to do this. In particular, these blocks are dictated by the application in which the matrix arises. For example, the matrix

$$A = \begin{bmatrix} 1 & 3 & 4 & 2 \\ 0 & 5 & -1 & 6 \\ 1 & 0 & 3 & -1 \end{bmatrix}$$

can be written as

$$A = \left[\begin{array}{cc|cc} 1 & 3 & 4 & 2 \\ 0 & 5 & -1 & 6 \\ \hline 1 & 0 & 3 & -1 \end{array} \right]$$

The lines above partition A into an array of four blocks. The first row of the partition consists of the 2×2 matrices

$$\begin{bmatrix} 1 & 3 \\ 0 & 5 \end{bmatrix} \quad \text{and} \quad \begin{bmatrix} 4 & 2 \\ -1 & 6 \end{bmatrix}$$

and the second row consists of the 1×2 matrices $(1, 0)$ and $(3, -1)$. We can also partition A as

$$A = \begin{bmatrix} 1 & 3 & 4 & 2 \\ 0 & 5 & -1 & 6 \\ 1 & 0 & 3 & -1 \end{bmatrix}$$

In this case there is only one row and two columns. The blocks of this row are the 3×2 matrices

$$\begin{bmatrix} 1 & 3 \\ 0 & 5 \\ 1 & 0 \end{bmatrix} \quad \text{and} \quad \begin{bmatrix} 4 & 2 \\ -1 & 6 \\ 3 & -1 \end{bmatrix}$$

We occasionally partition matrices into blocks, regarding the matrices in *block form*, to facilitate matrix multiplication. If two matrices are partitioned in the correct way, we can multiply them by regarding the blocks as individual entries. For example, suppose that matrices A and B are represented in block form

$$A = \begin{bmatrix} A_1 & A_2 \\ A_3 & A_4 \end{bmatrix} \quad \text{and} \quad B = \begin{bmatrix} B_1 & B_2 \\ B_3 & B_4 \end{bmatrix}$$

Then

$$AB = \begin{bmatrix} A_1 & A_2 \\ A_3 & A_4 \end{bmatrix} \begin{bmatrix} B_1 & B_2 \\ B_3 & B_4 \end{bmatrix}$$

$$= \begin{bmatrix} A_1 B_1 + A_2 B_3 & A_1 B_2 + A_2 B_4 \\ A_3 B_1 + A_4 B_3 & A_3 B_2 + A_4 B_4 \end{bmatrix}$$

As an illustration of the benefits of such a technique, suppose that the matrices A and B above are $2n \times 2n$ matrices for some large integer n, and suppose that each of the blocks (the A_i's and B_j's) is of size $n \times n$. In the case that $A_1 = A_3 = A_4 = 0$ and $A_2 = I_n$, we have that

$$AB = \begin{bmatrix} 0 & I_n \\ 0 & 0 \end{bmatrix} \begin{bmatrix} B_1 & B_2 \\ B_3 & B_4 \end{bmatrix}$$

$$= \begin{bmatrix} B_3 & B_4 \\ 0 & 0 \end{bmatrix}$$

Example 2   Let A and B be the matrices in block form

$$A = \begin{bmatrix} 1 & 3 & 4 & 2 \\ 0 & 5 & -1 & 6 \\ 1 & 0 & 3 & -1 \end{bmatrix} \quad \text{and} \quad B = \begin{bmatrix} 1 & 0 & 3 \\ 1 & 2 & 0 \\ 2 & -1 & 2 \\ 0 & 3 & 1 \end{bmatrix}$$

Then we can find the entries in the upper left block of AB using the method of block multiplication, by computing

$$\begin{bmatrix} 1 & 3 \\ 0 & 5 \end{bmatrix}\begin{bmatrix} 1 & 0 \\ 1 & 2 \end{bmatrix} + \begin{bmatrix} 4 & 2 \\ -1 & 6 \end{bmatrix}\begin{bmatrix} 2 & -1 \\ 0 & 3 \end{bmatrix} = \begin{bmatrix} 12 & 8 \\ 3 & 29 \end{bmatrix}$$

Similarly, the upper right block of AB can be obtained by computing

$$\begin{bmatrix} 1 & 3 \\ 0 & 5 \end{bmatrix}\begin{bmatrix} 3 \\ 0 \end{bmatrix} + \begin{bmatrix} 4 & 2 \\ -1 & 6 \end{bmatrix}\begin{bmatrix} 2 \\ 1 \end{bmatrix} = \begin{bmatrix} 13 \\ 4 \end{bmatrix}$$

the lower left block of AB can be obtained by computing

$$(1 \quad 0)\begin{bmatrix} 1 & 0 \\ 1 & 2 \end{bmatrix} + (3 \quad -1)\begin{bmatrix} 2 & -1 \\ 0 & 3 \end{bmatrix} = (7 \quad -6)$$

and the lower right block of AB can be obtained by computing

$$(1 \quad 0)\begin{bmatrix} 3 \\ 0 \end{bmatrix} + (3 \quad -1)\begin{bmatrix} 2 \\ 1 \end{bmatrix} = (8)$$

Putting these blocks together, we have that

$$AB = \begin{bmatrix} 12 & 8 & 13 \\ 3 & 29 & 4 \\ 7 & -6 & 8 \end{bmatrix}$$

As a general principle we have the following rule.

BLOCK MULTIPLICATION

If two matrices A and B are represented in block form so that the number of blocks in each row of A is the same as the number of blocks in each column of B, then the matrices can be multiplied in the same way as if the blocks were scalars, provided that the individual products are defined.

Example 3   Suppose that A is an $m \times n$ matrix, B is an $n \times p$ matrix, and C is an $n \times q$ matrix. We can use B and C to produce an $n \times (p + q)$ matrix consisting of the columns of B followed by the columns of C. We denote this new matrix by $(B|C)$. Then $A(B|C) = (AB|AC)$. For example, if

$$A = \begin{bmatrix} 2 & -1 \\ 1 & 0 \end{bmatrix} \qquad B = \begin{bmatrix} 1 & 3 \\ 0 & 2 \end{bmatrix} \qquad C = \begin{bmatrix} 2 & 0 & 1 \\ 1 & 0 & 3 \end{bmatrix}$$

then

$$A(B|C) = \begin{bmatrix} 2 & -1 \\ 1 & 0 \end{bmatrix}\begin{bmatrix} 1 & 3 & 2 & 0 & 1 \\ 0 & 2 & 1 & 0 & 3 \end{bmatrix}$$

$$= \begin{bmatrix} 2 & 4 & 3 & 0 & -1 \\ 1 & 3 & 2 & 0 & 1 \end{bmatrix}$$

$$= (AB|AC)$$

For any $m \times n$ matrix A, and any $n \times p$ matrix B, if we represent B as an array of column vectors,

$$B = ({}^1B, {}^2B, \ldots, {}^pB)$$

then

$$AB = (A\,{}^1B, A\,{}^2B, \ldots, A\,{}^pB)$$

Example 4   Suppose that A is an $m \times n$ matrix, and $\mathbf{x}$ is an $n \times 1$ column vector. If we represent A as an array of columns, then A is partitioned with individual blocks consisting of column vectors. Similarly, $\mathbf{x}$ can be partitioned with 1×1 blocks consisting of individual rows. Each such row is a square matrix of order 1. Thus, for

$$A = ({}^1A, {}^2A, \ldots, {}^nA) \qquad \text{and} \qquad \mathbf{x} = \begin{bmatrix} x_1 \\ x_2 \\ \vdots \\ x_n \end{bmatrix}$$

$$A\mathbf{x} = ({}^1A, {}^2A, \ldots, {}^nA)\begin{bmatrix} x_1 \\ x_2 \\ \vdots \\ x_n \end{bmatrix}$$

$$= x_1\,{}^1A + x_2\,{}^2A + \cdots + x_n\,{}^nA$$

Technically, each x_i is a 1×1 matrix and, as such, should be written after ^iA. However, a square matrix of order 1 can be identified with a scalar, and each ^iA can be viewed as a vector, so the result is a linear combination of the column vectors of A. For example, suppose that

$$A = \begin{bmatrix} 1 & 2 & 3 \\ 1 & -1 & 2 \end{bmatrix} \quad \text{and} \quad \mathbf{x} = \begin{bmatrix} -1 \\ 1 \\ 3 \end{bmatrix}$$

Then

$$Ax = (-1)\begin{bmatrix} 1 \\ 1 \end{bmatrix} + (1)\begin{bmatrix} 2 \\ -1 \end{bmatrix} + (3)\begin{bmatrix} 3 \\ 2 \end{bmatrix}$$

$$= \begin{bmatrix} 10 \\ 4 \end{bmatrix}$$

Example 5   Another way to treat the $m \times n$ matrix in Example 4 is to represent it as an array of m row vectors

$$A = \begin{bmatrix} _1A \\ _2A \\ \vdots \\ _mA \end{bmatrix}$$

Then, using block multiplication,

$$Ax = \begin{bmatrix} _1Ax \\ _2Ax \\ \vdots \\ _mAx \end{bmatrix}$$

Thus, for the matrices

$$A = \begin{bmatrix} 1 & 2 & 3 \\ 1 & -1 & 2 \end{bmatrix} \quad \text{and} \quad \mathbf{x} = \begin{bmatrix} -1 \\ 1 \\ 3 \end{bmatrix}$$

of Example 4,

$$Ax = \begin{bmatrix} 1 & 2 & 3 \\ \hline 1 & -1 & 2 \end{bmatrix} \begin{bmatrix} -1 \\ 1 \\ 3 \end{bmatrix}$$

$$= \begin{bmatrix} (1, 2, 3) \begin{bmatrix} -1 \\ 1 \\ 3 \end{bmatrix} \\ (1, -1, 2) \begin{bmatrix} -1 \\ 1 \\ 3 \end{bmatrix} \end{bmatrix}$$

$$= \begin{bmatrix} 10 \\ 4 \end{bmatrix}$$

The following is a summary of the important properties of block multiplication covered in this section. We assume that all matrix sizes are compatible.

1. $A(B|C) = (AB|AC)$

2. $A(^1B, {}^2B, \ldots, {}^pB) = (A\,{}^1B, A\,{}^2B, \ldots, A\,{}^pB)$

3. $({}^1A, {}^2A, \ldots, {}^nA) \begin{bmatrix} x_1 \\ x_2 \\ \vdots \\ x_n \end{bmatrix} = x_1\,{}^1A + x_2\,{}^2A + \cdots + x_n\,{}^nA$

Exercises

Exercises 1 and 2 use the following matrices:

$$A = \begin{bmatrix} 1 & -2 \\ 3 & 4 \end{bmatrix} \qquad B = \begin{bmatrix} 3 & 0 & -1 \\ 1 & 2 & 2 \end{bmatrix}$$

$$C = \begin{bmatrix} 1 & -1 & 2 \\ 0 & 3 & 1 \end{bmatrix} \qquad D = \begin{bmatrix} 2 & 1 \\ 0 & 1 \\ 1 & 3 \end{bmatrix}$$

1. Verify:
(a) $AI = IA = A$
(b) $3(AB) = (3A)B = A(3B)$
(c) $A(B + C) = AB + AC$

2. Verify that $A(BD) = (AB)D$.

3. Compute the products in block form of

(a) $\begin{bmatrix} 2 & 1 & 0 & 2 \\ 1 & 1 & 1 & 3 \\ \hline -1 & 2 & -2 & 0 \end{bmatrix}$ and

$\begin{bmatrix} 1 & 1 & 3 & 0 \\ -1 & 2 & -4 & 1 \\ \hline 3 & 1 & 1 & 1 \\ 2 & 0 & 2 & -1 \end{bmatrix}$

(b) $\begin{bmatrix} 2 & 0 \\ 3 & 1 \\ -1 & 5 \\ 1 & 2 \end{bmatrix} \begin{bmatrix} -1 & 2 & 3 & | & 0 \\ 2 & 0 & -1 & | & 2 \end{bmatrix}$

(c) $(-1, 3, 1) \begin{bmatrix} 1 & | & 2 \\ -1 & | & 1 \\ 0 & | & 1 \end{bmatrix}$

(d) $\begin{bmatrix} 1 & | & -1 & 0 \\ 0 & | & 1 & 2 \end{bmatrix} \begin{bmatrix} 1 \\ 3 \\ 2 \end{bmatrix}$

(e) $\begin{bmatrix} 1 & -1 & 0 \\ 0 & 1 & 2 \end{bmatrix} \begin{bmatrix} 1 \\ 3 \\ 2 \end{bmatrix}$

4. Let A be given in block form as

$$A = \begin{bmatrix} 3 & 2 & | & 4 & 1 \\ 0 & 1 & | & 2 & 4 \\ -1 & 3 & | & 5 & 6 \end{bmatrix} = \begin{bmatrix} C & D \\ E & F \end{bmatrix}$$

Verify that

$$A^t = \begin{bmatrix} C^t & E^t \\ D^t & F^t \end{bmatrix}$$

5. Suppose that A and B are diagonal matrices of the same size. Prove that the following matrices are also diagonal matrices.
   (a) $A + B$
   (b) cA, where c is a scalar
   (c) A^2
   (d) AB
   (e) A^k, where k is any positive integer

6. Prove that if A and B are diagonal matrices, then $(AB)_{ii} = (BA)_{ii} = A_{ii}B_{ii}$ for all i. Now use Exercise 5(d) to conclude that $AB = BA$.

7. Suppose that A has the block form

$$A = \begin{bmatrix} I_m & 0 \\ 0 & B \end{bmatrix}$$

where B is an $m \times m$ matrix. Prove that

$$A^k = \begin{bmatrix} I_m & 0 \\ 0 & B^k \end{bmatrix}$$

8. Prove parts (a), (c), and (e) of Theorem 2.2.1.

9. Suppose that A is an $m \times n$ matrix, B is an $n \times p$ matrix, and C is a $p \times q$ matrix. To prove part (f) of Theorem 2.2.1 directly, first prove that

$$(A(BC))_{ij} = \sum_{k=1}^{n} \sum_{l=1}^{p} A_{ik} B_{kl} C_{lj}$$

Now by reversing the order of summation, show that $((AB)C)_{ij}$ yields the same result.

10. Let B be a symmetric $n \times n$ matrix and let C be any $m \times n$ matrix. Prove that CBC^t is symmetric.

11. A square matrix A is called **upper triangular** if $A_{ij} = 0$ whenever $i > j$, that is, whenever all the entries below the (main) diagonal are zero. Suppose that A and B are upper triangular matrices of the same size. Prove that the following matrices are also upper triangular:
    (a) $A + B$
    (b) AB
    (c) Any diagonal matrix

12. A square matrix A is called **strictly upper triangular** if $A_{ij} = 0$ whenever $i \geq j$. Prove that for any strictly upper triangular matrix A of order n, we have that $A^n = 0$.

13. Use the transition matrix A and the population vector **p** in the application of this section to answer the following questions.
    (a) How many people will be living in the city in 3 years?
    (b) What proportion of the people now living in the city will be living in the city in 3 years?
    *(c) Use the program MATRIX to compute the proportion of people now living in the city who will be living in the suburbs in 5 years, in 10 years, in 20 years.

*14. Use the program MATRIX and the matrices

$$A = \begin{bmatrix} 1 & 2 & 1 & 3 \\ 1 & -1 & 2 & 4 \\ 0 & 2 & 8 & -5 \\ 1 & -3 & 2 & 4 \end{bmatrix}$$

$$B = \begin{bmatrix} 2 & 1 & -3 \\ 4 & 2 & 1 \\ 1 & 3 & -1 \\ 2 & 1 & 4 \end{bmatrix}$$

$$C = \begin{bmatrix} 2 & 3 & 4 \\ 1 & 2 & 1 \\ 5 & -6 & 8 \\ -1 & -1 & 2 \end{bmatrix}$$

$$D = \begin{bmatrix} 2 & 3 & 1 & -1 \\ -1 & 0 & 2 & 4 \\ 3 & 7 & 7 & 2 \end{bmatrix}$$

to compute:

(a) $B + C$ (b) CD

(c) AB (d) DA

(e) A^2C (f) DA^2

(g) $(AB)D$ (h) $A(BD)$

* This problem should be solved by using one of the programs noted in Appendix B.

2.3 THREE APPLICATIONS OF MATRIX MULTIPLICATION

We consider three applications of matrix multiplication.

The Leslie Matrix and Population Change

The population of an isolated colony of animals depends on the fertility and mortality rates for the various age groups of the colony. Fertility rates can be measured precisely only if observations are limited to females (since they are the ones who give birth). Since a relationship between the number of males and the number of females can usually be established, this restriction is not serious.

    Let us consider a hypothetical example to illustrate these ideas. Suppose that the members of a certain colony of mammals have a life span of under 3 years. We shall divide the females into three age groups: the first consists of those of age less than 1, the second consists of those of age 1, and the third consists of those of age 2. Suppose that of newborn females, 40% survive to age 1, and that 50% of these survive to age 2. Thus, we have a picture of the survival rates of these animals. In terms of reproductive rates, suppose that the females under 1 year of age do not give birth, that those of age 1 have an average of two female offspring, and that those of age 2 have an average of one female offspring. Let x_1, x_2, and x_3 be the numbers of females in the first, second, and third age groups, respectively. The vector $\mathbf{x} = (x_1, x_2, x_3)$ is called the *population distribution* for the female population of the colony. The information above can be used to predict the population distribution for the next year. Let $\mathbf{y} = (y_1, y_2, y_3)$ be next year's distribution. Since y_1 is the total number of female offspring born during the year, and since these offspring come from females of the second and third age groups, we have that $y_1 = 2x_2 + x_3$. The number y_2 is the total number of females in the second age group for the next year. Because these females are in the first age group this year, and because only 40% of these will survive to the next year, we have that $y_2 = 0.4x_1$. Similarly, $y_3 = 0.5x_2$. Collecting these results, we have

$$
\begin{aligned}
y_1 &= \qquad 2.0x_2 + 1.0x_3 \\
y_2 &= 0.4x_1 \\
y_3 &= \qquad 0.5x_2
\end{aligned}
$$

These three equations can be represented by the single matrix equation $\mathbf{y} = A\mathbf{x}$, where $\mathbf{x}$ and $\mathbf{y}$ are population distributions as defined above and A is the 3×3 matrix

$$
A = \begin{bmatrix} 0.0 & 2.0 & 1.0 \\ 0.4 & 0.0 & 0.0 \\ 0.0 & 0.5 & 0.0 \end{bmatrix}
$$

For example, suppose that $\mathbf{x} = (1000, 1000, 1000)$; that is, there are 1000 females in each age group. Then

$$
\begin{aligned}
\mathbf{y} &= A\mathbf{x} \\
&= \begin{bmatrix} 0.0 & 2.0 & 1.0 \\ 0.4 & 0.0 & 0.0 \\ 0.0 & 0.5 & 0.0 \end{bmatrix} \begin{bmatrix} 1000 \\ 1000 \\ 1000 \end{bmatrix} \\
&= \begin{bmatrix} 3000 \\ 400 \\ 500 \end{bmatrix}
\end{aligned}
$$

So one year later there are 3000 females under 1 year of age, 400 females who are 1 year old, and 500 females who are 2 years old.

For each positive integer k, let $\mathbf{x}(k)$ represent the population distribution after k years, and let $\mathbf{x}(0) = (x_1, x_2, x_3)$ represent the current population distribution. In the example above,

$$
\mathbf{x}(0) = \begin{bmatrix} 1000 \\ 1000 \\ 1000 \end{bmatrix} \qquad \text{and} \qquad \mathbf{x}(1) = \begin{bmatrix} 3000 \\ 400 \\ 500 \end{bmatrix}
$$

Then for any positive integer k, we have $\mathbf{x}(k) = A\mathbf{x}(k-1)$. Thus,

$$
\mathbf{x}(k) = A\mathbf{x}(k-1) = A^2\mathbf{x}(k-2) = \cdots = A^k\mathbf{x}(0)
$$

In this way we may predict population trends over the long term. For example, suppose that the current population distribution is $\mathbf{x}(0) = (1000, 1000, 1000)$, and that we want to predict the distribution in 10 years; that is, we want to find $\mathbf{x}(10) = A^{10}\mathbf{x}(0)$. We first compute A^{10}. If we round off the entries of A^{10} to three places after the decimal, we have

$$
A^{10} = \begin{bmatrix} 0.481 & 1.088 & 0.418 \\ 0.167 & 0.481 & 0.202 \\ 0.101 & 0.209 & 0.077 \end{bmatrix}
$$

Then

$$
\mathbf{x}(10) = A^{10}\mathbf{x}(0)
$$

$$
= \begin{bmatrix} 0.481 & 1.088 & 0.418 \\ 0.167 & 0.481 & 0.202 \\ 0.101 & 0.209 & 0.077 \end{bmatrix} \begin{bmatrix} 1000 \\ 1000 \\ 1000 \end{bmatrix}
$$

$$
= \begin{bmatrix} 1987 \\ 851 \\ 387 \end{bmatrix}
$$

where each entry is rounded off to the nearest whole number. If we continue this process in increments of 10 years, we find that (rounding to whole numbers)

$$
\mathbf{x}(20) = \begin{bmatrix} 2043 \\ 819 \\ 408 \end{bmatrix} \quad \text{and} \quad \mathbf{x}(30) = \mathbf{x}(40) = \begin{bmatrix} 2045 \\ 818 \\ 409 \end{bmatrix}
$$

It appears that the population stabilizes after 30 years. In fact, for the vector $\mathbf{z} = (2045, 818, 409)$, we have that $A\mathbf{z} = \mathbf{z}$ precisely. Under this circumstance, the population distribution is **stable**; that is, it does not change from year to year. The fact that there is a stable population distribution for this colony depends on the nature of the survival and reproductive rates of the age groups. Exercise 2 gives an example of a population model for which no stable population distribution exists.

We can generalize this situation to an arbitrary colony of animals. Suppose that we divide the females of the colony into n age groups, where x_i is the number of members of the ith group. The duration of time in an individual age group need not be a year. Let $\mathbf{x} = (x_1, x_2, \ldots, x_n)$ be the population distribution of the females of the colony. Let p_i be the probability that a female of the ith group will survive to the $(i + 1)$st group, and let b_i be the average number of female offspring of a member of the ith age group. If $\mathbf{y} = (y_1, y_2, \ldots, y_n)$ is the population distribution for the next time period, then

$$y_1 = b_1 x_1 + b_2 x_2 + \cdots \qquad + b_n x_n$$

$$y_2 = p_1 x_1$$

$$y_3 = \qquad\quad p_2 x_2$$

$$\vdots$$

$$y_n = \qquad\qquad\qquad p_{n-1} x_{n-1}$$

Therefore, if

$$A = \begin{bmatrix} b_1 & b_2 & \cdots & & b_n \\ p_1 & 0 & \cdots & & 0 \\ 0 & p_2 & \cdots & & 0 \\ \vdots & \vdots & & & \vdots \\ 0 & 0 & \cdots & p_{n-1} & 0 \end{bmatrix}$$

then we have that

$$\mathbf{y} = A\mathbf{x}$$

The matrix A is called the **Leslie matrix** for the population. As in the earlier example, if $\mathbf{x}(k)$ is the population distribution after k time intervals, then

$$\mathbf{x}(k) = A^k \mathbf{x}(0)$$

Analysis of Traffic Flow

Figure 2.3.1 represents the flow of traffic on a maze of one-way streets. The arrows indicate the direction of traffic flow. The number on any street beyond an intersection is the portion of the traffic entering the street from that intersection. For example, 30% of the traffic leaving intersection P_1 goes to P_4, and the other 70% goes to P_2. Notice that all the traffic leaving P_5 goes to P_8.

Suppose that on a particular day, x_1 cars enter the maze from the left of P_1, and x_2 cars enter from the left of P_3. Let w_1, w_2, w_3, and w_4 represent the numbers leaving the maze along the exits to the right (see Figure 2.3.1). We wish to determine the values of the w_i's. At first glance, this seems nearly impossible since there are so many routes for the traffic. However, we employ the well-known mathematical strategy of decomposing the problem into several simpler ones, solving each one individually, and putting them together to obtain the grand solution.

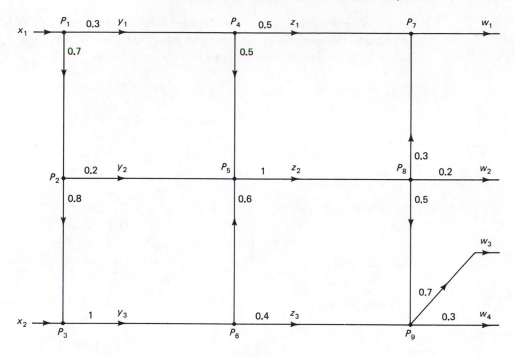

Figure 2.3.1

To begin with, consider only the portion of the traffic pattern involving intersections P_1, P_2, and P_3. Let y_1, y_2, and y_3 be the "expected" number of cars that exit along the three eastward routes. To find y_1, notice that all the cars entering the segment P_1P_4 come from the left of P_1, and that these constitute 30% of such cars. Therefore, $y_1 = 0.3x_1$. Also, $0.7x_1$ of the cars turn right at P_1, and of these, 20% enter P_2P_5. Because these are the only cars to do so, it follows that $y_2 = (0.2)(0.7)x_1 = 0.14x_1$. Furthermore, since 80% of the cars entering P_2 continue on to P_3P_6, we have that the number of such cars is $(0.8)(0.7)x_1 = 0.56x_1$. Finally, all the cars entering P_3 from the left enter P_3P_6, so $y_3 = 0.56x_1 + x_2$. Summarizing, we have

$$y_1 = 0.30x_1$$

$$y_2 = 0.14x_1$$

$$y_3 = 0.56x_1 + x_2$$

We can express this system of equations by the single matrix equation $\mathbf{y} = A\mathbf{x}$, where

$$\mathbf{y} = \begin{bmatrix} y_1 \\ y_2 \\ y_3 \end{bmatrix} \qquad A = \begin{bmatrix} 0.30 & 0 \\ 0.14 & 0 \\ 0.56 & 1 \end{bmatrix} \qquad \mathbf{x} = \begin{bmatrix} x_1 \\ x_2 \end{bmatrix}$$

Now consider the next set of intersections, P_4, P_5, and P_6. If we let z_1, z_2, and z_3 represent the numbers of cars that exit from the right of P_4, P_5, and P_6, respectively, then by a similar analysis we have that

$$z_1 = 0.5y_1$$
$$z_2 = 0.5y_1 + y_2 + 0.6y_3$$
$$z_3 = \qquad\qquad 0.4y_3$$

or $\mathbf{z} = B\mathbf{y}$, where

$$\mathbf{z} = \begin{bmatrix} z_1 \\ z_2 \\ z_3 \end{bmatrix} \quad \text{and} \quad B = \begin{bmatrix} 0.5 & 0 & 0 \\ 0.5 & 1 & 0.6 \\ 0 & 0 & 0.4 \end{bmatrix}$$

Finally, if we set

$$\mathbf{w} = \begin{bmatrix} w_1 \\ w_2 \\ w_3 \\ w_4 \end{bmatrix} \quad \text{and} \quad C = \begin{bmatrix} 1 & 0.30 & 0 \\ 0 & 0.20 & 0 \\ 0 & 0.35 & 0.7 \\ 0 & 0.15 & 0.3 \end{bmatrix}$$

then by the same argument, we have that $\mathbf{w} = C\mathbf{z}$. It follows that

$$\mathbf{w} = C\mathbf{z} = C(B\mathbf{y}) = (CB)(A\mathbf{x}) = (CBA)\mathbf{x}$$

Let $M = CBA$. Then

$$M = \begin{bmatrix} 1 & 0.30 & 0 \\ 0 & 0.20 & 0 \\ 0 & 0.35 & 0.7 \\ 0 & 0.15 & 0.3 \end{bmatrix} \begin{bmatrix} 0.5 & 0 & 0 \\ 0.5 & 1 & 0.6 \\ 0 & 0 & 0.4 \end{bmatrix} \begin{bmatrix} 0.30 & 0 \\ 0.14 & 0 \\ 0.56 & 1 \end{bmatrix}$$

$$= \begin{bmatrix} 0.3378 & 0.18 \\ 0.1252 & 0.12 \\ 0.3759 & 0.49 \\ 0.1611 & 0.21 \end{bmatrix}$$

For example, if 1000 cars enter the traffic pattern at P_1 and 2000 enter at P_3, then for $\mathbf{x} = (1000, 2000)$ we have

$$\mathbf{w} = M\mathbf{x}$$

$$= \begin{bmatrix} 0.3378 & 0.18 \\ 0.1252 & 0.12 \\ 0.3759 & 0.49 \\ 0.1611 & 0.21 \end{bmatrix} \begin{bmatrix} 1000 \\ 2000 \end{bmatrix}$$

$$= \begin{bmatrix} 697.8 \\ 365.2 \\ 1355.9 \\ 581.1 \end{bmatrix}$$

Naturally, the actual number of cars traveling on any path is a whole number, unlike the entries of $\mathbf{w}$. Since these calculations are based on probabilities, we cannot expect the answers to be exact. For example, approximately 698 cars exit the traffic pattern at P_7, and 365 cars exit the pattern at P_8.

This kind of analysis can be applied to other contexts, such as the flow of a fluid in a system of pipes or the exchange of money in an economy.

Application to Anthropology

In this application§ we shall see a fascinating use of matrix operations in the study of the marriage laws of the Natchez Indians.

Everyone in this tribe was a member of one of four classes: the Suns, the Nobles, the Honoreds, and the Stinkards. There were well-defined rules that determined class membership. The rules depended exclusively on the classes of the parents, and required that at least one of the parents be a Stinkard. Furthermore, the class of the child depended on the class of the other parent according to Table 2.3.1.

TABLE 2.3.1

Mother Is a Stinkard		Father Is a Stinkard	
Father	Child	Mother	Child
Sun	Noble	Sun	Sun
Noble	Honored	Noble	Noble
Honored	Stinkard	Honored	Honored
Stinkard	Stinkard	Stinkard	Stinkard

§ The material for this application may be found in the book *Introduction to Difference Equations*, John Wiley & Sons, Inc., New York, 1958, by Samuel Goldberg, who has kindly granted permission for its inclusion in this text. A Dover Publications, Inc. edition is scheduled to appear in 1986.

We are interested in determining the long-range distributions of these classes, that is, what the relative sizes of these classes will be in future generations. It is clear that there will be a problem of survival if the class of Stinkards becomes too small. To simplify matters, we make three assumptions:

1. In every generation each class is divided equally between males and females.
2. Each adult marries exactly once.
3. Each pair of parents has exactly one son and one daughter.

Because of assumption 1, we need only keep track of the number of males in each class for every generation. To do this we introduce the following notation:

$$s(k) = \text{number of male Suns in the } k\text{th generation}$$

$$n(k) = \text{number of male Nobles in the } k\text{th generation}$$

$$h(k) = \text{number of male Honoreds in the } k\text{th generation}$$

$$st(k) = \text{number of male Stinkards in the } k\text{th generation}$$

Our immediate goal is to find a relationship between the kth and $(k-1)$st generations in terms of the numbers of members in each class. Since every Sun male must have a Sun mother (and vice versa), we obtain the equation

$$s(k) = s(k-1)$$

The fact that every Noble male must have a Sun father or a Noble mother yields

$$n(k) = s(k-1) + n(k-1)$$

In addition, every Honored male must have a Noble father or an Honored mother. Thus,

$$h(k) = n(k-1) + h(k-1)$$

Finally, assumption 3 guarantees that the total number of males (and females) remains the same for each generation. Thus,

$$s(k) + n(k) + h(k) + st(k) = s(k-1) + n(k-1) + h(k-1) + st(k-1)$$

By using the previous relationships, we have

$$s(k-1) + [s(k-1) + n(k-1)] + [n(k-1) + h(k-1)] + st(k)$$
$$= s(k-1) + n(k-1) + h(k-1) + st(k-1)$$

or

$$st(k) = -s(k-1) - n(k-1) + st(k-1)$$

If we let

$$\mathbf{x}(k) = \begin{bmatrix} s(k) \\ n(k) \\ h(k) \\ st(k) \end{bmatrix} \quad \text{and} \quad A = \begin{bmatrix} 1 & 0 & 0 & 0 \\ 1 & 1 & 0 & 0 \\ 0 & 1 & 1 & 0 \\ -1 & -1 & 0 & 1 \end{bmatrix}$$

we may represent all our relationships by the single matrix equation

$$\mathbf{x}(k) = A\mathbf{x}(k-1)$$

Because this equation must hold for *all* k, we have

$$\mathbf{x}(k) = A\mathbf{x}(k-1) = AA\mathbf{x}(k-2) = \cdots = A^k\mathbf{x}(0)$$

So to find a convenient form for $\mathbf{x}(k)$, we need to evaluate A^k. We shall leave it to the exercises to show that if we write $A = I + B$, where

$$B = \begin{bmatrix} 0 & 0 & 0 & 0 \\ 1 & 0 & 0 & 0 \\ 0 & 1 & 0 & 0 \\ -1 & -1 & 0 & 0 \end{bmatrix}$$

then $A^k = I + kB + [k(k-1)/2]B^2$. Thus, carrying out the matrix multiplication, we obtain

$$\mathbf{x}(k) = \begin{bmatrix} s(0) \\ n(0) + ks(0) \\ h(0) + kn(0) + \dfrac{k(k-1)}{2}s(0) \\ st(0) - kn(0) - \dfrac{k(k+1)}{2}s(0) \end{bmatrix} \tag{1}$$

It is easy to see from equation (1) that if there are initially no Suns or Nobles [i.e., $n(0) = s(0) = 0$], the number of members in each class will remain the same from generation to generation.

On the other hand, consider the last entry of $\mathbf{x}(k)$. We can conclude that unless $n(0) = s(0) = 0$, the number of Stinkards will decrease to the point where there are insufficiently many of them to allow the other members to marry. At this point the social order will cease to exist.

Exercises

1. By observing a certain colony of mice it was found that all the animals die within 3 years. Of those offspring that are females, 60% live for at least 1 year. Of these, 20% reach their second birthday. The females who are under 1 year of age have an average of three female offspring. Those females between 1 and 2 years of age have an average of two female offspring while they are in this age group. None of the females of age 2 give birth.

   (a) Write down the Leslie matrix that describes this situation.
   (b) If the current population distribution for the females is given by the vector $(100, 60, 30)$, find the population distribution for the next year. Also, find the population distribution 4 years from now.
   (c) Show that there is no nontrivial stable population distribution for this colony of mice. [*Hint:* Let A be the Leslie matrix and suppose that $\mathbf{z} = (x, y, z)$ is a stable population distribution. Then $A\mathbf{z} = \mathbf{z}$. This is equivalent to $(A - I)\mathbf{z} = 0$. Now solve this homogeneous system of linear equations.]
   *(d) Use the program MATRIX to compute the population distribution in 10 years.

2. Suppose that the females of a certain colony of animals are divided into two age groups, and suppose that the Leslie matrix for this population is

$$A = \begin{bmatrix} 0 & 1 \\ 1 & 0 \end{bmatrix}$$

   (a) What proportion of the females of the first age group survive to the second age group?

   * This problem should be solved by using one of the programs noted in Appendix B.

   (b) How many female offspring do the females of each age group average?
   (c) If $\mathbf{x} = (a, b)$ is the current population distribution for the females of the colony, describe all future population distributions.

3. Water is pumped into a system of pipes at points A and B in Figure 2.3.2. At each of the junctions C, D, E, F, G, and H, the pipes are split and water flows according to the portions indicated in the diagram. If water flows into A and B at p and q gallons per minute, respectively, express the outputs at I, J, and K as a product of the column vector with entries p and q and two matrices.

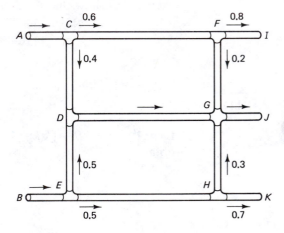

Figure 2.3.2

4. A certain medical foundation receives money from two sources: donations and interest earned on endowments. Of the donations received, 30% is used to defray the costs of raising money; only 10% of the interest is used to defray the cost of managing the

endowment funds. Of the rest of the money, the net income, 40% is used for research, and 60% is used to maintain medical clinics. Of the three expenses, research, clinics, and money raising, the portions going to materials and personnel are divided according to Table 2.3.2. Find a matrix M such that if p is the

TABLE 2.3.2

	Research	Clinics	Money Raising
material costs	80%	50%	70%
personnel costs	20%	50%	30%

value of donations and q is the value of interest, the product of M and the column vector with entries p and q yields the amounts of material and personnel costs of the foundation.

5. Recall the *binomial formula* for scalars a and b:

$$(a + b)^k = \sum_{i=0}^{k} \frac{k! a^{k-i} b^i}{i!(k-i)!}$$

(a) Suppose that A and B are $m \times m$ matrices which commute; that is, $AB = BA$. Prove that the binomial formula holds for $k = 2$, 3. That is, prove that

$$(A + B)^2 = A^2 + 2AB + B^2$$

and

$$(A + B)^3 = A^3 + 3A^2B + 3AB^2 + B^3$$

(b) Use mathematical induction to extend the results in part (a) to any positive integer k.

6. Use Exercise 5 to prove that if B is a square matrix such that $B^3 = 0$, then

$$(I + B)^k = I + kB + \frac{k(k-1)}{2} B^2$$

7. In reference to the application dealing with the Natchez Indians, suppose that initially there are 100 Sun males, 200 Noble males, 300 Honored males, and 8000 Stinkard males.
   (a) How many males will there be in each class in $k = 1, 2, 3$ generations?
   (b) How many generations will it take for there to be insufficiently many Stinkards to allow the other members to marry?

2.4 LINEAR TRANSFORMATIONS FROM R^n TO R^m

In Section 2.1 we saw that if we begin with an $m \times n$ matrix A and an $n \times 1$ (column) vector $\mathbf{x}$, the product of A and $\mathbf{x}$ is the $m \times 1$ (column) vector $A\mathbf{x}$. Thus, multiplication by A assigns to each vector in R^n a vector in R^m when the vectors of these spaces are identified with column vectors. This particular association is an example of a function from R^n to R^m. In fact, it is this special type of function that is the main object of study in linear algebra. In the next section we establish many of the important properties of such functions using the results that we have learned thus far.

As indicated in previous sections, it will sometimes be convenient to consider the vectors of R^n as column vectors as well as row vectors.

Definition   A *function* or *mapping* f from R^n to R^m, written $f: R^n \to R^m$, is a rule which assigns to each vector $\mathbf{x}$ in R^n a unique vector, denoted $f(\mathbf{x})$, in R^m. The vector $f(\mathbf{x})$ is called the *image* of $\mathbf{x}$ (under f). The space R^n is called the *domain* of f, and the set of all images $f(\mathbf{x})$ in R^m is called the *range* of f.

Notice that the definitions of function, domain, and range agree with the ones that were encountered in earlier courses where the domain and range were both subsets of R.

Example 1   Let $f: R^3 \to R^2$ be defined by the rule

$$f(x, y, z) = (x + y + z, x^2)$$

Then f is a function whose domain is R^3. Since $f(0, 1, 1) = (2, 0)$, the vector $(2, 0)$ is the image of $(0, 1, 1)$ under f. The vector $(2, 0)$ is also the image of $(0, 2, 0)$ under f. Actually, it can be shown that $(2, 0)$ is the image of infinitely many vectors in R^3. Not every vector in R^2 is the image of a vector under f because the second coordinate of any image is nonnegative.

Example 2   Let A be the 3×2 matrix

$$A = \begin{bmatrix} 1 & 0 \\ 2 & 1 \\ 1 & -1 \end{bmatrix}$$

Define the function $T: R^2 \to R^3$ by

$$T(\mathbf{x}) = A\mathbf{x}$$

Notice that the product of A with a column vector $\mathbf{x}$ in R^2 yields a column vector in R^3. For example,

$$T\begin{bmatrix} 1 \\ 3 \end{bmatrix} = \begin{bmatrix} 1 & 0 \\ 2 & 1 \\ 1 & -1 \end{bmatrix}\begin{bmatrix} 1 \\ 3 \end{bmatrix} = \begin{bmatrix} 1 \\ 5 \\ -2 \end{bmatrix}$$

So the vector $(1, 5, -2)$ is the image of $(1, 3)$ under the mapping T. Because

$$T\begin{bmatrix} x \\ y \end{bmatrix} = \begin{bmatrix} 1 & 0 \\ 2 & 1 \\ 1 & -1 \end{bmatrix}\begin{bmatrix} x \\ y \end{bmatrix} = \begin{bmatrix} x \\ 2x + y \\ x - y \end{bmatrix}$$

we may write the equation

$$T(x, y) = (x, 2x + y, x - y)$$

for any vector (x, y) in R^2.

Example 3

Using the function T defined in Example 2, determine whether each of the vectors (1, 1, 2) and (1, 3, 1) is an image under T.

The vector (1, 1, 2) is the image of a vector (x, y) under T if and only if there exists a vector $\mathbf{x} = (x, y)$ such that $T(\mathbf{x}) = (1, 1, 2)$, or

$$\begin{bmatrix} 1 & 0 \\ 2 & 1 \\ 1 & -1 \end{bmatrix} \begin{bmatrix} x \\ y \end{bmatrix} = \begin{bmatrix} x \\ 2x + y \\ x - y \end{bmatrix} = \begin{bmatrix} 1 \\ 1 \\ 2 \end{bmatrix}$$

If we identify the corresponding coordinates of the left and right sides of this equation, we obtain the system of linear equations

$$\begin{aligned} x & = 1 \\ 2x + y &= 1 \\ x - y &= 2 \end{aligned}$$

It can be shown that this sytem has the (unique) solution $x = 1$ and $y = -1$. Thus, (1, 1, 2) is the image of (1, −1) under T.

In a similar manner we can show that (1, 3, 1) is the image of the vector (x, y) under T if and only if

$$\begin{aligned} x & = 1 \\ 2x + y &= 3 \\ x - y &= 1 \end{aligned}$$

However, it can be shown that this system is inconsistent. Therefore, (1, 3, 1) is not the image of any vector under T.

The function T defined in Example 3 illustrates the use of a matrix to define a function by means of left multiplication.

Definition

For an $m \times n$ matrix A we define the ***left-multiplication function***, $L_A: R^n \to R^m$, by

$$L_A(\mathbf{x}) = A\mathbf{x}$$

for all $\mathbf{x}$ in R^n.

In addition to the function defined in Example 2, we have already seen that the functions representing rotations, reflections, and projections (Exercises 8, 9, and 10 of Section 2.1) may be expressed as left-multiplication functions for the proper choices of A. The next theorem is essentially a restatement of Theorem 2.1.2. However, it establishes the two most important properties of the function L_A.

Theorem 2.4.1 *Let A be an m × n matrix and let* **x** *and* **y** *be column vectors in* R^n. *Then:*

(a) $L_A(\mathbf{x} + \mathbf{y}) = L_A(\mathbf{x}) + L_A(\mathbf{y})$

(b) $L_A(c\mathbf{x}) = c(L_A(\mathbf{x}))$ *for any scalar c*

Proof

We shall prove part (a) and leave part (b) as an exercise.

(a) By Theorem 2.1.2 we have

$$L_A(\mathbf{x} + \mathbf{y}) = A(\mathbf{x} + \mathbf{y})$$
$$= A\mathbf{x} + A\mathbf{y}$$
$$= L_A(\mathbf{x}) + L_A(\mathbf{y}) \qquad \blacksquare$$

In particular, the three geometric functions mentioned above all share these two properties. For example, the rotation of a sum of vectors is the sum of the rotated vectors. To see this geometrically, notice that the parallelogram whose sides are given by the vectors is rotated by the same angle as the vectors. Hence, the diagonal of the parallelogram that represents the sum of the vectors is also rotated by the same angle (see Figure 2.4.1).

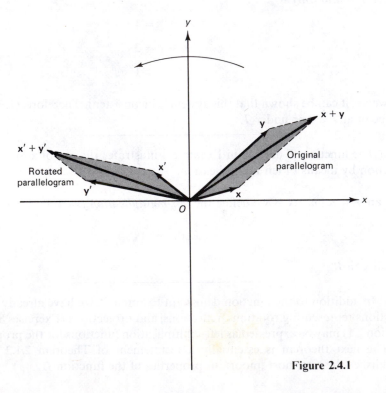

Figure 2.4.1

Theorem 2.4.1 tells us that the function L_A "preserves" the operations of vector addition and scalar multiplication. Functions that possess these two properties are called "linear transformations."

Definition A function $T: R^n \to R^m$ is called a ***linear transformation*** if for all vectors **x** and **y** in R^n and scalars c, we have

(a) $T(\mathbf{x} + \mathbf{y}) = T(\mathbf{x}) + T(\mathbf{y})$

(b) $T(c\mathbf{x}) = c(T(\mathbf{x}))$

For brevity, we shall often use the term ***linear*** instead of "linear transformation."

Linear transformations arise in a variety of contexts. For example, consider the operation of dot product, which was defined in Chapter 1. For a fixed vector **y** in R^n define $T: R^n \to R$ by $T(\mathbf{x}) = \mathbf{y} \cdot \mathbf{x}$. Theorem 1.6.1 allows us to conclude immediately that T is linear. It will be shown in Chapter 8 that many familiar operations such as differentiation and integration also produce linear transformations in a context more general than R^n.

We shall see shortly that all linear transformations are given by left-multiplication functions. First, however, it will be necessary to develop a number of simple properties of linear transformations.

Theorem 2.4.2 *Let $T: R^n \to R^m$ be a linear transformation. Then for any vectors **x** and **y** in R^n, we have:*

(a) $T(\mathbf{0}) = \mathbf{0}$

(b) $T(-\mathbf{x}) = -T(\mathbf{x})$

(c) $T(a\mathbf{x} + b\mathbf{y}) = aT(\mathbf{x}) + bT(\mathbf{y})$ *for any scalars a and b*

(d) $T(\sum_{i=1}^{k} a_i\mathbf{x}_i) = \sum_{i=1}^{k} a_i T(\mathbf{x}_i)$ *for any vectors $\mathbf{x}_1, \ldots, \mathbf{x}_k$ in R^n and scalars $a_1, \ldots, a_k$*

Proof

We shall prove parts (a) and (c) and leave the remaining parts for the exercises.

(a) By part (b) of the definition of linearity we have that

$$T(\mathbf{0}) = T(0\mathbf{0}) = 0T(\mathbf{0}) = \mathbf{0}$$

(c) For any vectors **x** and **y** and scalars a and b, we have

$$T(a\mathbf{x} + b\mathbf{y}) = T(a\mathbf{x}) + T(b\mathbf{y})$$
$$= aT(\mathbf{x}) + bT(\mathbf{y}) \qquad\blacksquare$$

Notice that part (d) allows us to "slide" T past the Σ sign and the scalars a_i. This symbolic manipulation will be used frequently throughout the remainder of the text.

We may use part (a) to conclude that the function $T: R \to R$ defined by $T(x) = 3x + 4$ is not linear because $T(0) = 4 \neq 0$. Likewise, we may use (b) to show that the function of Example 1 is not linear. The next example illustrates how to prove that a given function is linear.

Example 4   Define $T: R^2 \to R^2$ by $T(x, y) = (2x - y, x)$. To show that T is linear, we must verify properties (a) and (b) of the definition. For this purpose, let $\mathbf{x} = (a_1, a_2)$, $\mathbf{y} = (b_1, b_2)$, and let c be any scalar. Then

$$T(\mathbf{x} + \mathbf{y}) = T((a_1, a_2) + (b_1, b_2))$$
$$= T(a_1 + b_1, a_2 + b_2)$$
$$= (2(a_1 + b_1) - (a_2 + b_2), a_1 + b_1)$$

On the other hand,

$$T(\mathbf{x}) + T(\mathbf{y}) = T(a_1, a_2) + T(b_1, b_2)$$
$$= (2a_1 - a_2, a_1) + (2b_1 - b_2, b_1)$$
$$= (2(a_1 + b_1) - (a_2 + b_2), a_1 + b_1)$$

Thus, $T(\mathbf{x} + \mathbf{y}) = T(\mathbf{x}) + T(\mathbf{y})$, verifying property (a). Property (b) is verified similarly.

There are many ways to combine functions to form new functions. Among the more useful of these are the operations of "composition" and "sum" of functions. Although these operations are defined in general, we restrict our investigations to linear transformations.

Definition   Let $T: R^n \to R^m$ and $U: R^m \to R^p$ be linear transformations. The ***composition*** UT is the function $UT: R^n \to R^p$ defined by $UT(\mathbf{x}) = U(T(\mathbf{x}))$ for all $\mathbf{x}$ in R^n.

Example 5   Define the linear transformations $T: R^2 \to R^2$ and $U: R^2 \to R^3$ as follows:

$$T(x, y) = (2x + y, 3x)$$

and

$$U(x, y) = (2y, x + y, x - y)$$

Then

$$UT(x, y) = U(2x + y, 3x)$$
$$= (2(3x), (2x + y) + 3x, (2x + y) - 3x)$$
$$= (6x, 5x + y, -x + y)$$

Theorem 2.4.3 *Let* $T: R^n \to R^m$ *and* $U: R^m \to R^p$ *be linear transformations. Then the composition* $UT: R^n \to R^p$ *is linear.*

Proof
For $\mathbf{x}$ and $\mathbf{y}$ in R^n,

$$UT(\mathbf{x} + \mathbf{y}) = U(T(\mathbf{x} + \mathbf{y}))$$
$$= U(T(\mathbf{x}) + T(\mathbf{y}))$$
$$= U(T(\mathbf{x})) + U(T(\mathbf{y}))$$
$$= UT(\mathbf{x}) + UT(\mathbf{y})$$

This verifies part (a) of the definition of a linear transformation. Part (b) is verified similarly.   ■

It follows from Theorem 2.4.3 that function UT of Example 5 is linear.

If $T: R^n \to R^n$, the composition of T with itself is defined. We define T^2 as TT, and in general, T^n is defined by $T^{(n-1)}T$ for $n \geq 2$.

Definition   If T and U are linear transformations from R^n to R^m, then the **sum**, $T + U$, is the function from R^n to R^m defined by $(T + U)(\mathbf{x}) = T(\mathbf{x}) + U(\mathbf{x})$ for all $\mathbf{x}$ in R^n. For any scalar a, the **product** of T by a, aT, is the function from R^n to R^m defined by $(aT)(\mathbf{x}) = a(T(\mathbf{x}))$ for all $\mathbf{x}$ in R^n.

Example 6   Let T and U be linear transformations from R^3 to R^2 defined by

$$T(x, y, z) = (x + y, x - z)$$

and

$$U(x, y, z) = (y - z, x + y + z)$$

We shall compute $(T + U)(x, y, z)$ and $3T(x, y, z)$.

$$(T + U)(x, y, z) = T(x, y, z) + U(x, y, z)$$
$$= (x + y, x - z) + (y - z, x + y + z)$$
$$= (x + 2y - z, 2x + y)$$

Next we compute $3T(x, y, z)$.

$$3T(x, y, z) = 3(T(x, y, z))$$
$$= 3(x + y, x - z)$$
$$= (3x + 3y, 3x - 3z)$$

As we shall see (Corollary 2.4.7), the sum of linear transformations is linear, and the product of a linear transformation and a scalar is linear.

There are two special linear transformations that will appear frequently in the remainder of this text. The first is the *zero transformation*, $T_0: R^n \to R^m$, defined by $T_0(\mathbf{x}) = \mathbf{0}$ for all $\mathbf{x}$ in R^n. The second is the *identity transformation*, $I: R^n \to R^n$, defined by $I(\mathbf{x}) = \mathbf{x}$ for all $\mathbf{x}$ in R^n. Notice that we use the same symbol for the identity transformation and the identity matrix. The context should make it clear which is the correct interpretation. We shall leave it as an exercise to prove that both T_0 and I are linear.

One of the most important properties of linear transformations is that they are completely determined by their *action* on the standard vectors, $\mathbf{e}_1, \ldots, \mathbf{e}_n$. We state this result more formally below.

Theorem 2.4.4 *Let T and U be linear transformations defined from R^n to R^m. If $T(\mathbf{e}_j) = U(\mathbf{e}_j)$ for $j = 1, \ldots, n$, then $T = U$; that is, $T(\mathbf{x}) = U(\mathbf{x})$ for every $\mathbf{x}$ in R^n.*

Proof

Let $\mathbf{x} = (x_1, \ldots, x_n)$. We may write $\mathbf{x} = x_1 \mathbf{e}_1 + \cdots + x_n \mathbf{e}_n$. By appealing twice to Theorem 2.4.2(d), we have

$$T(\mathbf{x}) = T\left(\sum_{j=1}^{n} x_j \mathbf{e}_j\right)$$

$$= \sum_{j=1}^{n} x_j T(\mathbf{e}_j)$$

$$= \sum_{j=1}^{n} x_j U(\mathbf{e}_j)$$

$$= U\left(\sum_{j=1}^{n} x_j \mathbf{e}_j\right)$$

$$= U(\mathbf{x}) \qquad\blacksquare$$

It is interesting to note that the linear transformation T given in Example 4 may be represented as a left-multiplication function L_A, where A is given by

$$A = \begin{bmatrix} 2 & -1 \\ 1 & 0 \end{bmatrix}$$

since

$$\begin{bmatrix} 2 & -1 \\ 1 & 0 \end{bmatrix} \begin{bmatrix} x \\ y \end{bmatrix} = \begin{bmatrix} 2x - y \\ x \end{bmatrix}$$

We can therefore conclude that T is linear since L_A is linear by Theorem 2.4.1.

It is no coincidence that T can be represented as a left-multiplication. The next theorem states that all linear transformations are left-multiplication functions. As noted earlier, we shall allow a dual interpretation of the vectors of R^n; they may be considered either as row or column vectors, depending on the context in which they are used.

Theorem 2.4.5 *Let $T: R^n \to R^m$ be a linear transformation. Then there exists a unique $m \times n$ matrix A such that $T(\mathbf{x}) = L_A(\mathbf{x}) = A\mathbf{x}$ for all vectors $\mathbf{x}$ in R^n. That is, $T = L_A$.*

Proof

Let A be the $m \times n$ matrix whose jth column ${}^jA$ is given by

$$^jA = T(\mathbf{e}_j)$$

for $1 \le j \le n$. This definition makes sense if we consider the image $T(\mathbf{e}_j)$ as a column vector in R^m. By Theorem 2.4.1 L_A is linear. By Theorem 2.1.2(a) we have that

$$L_A(\mathbf{e}_j) = A\mathbf{e}_j = {}^jA = T(\mathbf{e}_j)$$

for $1 \le j \le n$. So, by Theorem 2.4.4, we have that T and L_A are equal.

To prove that A is unique, we must show that if B is an $m \times n$ matrix such that $L_A = L_B$, then $A = B$. However, the condition that $L_A = L_B$ is equivalent to the condition that $A\mathbf{x} = B\mathbf{x}$ for all vectors $\mathbf{x}$ in R^n, and so, by Corollary 2.1.3, we have that $A = B$. ∎

Example 7 If we consider the linear transformation T defined in Example 4, we have

$$T(1, 0) = \begin{bmatrix} 2 \\ 1 \end{bmatrix} \quad \text{and} \quad T(0, 1) = \begin{bmatrix} -1 \\ 0 \end{bmatrix}$$

Consequently, for

$$A = \begin{bmatrix} 2 & -1 \\ 1 & 0 \end{bmatrix}$$

we have that $T = L_A$.

Definition Let $T: R^n \to R^m$ be a linear transformation. We call the $m \times n$ matrix A such that $T = L_A$, the **standard matrix** of T.

Example 8   Define $T: R^3 \to R^2$ by

$$T(x, y, z) = (2x - 3y + 7z, 5x + 4y)$$

We determine the standard matrix of T. Since $T(\mathbf{e}_1) = (2, 5)$, $T(\mathbf{e}_2) = (-3, 4)$, and $T(\mathbf{e}_3) = (7, 0)$, we have that the matrix

$$A = \begin{bmatrix} \overset{\displaystyle T(\mathbf{e}_1)}{\downarrow} & \overset{\displaystyle T(\mathbf{e}_2)}{\downarrow} & \overset{\displaystyle T(\mathbf{e}_3)}{\downarrow} \\ 2 & -3 & 7 \\ 5 & 4 & 0 \end{bmatrix}$$

is the standard matrix of T.

Theorem 2.4.5 provides us with a correspondence between matrices and linear transformations. To make this correspondence more useful, we shall show that it preserves sums, products, and multiplication by scalars.

Theorem 2.4.6   *Let A and B be matrices such that the operations below are defined. Then:*
     (a) $L_{A+B} = L_A + L_B$
     (b) $L_{cA} = cL_A$ for all scalars c
     (c) $L_{AB} = L_A L_B$, where $L_A L_B$ denotes the composition of L_A and L_B

Proof
We shall prove parts (a) and (c) and leave part (b) as an exercise.
(a) For any vector $\mathbf{x}$ in R^n, we have

$$
\begin{aligned}
L_{A+B}(\mathbf{x}) &= (A + B)\mathbf{x} \\
&= A\mathbf{x} + B\mathbf{x} \\
&= L_A(\mathbf{x}) + L_B(\mathbf{x}) \\
&= (L_A + L_B)(\mathbf{x})
\end{aligned}
$$

Therefore, $L_{A+B} = L_A + L_B$.
(c) By Theorem 2.4.1, we have that L_A, L_B, and L_{AB} are linear. So, by Theorem 2.4.3, we have that $L_A L_B$ is linear. Now we may apply Theorem 2.1.2(a) and property 2 at the end of Section 2.2 to obtain

$$
\begin{aligned}
L_{AB}(\mathbf{e}_j) &= (AB)(\mathbf{e}_j) \\
&= {}^j(AB) \\
&= A({}^jB)
\end{aligned}
$$

On the other hand, we have

$$
\begin{aligned}
L_A L_B(\mathbf{e}_j) &= L_A(L_B(\mathbf{e}_j)) \\
&= L_A(B\mathbf{e}_j) \\
&= L_A({}^jB) \\
&= A({}^jB)
\end{aligned}
$$

So, by Theorem 2.4.4, $L_{AB} = L_A L_B$.   ■

Corollary 2.4.7

If T and U are linear transformations from R^n to R^m and a is a scalar, then $T + U$ and aT are also linear transformations from R^n to R^m.

Proof

By Theorem 2.4.5 there exist matrices A and B such that $T = L_A$ and $U = L_B$. So by Theorem 2.4.6(a), we have

$$
T + U = L_A + L_B = L_{(A+B)}
$$

which is linear by Theorem 2.4.1. The remainder of the proof is left as an exercise.   ■

Example 9

To illustrate parts (a) and (c) of Theorem 2.4.6, let T and U be the linear transformations as defined in Example 5. Define a third linear transformation $T': R^2 \to R^2$ by $T'(x, y) = (x - 2y, x + y)$. Let A, B, and C be the standard matrices of T, U, and T', respectively. Then

$$
A = \begin{bmatrix} 2 & 1 \\ 3 & 0 \end{bmatrix} \qquad B = \begin{bmatrix} 0 & 2 \\ 1 & 1 \\ 1 & -1 \end{bmatrix} \qquad C = \begin{bmatrix} 1 & -2 \\ 1 & 1 \end{bmatrix}
$$

So

$$
A + C = \begin{bmatrix} 3 & -1 \\ 4 & 1 \end{bmatrix} \qquad \text{and} \qquad BA = \begin{bmatrix} 6 & 0 \\ 5 & 1 \\ -1 & 1 \end{bmatrix}
$$

are the standard matrices of $T + T'$ and UT, respectively.

We are now free to choose either matrices or linear transformations to establish useful results. For example, in Exercise 9 of Section 2.2 the proof of the associativity of matrix multiplication was sketched. A very simple proof may now be obtained from Theorems 2.4.5 and 2.4.6. For, if A, B, and C are matrices such that the products $A(BC)$ and $(AB)C$ are defined, we have

$$L_{A(BC)} = L_A L_{BC} = L_A (L_B L_C) = (L_A L_B) L_C = L_{AB} L_C = L_{(AB)C}$$

By Theorem 2.4.5 we may conclude that $A(BC) = (AB)C$. Notice that we used the fact that the composition of functions is associative.

Exercises

Exercises 1–5 use the following matrices:

$$A = \begin{bmatrix} 1 & 2 & 3 \\ 3 & 0 & 1 \end{bmatrix} \quad B = \begin{bmatrix} 0 & -1 & 2 \\ 1 & 1 & 0 \end{bmatrix}$$

$$C = \begin{bmatrix} 2 & 1 \\ 0 & 3 \\ 1 & -1 \end{bmatrix}$$

1. For the transformation $L_A: R^n \to R^m$, what are the values of n and m?

2. For the transformation $L_B: R^n \to R^m$, what are the values of n and m?

3. For the transformation $L_C: R^n \to R^m$, what are the values of n and m?

4. (a) Compute $L_A(1, 0, 1)$ and $L_B(1, 0, 1)$.
   (b) Compute $A + B$.
   (c) Verify that $L_{A+B}(1, 0, 1) = L_A(1, 0, 1) + L_B(1, 0, 1)$.

5. (a) Compute AC.
   (b) Compute $L_{AC}(1, -1)$ and $L_C(1, -1)$.
   (c) Verify that $L_{AC}(1, -1) = (L_A L_C)(1, -1)$.

6. For each of the following, determine whether or not T is a linear transformation. Justify your answer.
   (a) $T(x, y) = (2x, y^2)$
   (b) $T(x, y, z) = x + y + z$
   (c) $T(x, y) = (0, x)$
   (d) $T(x, y) = (1, x)$
   (e) $T(x, y, z) = (y, z, x)$
   (f) $T(x) = (x, 2x, 3x)$
   (g) $T(x, y) = (x + y, x - y)$
   (h) $T(x, y) = (x + y, x - 2y, 0)$
   (i) $T(x, y) = (|x + y|, x + y)$

7. For the linear transformation $T: R^n \to R^m$ and the vector $\mathbf{y}$, determine whether or not $\mathbf{y}$ is the image of a vector in R^n.
   (a) $T(x, y) = (x + y, x - y)$, $\mathbf{y} = (2, 0)$
   (b) $T(x, y) = x + y$, $\mathbf{y} = 3$
   (c) $T(x, y) = (x, y, x + y)$, $\mathbf{y} = (1, 2, 1)$
   (d) $T(x, y, z) = (x, y)$, $\mathbf{y} = (1, 2)$
   (e) $T(x, y, z) = (x + y, y + z, x - z)$, $\mathbf{y} = (2, 1, 1)$
   (f) $T(x, y, z) = (x + y, y + z, x - z)$, $\mathbf{y} = (1, 1, 1)$

8. For the linear transformations T and U, the scalar c, and the vector $\mathbf{x}$ compute the image of $\mathbf{x}$ under $T + U$ and under cT.
   (a) $T(x, y) = (x - y, x)$, $U(x, y) = (y, x)$, $c = 4$, and $\mathbf{x} = (3, -2)$
   (b) $T(x, y) = x - 2y$, $U(x, y) = y - x$, $c = -3$, and $\mathbf{x} = (1, 4)$
   (c) $T(x, y, z) = (x + y - z, z - x, 2x + y)$, $U(x, y, z) = (3y - x, x + y, 2z)$, $c = 4$, and $\mathbf{x} = (0, -1, 3)$

9. For the linear transformations U and T find an expression for the image of an arbitrary vector $\mathbf{x}$ under the composition UT.
   (a) $T(x, y) = (x - y, x + y)$ and $U(x, y) = x + 2y$
   (b) $T(x, y, z) = (z - x, y + z)$ and $U(x, y) = (x - y, 2x + y)$
   (c) $T(x, y) = (x + y, x - y, 2x)$ and $U(x, y, z) = (x, 2y, 3z)$

10. For each part of Exercise 8:
    (a) Find the standard matrices for $T, U, T + U$, and cT.
    (b) Verify Theorems 2.4.5 and 2.4.6(a) and (b) for these matrices.

11. For each part of Exercise 9:
    (a) Find the standard matrices for T, U, and UT.
    (b) Verify Theorems 2.4.5 and 2.4.6(c) for these matrices.

12. Give a *direct* proof of Corollary 2.4.7 using only the definition of linear transformation.

13. Prove Theorem 2.4.2(b) and (d).

14. Prove Theorem 2.4.6(b).

15. Find the standard matrices of the given linear transformations.
    (a) $T(x, y) = (y, 0)$
    (b) $T(x, y) = (x + 2y, x - 3y)$
    (c) $T(x, y, z) = (x + y + z, 2x)$
    (d) $T(x, y) = (3y, 2x - y, x + y)$
    (e) $T(x, y, z) = (x, y, z)$
    (f) $T(x, y, z) = (0, 0, 0)$
    (g) $T: R^3 \to R^3$ defined by $T(\mathbf{x}) = 4\mathbf{x}$
    (h) $T: R^3 \to R$ defined by $T(\mathbf{x}) = \mathbf{y} \cdot \mathbf{x}$, where $\mathbf{y} = (2, -1, 1)$

16. Prove the converse of Theorem 2.4.2(c): For any mapping $T: R^n \to R^m$, if $T(a\mathbf{x} + b\mathbf{y}) = aT(\mathbf{x}) + bT(\mathbf{y})$ for any scalars a and b and any vectors $\mathbf{x}$ and $\mathbf{y}$ in R^n, then T is linear.

17. Let $T: R^2 \to R^3$ be the linear transformation such that $T(\mathbf{e}_1) = (1, 1, 3)$ and $T(\mathbf{e}_2) = (2, 0, 4)$. Determine $T(3, 2)$ and $T(2, 1)$.

18. Prove that for any linear transformation $T: R^2 \to R^2$, there are unique scalars a, b, c, and d such that

$$T(x, y) = (ax + by, cx + dy)$$

for every (x, y) in R^2.

19. (a) State and prove a generalization of the result of Exercise 18 for a linear transformation $T: R^n \to R^m$.
    (b) Apply this result to redo Exercise 6.

20. Prove that T_0 and I are linear transformations.

21. Prove:
    (a) $L_O = T_0$
    (b) $L_I = I$ (This result provides some justification for the use of I to denote both the identity matrix and the identity transformation.)

22. Let $T: R^2 \to R^2$ denote the linear transformation that reflects a vector about the y-axis.
    (a) Prove that $T^2 = I$.
    (b) Use part (a) to find a 2×2 matrix A such that $A^2 = I$.

23. Let $T: R^3 \to R^3$ denote the linear transformation that projects a vector on the xy-plane.
    (a) Prove that $T^2 = T$.
    (b) Use part (a) to find a 3×3 matrix A such that $A^2 = A$.

24. In Section 2.2 we saw that rotation matrices satisfy the equation $A_{\theta + \phi} = A_\theta A_\phi$ for any angles θ and ϕ.
    (a) Prove that $I = A^n_{(2\pi/n)}$ for any positive integer n.
    (b) Use Theorem 2.4.6 and part (a) to show that for any positive integer n, there exists a linear transformation $T: R^2 \to R^2$ such that $T^n = I$.

25. Let T and U be linear transformations from R^n to R^m. Use the definition of linear transformation to prove that for any scalar c the function $cT + U$ is also a linear transformation.

26. Complete the proof of Corollary 2.4.7.

27. Complete the proof of Theorem 2.4.3.

28. For an $m \times n$ matrix A define the mapping $R_A: R^m \to R^n$ by

$$R_A(\mathbf{x}) = \mathbf{x}A$$

where $\mathbf{x}$ is a $1 \times m$ row vector.
    (a) Show that R_A is a linear transformation.
    (b) Show that for any matrices A and B for which the product AB is defined,

$$R_{AB} = R_B R_A$$

    (c) Prove that $R_A = L_{(A^t)}$.

29. Prove Theorem 2.4.1(b).

†30. Prove the following generalization of Theorem 2.4.4: Let T and U be linear transformations from R^n to R^m, and suppose that R^n is the span of the subset $\{x_1, x_2, \ldots, x_k\}$. If $T(x_i) = U(x_i)$ for all i, then $T = U$.

31. Let y be a fixed vector in R^3. Let T be the linear transformation on R^3 defined by $T(x) = x \times y$, the cross product of x and y, that is defined in Exercise 18 of Section 1.6. Prove that T is linear.

*32. Let $T: R^4 \to R^3$ be defined by $T(x, y, z, w) = (x + 3y + 2w, x - z + w, x - y + w)$, and let $U: R^3 \to R^4$ be defined by $U(x, y, z) =$

$(x, x + y, y - z, x + 2y + z)$. Use the program MATRIX to compute

(a) $UT\begin{bmatrix} 1 \\ 4 \\ 6 \\ 8 \end{bmatrix}$

(b) $UTUT\begin{bmatrix} 2 \\ 1 \\ 0 \\ 1 \end{bmatrix}$

(c) $TU\begin{bmatrix} 3 \\ 1 \\ 7 \end{bmatrix}$

(d) $TUTU\begin{bmatrix} 1 \\ -5 \\ 3 \end{bmatrix}$

† This exercise will be used in a subsequent section.

* This problem should be solved by using one of the programs in Appendix B.

2.5 NULL SPACE AND RANGE

In this final section of Chapter 2 we shall see an interplay between all the major ideas that have been developed so far: vectors, subspaces, matrices, systems of equations, and linear transformations. The main strength of linear algebra is that it allows us to choose from among all these powerful tools to solve a wide variety of problems.

Consider the following simple homogeneous system of linear equations:

$$x + y + z = 0$$
$$2x + y \quad\quad = 0$$

This system may be represented as a single matrix equation: $Ax = 0$, where A is the coefficient matrix of the system. The set of solutions to this system is a subspace of R^3 (in fact, we can see that the solution space is the intersection of two planes through the origin). If we define the linear transformation $T = L_A: R^3 \to R^2$, the solution space may be identified with the set of vectors x in R^3 such that $T(x) = 0$. Sets that arise in this manner have a special significance and are defined below.

Definition   Let $T: R^n \to R^m$ be a linear transformation. We call the set of those x in R^n such that $T(x) = 0$, denoted by $N(T)$, the ***null space*** of T.

In regard to the system above, we saw that the null space was the solution space of a homogeneous system of linear equations and was a subspace. The next theorem tells us that null spaces are always subspaces.

Theorem 2.5.1 *Let $T: R^n \to R^m$ be a linear transformation. Then $N(T)$ is a subspace of R^n.*

Proof

Clearly, $N(T)$ is a subset of R^n. Since $T(\mathbf{0}) = \mathbf{0}$ by Theorem 2.4.2(a), we have that $N(T)$ is not empty. To prove that $N(T)$ is a subspace, we first consider any vectors $\mathbf{x}$ and $\mathbf{y}$ in $N(T)$. Then

$$T(\mathbf{x} + \mathbf{y}) = T(\mathbf{x}) + T(\mathbf{y})$$
$$= \mathbf{0} + \mathbf{0}$$
$$= \mathbf{0}$$

and hence $\mathbf{x} + \mathbf{y}$ lies in $N(T)$. Similarly, for any scalar a, $a\mathbf{x}$ lies in $N(T)$. We conclude that $N(T)$ is a subspace of R^n. ■

Corollary 2.5.2 *The solution space of a homogeneous system of linear equations in n unknowns is a subspace of R^n.*

The proof is left as an exercise.

Suppose that we define $T: R^3 \to R$ by $T(\mathbf{x}) = \mathbf{e}_3 \cdot \mathbf{x}$. In Section 2.4 we saw that a function of this form is a linear transformation. It is easy to see that $N(T)$ is the set of all vectors that are orthogonal to $\mathbf{e}_3$. Geometrically, this represents the xy-plane, which is clearly a subspace of R^3. However, it is not always easy to discover the null space geometrically. The next example illustrates how the tools of Chapter 1 may be used for this purpose.

Example 1 Define $T: R^4 \to R^3$ by

$$T(x, y, z, w) = (2x + y + z - w, x + y + 2z + 2w, x - z - 3w)$$

To find $N(T)$ we set $T(x, y, z, w) = (0, 0, 0)$. This equation is equivalent to the system of linear equations

$$2x + y + \ z - \ w = 0$$
$$x + y + 2z + 2w = 0$$
$$x \qquad - \ z - 3w = 0$$

If we solve this system by Gaussian elimination, we obtain the solution

$$(x, y, z, w) = s(1, -3, 1, 0) + t(3, -5, 0, 1)$$

where s and t are arbitrary scalars. It follows that $N(T)$ is the span of the set $\{(1, -3, 1, 0), (3, -5, 0, 1)\}$.

Of particular interest when solving a system of linear equations, $A\mathbf{x} = \mathbf{b}$, is whether or not the system has a *unique* solution. That is, if $\mathbf{x}$ and $\mathbf{y}$ are solutions, does

$\mathbf{x} = \mathbf{y}$? Stated in terms of linear transformations, does the condition $L_A(\mathbf{x}) = L_A(\mathbf{y})$ imply that $\mathbf{x} = \mathbf{y}$?

If f is any function such that $f(\mathbf{x}) = f(\mathbf{y})$ implies that $\mathbf{x} = \mathbf{y}$, that is, distinct elements of the domain have distinct images in the range, then f is said to be **one-to-one**. In earlier courses you may recall that monotonic functions were one-to-one. For linear transformations the next theorem provides an equivalent condition in terms of the null space.

Theorem 2.5.3 *A linear transformation T is one-to-one if and only if $N(T)$ is the zero subspace.*

Proof

Suppose that T is one-to-one and that $\mathbf{x}$ is in $N(T)$. Then the relations $T(\mathbf{x}) = \mathbf{0} = T(\mathbf{0})$ imply that $\mathbf{x} = \mathbf{0}$. Since $\mathbf{0}$ is clearly in $N(T)$, we have that $N(T) = \{\mathbf{0}\}$.

Now assume that $N(T) = \{\mathbf{0}\}$ and that $T(\mathbf{x}) = T(\mathbf{y})$. Then $\mathbf{0} = T(\mathbf{x}) - T(\mathbf{y}) = T(\mathbf{x} - \mathbf{y})$, so $\mathbf{x} - \mathbf{y}$ is in $N(T)$. Therefore, $\mathbf{x} - \mathbf{y} = \mathbf{0}$ or $\mathbf{x} = \mathbf{y}$, so T is one-to-one. ∎

It was shown that the subspace $N(T)$ of Example 1 is not trivial. Therefore, the transformation T of Example 1 is not one-to-one. Although it is quite easy to show directly that the zero transformation T_0 is not one-to-one, it also follows from Theorem 2.5.3 because $N(T_0)$ is the entire domain of T_0. Similarly, the identity transformation I is one-to-one because $N(I) = \{\mathbf{0}\}$.

Perhaps the most important question about a system of linear equations $A\mathbf{x} = \mathbf{b}$ is whether or not it has *any* solutions. In terms of transformations, we want to know if $\mathbf{b}$ is the image of a vector $\mathbf{x}$ under the transformation L_A. Recall that the set of all images of a transformation is called the range of the transformation. So the question of the existence of a solution to the equation $A\mathbf{x} = \mathbf{b}$ is equivalent to the question of whether or not $\mathbf{b}$ lies in the range of L_A. The range of a linear transformation plays an important role in linear algebra. We shall denote the range of T by $R(T)$.

To illustrate these ideas, consider the system of linear equations

$$x + y + z = 2$$
$$2x + y \quad\;\; = 3$$

If we define $T = L_A$, where A is the coefficient matrix of the system, it is easy to see that this system has a solution if and only if the vector $(2, 3)$ is in the range of T, in which case, $(2, 3)$ is the image of a solution. Notice that $\mathbf{x} = (1, 1, 0)$ is one such solution.

The next theorem provides a theoretical condition that will allow us to determine if a vector is contained in the range of a linear transformation.

Theorem 2.5.4 *Let $T: R^n \to R^m$ be a linear transformation. Suppose that $\{\mathbf{x}_1, \mathbf{x}_2, \ldots, \mathbf{x}_k\}$ spans R^n. Then a vector is in the range of T if and only if it is a linear combination of the vectors $T(\mathbf{x}_1)$, $T(\mathbf{x}_2), \ldots, T(\mathbf{x}_k)$, that is,*

$$R(T) = \text{span } \{T(\mathbf{x}_1), T(\mathbf{x}_2), \ldots, T(\mathbf{x}_k)\}$$

Proof

A vector $\mathbf{y}$ is in $R(T)$ if and only if $\mathbf{y} = T(\mathbf{x})$ for some $\mathbf{x}$ in R^n. Since $\{\mathbf{x}_1, \mathbf{x}_2, \ldots, \mathbf{x}_k\}$ spans R^n, any vector $\mathbf{x}$ in R^n is a linear combination of the $\mathbf{x}_i$'s. That is,

$$\mathbf{x} = a_1\mathbf{x}_1 + a_2\mathbf{x}_2 + \cdots + a_k\mathbf{x}_k$$

where the a_i's are scalars. Consequently, $\mathbf{y}$ is in $R(T)$ if and only if

$$\begin{aligned}
\mathbf{y} &= T(\mathbf{x}) \\
&= T(a_1\mathbf{x}_1 + a_2\mathbf{x}_2 + \cdots + a_k\mathbf{x}_k) \\
&= a_1 T(\mathbf{x}_1) + a_2 T(\mathbf{x}_2) + \cdots + a_k T(\mathbf{x}_k)
\end{aligned}$$

Thus, $\mathbf{y}$ is in $R(T)$ if and only if $\mathbf{y}$ is a linear combination of the $T(\mathbf{x}_i)$'s.   ■

Because the span of any set is a subspace we have:

Corollary 2.5.5

The range of a linear transformation from R^n to R^m is a subspace of R^m.

Example 2

Define $T: R^3 \to R^2$ by

$$T(x, y, z) = (2x + y - 4z, x - y)$$

We shall find the range of T, and relate it to solutions of a system of linear equations. Recall that $\{\mathbf{e}_1, \mathbf{e}_2, \mathbf{e}_3\}$ spans R^3. Because $T(\mathbf{e}_1) = (2, 1)$, $T(\mathbf{e}_2) = (1, -1)$, and $T(\mathbf{e}_3) = (-4, 0)$, we have that $\{(2, 1), (1, -1), (-4, 0)\}$ spans $R(T)$. Since $(2, 1)$ and $(1, -1)$ are not parallel, they span R^2. Thus, by Theorem 2.5.4, $R(T) = R^2$. In terms of a system of linear equations, we may conclude that the system

$$\begin{aligned}
2x + y - 4z &= a \\
x - y \qquad &= b
\end{aligned}$$

has a solution for any choice of scalars a and b.

Notice that the computation in Example 2 is the same as that used to compute the columns of the standard matrix of T. From this we may deduce additional corollaries.

Corollary 2.5.6

If A is the standard matrix of T, the range of T is the span of the columns of A.

Corollary 2.5.7

The system $A\mathbf{x} = \mathbf{b}$ has a solution if and only if $\mathbf{b}$ is a linear combination of the columns of A.

Example 3   Consider the system of linear equations $A\mathbf{x} = \mathbf{b}$ (written in matrix form)

$$\begin{bmatrix} 1 & 1 \\ 1 & -1 \end{bmatrix} \begin{bmatrix} x \\ y \end{bmatrix} = \begin{bmatrix} 2 \\ 0 \end{bmatrix}$$

Because

$$\mathbf{b} = \begin{bmatrix} 2 \\ 0 \end{bmatrix} = 1 \begin{bmatrix} 1 \\ 1 \end{bmatrix} + 1 \begin{bmatrix} 1 \\ -1 \end{bmatrix}$$

the vector $\mathbf{b}$ is a linear combination of the columns of A. So, by Corollary 2.5.7, the system has a solution.

As we have seen, there are two important subspaces associated with a linear transformation $T: R^n \to R^m$, namely, $N(T)$ and $R(T)$. We must remember that the null space, $N(T)$, is a subspace of R^n, and that the range, $R(T)$, is a subspace of R^m. This situation is depicted graphically in Figure 2.5.1.

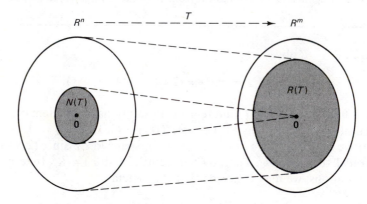

Figure 2.5.1

Notice that if A is an $m \times n$ matrix, then $A\mathbf{x} = \mathbf{b}$ has a solution for every $\mathbf{b}$ if and only if $R(L_A)$ is all of R^m, that is, if and only if the linear transformation $T = L_A$ is "onto." Recall that if f is any function from R^n to R^m, then f is ***onto*** if and only if the range of f is R^m. Consequently, we have the following corollary.

Corollary 2.5.8   *If the $m \times n$ matrix A is the standard matrix of T, then T is onto if and only if the columns of A span R^m.*

We leave it as an exercise to show that I is onto but T_0 is not.

Example 4   Consider any system of linear equations $A\mathbf{x} = \mathbf{b}$, where A is the 3×3 matrix

$$A = \begin{bmatrix} 1 & 0 & 1 \\ 0 & 1 & 1 \\ 1 & 1 & 0 \end{bmatrix}$$

It is easy to show that the columns of A span R^3 (we omit the details). Consequently, in view of Corollary 2.5.8, we have that any system of linear equations of the form

$$x + \qquad z = b_1$$
$$y + z = b_2$$
$$x + y \qquad = b_3$$

has a solution.

In summary, there are two ways in which we can view the system of linear equations given by $A\mathbf{x} = \mathbf{b}$. From one point of view, we want a practical method of finding solutions, for example, Gaussian elimination. From another point of view, we would like some more theoretical information before we attempt to find a solution, for example, the size (we shall see later that the appropriate word is "dimension") and form of the solution set. The search for answers to these more theoretical questions is frequently made easier when the problems are stated in terms of linear transformations. If we set $T = L_A$, we can formulate statements about solutions in terms of the linear transformation T (see Exercises 13 and 14). We summarize these in Table 2.5.1.

TABLE 2.5.1 $T: R^n \to R^m$

Property of Subspaces	Property of Transformations	Property of Systems
$N(T) = \{\mathbf{0}\}$	T is one-to-one	for any vector $\mathbf{b}$ in R^m, $A\mathbf{x} = \mathbf{b}$ has at most one solution
$R(T) = R^m$	T is onto	for any vector $\mathbf{b}$ in R^m, $A\mathbf{x} = \mathbf{b}$ has a solution

Exercises

1. For each of the linear transformations $T: R^2 \to R^2$:
   (i)   Describe the null space geometrically.
   (ii) Describe the range geometrically.
   (iii) Use Theorem 2.5.3 and part (i) to determine if T is one-to-one.
   (iv) Use (ii) to determine if T is onto.
   (a) The projection on the x-axis: $T(x, y) = (x, 0)$.
   (b) The reflection about the x-axis: $T(x, y) = (x, -y)$.
   (c) The rotation by an angle θ.

2. For each of the linear transformations T:
   (i) Find a set that spans the null space.

   (ii) Find a set that spans the range (use Theorem 2.5.4 applied to the standard vectors).
   (a) $T: R^2 \to R^2$ defined by $T(x, y) = (2x - y, x + y)$.
   (b) $T: R^3 \to R^2$ defined by $T(x, y, z) = (2x - y, x + y)$.
   (c) $T: R^3 \to R^3$ defined by $T(x, y, z) = (x + y + z, 0, 2x + z)$.
   (d) $T: R^2 \to R^4$ defined by $T(x, y) = (x + y, x - y, x, y)$.
   (e) $T: R^3 \to R^4$ defined by $T(x, y, z) = (x + y, y + z, x - z, x + 2y + z)$.
   (f) $T: R^3 \to R$ defined by $T(\mathbf{x}) = \mathbf{e}_1 \cdot \mathbf{x}$.

(g) $T: R^3 \rightarrow R$ defined by $T(\mathbf{x}) = (1, 1, 1) \cdot \mathbf{x}$

(h) T is the projection on the xy-plane:
$$T(x, y, z) = (x, y, 0)$$

(i) T is the reflection about the xy-plane:
$$T(x, y, z) = (x, y, -z)$$

3. Define $T: R^3 \rightarrow R^3$ by

$$T(x, y, z) = (x + y - z, x + 2y, 2x + z)$$

(a) Prove that $\mathbf{y} = (1, 3, 3)$ is in the range of T.

(b) Find a vector $\mathbf{x}$ whose image is the vector $\mathbf{y}$ given in part (a).

4. Prove that the identity transformation I is onto but that the zero transformation T_0 is not.

5. Let $\mathbf{y}$ be a fixed vector in R^2. Define $T: R^2 \rightarrow R$ by $T(\mathbf{x}) = \mathbf{y} \cdot \mathbf{x}$ for all $\mathbf{x}$ in R^2. Prove that T is onto if and only if $\mathbf{y} \neq \mathbf{0}$.

6. Verify all the statements in Example 1.

7. Suppose that U and T are linear transformations which are both one-to-one and such that UT is defined. Prove that UT is one-to-one.

8. Determine if the result of Exercise 7 is still valid if we assume that only one of the linear transformations is one-to-one.

9. Suppose that U and T are linear transformations which are both onto and such that UT is defined. Prove that UT is onto.

10. Determine if the result of Exercise 9 is still valid if we assume that only one of the linear transformations is onto.

11. Use the definition of subspace to prove that $R(T)$ is a subspace for any linear transformation T.

12. Prove Corollary 2.5.2.

13. Let A be an $m \times n$ matrix. Prove that the system of linear equations $A\mathbf{x} = \mathbf{b}$ has at most one solution for every $\mathbf{b}$ if and only if L_A is one-to-one.

14. Let A be an $m \times n$ matrix. Prove that the system of linear equations $A\mathbf{x} = \mathbf{b}$ has a solution if and only if $\mathbf{b}$ lies in $R(L_A)$.

15. Complete the proof of Theorem 2.5.1 by showing that $N(T)$ is closed under products by scalars.

16. Let $T: R^n \rightarrow R^m$ be linear. Prove that for any subspace V of R^n, the subset

$$T(V) = \{T(\mathbf{x}) : \mathbf{x} \text{ is a vector in } V\}$$

is a subspace of R^m.

*17. Use the program LINSYST to do Exercise 2(c), (d), and (e).

* This problem should be solved by using one of the programs noted in Appendix B.

Key Words

3

Dimension and Rank

In Chapter 1 we used linear combinations of vectors to produce new vectors. This process led to the definition of the span of a finite set. It was shown that the span of a finite set is a subspace of R^n.

In this chapter we take the opposite approach. Starting with a subspace V of R^n, we find finite subsets of V which span V. Furthermore, we seek the *best* way of choosing such a spanning set. For example, if different sets span the same subspace, it is usually better to choose the one with the fewest vectors. This choice minimizes the number of calculations.

These considerations lead to the concept of "dimension," which in turn leads to a deeper understanding of matrices and systems of linear equations.

3.1 LINEAR DEPENDENCE, INDEPENDENCE, AND BASES

We begin this section by examining particular spanning sets for a subspace of R^3.

Example 1 It can be verified that the span of the set

$$S = \{(1, 0, 0), (0, 1, 0), (1, -1, 0)\}$$

is the xy-plane $\{(x, y, 0): x \text{ and } y \text{ are in } R\}$. We denote this subspace by V. Notice that the vector $(1, -1, 0)$ of S is a linear combination of the other vectors of S. As we shall

129

see by Theorem 3.1.1, $(1, -1, 0)$ can be deleted, and we have a smaller set $S' = \{(1, 0, 0), (0, 1, 0)\}$ which also spans V. Since

$$(1, 0, 0) = (0, 1, 0) + (1, -1, 0)$$

we could also have eliminated $(1, 0, 0)$ from S to conclude that $\{(0, 1, 0), (1, -1, 0)\}$ spans V.

Example 1 illustrates a technique for reducing the size of a spanning set. This technique is summarized in the following theorem.

Theorem 3.1.1   *Let $S = \{\mathbf{x}_1, \mathbf{x}_2, \ldots, \mathbf{x}_k\}$ be a subset of R^n, and suppose that for some i, $\mathbf{x}_i$ is a linear combination of the other vectors of S. If S' denotes the subset of S obtained by deleting $\mathbf{x}_i$, then the span of S' is equal to the span of S.*

Proof

Suppose that $\mathbf{x}_i$ is a linear combination of the other vectors of S. For notational convenience (since it is not a substantive part of the proof), we suppose that $i = k$. We want to show that $S' = \{\mathbf{x}_1, \mathbf{x}_2, \ldots, \mathbf{x}_{k-1}\}$ has the same span as S. Since S' is a subset of S, the span of S' is contained in the span of S. So we must show that the span of S is contained in the span of S'. Since $\mathbf{x}_k$ is a linear combination of the other vectors of S, we have that

$$\mathbf{x}_k = a_1 \mathbf{x}_1 + a_2 \mathbf{x}_2 + \cdots + a_{k-1} \mathbf{x}_{k-1}$$

for some scalars $a_1, a_2, \ldots, a_{k-1}$. Now let $\mathbf{x}$ be any vector in the span of S. Since $\mathbf{x}$ is a linear combination of the vectors of S, there are scalars $b_1, b_2, \ldots, b_k$ such that

$$\mathbf{x} = b_1 \mathbf{x}_1 + b_2 \mathbf{x}_2 + \cdots + b_k \mathbf{x}_k$$

$$= b_1 \mathbf{x}_1 + b_2 \mathbf{x}_2 + \cdots + b_{k-1} \mathbf{x}_{k-1} + b_k \left(\sum_{j=1}^{k-1} a_j \mathbf{x}_j \right)$$

$$= \sum_{j=1}^{k-1} (b_j + b_k a_j) \mathbf{x}_j$$

Therefore, $\mathbf{x}$ is contained in the span of S'. Thus, the span of S is contained in the span of S'. We conclude that S and S' have the same span. ∎

Example 2   Let V be the subspace of R^4 spanned by the set

$$S = \{(1, 2, 1, 3), (1, 0, 1, 4), (0, 2, 0, -1), (1, 4, 1, 2)\}$$

Use Theorem 3.1.1 to eliminate as many vectors in S as possible to obtain a "smallest" subset of S that still spans V.

We first find a vector in S which is a linear combination of the other vectors of S. Since we have no way of choosing such a vector, we pick one arbitrarily and then test to see whether it is a linear combination of the other vectors of the set. Consider the vector $(1, 2, 1, 3)$. We wish to find scalars x, y, and z so that

$$x(1, 0, 1, 4) + y(0, 2, 0, -1) + z(1, 4, 1, 2) = (1, 2, 1, 3)$$

Combining the terms on the left side of the equation, and equating corresponding coordinates of the left and right sides, we obtain

$$
\begin{aligned}
x \quad\;\; + \;\; z &= 1 \\
2y + 4z &= 2 \\
x \quad\;\; + \;\; z &= 1 \\
4x - \;\; y + 2z &= 3
\end{aligned}
$$

Using the methods of Chapter 1, this system can be simplified to the equivalent system

$$
\begin{aligned}
x \quad\;\; + \;\; z &= 1 \\
y + 2z &= 1
\end{aligned}
$$

There are many solutions to this system, for example, $x = 0$, $y = -1$, and $z = 1$. Thus,

$$(1, 2, 1, 3) = -(0, 2, 0, -1) + (1, 4, 1, 2)$$

So, by Theorem 3.1.1, $(1, 2, 1, 3)$ can be eliminated, and we have that

$$S' = \{(1, 0, 1, 4), (0, 2, 0, -1), (1, 4, 1, 2)\}$$

is also a set that spans V.

S' is not the smallest subset that spans V. Using the same technique as above, it can be shown that $(1, 0, 1, 4)$ is a linear combination of $(0, 2, 0, -1)$ and $(1, 4, 1, 2)$. Reapplying Theorem 3.1.1, we have that the set

$$S'' = \{(0, 2, 0, -1), (1, 4, 1, 2)\}$$

spans the subspace V. If either vector is eliminated from S'', the new set will not span V since neither vector is a linear combination of the other.

———————————

These examples illustrate the need to recognize when several vectors are related in such a way that one of them is a linear combination of the others. The following definition of "linear dependence" is equivalent to this condition, but is easier to use (see Theorem 3.1.2).

Definition

A set of vectors $\{\mathbf{x}_1, \mathbf{x}_2, \ldots, \mathbf{x}_k\}$ is *linearly dependent* if there are scalars c_1, $c_2, \ldots, c_k$—not all zero—such that

$$c_1\mathbf{x}_1 + c_2\mathbf{x}_2 + \cdots + c_k\mathbf{x}_k = \mathbf{0}$$

Notice that the zero vector $\mathbf{0}$ of R^n can be represented as a linear combination of any nonempty finite subset $\{\mathbf{x}_1, \mathbf{x}_2, \ldots, \mathbf{x}_k\}$ of R^n,

$$0\mathbf{x}_1 + 0\mathbf{x}_2 + \cdots + 0\mathbf{x}_k = \mathbf{0}$$

We shall call such a representation *trivial*. It follows that a set is linearly dependent if and only if there exists a *nontrivial* representation of $\mathbf{0}$ as a linear combination of the vectors of the set.

For example, if $S = \{(1, 1, -1), (-2, 1, 1), (0, -3, 1), (5, 6, 8)\}$, there is a nontrivial representation of $\mathbf{0}$ as a linear combination of the vectors of S, namely,

$$2(1, 1, -1) + 1(-2, 1, 1) + 1(0, -3, 1) + 0(5, 6, 8) = \mathbf{0}$$

The representation is nontrivial because *at least one* of the scalars above is not zero. Thus S is linearly dependent.

Example 3

Determine whether or not the set

$$\{(1, 2, 1), (1, 0, 1), (1, 4, 1), (1, 2, 3)\}$$

is linearly dependent.

We want to determine if there exists a set of scalars c_1, c_2, c_3, and c_4 such that at least one of them is not zero and

$$c_1(1, 2, 1) + c_2(1, 0, 1) + c_3(1, 4, 1) + c_4(1, 2, 3) = (0, 0, 0)$$

This vector equation is equivalent to the linear system

$$
\begin{aligned}
c_1 + c_2 + c_3 + c_4 &= 0 \\
2c_1 \phantom{{}+ c_2} + 4c_3 + 2c_4 &= 0 \\
c_1 + c_2 + c_3 + 3c_4 &= 0
\end{aligned}
$$

By the method of Gaussian elimination it can be shown that the solutions to this system are of the form $c_1 = -2s, c_2 = s, c_3 = s$, and $c_4 = 0$, for arbitrary values of s. If we set $s = 1$, then we have that $c_1 = -2, c_2 = 1, c_3 = 1$, and $c_4 = 0$. So

$$(-2)(1, 2, 1) + 1(1, 0, 1) + 1(1, 4, 1) + 0(1, 2, 3) = (0, 0, 0)$$

Since this representation of $\mathbf{0}$ is nontrivial, we have that S is linearly dependent.

A careful examination of the last equation in Example 3 shows how linear dependence is related to the problem of representing a vector as a linear combination of other vectors. Notice that we can solve this equation for $(1, 2, 1)$ to obtain

$$(1, 2, 1) = \left(\frac{1}{2}\right)(1, 0, 1) + \left(\frac{1}{2}\right)(1, 4, 1) + 0(1, 2, 3)$$

So $(1, 2, 1)$ is a linear combination of the other vectors of S. Similarly, it can be shown that $(1, 4, 1)$ is a linear combination of the other vectors of S.

The next theorem justifies the relation between linear dependence and the representation of a vector as a linear combination of other vectors.

Theorem 3.1.2 *Let $S = \{\mathbf{x}_1, \mathbf{x}_2, \ldots, \mathbf{x}_k\}$ be a subset of R^n. Then S is linearly dependent if and only if at least one of the vectors in S is a linear combination of the other vectors in S.*

Proof

Suppose that S is linearly dependent. Then there exist scalars $c_1, c_2, \ldots, c_k$, not all zero, such that

$$c_1\mathbf{x}_1 + c_2\mathbf{x}_2 + \cdots + c_k\mathbf{x}_k = \mathbf{0}$$

Without loss of generality, suppose that $c_k \neq 0$. Moving the first $k - 1$ terms to the right side of the equation and dividing by c_k, we obtain

$$\mathbf{x}_k = \left(\frac{-c_1}{c_k}\right)\mathbf{x}_1 + \left(\frac{-c_2}{c_k}\right)\mathbf{x}_2 + \cdots + \left(\frac{-c_{k-1}}{c_k}\right)\mathbf{x}_{k-1}$$

So $\mathbf{x}_k$ is a linear combination of the other vectors of the set.

We now prove the converse. Suppose that at least one of the vectors in S is a linear combination of the others. Without loss of generality, suppose that $\mathbf{x}_k$ is a linear combination of $\mathbf{x}_1, \mathbf{x}_2, \ldots, \mathbf{x}_{k-1}$. Then there exist scalars $c_1, c_2, \ldots, c_{k-1}$ such that

$$c_1\mathbf{x}_1 + c_2\mathbf{x}_2 + \cdots + c_{k-1}\mathbf{x}_{k-1} = \mathbf{x}_k$$

Therefore,

$$c_1\mathbf{x}_1 + c_2\mathbf{x}_2 + \cdots + c_{k-1}\mathbf{x}_{k-1} + (-1)\mathbf{x}_k = \mathbf{0}$$

Since the last coefficient is not zero, S is linearly dependent. ∎

Example 4 Determine whether or not the set $S = \{(1, 2, 1), (2, 2, 3), (1, 0, 1)\}$ is linearly dependent.

We wish to determine if there are scalars $c_1, c_2,$ and c_3 not all zero so that

$$c_1(1, 2, 1) + c_2(2, 2, 3) + c_3(1, 0, 1) = (0, 0, 0)$$

Converting to a system of linear equations, we have

$$c_1 + 2c_2 + c_3 = 0$$
$$2c_1 + 2c_2 = 0$$
$$c_1 + 3c_2 + c_3 = 0$$

By the method of Gaussian elimination, we obtain that $c_1 = c_2 = c_3 = 0$. Since all these scalars are zero, we conclude that S is not linearly dependent.

In the example above we verified that the given set is not linearly dependent by appealing directly to the definition of linear dependence. Later, we shall use less direct but computationally simpler methods of doing this. These methods involve the use of elementary row operations on matrices.

Finite sets that are not linearly dependent are at least as important as linearly dependent sets.

Definition   A finite subset S of R^n is ***linearly independent*** if S is not linearly dependent.

In terms of representations of **0**, we see that *a finite set S is linearly independent if and only if the only representation of 0 as a linear combination of the vectors of S is the trivial one.* To show that a set $\{\mathbf{x}_1, \mathbf{x}_2, \ldots, \mathbf{x}_k\}$ is linearly independent, we set

$$\sum_{i=1}^{k} c_i \mathbf{x}_i = \mathbf{0}$$

and then prove that $c_i = 0$ for each i. Using this argument, we conclude that the set S of Example 4 is linearly independent.

We shall show that the set $\{\mathbf{e}_1, \mathbf{e}_2, \ldots, \mathbf{e}_n\}$ of standard vectors is linearly independent. Consider any scalars $c_1, c_2, \ldots, c_n$ for which

$$c_1\mathbf{e}_1 + c_2\mathbf{e}_2 + \cdots + c_n\mathbf{e}_n = \mathbf{0}$$

Combining the terms on the left side of this equation, we obtain

$$(c_1, c_2, \ldots, c_n) = (0, 0, \ldots, 0)$$

Equating the corresponding components, we have that $c_i = 0$ for all i. Since there are no nonzero scalars, we have that the set of standard vectors is linearly independent.

By convention, *the empty subset of R^n is linearly independent.* Logicians justify this convention by the following argument. If a set is linearly dependent, there exists a nontrivial representation of **0** as a linear combination of the vectors of the set. But the empty set contains no vectors. So no such representation exists. Therefore, the empty set is not linearly dependent. So it must be linearly independent.

By repeated application of Theorems 3.1.1 and 3.1.2, we obtain:

Corollary 3.1.3

If a subspace V of R^n is spanned by a finite subset S, then there is a linearly independent subset S' of S which also spans V.

Intuitively, this corollary states that if S is linearly dependent, we can "throw away" (remove) redundant vectors until we are left with a linearly independent set which also spans V.

Notice that the set S'' of Example 2 is a linearly independent subset of S which also spans V.

As we have seen, $\mathbf{0}$ can be represented in only one way as a linear combination of a linearly independent set. This property of the uniqueness of the representation of $\mathbf{0}$ extends to other vectors.

Theorem 3.1.4   *Let $S = \{\mathbf{x}_1, \mathbf{x}_2, \ldots, \mathbf{x}_k\}$ be a nonempty, linearly independent subset of R^n. Any vector $\mathbf{x}$ in the span of S can be expressed as a linear combination of the vectors in S in only one way. That is, if*

$$\mathbf{x} = a_1\mathbf{x}_1 + a_2\mathbf{x}_2 + \cdots + a_k\mathbf{x}_k$$

then the coefficients $a_1, a_2, \ldots, a_k$ are unique.

Proof

Suppose that a vector $\mathbf{x}$ in the span of S can be represented as a linear combination of the vectors of S in two ways. That is,

$$\mathbf{x} = a_1\mathbf{x}_1 + a_2\mathbf{x}_2 + \cdots + a_k\mathbf{x}_k$$

and

$$\mathbf{x} = b_1\mathbf{x}_1 + b_2\mathbf{x}_2 + \cdots + b_k\mathbf{x}_k$$

We wish to show that $a_i = b_i$ for all i. Since

$$a_1\mathbf{x}_1 + a_2\mathbf{x}_2 + \cdots + a_k\mathbf{x}_k = b_1\mathbf{x}_1 + b_2\mathbf{x}_2 + \cdots + b_k\mathbf{x}_k$$

we have

$$(a_1 - b_1)\mathbf{x}_1 + (a_2 - b_2)\mathbf{x}_2 + \cdots + (a_k - b_k)\mathbf{x}_k = \mathbf{0}$$

Finally, because S is linearly independent, the coefficients $a_i - b_i$ must all be zero. We conclude that $a_i = b_i$ for all i. ∎

It should be noted that the converse of Theorem 3.1.4 is also valid (see Exercise 19).

Example 5 Consider the set $\{(1, 0, 0), (0, 1, 0), (1, -1, 0)\}$ of Example 1. This set is not linearly independent. Notice that the vector $(2, 2, 0)$ can be represented as a linear combination of the vectors in the set in more than one way:

$$(2, 2, 0) = 2(1, 0, 0) + 2(0, 1, 0) + 0(1, -1, 0)$$

and

$$(2, 2, 0) = 1(1, 0, 0) + 3(0, 1, 0) + 1(1, -1, 0)$$

Spanning sets are used to describe subspaces of R^n. By Theorem 3.1.4 the representation of any vector as a linear combination of vectors of a linearly independent set is unique. For this reason linearly independent spanning sets play an important role in linear algebra.

Definition A subset S of a subspace V of R^n is called a *basis* for V if S is linearly independent and spans V.

We can use this definition to restate Corollary 3.1.3.: *If a subspace V of R^n is spanned by a finite subset S, then there is a subset S' of S which is a basis for V.*

Any linearly independent subset of R^n is a basis for a subspace of R^n, namely, its span.

Earlier in this section it was shown that the subset $\{e_1, e_2, \ldots, e_n\}$ of standard vectors of R^n is linearly independent. In Section 1.3 it was shown that this set spans R^n. Therefore, this set is a basis for R^n. We shall call this basis the *standard basis* of R^n.

Example 6 Let $S = \{(1, 1), (1, -1)\}$. Show that S is a basis for R^2.

Consider any scalars x and y such that

$$x(1, 1) + y(1, -1) = (0, 0)$$

Then

$$x + y = 0$$
$$x - y = 0$$

This system has only the trivial solution $x = 0$ and $y = 0$. Therefore, S is linearly independent.

Although we have seen a geometric argument that two nonparallel vectors in R^2 span a plane, and hence all of R^2, we now provide an algebraic argument that S spans R^2. Let (a, b) be any vector in R^2. We wish to find scalars x and y such that

$$x(1, 1) + y(1, -1) = (a, b)$$

This equation is equivalent to

$$x + y = a$$
$$x - y = b$$

This system has the solution

$$x = \frac{1}{2}(a + b) \quad \text{and} \quad y = \frac{1}{2}(a - b)$$

Therefore, any vector in R^2 is a linear combination of $(1, 1)$ and $(1, -1)$. We conclude that S is a basis for R^2.

There is an easier way to check if a set of two vectors is linearly independent. By Exercise 8, a set of two vectors is linearly dependent if and only if one of the vectors is a scalar multiple of the other. Since this is not the case with the vectors $(1, 1)$ and $(1, -1)$, S is linearly independent.

By the arguments of Example 6, we see that *any* two nonparallel vectors in R^2 form a basis for R^2. In general, subspaces have many bases.

Example 7   Show that $S = \{(1, -1, 0, 0), (1, 0, -1, 0), (1, 0, 0, -1)\}$ is a basis for the subspace

$$V = \{(a, b, c, d) : a + b + c + d = 0\}$$

of R^4.

It should be noted that V is indeed a subspace of R^4. This can be shown by verifying that V satisfies the definition of subspace given in Section 1.3. To prove that S is linearly independent, consider any scalars x, y, and z such that

$$x(1, -1, 0, 0) + y(1, 0, -1, 0) + z(1, 0, 0, -1) = (0, 0, 0, 0)$$

Then

$$x + y + z = 0$$
$$-x \qquad\qquad = 0$$
$$-y \qquad = 0$$
$$-z = 0$$

The only solution to this system is $x = 0$, $y = 0$, $z = 0$. Therefore, S is linearly independent.

To prove that S spans V, first notice that S is a subset of V since the sum of the components of each vector of S equals zero. Therefore, the span of S is contained in V

because V is a subspace of R^4. Now consider any vector (a, b, c, d) of V. Since $a + b + c + d = 0$, we have $a = -b - c - d$. Therefore,

$$(a, b, c, d) = (-b - c - d, b, c, d)$$
$$= (-b, b, 0, 0) + (-c, 0, c, 0) + (-d, 0, 0, d)$$
$$= -b(1, -1, 0, 0) - c(1, 0, -1, 0) - d(1, 0, 0, -1)$$

So (a, b, c, d) is a linear combination of the vectors of S. We conclude that V is the span of S. Therefore, S is a basis for V.

Example 8   Find a basis for the zero subspace of R^n.

The span of the empty subset of R^n is $\{0\}$, by definition, as given in Section 1.3. As noted earlier, the empty set is linearly independent. Therefore, the empty set is a basis for $\{0\}$. Actually, the empty set is the only basis for $\{0\}$. Notice that the set $\{0\}$ is linearly dependent, and in fact, any set containing the zero vector is linearly dependent (see Exercise 5).

Example 9   Let $S = \{(1, 2, 1), (1, 1, 3), (1, 0, 5)\}$. Let V be the span of S. Find a basis for V.

By Corollary 3.1.3 there exists a linearly independent subset of S which spans V. This subset must be a basis for V. Consider scalars x, y, and z such that

$$x(1, 2, 1) + y(1, 1, 3) + z(1, 0, 5) = (0, 0, 0)$$

Then

$$x + y + z = 0$$
$$2x + y \quad\quad = 0$$
$$x + 3y + 5z = 0$$

This system is equivalent to

$$x + y + z = 0$$
$$y + 2z = 0$$

which has nontrivial solutions, for example, $z = 1$, $y = -2$, and $x = 1$. Therefore,

$$1(1, 2, 1) - 2(1, 1, 3) + 1(1, 0, 5) = (0, 0, 0)$$

So we can eliminate one of the vectors of S, say $(1, 0, 5)$, to obtain a set $S' = \{(1, 2, 1), (1, 1, 3)\}$ which also spans V. The set S' is linearly independent since the vectors of S' are not scalar multiples of each other. Therefore, S' is a basis for V.

Matrix Approach to Constructing a Basis from a Spanning Set ___

Let S be a subset of R^n that spans a subspace V. We have seen how to use Theorems 3.1.1 and 3.1.2 to eliminate vectors from S to obtain a basis for V. We now consider another method which is more mechanical and less tedious. It applies elementary row operations to the matrix whose rows are the vectors of S. The idea is to perform elementary row operations on this matrix to simplify its form. As we shall see, the rows of the new matrix span the same subspace of R^n.

Theorem 3.1.5   *Let A be an $m \times n$ matrix. Suppose that B is a matrix obtained from A by means of a sequence of elementary row operations. Then the rows of B and the rows A span the same subspace of R^n.*

Proof

If B is obtained from A by a single elementary row operation of type 1, then these two matrices have the same rows, but in a different order. Clearly, they span the same subspace.

We leave it as an exercise to prove that if B is obtained from A by means of a single elementary row operation of type 2, then the rows of A and the rows B span the same subspace.

Now suppose that B is obtained from A by a single elementary row operation of type 3. Let V be the subspace of R^n spanned by the rows of A, and let W be the subspace of R^n spanned by the rows of B. We show that $V = W$. Let $\mathbf{x}_1, \mathbf{x}_2, \ldots, \mathbf{x}_m$ denote the rows of A and let $\mathbf{y}_1, \mathbf{y}_2, \ldots, \mathbf{y}_m$ denote the rows of B. Then there are integers j and k, $j \neq k$, and a scalar c, such that

$$\mathbf{y}_i = \mathbf{x}_i \qquad \text{for } i \neq k$$

and

$$\mathbf{y}_k = c\mathbf{x}_j + \mathbf{x}_k$$

Therefore, any row of B is either a row of A or a linear combination of two rows of A. We conclude that the rows of B are contained in V. Therefore, W is contained in V. Since

$$\mathbf{x}_k = \mathbf{y}_k - c\mathbf{x}_j$$
$$= \mathbf{y}_k - c\mathbf{y}_j$$

we have that A can be obtained from B by an elementary row operation of type 3. By a similar argument, V is contained in W. We conclude that $V = W$.

If B is obtained from A by a sequence of more than one elementary row operation, then we can apply the results above several times to conclude that the rows of B and the rows of A span the same subspace of R^n.   ■

Example 10   We rework Example 9, applying Theorem 3.1.5. Let A denote the 3×3 matrix whose rows are the vectors of the set S of Example 9. Let V denote the space spanned by S. Then

$$A = \begin{bmatrix} 1 & 2 & 1 \\ 1 & 1 & 3 \\ 1 & 0 & 5 \end{bmatrix}$$

By means of three elementary row operations A can be transformed into a matrix

$$B = \begin{bmatrix} 1 & 2 & 1 \\ 0 & -1 & 2 \\ 0 & 0 & 0 \end{bmatrix}$$

Applying Theorem 3.1.5, we see that the rows of B span V. We can ignore the third row of B since it is zero. Therefore, the set $\{(1, 2, 1), (0, -1, 2)\}$ spans V. Since this set is linearly independent, it is a basis for V.

As is clear from Example 10, we may apply Theorem 3.1.5 to transform a matrix A to a new matrix B so that a basis for the span of the rows of A can be obtained by simple observation. It turns out that if B is in triangular form, the task is especially easy.

Theorem 3.1.6   *Let A be a matrix in triangular form. Then the nonzero rows of A are linearly independent.*

We omit the proof of this theorem. However, we can gain some insight into why Theorem 3.1.6 works from the next example.

Example 11   Consider the matrix

$$\begin{bmatrix} 3 & 2 & 4 & 4 & 1 \\ 0 & 0 & 2 & 2 & 3 \\ 0 & 0 & 0 & 1 & 7 \\ 0 & 0 & 0 & 0 & 0 \end{bmatrix}$$

which is in triangular form. We will show that the first three rows of this matrix are linearly independent. Let $\mathbf{x}_1$, $\mathbf{x}_2$, and $\mathbf{x}_3$ denote the first three rows of the matrix. Suppose that

$$c_1 \mathbf{x}_1 + c_2 \mathbf{x}_2 + c_3 \mathbf{x}_3 = (0, 0, 0, 0, 0)$$

That is,

$$(3c_1, 2c_1, 4c_1 + 2c_2, 4c_1 + 2c_2 + c_3, c_1 + 3c_2 + 7c_3) = (0, 0, 0, 0, 0)$$

Since the first component of this linear combination is $3c_1$, we have that $c_1 = 0$. Therefore, the equation above becomes

$$c_2 \mathbf{x}_2 + c_3 \mathbf{x}_3 = (0, 0, 0, 0, 0)$$

The third component of this linear combination is $2c_2$. Therefore, $c_2 = 0$. So the equation reduces to

$$c_3 \mathbf{x}_3 = (0, 0, 0, 0, 0)$$

Since $\mathbf{x}_3$ is not the zero vector, we have that $c_3 = 0$. We conclude that the nonzero rows of the matrix are linearly independent.

The technique for finding a basis for a subspace with a given spanning set is straightforward. Suppose that the set $\{\mathbf{x}_1, \mathbf{x}_2, \ldots, \mathbf{x}_k\}$ spans the subspace V of R^n. First construct the $k \times n$ matrix A whose rows are given by the $\mathbf{x}_i$'s. Next, by means of elementary row operations, transform A into a matrix B in triangular form. Then the rows of B span V. So the nonzero rows of B span V also. But this set of rows is linearly independent. Thus the set of nonzero rows of B is a basis for V.

Example 12   Let V be the subspace of R^4 spanned by

$$S = \{(1, 2, 1, 3), (2, 1, -1, 0), (1, -1, -2, -3), (1, 5, 4, 9)\}$$

Find a basis for V.

First, form the matrix A whose rows are the vectors of S:

$$A = \begin{bmatrix} 1 & 2 & 1 & 3 \\ 2 & 1 & -1 & 0 \\ 1 & -1 & -2 & -3 \\ 1 & 5 & 4 & 9 \end{bmatrix}$$

Reducing A to triangular form, we obtain the matrix B:

$$B = \begin{bmatrix} 1 & 2 & 1 & 3 \\ 0 & -3 & -3 & -6 \\ 0 & 0 & 0 & 0 \\ 0 & 0 & 0 & 0 \end{bmatrix}$$

Therefore, $\{(1, 2, 1, 3), (0, -3, -3, -6)\}$, the set of nonzero rows of B, is a basis for V.

Exercises

1. For each of the following parts, determine whether the given set is linearly dependent or linearly independent.
   (a) $\{(1, 2, 3)\}$
   (b) $\{(1, 1), (2, 2)\}$
   (c) $\{(1, 0, 1), (1, 0, -1)\}$
   (d) $\{(1, 2, 1), (1, 1, 1), (0, 1, 0)\}$
   (e) $\{(1, 1, 1), (1, -1, 1), (1, 0, 1)\}$
   (f) $\{(1, 0, 0, 0), (1, 1, 0, 0), (1, 1, 1, 0), (1, 1, 1, 1)\}$
   (g) $\{(1, 2, 1, 1), (2, 3, 1, -1), (1, 1, 0, -2), (1, 3, 2, 4)\}$

2. For each of the following sets S, find a subset that is a basis for the span of S.
   (a) $\{(1, 2), (2, 4)\}$
   (b) $\{(1, 1, 2)\}$
   (c) $\{(1, 1, 1), (1, 1, -1), (1, 1, 0)\}$
   (d) $\{(1, 1, 0), (1, 0, 1), (0, 1, 1)\}$
   (e) $\{(1, 2, 1), (-1, -2, -1), (3, 6, 3)\}$
   (f) $\{(1, -1, 1), (-2, 3, 0), (-3, 2, 2), (4, 1, -4)\}$
   (g) $\{(1, 2, 3, 4), (1, 1, 1, 1), (0, 1, 2, 3), (0, 0, 1, 2)\}$
   (h) $\{(1, 2, 3, 1), (1, 3, 4, 3), (2, 4, 6, 0), (1, 2, 4, 2)\}$

3. For each set S of Exercise 2, use matrix manipulations to find a basis for the span of S. Notice that in contrast to Exercise 2, the basis that you obtain is usually not a subset of S.

4. For each set S below, determine whether or not S is a basis for the given subspace V.
   (a) $S = \{(1, 2), (2, 3)\}$ and $V = R^2$
   (b) $S = \{(1, 1)\}$ and $V = \{(x, y)$ in $R^2 : x = y\}$
   (c) $S = \{(1, 2, 3), (4, 5, 6), (7, 8, 9)\}$ and $V = R^3$
   (d) $S = \{(1, -1)\}$ and $V = R^2$
   (e) $S = \{(2, -1, 0), (1, 0, -1)\}$ and $V = \{(x, y, z)$ in $R^3 : x + 2y + z = 0\}$
   (f) $S = \{(1, 1, 2, -4), (1, -1, 0, 0)\}$ and $V = \{(x, y, z, w)$ in $R^4 : x + y + z + w = 0\}$
   (g) $S = \{(1, 1, 2, -4), (1, -1, 0, 0), (2, -2, 0, 0)\}$ and $V = \{(x, y, z, w)$ in $R^4 : x + y + z + w = 0\}$

   (h) $S = \{(1, 1, 2, -4), (1, -1, 0, 0), (0, 0, 1, -1)\}$ and $V = \{(x, y, z, w)$ in $R^4 : x + y + z + w = 0\}$

†5. Prove that any finite subset of R^n which contains $\mathbf{0}$ is linearly dependent.

6. Show that for any nonzero vector $\mathbf{x}$ in R^n the set $\{\mathbf{x}\}$ is linearly independent.

†7. Show that for any vector $\mathbf{x}$ in R^n and for any scalar c ($c \neq 1$), the set $\{\mathbf{x}, c\mathbf{x}\}$ is linearly dependent.

8. Prove that for any two distinct vectors $\mathbf{x}$ and $\mathbf{y}$ in R^n, the set $S = \{\mathbf{x}, \mathbf{y}\}$ is linearly dependent if and only if $\mathbf{x}$ is a scalar multiple of $\mathbf{y}$, or $\mathbf{y}$ is a scalar multiple of $\mathbf{x}$.

9. Show that if $\{\mathbf{x}, \mathbf{y}\}$ is a linearly independent subset of R^n, then $\{\mathbf{x} + \mathbf{y}, \mathbf{x} - \mathbf{y}\}$ is also linearly independent.

†10. Let S' be a finite subset of R^n, and let S be a subset of S'. Prove that:
   (a) If S is linearly dependent, then S' is linearly dependent.
   (b) If S' is linearly independent, then S is linearly independent.

11. Prove that for any linearly independent set of vectors $\{\mathbf{x}_1, \mathbf{x}_2, \ldots, \mathbf{x}_k\}$, and any nonzero scalars $c_1, c_2, \ldots, c_k$ (not necessarily distinct), the set $\{c_1\mathbf{x}_1, c_2\mathbf{x}_2, \ldots, c_k\mathbf{x}_k\}$ is linearly independent.

12. Prove Theorem 3.1.5 for the case that B is obtained from A by an elementary row operation of type 2.

†13. Suppose that $T: R^n \to R^m$ is a linear transformation, and that $\{\mathbf{y}_1, \mathbf{y}_2, \ldots, \mathbf{y}_k\}$ is a linearly independent subset of R^m. Prove that if $\mathbf{z}_1, \mathbf{z}_2, \ldots, \mathbf{z}_k$ are vectors in R^n such that $T(\mathbf{z}_i) = \mathbf{y}_i$ for each i, then $\{\mathbf{z}_1, \mathbf{z}_2, \ldots, \mathbf{z}_k\}$ is linearly independent.

†14. Suppose that A is an $m \times n$ matrix whose rows are distinct and form a linearly independent subset of R^n. Let B be any matrix obtained from A by means of a finite sequence of elementary row operations. Prove

---

† This exercise will be used in a subsequent section.

that the rows of B form a linearly independent set. (*Hint*: First establish the result for the case that B is obtained from A by one elementary row operation.)

†15. Use Exercise 14 to derive the following test for linear independence: Let $S = \{\mathbf{x}_1, \mathbf{x}_2, \ldots, \mathbf{x}_m\}$ be a set of m vectors in R^n and let A be the $m \times n$ matrix whose ith row is $\mathbf{x}_i$. Transform A into a matrix B in triangular form by means of elementary row operations. Then S is linearly independent if and only if B has no zero rows.

16. Apply the test of Exercise 15 to the problems of Exercise 1.

17. Suppose that $T: R^n \to R^m$ is a linear transformation which is one-to-one. Prove that if $\{\mathbf{x}_1, \mathbf{x}_2, \ldots, \mathbf{x}_k\}$ is a linearly independent subset of R^n, then $\{T(\mathbf{x}_1), T(\mathbf{x}_2), \ldots, T(\mathbf{x}_k)\}$ is a linearly independent subset of R^m.

†18. Modify the proof of Theorem 3.1.2 to prove the following: Let S be a linearly independent subset of R^n, and let $\mathbf{x}$ be a vector in R^n that is not in the span of S. Then $S \cup \{\mathbf{x}\}$ is linearly independent.

19. Prove the converse of Theorem 3.1.4: Let S be a nonempty finite subset of R^n. If every vector in the span of S can be expressed as a linear combination of the vectors in S in only one way, then S is linearly independent.

*20. Use the program LINSYST to test each set of Exercise 1 to determine whether the set is linearly dependent or linearly independent.

*21. For each of the following sets, use the program ROW REDUCTION to find a basis for the span of the set.
   (a) $\{(1, 3, 7), (4, 2, 1), (1, 1, 9), (3, 5, 2)\}$
   (b) $\{(1, 3, 4, 2), (2, 1, 5, 1), (-1, 7, 2, 4),$ $(3, 4, 9, 3)\}$
   (c) $\{(1, 1, 4, 5, 6), (2, 3, 3, 4, 1), (1, 0, 9, 11, 17),$ $(2, 0, 1, 7, 7)\}$

* This problem should be solved by using one of the programs noted in Appendix B.

3.2 DIMENSION OF SUBSPACES OF R^n

It was shown in Section 3.1 that R^n has a basis consisting of n vectors. Since R^n is the largest subspace of itself, it seems reasonable to expect that bases for other subspaces of R^n contain fewer than n vectors. As we shall see, this is indeed the case. This section is concerned with the number of vectors in a basis for a subspace of R^n. We shall show that this number is unique and that it is an intrinsic property of the subspace. It is called the *dimension* of the subspace.

As a preliminary result we show that the number n is a limit to the size of any linearly independent subset of R^n. This is the content of Theorem 3.2.1. The following example serves to illustrate the ideas behind the proof of the theorem.

Example 1 Use Exercise 15 of Section 3.1 to show that $S = \{(1, 2, 3), (1, 3, 2), (2, 1, 1), (3, 0, 2)\}$ is linearly dependent.

Form the matrix A whose rows are the vectors of S. By means of elementary row operations transform A into a matrix B in triangular form.

$$A = \begin{bmatrix} 1 & 2 & 3 \\ 1 & 3 & 2 \\ 2 & 1 & 1 \\ 3 & 0 & 2 \end{bmatrix} \to \begin{bmatrix} 1 & 2 & 3 \\ 0 & -1 & -1 \\ 0 & -3 & -5 \\ 0 & -6 & -7 \end{bmatrix} \to \begin{bmatrix} 1 & 2 & 3 \\ 0 & -1 & -1 \\ 0 & 0 & -2 \\ 0 & 0 & -1 \end{bmatrix} \to \begin{bmatrix} 1 & 2 & 3 \\ 0 & -1 & -1 \\ 0 & 0 & -2 \\ 0 & 0 & 0 \end{bmatrix} = B$$

Notice that B has a zero row. We conclude by Exercise 15 that S is not linearly independent.

The proof of Theorem 3.2.1 makes use of Exercise 14 of Section 3.1, rather than Exercise 15 as in Example 1. However, Exercise 15 is really just a special application of Exercise 14.

Theorem 3.2.1   *Any linearly independent subset of R^n contains at most n vectors.*

Proof

Let $S = \{\mathbf{x}_1, \mathbf{x}_2, \ldots, \mathbf{x}_p\}$ be a linearly independent subset of R^n consisting of p vectors. We shall show that $p \leq n$. Let A be the $p \times n$ matrix whose ith row is $\mathbf{x}_i$. By means of several elementary row operations, we may transform A into a matrix B in triangular form. By Exercise 14 of Section 3.1, we have that the rows of B are linearly independent. Therefore, each row of B has a nonzero entry. Since B is in triangular form, the leading nonzero entry of any row of B is to the right of the leading nonzero entry of the preceding row of B. So these leading nonzero entries must occur in separate columns of B. Because there are p such entries, and because B has n columns, it must be the case that $p \leq n$. ∎

Another way of stating Theorem 3.2.1 is now given.

Corollary 3.2.2   *Any finite subset of R^n containing at least $n + 1$ vectors is linearly dependent.*

Example 2   Show without calculations that the set $\{(1, 2, 4), (1, -1, 2), (1, 1, 1), (-2, 3, 8)\}$ is linearly dependent.

Since this subset of R^3 contains four vectors, the set is linearly dependent by Corollary 3.2.2.

For any subspace V of R^n, a linearly independent subset S of V is also a linearly independent subset of R^n. Therefore, S cannot contain more than n vectors. Actually, depending on the choice of the subspace V, it may be the case that S contains far fewer than n vectors. In fact, as we shall see, the only subspace of R^n that contains a linearly independent subset of n vectors is R^n itself.

Example 3   Let V be the span of $\{(1, 2, 3)\}$. Then V is the set of all scalar multiples $(1, 2, 3)$. We can visualize V as the line in R^3 containing the points $(0, 0, 0)$ and $(1, 2, 3)$. Since the set $\{(1, 2, 3)\}$ is linearly independent, it serves as a basis for V. We show that any subset of V consisting of at least two vectors is linearly dependent. In view of Exercise 10(a) of Section 3.1 we need only show that any subset of V consisting of two vectors is linearly dependent. Let $S = \{\mathbf{x}, \mathbf{y}\}$ be such a set. If $\mathbf{x}$ or $\mathbf{y}$ is $\mathbf{0}$, then S is linearly dependent by Exercise 5 of Section 3.1. Otherwise, since both $\mathbf{x}$ and $\mathbf{y}$ are nonzero multiples of

(1, 2, 3), **y** is a multiple of **x**. Therefore, by Exercise 7 of Section 3.1, S is linearly dependent. So the largest number of vectors in any linearly independent subset of V is one.

The fact that there is a limit to the size of a linearly independent subset of a subspace of R^n is a key to the construction of a basis for the subspace.

Theorem 3.2.3   *Let V be a subspace of R^n. Let S be any linearly independent subset of V. Then there exists a basis for V which contains S.*

Proof

Let V be a subspace of R^n and let S be a linearly independent subset of V. If span of $S = V$, then S is a basis for V. Otherwise, there exists a vector $\mathbf{x}_1$ in V such that $\mathbf{x}_1$ is not in the span of S. By Exercise 18 of Section 3.1, the set $S_1 = S \cup \{\mathbf{x}_1\}$ is linearly independent. If span of $S_1 = V$, then S_1 is a basis for V which contains S. If not, then, as above, we find a vector $\mathbf{x}_2$ in V such that $S_2 = S_1 \cup \{\mathbf{x}_2\}$ is linearly independent. If S_2 spans V, then S_2 is a basis for V which contains S. We continue this process of selecting larger and larger linearly independent subsets of V. By Theorem 3.2.1, this process must terminate because no linearly independent subset of V can contain more than n vectors. Thus, we eventually obtain a subset of V which contains S and is a basis for V. ∎

In view of Theorem 3.2.3 we have the following.

Corollary 3.2.4   *Every subspace of R^n has a basis.*

Proof

Let V be a subspace of R^n. Choose any linearly independent subset of V, for example, the empty set, and apply Theorem 3.2.3. ∎

Corollary 3.2.5   *Every linearly independent subset of R^n consisting of n vectors is a basis for R^n.*

Proof

Let S be a linearly independent subset of R^n consisting of n vectors. By Theorem 3.2.3, S can be extended to a basis S' of R^n. By Theorem 3.2.1, S' has at most n vectors. Therefore, $S' = S$. So S is a basis for R^n. ∎

Example 4   We shall use the argument in the proof of Theorem 3.2.3 to show that the set $S = \{(1, 2, 0), (3, 5, 0), (7, 8, 9)\}$ is a basis for R^3. First notice that the subset $\{(1, 2, 0), (3, 5, 0)\}$ is linearly independent and lies in the xy-plane, a subspace of R^3. The third vector of S, $(7, 8, 9)$, does not lie in the xy-plane. Therefore, by Exercise 18 of Section 3.1, S is linearly independent. Hence, by Corollary 3.2.5, S is a basis for R^3.

The following example illustrates the process of extending a linearly independent subset of R^n to a basis for R^n, as used in the proof of Theorem 3.2.3. Its purpose is to aid in the understanding of the proof of the theorem. A more practical method of extending a basis will be given in Example 8 at the end of this section.

Example 5   Extend $S = \{(1, 1, 1)\}$ to a basis for R^3. Notice that S is linearly independent. Choose any vector in R^3 which is not a linear combination of the vector of S. Since S contains only one vector, it suffices to choose any vector that is not a multiple of $(1, 1, 1)$. For example, choose $(1, 0, 0)$. Then $S' = \{(1, 1, 1), (1, 0, 0)\}$ is a linearly independent subset of R^3. We obtained S' from S by adjoining a standard vector. Adjoin another standard vector, say $(0, 1, 0)$, to S' to obtain a set $S'' = \{(1, 1, 1), (1, 0, 0), (0, 1, 0)\}$. To show that S'' is linearly independent, we first suppose that

$$x(1, 1, 1) + y(1, 0, 0) + z(0, 1, 0) = (0, 0, 0)$$

Then

$$
\begin{aligned}
x + y & \phantom{{}+z} = 0 \\
x & + z = 0 \\
x & \phantom{{}+z} = 0
\end{aligned}
$$

Thus, $x = y = z = 0$. So S'' is linearly independent. By Corollary 3.2.5, S'' is a basis for R^3.

_____

Example 5 is interesting in that the task of showing that S'' is a basis for R^3 does not involve verifying that S'' spans R^3. It is the number of vectors in S'' which is significant here. In general, the number of vectors in a basis for a subspace does not depend on the choice of basis, but is an intrinsic property of the subspace.

Theorem 3.2.6   *Any two bases for a subspace V of R^n contain the same number of vectors.*

Proof
Let S and S' be bases for a subspace V of R^n. If $V = \{\mathbf{0}\}$, then the only basis for V is the empty set, so $S = S'$. Therefore, we assume that $V \neq \{\mathbf{0}\}$. Suppose that $S = \{\mathbf{x}_1, \mathbf{x}_2, \ldots, \mathbf{x}_k\}$ and $S' = \{\mathbf{y}_1, \mathbf{y}_2, \ldots, \mathbf{y}_m\}$. We must show that $k = m$.
Define a mapping $T: R^k \to R^n$ by

$$T(c_1, c_2, \ldots, c_k) = c_1\mathbf{x}_1 + c_2\mathbf{x}_2 + \cdots + c_k\mathbf{x}_k$$

Then:
   (a) T is a linear transformation.
   (b) The range of T is V.

The verifications of (a) and (b) are straightforward and are left as exercises. Since V is the range of T, and S' is contained in V, for each $i = 1, 2, \ldots, m$, there is a vector $\mathbf{z}_i$ in R^k such that $T(\mathbf{z}_i) = \mathbf{y}_i$. Since S' is linearly independent, the set $\{\mathbf{z}_1, \mathbf{z}_2, \ldots, \mathbf{z}_m\}$ is a linearly independent subset of R^k by Exercise 13 of Section 3.1. Therefore, we have that $m \leq k$ by Theorem 3.2.1. If we reverse the roles of S and S' and repeat the argument above, we have that $k \leq m$. Therefore, $k = m$, and S and S' have the same number of vectors. ∎

In view of this theorem, the number of vectors in any basis for a subspace V of R^n depends only on V.

Definition   The unique number of vectors in any basis for a subspace V of R^n is called the *dimension* of V. We denote this number by *dim V*.

For example, because R^n has a basis consisting of n vectors, it has dimension n; that is, dim $R^n = n$. At the other extreme, the dimension of the zero subspace is zero because its only basis is the empty set which has zero members.

Theorem 3.2.7   *Let V be a subspace of R^n. Let W be any subspace of R^n which is contained in V. Then:*
(a) *dim $W \leq$ dim V.*
(b) *If dim $W =$ dim V, then $W = V$.*

Proof

(a) Let S be a basis for W. Then S is a linearly independent subset of V. So, by Theorem 3.2.3, we can extend S to a basis S' for V. Since the number of vectors in S is less than or equal to the number of vectors in S', we have (a).

(b) Suppose that dim $W =$ dim V. Let S be a basis for W. As above, extend S to a basis S' for V. Since S is a subset of S', and since S and S' contain the same number of vectors, we have $S = S'$. Therefore, $W = V$. ∎

The following result generalizes Corollary 3.2.5.

Corollary 3.2.8   *Let V be a subspace of R^n of dimension k. Any linearly independent subset of V consisting of k vectors is a basis for V.*

Proof

Let S be a linearly independent subset of V consisting of k vectors. Let W be the span of S. Then dim $W =$ dim $V = k$, and W is contained in V. So, by Theorem 3.2.7, $W = V$. Therefore, S is a basis for V. ∎

Compare the following corollary with Theorem 3.2.1.

Corollary 3.2.9   *Let V be a subspace of R^n of dimension equal to k. Any linearly independent subset of V contains at most k vectors.*

Proof

Let S be a linearly independent subset of V. Let W be the span of S. Then S is a basis for W, and W is contained in V. So, by Theorem 3.2.7, the number of vectors of S is at most k. ∎

Example 6   Describe all subspaces of R^3.

We can classify the subspaces of R^3 by dimension. By Theorem 3.2.7 the dimension of any subspace of R^3 is at most 3.

Subspaces of dimension 0: The only subspace of dimension 0 is $\{(0, 0, 0)\}$.

Subspaces of dimension 1: Any subspace of dimension 1 is spanned by a single nonzero vector. Geometrically, this subspace is a line that passes through the origin (see Section 1.2).

Subspaces of dimension 2: Any subspace V of dimension 2 is spanned by a linearly independent set $\{\mathbf{x}, \mathbf{y}\}$ of two vectors. So

$$V = \{s\mathbf{x} + t\mathbf{y} : s, t \text{ are in } R\}$$

Geometrically, this set is a plane that contains the origin.

Subspaces of dimension 3: By Theorem 3.2.7, the only subspace of R^3 of dimension equal to 3 is R^3 itself.

Example 7   Apply the preceding theorems on dimension to show that $S = \{(1, 2, 0), (1, 3, 1)\}$ is a basis for the subspace

$$V = \{(x, y, z) : 2x - y + z = 0\}$$

Clearly, the vectors of S lie in V and are linearly independent. Since the equation defining V is the equation of a plane, V has dimension 2. Thus by Corollary 3.2.8, S is a basis for V.

We have used Corollary 3.2.8 to show that a linearly independent set is a basis for a subspace. The next corollary considers the same problem for spanning sets.

Corollary 3.2.10

Let V be a subspace of R^n of dimension k. Any spanning set for V consisting of k vectors is a basis for V.

Proof

Suppose that a set S which consists of k vectors spans V. By Corollary 3.1.3 there exists a subset S' of S such that S' is a basis for V. Therefore, since dim $V = k$, we have that S' contains k vectors. But S consists of k vectors, and S' is a subset of S. Therefore, $S' = S$. So S is a basis for V. ∎

Example 8   Extend

$$S = \{(0, 1, 2, 1, 3), (0, 1, 2, 1, 4), (0, 2, 3, 1, -1)\}$$

to a basis for R^5.

First form the matrix A whose rows are the vectors of S,

$$A = \begin{bmatrix} 0 & 1 & 2 & 1 & 3 \\ 0 & 1 & 2 & 1 & 4 \\ 0 & 2 & 3 & 1 & -1 \end{bmatrix}$$

By Theorem 3.1.5, the space spanned by the rows of any matrix obtained from A by elementary row operations is identical to the span of S. Therefore, transforming A into a matrix B in triangular form, we have that the rows of

$$B = \begin{bmatrix} 0 & 1 & 2 & 1 & 3 \\ 0 & 0 & -1 & -1 & -7 \\ 0 & 0 & 0 & 0 & 1 \end{bmatrix}$$

span the same space that is spanned by S. By Theorem 3.1.6, the rows of B are linearly independent. We wish to extend B to a larger 5×5 matrix whose rows are linearly independent. Theorem 3.1.6 provides us with a way to do this. Simply adjoin two more nonzero rows to B to obtain a matrix which is also in triangular form. In this case the rows $(1, 0, 0, 0, 0)$ and $(0, 0, 0, 1, 0)$ will do. Thus,

$$B' = \begin{bmatrix} 1 & 0 & 0 & 0 & 0 \\ 0 & 1 & 2 & 1 & 3 \\ 0 & 0 & -1 & -1 & -7 \\ 0 & 0 & 0 & 1 & 0 \\ 0 & 0 & 0 & 0 & 1 \end{bmatrix}$$

is in triangular form. By Theorem 3.1.6, these rows are linearly independent. Therefore, by Corollary 3.2.5, the rows of B' form a basis for R^5. This basis was obtained by adjoining the vectors $(1, 0, 0, 0, 0)$ and $(0, 0, 0, 1, 0)$ to the three rows of B. Since the rows of B were obtained by performing elementary row operations on A, it follows that the 5×5 matrix whose rows consist of the vectors of S and the vectors $(1, 0, 0, 0, 0)$ and $(0, 0, 0, 1, 0)$ can be transformed by means of elementary row operations into B'. Therefore, the set

$$\{(0, 1, 2, 1, 3), (0, 1, 2, 1, 4), (0, 2, 3, 1, -1), (1, 0, 0, 0, 0), (0, 0, 0, 1, 0)\}$$

also spans R^5. So, by Corollary 3.2.10, this set is a basis for R^5.

Compare the next corollary with Corollary 3.2.9.

Corollary 3.2.11

Let V be a subspace of R^n of dimension equal to k. Then any subset of V which spans V contains at least k vectors.

The proof of Corollary 3.2.11 is omitted and left as an exercise.

Exercises

1. Without doing any computation, give reasons why:
   (a) $\{(1, 2, 1, 2), (1, 2, 3, 5), (2, 4, 5, -1)\}$ does not span R^4.
   (b) $\{(1, 3, 1, 1), (1, 2, 5, 1), (0, 1, 0, 2),$ $(3, 1, -1, 1), (2, 4, -6, 8)\}$ is not linearly independent.
   (c) $\{(1, 2, 1), (1, 0, 3), (2, 3, 6), (8, 2, 5)\}$ is not a basis for R^3.

2. For each linearly independent subset S, extend S to a basis for the subspace V.
   (a) $S = \{(1, 2, 1), (1, -1, 1)\}$, $V = R^3$
   (b) $S = \{(1, 1, 1)\}$, $V = \{(x, y, z) : x - 2y + z = 0\}$
   (c) $S = \{(1, -1)\}$, $V = \{(x, y) : x + y = 0\}$
   (d) $S = \varnothing$, the empty set, $V = R^5$
   (e) $S = \{(1, 1, 2, 1), (0, 1, 1, 0)\}$, $V = R^4$
   (f) $S = \{(1, 1, 2, 1), (0, 1, 1, 0)\}$, $V = \{(x, y, z, w) : x = w\}$

3. For each of the parts of Exercise 4 of Section 3.1
   (i) Find the dimension of the span of S.
   (ii) Find the dimension of V.
   (iii) Use parts (i) and (ii) to do Exercise 4 of Section 3.1.

4. Find a basis for and the dimension of each subspace V.
   (a) $V = \{(x, y, z) : x + y - 3z = 0\}$
   (b) $V = \{(x, y, z) : x + y + z = 0$ and $x + 2y + 3z = 0\}$
   (c) $V = \{(x, y, z, w) : x + 2y - z + 3w = 0\}$
   (d) $V = \{(x, y, z, w) : x = 2y,$ $y = 3z,$ and $z = 4w\}$
   (e) $V = \{(x, y, z, w, t) : x = t$ and $z = w\}$

5. For any nonzero vector $\mathbf{y} = (a_1, a_2, \ldots, a_n)$ in R^n, let $V = \{\mathbf{x}$ in $R^n : \mathbf{y} \cdot \mathbf{x} = 0\}$.

   (a) Show that V is a subspace of R^n.
   (b) Prove that dim $V = n - 1$.
   (c) Find a basis for V.

6. Let W be a subspace of R^n. Consider any vector $\mathbf{y}$ in R^n but not in W. Let $V = \{s\mathbf{y} + \mathbf{x} : \mathbf{x}$ is in W and s is a scalar$\}$.
   (a) Prove that V is a subspace of R^n.
   (b) Prove that dim $V = \dim W + 1$.

7. Prove Corollary 3.2.11.

8. For the mapping $T : R^k \to R^n$ as defined in the proof of Theorem 3.2.6:
   (a) Show that T is a linear transformation.
   (b) Show that the range of T is V.

9. Let $T : R^n \to R^m$ be a linear transformation and suppose that V is a subspace of R^n. Let $T(V) = \{T(\mathbf{x}) : \mathbf{x}$ is in $V\}$. Recall from Exercise 16 of Section 2.5 that $T(V)$ is a subspace of R^m. Prove that if T is one-to-one, then
   (a) If $\{\mathbf{x}_1, \mathbf{x}_2, \ldots, \mathbf{x}_k\}$ is a basis for the subspace V of R^n, then $\{T(\mathbf{x}_1), T(\mathbf{x}_2), \ldots, T(\mathbf{x}_k)\}$ is a basis for $T(V)$. (*Hint*: See Exercise 17 of Section 3.1.)
   (b) dim $T(V) = \dim V$.

*10. Use the program ROW REDUCTION and the method of Example 8 to extend the given linearly independent set to a basis for R^n.
   (a) $\{(1, 0, 4)\}$
   (b) $\{(1, 3, 2, 1, 1), (2, 6, -1, 2, 0)\}$
   (c) $\{(3, 2, 1, 1, 5), (4, 1, 1, 2, 1), (1, 0, 1, 1, 1)\}$

* This problem should be solved by using one of the programs noted in Appendix B.

3.3 RANK AND NULLITY OF MATRICES AND LINEAR TRANSFORMATIONS

In this section the concept of dimension is applied to linear transformations and matrices. In Chapter 2 we studied two special kinds of subspaces: null spaces and ranges. The dimensions of these subspaces give useful information about linear transformations and matrices.

Definition   Let $T: R^n \to R^m$ be a linear transformation. We define the **rank** of T, denoted **rank** T, as the dimension of the range of T. Symbolically, we write

$$\text{rank } T = \dim R(T)$$

Definition   Let A be an $m \times n$ matrix. We define the **rank** of A, denoted **rank** A, as the rank of the linear transformation L_A, the transformation determined by left-multiplication by A. Symbolically, we write

$$\text{rank } A = \text{rank } L_A$$

There is an easy way to describe the rank of a linear transformation or matrix. If A is the standard matrix of a linear transformation $T: R^n \to R^m$, then, by Corollary 2.5.6, we have that $R(T)$ is the subspace of R^m consisting of the span of the columns of A. So the rank of T or A is the dimension of this subspace. We state this result formally.

Theorem 3.3.1   *For an $m \times n$ matrix A, the rank of A is the dimension of the subspace of R^m spanned by the columns of A.*

Example 1   Let $T: R^4 \to R^3$ be defined by

$$T(x, y, z, w) = (x + 2y + z + w, 2x - y + z + 3w, x - 3y + 2w)$$

Find the rank of T.

The standard matrix of T is

$$A = \begin{bmatrix} 1 & 2 & 1 & 1 \\ 2 & -1 & 1 & 3 \\ 1 & -3 & 0 & 2 \end{bmatrix}$$

So $R(T)$ is the span of the columns of A, that is, the span of the set

$$\{(1, 2, 1), (2, -1, -3), (1, 1, 0), (1, 3, 2)\}$$

It is easy to show that $\{(1, 2, 1), (0, 1, 1)\}$ is a basis for this span. Therefore, $R(T)$ has dimension 2. Hence, rank T = rank A = 2.

Example 2 Let $T_0 : R^n \to R^m$ be the zero transformation, that is, $T_0(\mathbf{x}) = \mathbf{0}$ for all $\mathbf{x}$ in R^n. Then $R(T_0)$ is the zero subspace of R^n. Thus, rank $T_0 = 0$.

Example 3 Let $I : R^n \to R^n$ be the identity transformation, defined by $I(\mathbf{x}) = \mathbf{x}$ for all $\mathbf{x}$ in R^n. Then $R(I) = R^n$. Therefore, rank $I = \dim R^n = n$.

The other subspace associated with a linear transformation is its null space.

Definition Let $T : R^n \to R^m$ be a linear transformation. We define the **nullity** of T, denoted **nullity** T, as the dimension of the null space of T. Symbolically,

$$\text{nullity } T = \dim N(T)$$

Definition Let A be a matrix. We define the **nullity** of A, denoted **nullity** A, as the nullity of L_A. Symbolically,

$$\text{nullity } A = \text{nullity } L_A$$

From this definition it follows that the nullity of A is the dimension of the subspace $\{\mathbf{x} : A\mathbf{x} = \mathbf{0}\}$.

Example 4 Consider the linear transformation $T : R^4 \to R^3$ defined in Example 1.

$$T(x, y, z, w) = (x + 2y + z + w, 2x - y + z + 3w, x - 3y + 2w)$$

Then $N(T) = \{(x, y, z, w) : T(x, y, z, w) = \mathbf{0}\}$. We will find the nullity of T by first obtaining a basis for $N(T)$. If we set $T(x, y, z, w) = \mathbf{0}$, we obtain the homogeneous system of linear equations

$$x + 2y + z + w = 0$$
$$2x - y + z + 3w = 0$$
$$x - 3y + 2w = 0$$

Using Gaussian elimination to solve this system, we have that a vector (x, y, z, w) is a solution to this system if and only if it is of the form

$$(x, y, z, w) = \left(-\frac{3}{5}s - \frac{7}{5}t, \ -\frac{1}{5}s + \frac{1}{5}t, \ s, \ t \right)$$

$$= s\left(-\frac{3}{5}, -\frac{1}{5}, 1, 0 \right) + t\left(-\frac{7}{5}, \frac{1}{5}, 0, 1 \right)$$

Thus, the null space of T is the span of

$$\left\{ \left(-\frac{3}{5}, -\frac{1}{5}, 1, 0 \right), \left(-\frac{7}{5}, \frac{1}{5}, 0, 1 \right) \right\}$$

Since this set is linearly independent, it is a basis for $N(T)$. Therefore,

$$\text{nullity } T = \dim N(T)$$

$$= 2$$

Example 5   Let $T_0: R^n \to R^m$ be the zero transformation, as in Example 2. Then $N(T) = R^n$, and therefore nullity $T = n$.

Example 6   Let $I: R^n \to R^n$ be the identity transformation. Then $N(T) = \{\mathbf{0}\}$, and therefore nullity $T = 0$.

The rank and nullity of a linear transformation T can be used to measure the extent to which T is onto or one-to-one.

Theorem 3.3.2   *Let $T: R^n \to R^m$ be a linear transformation. Then,*
(a) *T is one-to-one if and only if nullity $T = 0$.*
(b) *T is onto if and only if rank $T = m$.*

Proof
(a) By Theorem 2.5.3, T is one-to-one if and only if $N(T) = \{\mathbf{0}\}$. Since $\{\mathbf{0}\}$ is the only subspace of R^n of dimension zero, we have the result.
(b) By definition, T is onto if and only if $R(T) = R^m$. Since the only subspace of R^m of dimension m is R^m itself, we have the result. ■

For example, the linear transformation T of Examples 1 and 4 has rank 2. Therefore, since $m = 3$, we have that T is not onto. Since nullity $T = 2$, we have that T is not one-to-one.

There is a relationship between the rank and the nullity of a linear transformation which allows us to compute one of these if the other one is known. The nullity of a linear transformation is a measure of how much is "lost" (carried to zero) when the transformation is applied. The rank is a measure of how much "survives." One might conjecture that the sum of the rank and nullity is a measure of the original size (the dimension) of the space. Consider the following example.

Example 7   Let $T: R^3 \to R^2$ be defined by $T(x, y, z) = (0, z)$. It is easy to verify that T is linear, the range of T is the y-axis in R^2, and the null space of T is the xy-plane in R^3 (see Figure 3.3.1). Clearly, rank $T = 1$ and nullity $T = 2$. Notice that 1 and 2 sum to 3, the dimension of R^3.

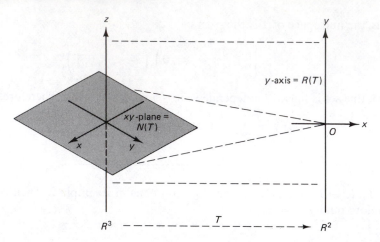

Figure 3.3.1

Theorem 3.3.3 *(Dimension theorem for linear transformations)* Let $T: R^n \to R^m$ be a linear transformation. Then

$$\text{rank } T + \text{nullity } T = n$$

Corollary 3.3.4 *(Dimension theorem for matrices)* For any $m \times n$ matrix A

$$\text{rank } A + \text{nullity } A = n$$

Proof of Theorem 3.3.3
Let $\{\mathbf{x}_1, \mathbf{x}_2, \ldots, \mathbf{x}_k\}$ be a basis for $N(T)$. By Theorem 3.2.3, this set can be extended to a basis

$$\{\mathbf{x}_1, \mathbf{x}_2, \ldots, \mathbf{x}_k, \mathbf{x}_{k+1}, \ldots, \mathbf{x}_n\}$$

for R^n. By Theorem 2.5.4, the set of images

$$\{T(\mathbf{x}_1), T(\mathbf{x}_2), \ldots, T(\mathbf{x}_k), T(\mathbf{x}_{k+1}), \ldots, T(\mathbf{x}_n)\}$$

spans $R(T)$. But $T(\mathbf{x}_1) = T(\mathbf{x}_2) = \cdots = T(\mathbf{x}_k) = \mathbf{0}$, and therefore, the set $S = \{T(\mathbf{x}_{k+1}), \ldots, T(\mathbf{x}_n)\}$ spans $R(T)$. We shall show that S is linearly independent. Consider any scalars $c_{k+1}, c_{k+2}, \ldots, c_n$ such that

$$c_{k+1} T(\mathbf{x}_{k+1}) + \cdots + c_n T(\mathbf{x}_n) = \mathbf{0}$$

Since T is linear, we have

$$T(c_{k+1}\mathbf{x}_{k+1} + \cdots + c_n\mathbf{x}_n) = \mathbf{0}$$

Therefore, the vector $c_{k+1}\mathbf{x}_{k+1} + \cdots + c_n\mathbf{x}_n$ lies in $N(T)$. Since $\{\mathbf{x}_1, \mathbf{x}_2, \ldots, \mathbf{x}_k\}$ is a basis for $N(T)$, there are scalars $c_1, c_2, \ldots, c_k$ such that

$$c_1\mathbf{x}_1 + \cdots + c_k\mathbf{x}_k = c_{k+1}\mathbf{x}_{k+1} + \cdots + c_n\mathbf{x}_n$$

So

$$c_1\mathbf{x}_1 + \cdots + c_k\mathbf{x}_k - c_{k+1}\mathbf{x}_{k+1} - \cdots - c_n\mathbf{x}_n = \mathbf{0}$$

Since $\{\mathbf{x}_1, \mathbf{x}_2, \ldots, \mathbf{x}_n\}$ is a basis for R^n, the coefficients must all be zero. Therefore,

$$c_1 = c_2 = \cdots = c_k = -c_{k+1} = \cdots = -c_n = 0$$

In particular, $c_i = 0$ for $i = k + 1, \ldots, n$. Hence, S is linearly independent. It follows that $S = \{T(\mathbf{x}_{k+1}), \ldots, T(\mathbf{x}_n)\}$ is a basis for $R(T)$. Notice that this set consists of $n - k$ vectors. So we have that

$$\text{rank } T = \dim R(T)$$

$$= n - k$$

Thus,

$$\text{rank } T + \text{nullity } T = (n - k) + k$$

$$= n \qquad \blacksquare$$

If we apply Theorem 3.3.3 to the linear transformation L_A, we then obtain Corollary 3.3.4.

For an additional illustration of this result, consider the linear transformation T of Examples 1 and 4. In this case we have that rank $T = 2$, nullity $T = 2$, and $n = 4$. So rank T + nullity $T = n$.

There are some important consequences of this theorem which are easy to establish.

Corollary 3.3.5

Let $T: R^n \to R^m$ be a linear transformation. Then:

(a) If $m > n$, then T is not onto.

(b) If $m < n$, then T is not one-to-one.

(c) If $m = n$, then T is one-to-one if and only if T is onto.

Proof

(a) Suppose that $m > n$. Then

$$\text{rank } T \leq \text{rank } T + \text{nullity } T$$

$$= n$$

$$< m$$

Therefore, rank $T < m$, so we can conclude that T is not onto.

(b) Suppose that $m < n$. Since $R(T)$ is a subspace of R^m, we have that rank $T \leq m$. Therefore, $m - $ rank $T \geq 0$. So, by Theorem 3.3.3, we have

$$\text{nullity } T = n - \text{rank } T$$

$$> m - \text{rank } T$$

$$\geq 0$$

Therefore, nullity $T > 0$. Thus, by Theorem 3.3.2(a), we have that T is not one-to-one. (c) Suppose that $m = n$. Then T is onto if and only if rank $T = n$ if and only if nullity $T = 0$ (Theorem 3.3.3) if and only if T is one-to-one.   ∎

This corollary has a number of interesting consequences. For example, from part (a) we may conclude that there does not exist *any* linear transformation of R^2 onto R^3. That is, one cannot "cover" a three-dimensional space by a two-dimensional space. By part (b) we may conclude that there does not exist *any* one-to-one linear transformation from R^3 into R^2. That is, one cannot "squeeze" a three-dimensional vector space into a two-dimensional vector space without mapping distinct vectors onto a common image. These ideas are depicted visually in Figure 3.3.2.

Example 8   For the linear transformation $T: R^3 \to R^2$ defined by

$$T(x, y, z) = (x + 2y - z, x + y + z)$$

use the dimension theorem to obtain a basis for $N(T)$.

We first determine the dimension of $N(T)$. Since $T(1, 0, 0) = (1, 1)$ and $T(0, 1, 0) = (2, 1)$, and since $\{(1, 1), (2, 1)\}$ is linearly independent, we have that rank $T = 2$. By the dimension theorem it follows that nullity $T = 3 - 2 = 1$. Since $N(T)$

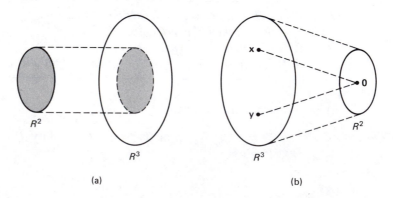

(a)   (b)

Figure 3.3.2

has dimension 1, any nonzero vector in $N(T)$ is a basis for $N(T)$. So we seek any nontrivial solution to the system of equations

$$x + 2y - z = 0$$
$$x + \ y + z = 0$$

By Gaussian elimination, we discover that $(-3, 2, 1)$ is a solution. Since this vector is a nonzero vector of $N(T)$, $\{(-3, 2, 1)\}$ is a basis for $N(T)$.

By Theorem 3.3.1, the rank of an $m \times n$ matrix A is the dimension of the subspace of R^m spanned by the columns of A. This subspace is called the *column space* of A. Notice that the column space of A coincides with the range of L_A. Similarly, the rows of A span a subspace of R^n called the *row space* of A. These two subspaces are quite distinct. Indeed, if $m \neq n$, the spaces have no vectors in common. However, as we shall see (Theorem 3.3.7), their dimensions are equal.

Example 9   Describe the column and row spaces of the matrices

$$A = \begin{bmatrix} 1 & 2 & 3 \\ 4 & 5 & 6 \end{bmatrix} \qquad B = (1 \quad 2 \quad 3 \quad 1) \qquad C = \begin{bmatrix} 1 & 0 & 0 \\ 0 & 1 & 0 \\ 0 & 0 & 1 \end{bmatrix}$$

Since the first two columns of A are linearly independent, the column space of A is all of R^2. Since the rows of A are linearly independent, the row space of A is the two-dimensional subspace of R^3 spanned by the rows.

Clearly, the columns of B span R, and the row of B spans a one-dimensional subspace of R^4.

The columns of C consist of the standard basis of R^3. Thus, the column space of C is R^3. The same argument for the rows of C leads to the same conclusion.

In what follows, we lay the groundwork for the proof of Theorem 3.3.7, which states that the row space and the column space of a matrix have the same dimension. Since the rows of a matrix A are the columns of A^t, the transpose operation plays an important role in establishing this result. We shall also take advantage of a simple and yet useful relationship between the transpose and the dot product. Let $\mathbf{x}$ and $\mathbf{y}$ be column vectors in R^n. Then $\mathbf{x}^t$ is a row vector, and the matrix product $(\mathbf{x}^t)\mathbf{y}$ is a 1×1 matrix. This matrix can be identified with its single entry and, as such, is simply the dot product $\mathbf{x} \cdot \mathbf{y}$ (see Exercise 8). Thus we have that

$$\mathbf{x} \cdot \mathbf{y} = \mathbf{x}^t \mathbf{y}$$

We begin with an application of this fact to establish a result that is useful in Chapter 7 as well as here.

Theorem 3.3.6 *Let A be an m × n matrix. Then:*
(a) $A\mathbf{x} \cdot \mathbf{y} = \mathbf{x} \cdot A^t\mathbf{y}$ *for all vectors* $\mathbf{x}$ *in* R^n *and* $\mathbf{y}$ *in* R^m.
(b) rank $A^t A$ = rank A.

Proof
Part (a) follows from the equations

$$(A\mathbf{x}) \cdot \mathbf{y} = (A\mathbf{x})^t\mathbf{y} = (\mathbf{x}^t A^t)\mathbf{y} = \mathbf{x}^t(A^t\mathbf{y}) = \mathbf{x} \cdot (A^t\mathbf{y})$$

(b) By the dimension theorem, part (b) is equivalent to the statement that nullity $A^t A$ = nullity A. To prove this, we shall show that for any vector $\mathbf{x}$ in R^n, we have that $A^t A\mathbf{x} = \mathbf{0}$ if and only if $A\mathbf{x} = \mathbf{0}$. It is clear that if $A\mathbf{x} = \mathbf{0}$, then $A^t A\mathbf{x} = \mathbf{0}$. So assume that $A^t A\mathbf{x} = \mathbf{0}$. From part (a) we have

$$0 = A^t A\mathbf{x} \cdot \mathbf{x} = A\mathbf{x} \cdot A^{tt}\mathbf{x} = A\mathbf{x} \cdot A\mathbf{x} = \|A\mathbf{x}\|^2$$

Therefore, $A\mathbf{x} = \mathbf{0}$. ■

We now apply Theorem 3.3.6 to compare the ranks of a matrix and its transpose.

Theorem 3.3.7 *For any matrix A,* rank A = rank A^t. *Hence,* rank A = dim (*row space of A*) = dim (*column space of A*)

Proof
By part (b) of Theorem 3.3.6, rank A = rank $A^t A$. By Exercise 14, we have that rank $A^t A \leq$ rank A^t. Therefore,

$$\text{rank } A \leq \text{rank } A^t$$

Since $A^{tt} = A$, we may apply this inequality to A^t to obtain

$$\text{rank } A^t \leq \text{rank } A$$

We conclude that rank A = rank A^t. Since the rows of A are the columns of A^t, we have that the row space of A is the column space of A^t. Therefore,

$$\text{dim (row space of } A) = \text{dim (column space of } A^t) = \text{rank } A^t = \text{rank } A$$
$$= \text{dim (column space of } A) \quad ■$$

By Theorem 3.1.5, the elementary row operations preserve the row space of a matrix. By Theorem 3.3.7, the dimension of the row space is the rank of the matrix. Therefore, we have the following result.

Corollary 3.3.8   *Elementary row operations preserve the rank of a matrix.*

Although elementary row operations preserve the dimension of the column space of a matrix, they do not necessarily preserve the column space itself. For example, for the 2×2 matrix

$$A = \begin{bmatrix} 1 & 0 \\ 1 & 0 \end{bmatrix}$$

can be obtained from the matrix

$$B = \begin{bmatrix} 1 & 0 \\ 0 & 0 \end{bmatrix}$$

by means of an elementary row operation. However, the column spaces of A and B are quite distinct.

Since the nonzero rows of a matrix in triangular form are linearly independent (Theorem 3.1.6), we have the following result.

Corollary 3.3.9
If a matrix is in triangular form, then the number of nonzero rows of the matrix is equal to its rank.

Corollaries 3.3.8 and 3.3.9 provide us with a method of computing the rank of a matrix. First, use elementary row operations to transform the matrix into triangular form. By Corollary 3.3.8, the transformed matrix has the same rank as the original one. Then, by Corollary 3.3.9, the number of nonzero rows of the transformed matrix equals the rank.

Example 10   Use elementary row operations to find the rank of the matrix

$$A = \begin{bmatrix} 0 & 0 & 1 & 3 & 3 \\ 2 & 3 & 1 & 5 & 2 \\ 4 & 6 & 1 & 7 & 2 \\ 4 & 6 & 1 & 7 & 1 \end{bmatrix}$$

By means of a succession of elementary row operations we transform A into a matrix in triangular form.

$$A = \begin{bmatrix} 0 & 0 & 1 & 3 & 3 \\ 2 & 3 & 1 & 5 & 2 \\ 4 & 6 & 1 & 7 & 2 \\ 4 & 6 & 1 & 7 & 1 \end{bmatrix} \rightarrow \begin{bmatrix} 2 & 3 & 1 & 5 & 2 \\ 0 & 0 & 1 & 3 & 3 \\ 4 & 6 & 1 & 7 & 2 \\ 4 & 6 & 1 & 7 & 1 \end{bmatrix} \rightarrow \begin{bmatrix} 2 & 3 & 1 & 5 & 2 \\ 0 & 0 & 1 & 3 & 3 \\ 0 & 0 & -1 & -3 & -2 \\ 0 & 0 & -1 & -3 & -3 \end{bmatrix}$$

$$\rightarrow \begin{bmatrix} 2 & 3 & 1 & 5 & 2 \\ 0 & 0 & 1 & 3 & 3 \\ 0 & 0 & 0 & 0 & 1 \\ 0 & 0 & 0 & 0 & 0 \end{bmatrix}$$

This last matrix is in triangular form. Since it has three nonzero rows, its rank is 3, and therefore, rank $A = 3$.

Using the matrix A of Example 10, we can draw some conclusions.

(a) Consider the transformation $L_A: R^5 \to R^4$. Since rank $L_A = 3$, it is not onto. Notice that we cannot use Corollary 3.3.5(a) to draw this conclusion.
(b) There is a vector $\mathbf{b}$ in R^4 for which the system $A\mathbf{x} = \mathbf{b}$ is inconsistent.
(c) Nullity of L_A equals 2.
(d) The maximum number of linearly independent vectors in the solution space of $A\mathbf{x} = \mathbf{0}$ is two.
(e) L_A is not one-to-one. [Notice that in contrast to part (a), we can use Corollary 3.3.5(b) to draw this conclusion.]

Example 11   Find the rank and nullity of the linear transformation $T: R^4 \to R^3$ defined by

$$T(x, y, z, w) = (x + 2y + z - w, \ x + 5y + 2z + w, \ -x + y + 3w)$$

Let A be the standard matrix of T. Then

$$A = \begin{bmatrix} 1 & 2 & 1 & -1 \\ 1 & 5 & 2 & 1 \\ -1 & 1 & 0 & 3 \end{bmatrix}$$

Then rank $T = $ rank A. We can use elementary row operations to transform A into the following matrix, which is in triangular form:

$$\begin{bmatrix} 1 & 2 & 1 & -1 \\ 0 & 3 & 1 & 2 \\ 0 & 0 & 0 & 0 \end{bmatrix}$$

Therefore, rank $A = 2$. We conclude that rank $T = 2$. So, by the dimension theorem, nullity $T = 4 - 2 = 2$.

We can define elementary column operations in a manner identical to elementary row operations. Compare the following definition to the one for elementary row operations given in Chapter 1.

Definition   An *elementary column operation* on a matrix is an operation that results in a new matrix in one of three ways:

1. Interchange two columns of the matrix.
2. Multiply a column of the matrix by a nonzero scalar.

3. Add a multiple of one of the columns of the matrix to another column of the matrix.

An elementary column operation is of *type 1*, *2*, or *3* depending on whether it is obtained by operation 1, 2, or 3.

Since the columns of a matrix are the same as the rows of its transpose, we can establish results analogous to Theorem 3.1.5 and Corollary 3.3.8 for columns. We state the result without proof (see Exercise 19).

Theorem 3.3.10   *If a matrix B is obtained from A by means of a sequence of elementary column operations, then B has the same column space as A. Therefore, A and B have the same rank.*

Exercises

1. Find the rank and nullity of each matrix.

   (a) $\begin{bmatrix} 1 & 2 & 1 \\ 2 & 1 & 3 \\ 1 & 2 & 4 \end{bmatrix}$

   (b) $\begin{bmatrix} 1 & 2 & 3 & 4 \\ 2 & 4 & 6 & 8 \end{bmatrix}$

   (c) $\begin{bmatrix} 1 & 2 & 1 & -1 \\ 2 & 1 & 1 & 0 \\ 1 & -7 & -2 & 5 \end{bmatrix}$

   (d) $\begin{bmatrix} 1 & 2 & 3 \\ 4 & 5 & 6 \\ 7 & 8 & 9 \end{bmatrix}$

   (e) $\begin{bmatrix} 1 & 2 & 3 & 4 \\ 5 & 6 & 7 & 8 \\ 9 & 10 & 11 & 12 \\ 13 & 14 & 15 & 16 \end{bmatrix}$

   (f) $\begin{bmatrix} 1 & 2 & 1 \\ 1 & 3 & -1 \\ 2 & 4 & 1 \\ 0 & 1 & 2 \end{bmatrix}$

2. Find the rank and nullity of each linear transformation T.

   (a) $T: R^3 \to R^2$, $T(x, y, z) = (x + y + z, x + y + z)$

   (b) $T: R^2 \to R^3$, $T(x, y) = (x + y, x - y, x)$

   (c) $T: R^3 \to R^3$, $T(x, y, z) = (x + 2y - z, x + y, y + z)$

   (d) $T: R^4 \to R^3$, $T(x, y, z, w) = (x + 2y, z - w, w)$

   (e) $T: R^3 \to R^3$, $T(x, y, z) = (x + y, y - 2z, x + 2z)$

   (f) $T: R^3 \to R$, $T(x, y, z) = x + 2y - z$

3. For each linear transformation T of Exercise 2, use your answer to decide if T is one-to-one or onto.

4. Find the rank and nullity of the linear transformation T from the information that is provided.

   (a) $T: R^n \to R^m$, and T is onto.

   (b) $T: R^n \to R^m$, and T is one-to-one.

   (c) $T: R^n \to R^{2n}$, and T is defined by $T(x_1, x_2, \ldots, x_n) = (x_1, x_2, \ldots, x_n, x_1, x_2, \ldots, x_n)$.

   (d) $T: R^n \to R^{n-1}$, and T is defined by $T(x_1, x_2, \ldots, x_n) = (x_1, x_2, \ldots, x_{n-1})$.

   (e) $T: R^n \to R^{n-1}$, and T is defined by $T(x_1, x_2, \ldots, x_n) = (x_1, x_1, \ldots, x_1)$.

   (f) $T: R^n \to R$, where $T(\mathbf{x}) = \mathbf{y} \cdot \mathbf{x}$, for some nonzero vector $\mathbf{y}$ in R^n.

   (g) $T: R^n \to R$, and T is defined by $T(x_1, x_2, \ldots, x_n) = x_1 + x_2 + \cdots + x_n$.

5. For the linear transformations in parts (c) through (g) of Exercise 4, use your answer to determine if T is one-to-one or onto.

6. (a) For any linear transformation $T: R^7 \to R^5$, deduce that T is not one-to-one.

   (b) For any linear transformation $T: R^5 \to R^8$, deduce that T is not onto.

7. Prove that for a linear transformation T from R^n to R^m:
   (a) If T is one-to-one, then rank $T = n$.
   (b) Use part (a) to argue that if T is one-to-one and onto, then $n = m$. [This is a partial converse to Corollary 3.3.5(c).]

8. Let $\mathbf{x}$ and $\mathbf{y}$ be $n \times 1$ column vectors. Show that the matrix product $\mathbf{x}^t\mathbf{y}$ is a 1×1 matrix whose entry is the dot product $\mathbf{x} \cdot \mathbf{y}$.

9. Prove that for any square matrix A, nullity $A =$ nullity A^t.

10. For any matrix A and any nonzero scalar c, prove that A and cA have the same rank.

11. Let $T: R^n \to R^m$ and $U: R^m \to R^p$ be linear transformations.
    (a) Prove that $N(T)$ is contained in $N(UT)$.
    (b) Use part (a) to prove that nullity $T \le$ nullity UT.
    (c) Use part (b) to prove that rank $UT \le$ rank T.

12. Use Exercise 11 to prove that if A and B are matrices such that AB is defined, then:
    (a) nullity $B \le$ nullity AB.
    (b) rank $AB \le$ rank B.
    (*Hint*: Apply Exercise 11, with $U = L_A$ and $T = L_B$.)

13. Let $T: R^n \to R^m$ and $U: R^m \to R^p$ be linear transformations.
    (a) Prove that $R(UT)$ is contained in $R(U)$.
    (b) Use part (a) to prove that rank $UT \le$ rank U.

14. Use Exercise 13 to prove that if A and B are matrices for which AB is defined, then rank $AB \le$ rank A.

15. Give examples of matrices A and B for which:
    (a) rank $(A + B) \ne$ (rank A) + (rank B).
    (b) rank $AB \ne$ (rank A)(rank B).

16. Show that the rank of any 2×2 rotation matrix is 2.

17. For an $m \times n$ matrix A recall the linear transformation $R_A: R^m \to R^n$ defined in Exercise 28 of Section 2.4. Prove that rank $R_A =$ rank A.

18. Prove that if matrices A and B have the same number of columns, then rank $A =$ rank B if and only if nullity $A =$ nullity B.

19. Prove Theorem 3.3.10.

*20. Use the program ROW REDUCTION to find the rank and nullity of each of the following matrices.

(a) $\begin{bmatrix} 2 & 4 & 5 & 7 & 8 \\ -1 & 2 & 4 & 6 & 9 \\ 6 & 4 & 2 & 2 & -2 \\ 3 & 2 & 1 & 1 & -1 \end{bmatrix}$

(b) $\begin{bmatrix} 0 & 1 & -1 & 3 & 7 & 2 \\ 2 & 1 & 6 & 4 & 2 & 0 \\ 3 & 1 & 0 & 9 & 9 & 2 \\ 3 & 4 & 16 & 9 & 11 & -2 \end{bmatrix}$

(c) $\begin{bmatrix} 1 & 2 & 3 & 4 & 5 \\ 6 & 7 & 8 & 9 & 10 \\ 11 & 12 & 13 & 14 & 15 \\ 16 & 17 & 18 & 19 & 20 \\ 21 & 22 & 23 & 24 & 25 \end{bmatrix}$

* This problem should be solved by using one of the programs noted in Appendix B.

3.4 INVERTIBLE MATRICES AND INVERTIBLE LINEAR TRANSFORMATIONS

For any real number $a \ne 0$ there is a unique scalar b, called the (multiplicative) inverse of a such that $ab = ba = 1$. For example, if $a = 2$, then $b = 0.5$. It is useful to consider this property in the context of matrices and linear transformations, where the matrix analog of the number 1 is the identity matrix I.

Definition   A square matrix A is **invertible** or **nonsingular** if there exists a square matrix B (of the same order as A) such that $AB = BA = I$.

Example 1   For the matrices

$$A = \begin{bmatrix} 1 & 2 \\ 3 & 5 \end{bmatrix} \quad \text{and} \quad B = \begin{bmatrix} -5 & 2 \\ 3 & -1 \end{bmatrix}$$

$$AB = \begin{bmatrix} 1 & 2 \\ 3 & 5 \end{bmatrix}\begin{bmatrix} -5 & 2 \\ 3 & -1 \end{bmatrix} = \begin{bmatrix} 1 & 0 \\ 0 & 1 \end{bmatrix}$$

and

$$BA = \begin{bmatrix} -5 & 2 \\ 3 & -1 \end{bmatrix}\begin{bmatrix} 1 & 2 \\ 3 & 5 \end{bmatrix} = \begin{bmatrix} 1 & 0 \\ 0 & 1 \end{bmatrix}$$

So A is invertible.

Example 2   Consider the matrix

$$A = \begin{bmatrix} 1 & 2 \\ 0 & 0 \end{bmatrix}$$

Then, for any matrix

$$B = \begin{bmatrix} a & b \\ c & d \end{bmatrix}$$

$$AB = \begin{bmatrix} 1 & 2 \\ 0 & 0 \end{bmatrix}\begin{bmatrix} a & b \\ c & d \end{bmatrix}$$

$$= \begin{bmatrix} a + 2c & b + 2d \\ 0 & 0 \end{bmatrix}$$

$$\neq \begin{bmatrix} 1 & 0 \\ 0 & 1 \end{bmatrix}$$

So A is not invertible.

As we have seen in Example 2, not every nonzero matrix is invertible. The next theorem tells us that when the matrix A is invertible, the matrix B in the definition is unique.

Theorem 3.4.1   *If A is an invertible $n \times n$ matrix, then there exists exactly one $n \times n$ matrix B such that $AB = BA = I$.*

Proof

Suppose that A is invertible and that there exist square matrices B and C such that $AB = BA = I$ and $AC = CA = I$. Then

$$B = BI = B(AC) = (BA)C = IC = C$$ ∎

In view of Theorem 3.4.1, we have the following definition.

Definition   Let A be an invertible matrix. Let B be the (unique) square matrix such that $AB = BA = I$. Then B is called the **inverse** of A, and it is denoted by A^{-1}.

Example 3   Recall the rotation matrix

$$A_\theta = \begin{bmatrix} \cos\theta & -\sin\theta \\ \sin\theta & \cos\theta \end{bmatrix}$$

For any nonzero vector $\mathbf{x}$ in R^2, we have that $A_\theta \mathbf{x}$ represents the vector obtained by rotating $\mathbf{x}$ counterclockwise by the angle θ. Furthermore, for any angle ϕ, we have

$$A_\theta A_\phi = A_{\theta+\phi} \qquad \text{and} \qquad A_0 = I$$

Therefore, for any angle θ

$$A_\theta A_{-\theta} = A_{\theta-\theta} = A_0 = I$$

Similarly,

$$A_{-\theta} A_\theta = I$$

Therefore,

$$A_{-\theta} = (A_\theta)^{-1}$$

For example, for $\theta = \pi/3$,

$$A_\theta = \begin{bmatrix} \frac{1}{2} & -\sqrt{3}/2 \\ \sqrt{3}/2 & \frac{1}{2} \end{bmatrix}$$

and

$$A_{-\theta} = \begin{bmatrix} \frac{1}{2} & \sqrt{3}/2 \\ -\sqrt{3}/2 & \frac{1}{2} \end{bmatrix}$$

As an exercise, verify that the product of these matrices in either order is I.

Notice that if a matrix A is invertible, then so is A^{-1}. In fact, A is the inverse of A^{-1}. That is, $(A^{-1})^{-1} = A$ (see Exercise 8). We list this result together with other elementary properties of matrix inverses.

Theorem 3.4.2   *Let A and B be invertible matrices of the same order. Then:*
   (a) *A^{-1} is invertible and $(A^{-1})^{-1} = A$.*
   (b) *AB is invertible and $(AB)^{-1} = B^{-1}A^{-1}$.*
   (c) *A^t is invertible and $(A^t)^{-1} = (A^{-1})^t$.*

Proof
The proof of part (a) is an exercise.
(b)

$$(AB)B^{-1}A^{-1} = A(BB^{-1})A^{-1}$$
$$= AIA^{-1}$$
$$= AA^{-1}$$
$$= I$$

Similarly, $B^{-1}A^{-1}(AB) = I$. Therefore, AB is invertible and has inverse $B^{-1}A^{-1}$.
(c) We must show that $(A^t)(A^{-1})^t = (A^{-1})^t(A^t) = I$. By Theorem 2.1.4 we have that

$$(A^t)(A^{-1})^t = (A^{-1}A)^t = I^t = I$$

Similarly, $(A^{-1})^t(A^t) = I$. Therefore, A^t is invertible, and $(A^{-1})^t = (A^t)^{-1}$ ∎

Example 4   From Example 1, we see that for the matrix

$$A = \begin{bmatrix} 1 & 2 \\ 3 & 5 \end{bmatrix}$$

we have

$$A^{-1} = \begin{bmatrix} -5 & 2 \\ 3 & -1 \end{bmatrix}$$

It is easy to verify that for the matrix

$$C = \begin{bmatrix} 1 & 1 \\ 1 & 3 \end{bmatrix}$$

we have

$$C^{-1} = \begin{bmatrix} 1.5 & -0.5 \\ -0.5 & 0.5 \end{bmatrix}$$

So

$$AC = \begin{bmatrix} 1 & 2 \\ 3 & 5 \end{bmatrix} \begin{bmatrix} 1 & 1 \\ 1 & 3 \end{bmatrix} = \begin{bmatrix} 3 & 7 \\ 8 & 18 \end{bmatrix}$$

and

$$C^{-1}A^{-1} = \begin{bmatrix} 1.5 & -0.5 \\ -0.5 & 0.5 \end{bmatrix} \begin{bmatrix} -5 & 2 \\ 3 & -1 \end{bmatrix} = \begin{bmatrix} -9 & 3.5 \\ 4 & -1.5 \end{bmatrix}$$

Finally, notice that

$$\begin{bmatrix} 3 & 7 \\ 8 & 18 \end{bmatrix} \begin{bmatrix} -9 & 3.5 \\ 4 & -1.5 \end{bmatrix} = \begin{bmatrix} 1 & 0 \\ 0 & 1 \end{bmatrix}$$

Similarly, the product in the reverse order is the identity matrix.

Part (b) of Theorem 3.4.2 can be extended to products of more than two matrices.

Corollary 3.4.3

Let $A_1, A_2, \ldots, A_k$ *be invertible matrices of the same order. Then their product* $A_1 A_2 \ldots A_k$ *is invertible and*

$$(A_1 A_2 \ldots A_k)^{-1} = (A_k)^{-1}(A_{k-1})^{-1} \cdots (A_1)^{-1}$$

Invertible Matrices and Systems of Linear Equations

Consider a system of n linear equations in n unknowns. Let A be the coefficient matrix for the system. Then the system can be represented as a single matrix equation

$$A\mathbf{x} = \mathbf{b}$$

If A is invertible, we can multiply both sides of the matrix equation above *on the left* by A^{-1} to solve for $\mathbf{x}$:

$$A^{-1}(A\mathbf{x}) = A^{-1}\mathbf{b}$$
$$(A^{-1}A)\mathbf{x} = A^{-1}\mathbf{b}$$
$$I\mathbf{x} = A^{-1}\mathbf{b}$$
$$\mathbf{x} = A^{-1}\mathbf{b}$$

We summarize this result.

Theorem 3.4.4 *Let $A\mathbf{x} = \mathbf{b}$ be a system of n equations in n unknowns, with an invertible coefficient matrix A. Then the system has the unique solution $\mathbf{x} = A^{-1}\mathbf{b}$.*

Example 5   Solve the system of linear equations

$$\begin{aligned} x + 2y + \ \ z &= 1 \\ x + \ \ y + 5z &= 2 \\ x + 2y + 2z &= 3 \end{aligned}$$

given the information that the matrix

$$A = \begin{bmatrix} 1 & 2 & 1 \\ 1 & 1 & 5 \\ 1 & 2 & 2 \end{bmatrix}$$

is invertible and that

$$A^{-1} = \begin{bmatrix} 8 & 2 & -9 \\ -3 & -1 & 4 \\ -1 & 0 & 1 \end{bmatrix}$$

The system can be represented by the single matrix equation

$$\begin{bmatrix} 1 & 2 & 1 \\ 1 & 1 & 5 \\ 1 & 2 & 2 \end{bmatrix} \begin{bmatrix} x \\ y \\ z \end{bmatrix} = \begin{bmatrix} 1 \\ 2 \\ 3 \end{bmatrix}$$

So, by Theorem 3.4.4, the system has the unique solution

$$\begin{bmatrix} x \\ y \\ z \end{bmatrix} = \begin{bmatrix} 8 & 2 & -9 \\ -3 & -1 & 4 \\ -1 & 0 & 1 \end{bmatrix} \begin{bmatrix} 1 \\ 2 \\ 3 \end{bmatrix}$$

$$= \begin{bmatrix} -15 \\ 7 \\ 2 \end{bmatrix}$$

Thus, $x = -15$, $y = 7$, and $z = 2$.

Of course, the method of Example 5 is useful only if the coefficient matrix is invertible. Furthermore, the inverse of the coefficient matrix must be available. This section is concerned with determining whether or not a matrix is invertible. The next theorem is a first step in this direction.

Theorem 3.4.5 *Let A be a (square) matrix of order n. If A is invertible, then* rank $A = n$.

Proof

Suppose that A is invertible. By the dimension theorem, we need only show that A has nullity equal to zero. Consider any vector $\mathbf{x}$ in R^n such that $A\mathbf{x} = \mathbf{0}$. Then

$$
\begin{aligned}
\mathbf{x} &= (A^{-1}A)\mathbf{x} \\
&= A^{-1}(A\mathbf{x}) \\
&= A^{-1}(\mathbf{0}) \\
&= \mathbf{0}
\end{aligned}
$$

Therefore, nullity $A = 0$. ∎

The converse of this theorem is also true. We shall establish its validity later in the section. However, we can apply Theorem 3.4.5 immediately to characterize 2×2 invertible matrices.

Theorem 3.4.6 *For a 2×2 matrix*

$$
A = \begin{bmatrix} a & b \\ c & d \end{bmatrix}
$$

A is invertible if and only if $ad - bc \neq 0$, in which case

$$
A^{-1} = \frac{1}{ad - bc} \begin{bmatrix} d & -b \\ -c & a \end{bmatrix}
$$

Proof

First suppose that $ad - bc \neq 0$. Then, for the matrix

$$
B = \frac{1}{ad - bc} \begin{bmatrix} d & -b \\ -c & a \end{bmatrix}
$$

we have that

$$
\begin{aligned}
BA &= \frac{1}{ad - bc} \begin{bmatrix} d & -b \\ -c & a \end{bmatrix} \begin{bmatrix} a & b \\ c & d \end{bmatrix} \\
&= \frac{1}{ad - bc} \begin{bmatrix} ad - bc & 0 \\ 0 & ad - bc \end{bmatrix} \\
&= I
\end{aligned}
$$

Similarly, $AB = I$. Therefore, A is invertible, and by the uniqueness of the inverse we have

$$A^{-1} = \frac{1}{ad - bc} \begin{bmatrix} d & -b \\ -c & a \end{bmatrix}$$

Now suppose that A is invertible. Then rank $A = 2$, by Theorem 3.4.5. Therefore, the rows of A form a linearly independent set. Since the columns are linearly independent also, the first column is not the zero vector. So one of its entries, a or c, is not zero. Since $\{(a, b), (c, d)\}$ is linearly independent, we must have

$$(0, 0) \neq a(c, d) - c(a, b) = (0, ad - bc)$$

Thus, $ad - bc \neq 0$. ∎

Example 6   In a particular elementary school, $\frac{1}{3}$ of the children who order a hot lunch on a school day will order a hot lunch on the next school day, while $\frac{2}{3}$ of them will bring a bag lunch. Furthermore, $\frac{1}{2}$ of the children who bring a bag lunch on a school day will bring a bag lunch on the next school day, while $\frac{1}{2}$ of them will order a hot lunch. If 40 children ordered hot lunches and 50 children brought bag lunches today, how many children ordered hot lunches and brought bag lunches yesterday?

Let x be the number of children who ordered hot lunches yesterday, and let y be the number of children who brought bag lunches yesterday. Then

$$\frac{1}{3} x + \frac{1}{2} y = 40$$

$$\frac{2}{3} x + \frac{1}{2} y = 50$$

This equation can be written in matrix form:

$$\begin{bmatrix} \frac{1}{3} & \frac{1}{2} \\ \frac{2}{3} & \frac{1}{2} \end{bmatrix} \begin{bmatrix} x \\ y \end{bmatrix} = \begin{bmatrix} 40 \\ 50 \end{bmatrix}$$

Let A be the coefficient matrix of the system. Then, by Theorem 3.4.6,

$$A^{-1} = \frac{1}{\frac{1}{6} - \frac{2}{6}} \begin{bmatrix} \frac{1}{2} & -\frac{1}{2} \\ -\frac{2}{3} & \frac{1}{3} \end{bmatrix}$$

$$= \begin{bmatrix} -3 & 3 \\ 4 & -2 \end{bmatrix}$$

So the system has the solution

$$\begin{bmatrix} x \\ y \end{bmatrix} = \begin{bmatrix} -3 & 3 \\ 4 & -2 \end{bmatrix} \begin{bmatrix} 40 \\ 50 \end{bmatrix}$$

$$= \begin{bmatrix} 30 \\ 60 \end{bmatrix}$$

Therefore, $x = 30$ and $y = 60$.

We now establish the converse of Theorem 3.4.5.

Theorem 3.4.7   *If A is a square matrix of order n, then A is invertible if and only if* rank $A = n$.

Proof
By Theorem 3.4.5, if A is invertible, then rank $A = n$. Now suppose that rank $A = n$. Then the linear transformation L_A is onto. Therefore, for each standard vector $\mathbf{e}_i$, there exists a vector $\mathbf{x}_i$ in R^n such that

$$A\mathbf{x}_i = L_A(\mathbf{x}_i) = \mathbf{e}_i$$

Let B be the $n \times n$ matrix whose ith column is $\mathbf{x}_i$. We show that B is the inverse of A. Using block multiplication, we have

$$\begin{aligned} AB &= A(\mathbf{x}_1, \mathbf{x}_2, \ldots, \mathbf{x}_n) \\ &= (A\mathbf{x}_1, A\mathbf{x}_2, \ldots, A\mathbf{x}_n) \\ &= (\mathbf{e}_1, \mathbf{e}_2, \ldots, \mathbf{e}_n) \\ &= I \end{aligned}$$

The proof that $BA = I$ is less direct. Since $\{\mathbf{e}_1, \mathbf{e}_2, \ldots, \mathbf{e}_n\}$ is linearly independent, so is $\{\mathbf{x}_1, \mathbf{x}_2, \ldots, \mathbf{x}_n\}$ by Exercise 13 of Section 3.1. Therefore, rank $B = n$. Repeating the argument above with B in place of A, we have an $n \times n$ matrix C such that $BC = I$. Thus,

$$A = AI = A(BC) = (AB)C = IC = C$$

Therefore, $BA = I$. Thus, A is invertible and $B = A^{-1}$. ∎

Corollary 3.4.8   *Let A be a square matrix. Then A is invertible if and only if the system of linear equations $A\mathbf{x} = \mathbf{0}$ has only the trivial solution.*

Proof
Suppose that A is an $n \times n$ matrix. Then A is invertible if and only if rank $A = n$ (Theorem 3.4.7). Rank $A = n$ if and only if nullity $A = 0$ (by the dimension theorem).

Finally, since nullity A is the dimension of the solution space of $A\mathbf{x} = \mathbf{0}$, we have that A is invertible if and only if this space is the zero subspace. ∎

Example 7   Determine which of the following matrices are invertible.

$$A = \begin{bmatrix} 1 & 2 & 3 & 4 \\ 1 & 5 & 5 & 5 \\ 2 & 1 & 0 & 6 \\ 0 & 3 & 6 & 2 \end{bmatrix} \qquad B = \begin{bmatrix} 1 & 2 & 1 \\ 1 & 3 & 2 \\ 1 & 1 & 1 \end{bmatrix}$$

Performing elementary row operations on A to put A into triangular form, we have

$$A = \begin{bmatrix} 1 & 2 & 3 & 4 \\ 1 & 5 & 5 & 5 \\ 2 & 1 & 0 & 6 \\ 0 & 3 & 6 & 2 \end{bmatrix} \rightarrow \begin{bmatrix} 1 & 2 & 3 & 4 \\ 0 & 3 & 2 & 1 \\ 0 & -3 & -6 & -2 \\ 0 & 3 & 6 & 2 \end{bmatrix} \rightarrow \begin{bmatrix} 1 & 2 & 3 & 4 \\ 0 & 3 & 2 & 1 \\ 0 & 0 & -4 & -1 \\ 0 & 0 & 4 & 1 \end{bmatrix}$$

$$\rightarrow \begin{bmatrix} 1 & 2 & 3 & 4 \\ 0 & 3 & 2 & 1 \\ 0 & 0 & -4 & -1 \\ 0 & 0 & 0 & 0 \end{bmatrix}$$

Since the last matrix is in triangular form, the rank of A is 3 by Corollary 3.3.9. Therefore, by Theorem 3.4.7, A is not invertible.

Performing elementary row operations on B, we have

$$B = \begin{bmatrix} 1 & 2 & 1 \\ 1 & 3 & 2 \\ 1 & 1 & 1 \end{bmatrix} \rightarrow \begin{bmatrix} 1 & 2 & 1 \\ 0 & 1 & 1 \\ 0 & -1 & 0 \end{bmatrix} \rightarrow \begin{bmatrix} 1 & 2 & 1 \\ 0 & 1 & 1 \\ 0 & 0 & 1 \end{bmatrix}$$

Thus, rank $B = 3$, since the final matrix is in triangular form and has three nonzero rows. Therefore, by Theorem 3.4.7, we have that B is invertible.

Since the matrix B in Example 7 is invertible, we can apply Corollary 3.4.8 to conclude that the system of linear equations $B\mathbf{x} = \mathbf{0}$ has only the trivial solution.

Additional Properties of Matrix Inverses

The next theorem tells us that for a square matrix A, we need only check that $AB = I$ or $BA = I$ (but not both) to conclude that B is the inverse of A.

Theorem 3.4.9   *Let A be a (square) matrix of order n. If there exists a square matrix B of order n such that $AB = I$ or $BA = I$, then A is invertible and B is the inverse of A.*

Proof

Suppose that $AB = I$. We argue that rank $A = n$. Consider any vector $\mathbf{y}$ in R^n. Let $\mathbf{x} = B\mathbf{y}$. Then

$$L_A(\mathbf{x}) = A\mathbf{x} = A(B\mathbf{y}) = (AB)\mathbf{y} = I\mathbf{y} = \mathbf{y}$$

Therefore, $\mathbf{y}$ lies in the $R(L_A)$. Hence, $R(L_A) = R^n$, and it follows that rank $A = n$. So, by Theorem 3.4.7 we have that A is invertible. Since $AB = I$,

$$A^{-1} = A^{-1}I = A^{-1}(AB) = (A^{-1}A)B = IB = B$$

For the case $BA = I$, we may use the argument above to conclude that B is invertible and that $B^{-1} = A$. Thus, $A^{-1} = (B^{-1})^{-1} = B$ by Theorem 3.4.2. ∎

The following corollary is of use in Chapter 5.

Corollary 3.4.10

Let A and B be square matrices of order n. If AB is invertible, then both A and B are invertible.

Proof

Suppose that AB is invertible. Let C be the inverse of AB. Then

$$A(BC) = (AB)C = I$$

So, by Theorem 3.4.9, A is invertible. By a similar argument, B is invertible. ∎

Invertible Linear Transformations

The concept of invertibility can be extended to linear transformations.

Definition

A linear transformation $T: R^n \to R^n$ is *invertible* if there exists a function $U: R^n \to R^n$ such that $TU = UT = I$, where I is the identity transformation on R^n.

It is left as an exercise (see Exercise 17) to show that the function U must be linear.

It is an elementary fact that a function is invertible if and only if it is one-to-one and onto.

Example 8

Let T and U be linear transformations from R^2 to R^2 defined by

$$T(x, y) = (2x + y, 3x + 2y)$$

and

$$U(x, y) = (2x - y, -3x + 2y)$$

Then, for any vector (x, y),

$$UT(x, y) = U(2x + y, 3x + 2y)$$

$$= (2(2x + y) - (3x + 2y), -3(2x + y) + 2(3x + 2y))$$

$$= (x, y)$$

$$= I(x, y)$$

Therefore, $UT = I$. Similarly, $TU = I$. Thus, T and U are invertible.

As with matrices, the transformation U in the definition above is unique.

Theorem 3.4.11   *For an invertible linear transformation T, the linear transformation U such that $TU = UT = I$ is unique.*

Since the proof is similar to the proof of Theorem 3.4.1, it is left as an exercise.

Definition   For an invertible linear transformation T, the unique linear transformation U for which $TU = UT = I$ is called the ***inverse*** of T, and it is denoted by T^{-1}.

Because of the correspondence between linear transformations and matrices, the results on matrix invertibility can be applied directly to linear transformations.

Theorem 3.4.12   *Let $T: R^n \to R^n$ be linear. Let A be the standard matrix of T. Then:*
   (a) T is invertible if and only if A is invertible.
   (b) If T is invertible, then $T^{-1} = L_{A^{-1}}$.

Proof
First suppose that T is invertible. Let A be the standard matrix of T, and let B be the standard matrix of T^{-1}. Then

$$L_I = I = TT^{-1} = L_A L_B = L_{AB}$$

Therefore, $AB = I$. By Theorem 3.4.9, we have that A is invertible, and $B = A^{-1}$. This proves part (b) and the first part of (a). Now suppose that A is invertible. We shall show that T is invertible. Let $U = L_{A^{-1}}$. Then

$$TU = L_A L_{A^{-1}} = L_{AA^{-1}} = L_I = I$$

Similarly, $UT = I$. So T is invertible.   ∎

Example 9   Determine which of the following linear transformations are invertible.
   (a) $T: R^3 \to R^3$ defined by $T(x, y, z) = (2x + y + z, 4x + y + z, 2x + 2y + z)$.
   (b) $U: R^3 \to R^3$ defined by $U(x, y, z) = (x + y + z, x - y - z, x)$.

Let A and B be the standard matrices for T and U, respectively. Then

$$A = \begin{bmatrix} 2 & 1 & 1 \\ 4 & 1 & 1 \\ 2 & 2 & 1 \end{bmatrix} \quad \text{and} \quad B = \begin{bmatrix} 1 & 1 & 1 \\ 1 & -1 & -1 \\ 1 & 0 & 0 \end{bmatrix}$$

It is easy to verify that rank $A = 3$ and rank $B = 2$. Therefore, T is invertible, but U is not invertible.

Example 10   For the linear transformation T on R^2 defined by

$$T(x, y) = (x + y, x + 2y)$$

find an expression for $T^{-1}(x, y)$.

Let A be the standard matrix of T. Then

$$A = \begin{bmatrix} 1 & 1 \\ 1 & 2 \end{bmatrix} \quad \text{and} \quad A^{-1} = \begin{bmatrix} 2 & -1 \\ -1 & 1 \end{bmatrix}$$

Therefore, by Theorem 3.4.12,

$$T^{-1}\begin{bmatrix} x \\ y \end{bmatrix} = L_{A^{-1}}\begin{bmatrix} x \\ y \end{bmatrix}$$

$$= A^{-1}\begin{bmatrix} x \\ y \end{bmatrix}$$

$$= \begin{bmatrix} 2 & -1 \\ -1 & 1 \end{bmatrix}\begin{bmatrix} x \\ y \end{bmatrix}$$

$$= \begin{bmatrix} 2x - y \\ -x + y \end{bmatrix}$$

or

$$T^{-1}(x, y) = (2x - y, -x + y)$$

We conclude this section with a theorem which summarizes the conditions that are equivalent to invertibility of a linear transformation.

Theorem 3.4.13   *Let $T: R^n \to R^n$ be linear. Then the following are equivalent:*
   (a) *T is invertible.*
   (b) *rank $T = n$.*
   (c) *T is onto.*
   (d) *T is one-to-one.*

Proof

First note that parts (b) and (c) are equivalent by Theorem 3.3.2, and that parts (c) and (d) are equivalent by Corollary 3.3.5. Therefore, parts (b), (c), and (d) are equivalent conditions. If T is invertible, then it is one-to-one and onto and hence part (a) implies parts (c) and (d) [and hence part (b)]. Conversely, if any of parts (b), (c), or (d) hold, then they all do. Therefore, T is one-to-one and onto. Thus, T is invertible. ∎

Exercises

1. For each of the following parts, determine whether or not the matrix is invertible.

   (a) $\begin{bmatrix} 1 & 2 \\ 6 & 5 \end{bmatrix}$

   (b) $\begin{bmatrix} 3 & 1 \\ -3 & -1 \end{bmatrix}$

   (c) $\begin{bmatrix} 2 & 1 \\ 1 & 2 \end{bmatrix}$

   (d) $\begin{bmatrix} 0 & 1 \\ 1 & 0 \end{bmatrix}$

   (e) $\begin{bmatrix} 1 & 2 & 1 \\ 1 & 3 & 2 \\ 1 & 4 & 3 \end{bmatrix}$

   (f) $\begin{bmatrix} 1 & 2 & 3 \\ 1 & 2 & 1 \\ 2 & 1 & 3 \end{bmatrix}$

   (g) $\begin{bmatrix} 1 & 1 & 1 \\ 2 & 2 & 2 \\ 3 & 3 & 3 \end{bmatrix}$

   (h) $\begin{bmatrix} 2 & 1 & -1 \\ 1 & 1 & 2 \\ 2 & 2 & 3 \end{bmatrix}$

2. Use invertible matrices to solve the linear systems.

   (a) $\begin{aligned} x + 3y &= 4 \\ 2x + 5y &= 7 \end{aligned}$

   (b) $\begin{aligned} 3x - 2y &= 1 \\ 4x - 3y &= -3 \end{aligned}$

3. For the matrices A and B, determine whether or not $B = A^{-1}$.

   (a) $A = \begin{bmatrix} 1 & 3 & 1 \\ 2 & 5 & 3 \\ 1 & 2 & 3 \end{bmatrix}$

   $B = \begin{bmatrix} -9 & 7 & -4 \\ 3 & -2 & 1 \\ 1 & -1 & 1 \end{bmatrix}$

   (b) $A = \begin{bmatrix} 1 & 4 & 2 \\ 1 & 1 & 1 \\ 2 & 1 & -1 \end{bmatrix}$

   $B = \begin{bmatrix} 1 & 2 & 2 \\ 1 & -1 & -1 \\ -2 & 1 & 1 \end{bmatrix}$

4. Use Exercise 3(a) to solve the system

   $$\begin{aligned} x + 3y + z &= 4 \\ 2x + 5y + 3z &= 1 \\ x + 2y + 3z &= -1 \end{aligned}$$

5. Determine which of the following linear transformations are invertible.

   (a) $T: R^2 \to R^2$, defined by $T(x, y) = (x + 4y, 2x - 3y)$.

   (b) $T: R^3 \to R^3$, defined by $T(x, y, z) = (y, z, x)$.

   (c) $T: R^3 \to R^3$, defined by $T(x, y, z) = (x + 2y + z, x - y, 2x + y + z)$.

   (d) $T: R^n \to R^n$ defined by $T(x_1, x_2, \ldots, x_n) = (x_1, x_2, \ldots, x_{n-1}, 0)$.

   (e) $T: R^n \to R^n$ defined by $T(x_1, x_2, \ldots, x_n) = (x_2, x_3, \ldots, x_n, x_1)$.

   (f) $T: R^n \to R^n$ defined by $T(x_1, x_2, \ldots, x_n) = (x_n, x_{n-1}, \ldots, x_2, x_1)$.

6. For each linear transformation $T: R^2 \to R^2$, find $T^{-1}(x, y)$.

   (a) $T(x, y) = (2x + 3y, 3x + 4y)$

   (b) $T(x, y) = (x + y, x)$

   (c) $T(x, y) = (x + 2y, 2x + 3y)$

7. For the linear transformation $T: R^2 \to R^2$ determine whether or not T is invertible.

   (a) T is a rotation.

   (b) T is a reflection.

   (c) T is a projection of R^2 on a line.

8. Prove Theorem 3.4.2(a).

9. For an invertible matrix A, prove that for any nonzero scalar c, cA is invertible and $(cA)^{-1} = (1/c)A^{-1}$.

10. Let

$$A = \begin{bmatrix} a_1 & 0 & \cdots & 0 \\ 0 & a_2 & \cdots & 0 \\ \cdot & \cdot & \cdots & 0 \\ 0 & \cdot & \cdots & a_n \end{bmatrix}$$

be a diagonal matrix with diagonal entries a_1, $a_2, \ldots, a_n$. Prove that A is invertible if and only if $a_i \neq 0$ for all i. If A is invertible, describe A^{-1}.

11. Give an example of invertible 2×2 matrices A and B such that $A + B$ is not invertible.

12. Complete the proof of Theorem 3.4.2. Prove that if A and B are invertible matrices, then:
    (a) $(B^{-1}A^{-1})(AB) = I$.
    (b) $(A^{-1})^t A^t = I$.

13. Complete the proof of Theorem 3.4.6 by verifying that $AB = I$.

14. Complete the proof of Corollary 3.4.10.

15. Prove that if T and U are linear transformations from R^n to R^n such that $TU = I$, then both T and U are invertible, $T^{-1} = U$, and $U^{-1} = T$.

16. Prove that if T and U are invertible linear transformations from R^n to R^n, then TU is invertible, and $(TU)^{-1} = U^{-1}T^{-1}$.

17. Let $T: R^n \to R^n$ be a linear transformation, and suppose that U is the inverse of T. Prove that U is linear. [*Hint*: To prove the additive property of U, let $\mathbf{x}$ and $\mathbf{y}$ be vectors in R^n, and let $\mathbf{x}' = U(\mathbf{x})$ and $\mathbf{y}' = U(\mathbf{y})$. Show that $\mathbf{x} = T(\mathbf{x}')$ and $\mathbf{y} = T(\mathbf{y}')$. Now use these equations to show that $U(\mathbf{x} + \mathbf{y}) = U(\mathbf{x}) + U(\mathbf{y})$.]

18. This exercise refers to the linear transformation R_A introduced in Exercise 28 of Section 2.4. For a square matrix A, prove that R_A is invertible if and only if A is invertible.

19. Use the inverse of the rotation matrix A_θ to verify the trigonometric identity

$$\sin^2 \theta + \cos^2 \theta = 1$$

20. Let

$$A = \begin{bmatrix} 1 & 2 & 3 \\ 1 & 1 & 1 \end{bmatrix} \quad \text{and} \quad B = \begin{bmatrix} 1 & 2 \\ 1 & 1 \\ 0 & 1 \end{bmatrix}$$

    (a) Compute AB.
    (b) Is AB invertible?
    (c) Does part (b) contradict Corollary 3.4.10? Justify your answer.

21. Let A and C be invertible matrices and let

$$M = \begin{bmatrix} A & O \\ O & C \end{bmatrix}$$

    where the zero matrices are chosen to be of sizes compatible with A and C. Prove that M is invertible, and that

$$M^{-1} = \begin{bmatrix} A^{-1} & O \\ O & C^{-1} \end{bmatrix}$$

22. Prove the following generalization of Exercise 21. Let A be an invertible matrix of order m, C be an invertible matrix of order n, and B, an arbitrary $m \times n$ matrix. Prove that the matrix

$$M = \begin{bmatrix} A & B \\ O & C \end{bmatrix}$$

    is invertible, and that

$$M^{-1} = \begin{bmatrix} A^{-1} & -A^{-1}BC^{-1} \\ O & C^{-1} \end{bmatrix}$$

*23. For each pair of matrices A and B, use the program MATRIX to determine whether or not $B = A^{-1}$.

    (a) $A = \begin{bmatrix} 1 & 2 & 1 & 3 \\ 2 & 5 & 6 & 7 \\ -1 & 1 & 10 & 2 \\ 3 & 8 & 9 & 16 \end{bmatrix}$

    $B = \begin{bmatrix} -213 & 71 & -33 & 13 \\ 143 & -47 & 22 & -9 \\ -33 & 11 & -5 & 2 \\ -13 & 4 & -2 & 1 \end{bmatrix}$

(b) $A = \begin{bmatrix} 1 & 2 & 1 & 3 \\ 3 & 1 & -1 & 1 \\ 0 & 1 & 0 & 2 \\ 1 & 1 & 2 & -1 \end{bmatrix}$

(a) $\begin{bmatrix} 1 & 2 & 5 & -1 \\ 0 & 3 & 2 & 1 \\ 0.5 & 1 & 5.5 & -0.5 \\ 1 & 3.5 & 6 & -0.5 \end{bmatrix}$

$B = \begin{bmatrix} 0.8 & 0.2 & -1.5 & -0.3 \\ -2.3 & 0.3 & 4.0 & 1.3 \\ 1.3 & -0.3 & -2.0 & -0.3 \\ 1.2 & -0.7 & -1.5 & -0.7 \end{bmatrix}$

(b) $\begin{bmatrix} 1 & 2 & 3 & 4 \\ 5 & 6 & 7 & 8 \\ 9 & 10 & 11 & 12 \\ 13 & 14 & 15 & 16 \end{bmatrix}$

(c) $A = \begin{bmatrix} 1 & 2 & 3 & 4 & 5 \\ 0 & 2 & 3 & 4 & 5 \\ 2 & 4 & 7 & 12 & 14 \\ 3 & 6 & 11 & 19 & 25 \\ 0 & 0 & 1 & 3 & 7 \end{bmatrix}$

(c) $\begin{bmatrix} 1 & 2 & 4 & 3 & 2 \\ -1 & 3 & -2 & 8 & 16 \\ 2 & 4 & 9 & 9 & 6 \\ 0 & 3 & -2 & 11 & 15 \\ -1 & 2 & -1 & 6 & 12 \end{bmatrix}$

$B = \begin{bmatrix} 1 & -1 & 0 & 0 & 0 \\ 10.5 & 0.5 & 18 & -16 & 11.5 \\ -10.0 & 0 & -19 & 16 & -12 \\ 1 & 0 & 4 & -3 & 2 \\ 1 & 0 & 1 & -1 & 1 \end{bmatrix}$

(d) $\begin{bmatrix} 1 & 2 & 1 & 3 & 4 \\ 3 & -2 & 2 & -3 & 1 \\ 3 & 3 & 1 & 2 & 5 \\ 2 & 2 & -1 & 2 & 1 \\ 1 & 2 & 1 & 3 & 4 \end{bmatrix}$

*24. Use the program ROW REDUCTION to determine whether or not each of the following matrices is invertible.

* This problem should be solved by using one of the programs noted in Appendix B.

3.5 ELEMENTARY MATRICES AND THE CONSTRUCTION OF MATRIX INVERSES

Now that we can recognize when a matrix or a linear transformation is invertible, we must develop a procedure for computing its inverse. In view of Theorem 3.4.12, we seek a procedure for computing the inverse of a matrix. The elementary row operations play a key role in this procedure. The idea is to transform the matrix into the identity matrix using elementary row operations. These elementary operations are used simultaneously to compute the inverse of the original matrix.

The next theorem assures us that an invertible matrix can be transformed into the identity matrix by means of elementary row operations.

Theorem 3.5.1 *Let A be an invertible matrix. Then A can be transformed into I by means of a finite number of elementary row operations.*

We shall not give a formal proof of this theorem. In its place we provide an example that illustrates the method by which an invertible matrix can be transformed into *I*.

Example 1 Use elementary row operations to transform the 3×3 matrix

$$A = \begin{bmatrix} 1 & 2 & 3 \\ 2 & 2 & 2 \\ 1 & 2 & 8 \end{bmatrix}$$

into I.

First transform A into the matrix in the triangular form given below.

$$\begin{bmatrix} 1 & 2 & 3 \\ 0 & -2 & -4 \\ 0 & 0 & 5 \end{bmatrix}$$

Next, using elementary row operations of type 2, divide each row by its first nonzero entry to obtain

$$\begin{bmatrix} 1 & 2 & 3 \\ 0 & 1 & 2 \\ 0 & 0 & 1 \end{bmatrix}$$

Finally, by elementary row operations of type 3, eliminate the nonzero entries above the diagonal entries to obtain

$$I = \begin{bmatrix} 1 & 0 & 0 \\ 0 & 1 & 0 \\ 0 & 0 & 1 \end{bmatrix}$$

Associated with elementary matrix operations are certain matrices called "elementary" matrices. They are used to convert elementary matrix operations into matrix products.

Definition An $n \times n$ matrix E is called an **elementary matrix** if E can be obtained from I by one elementary row or column operation.

For example, the matrix

$$E = \begin{bmatrix} 1 & 0 & 0 \\ 0 & 1 & 0 \\ 2 & 0 & 1 \end{bmatrix}$$

is an elementary matrix because E can be obtained from I by adding 2 times its first row to its third row. Notice that E can also be obtained from I by adding 2 times its third column to its first column. Thus, E can be obtained from I by means of an elementary row or column operation. This dual method of construction is common to all elementary matrices. We state this fact as a theorem without proof.

Theorem 3.5.2 *Any $n \times n$ elementary matrix E can be obtained from I by either an elementary row operation or an elementary column operation.*

The next theorem gives us a justification for considering elementary matrices. Its proof is omitted.

Theorem 3.5.3 *Let A and B be $n \times n$ matrices.*
    (a) *If B can be obtained from A by means of an elementary row operation, then $B = EA$, where E is the elementary matrix obtained from I by the same elementary row operation.*
    (b) *If B can be obtained from A by means of an elementary column operation, then $B = AE$, where E is the elementary matrix obtained from I by the same elementary column operation.*

Example 2   Consider the matrices

$$A = \begin{bmatrix} 1 & 2 & 1 \\ 2 & -1 & 4 \\ 1 & 0 & 1 \end{bmatrix} \quad \text{and} \quad B = \begin{bmatrix} 1 & 2 & 1 \\ 5 & 5 & 7 \\ 1 & 0 & 1 \end{bmatrix}$$

Notice that B can be obtained from A by adding 3 times the first row of A to the second row of A. This is an elementary row operation of type 3. Performing the same operation on I yields the elementary matrix

$$E = \begin{bmatrix} 1 & 0 & 0 \\ 3 & 1 & 0 \\ 0 & 0 & 1 \end{bmatrix}$$

Now notice that

$$EA = \begin{bmatrix} 1 & 0 & 0 \\ 3 & 1 & 0 \\ 0 & 0 & 1 \end{bmatrix} \begin{bmatrix} 1 & 2 & 1 \\ 2 & -1 & 4 \\ 1 & 0 & 1 \end{bmatrix}$$

$$= \begin{bmatrix} 1 & 2 & 1 \\ 5 & 5 & 7 \\ 1 & 0 & 1 \end{bmatrix}$$

$$= B$$

In view of Theorem 3.5.3 and the fact that any elementary matrix can be obtained by either an elementary row or column operation, we have:

Corollary 3.5.4

For any $n \times n$ matrix A and any $n \times n$ elementary matrix E:
(a) *EA can be obtained by applying an elementary row operation to A.*
(b) *AE can be obtained by applying an elementary column operation to A.*

The following theorem gives us information about inverses of elementary matrices.

Theorem 3.5.5 *Every elementary matrix is invertible. Furthermore, the inverse of an elementary matrix is also an elementary matrix.*

Proof

Let E be an $n \times n$ elementary matrix. In view of Theorem 3.5.2, we may assume that E can be obtained from I by an elementary row operation.

First suppose that the operation is of type 1. Then E can be obtained by interchanging two rows of I. By Theorem 3.5.3, the product EE is the result of interchanging the same rows of E, thus producing I again. Hence, $EE = I$. Therefore, E is invertible and has inverse E.

The proof for the case that the operation is of type 2 is left as an exercise.

Finally, suppose that the operation is of type 3. Suppose that E is obtained from I by adding c times its ith row to its jth row, where c is a scalar and $i \neq j$. Let F be the elementary matrix obtained from I by adding $-c$ times its ith row to its jth row. Then the product FE is the result of adding $-c$ times the ith row of E to the jth row of E. The result is, of course, I again. Therefore, $FE = I$. So E is invertible, and F is the inverse of E. ∎

Example 3

Let E be the elementary matrix

$$E = \begin{bmatrix} 1 & 0 & 0 \\ 0 & 1 & 0 \\ 3 & 0 & 1 \end{bmatrix}$$

We shall find E^{-1}. Notice that the result of adding -3 times the first row of E to the third row of E results in I. If F is the elementary matrix obtained from I by means of this same operation, then

$$F = \begin{bmatrix} 1 & 0 & 0 \\ 0 & 1 & 0 \\ -3 & 0 & 1 \end{bmatrix}$$

Therefore, by Theorem 3.5.3, $FE = I$. Thus, $E^{-1} = F$.

Theorem 3.5.6 *Every invertible matrix is a product of elementary matrices.*

Proof

Let A be an invertible $n \times n$ matrix. By Theorem 3.5.1 we have that A can be transformed into I by means of a finite number of elementary row operations. By Theorem 3.5.3, each such operation can be produced by left multiplication by an elementary matrix. Therefore, there exist elementary matrices $E_1, E_2, \ldots, E_p$ such that

$$E_p E_{p-1} \cdots E_1 A = I$$

Solving this equation for A, we have, by Corollary 3.4.3, that

$$A = (E_p E_{p-1} \cdots E_1)^{-1}$$
$$= (E_1)^{-1}(E_2)^{-1} \cdots (E_p)^{-1}$$

Therefore, since the inverse of an elementary matrix is an elementary matrix (Theorem 3.5.5), A is the product of elementary matrices.   ∎

Computation of the Matrix Inverse

Let A be an invertible $n \times n$ matrix. Let M be the $n \times 2n$ matrix whose first n columns are the columns of A, and whose last n columns are the columns of the $n \times n$ identity matrix I. That is,

$$M = (A \mid I)$$

For example, if

$$A = \begin{bmatrix} 1 & 2 \\ 3 & 4 \end{bmatrix}$$

then

$$M = \begin{bmatrix} 1 & 2 & 1 & 0 \\ 3 & 4 & 0 & 1 \end{bmatrix}$$

Recall, by Example 2 of Section 2.2, that for any $n \times n$ matrix C, the product CM can be computed using block multiplication of C on A and I. That is,

$$CM = (CA \mid CI) = (CA \mid C)$$

Theorem 3.5.7   *For any invertible $n \times n$ matrix A,*
  (a) *The matrix $(A \mid I)$ can be transformed into the $n \times 2n$ matrix $(I \mid A^{-1})$ by means of a finite number of elementary row operations.*
  (b) *If $(A \mid I)$ is transformed into $(I \mid B)$ by means of elementary row operations, then $B = A^{-1}$.*

Proof

(a) Suppose that A is an invertible $n \times n$ matrix. Then

$$A^{-1}(A|I) = (A^{-1}A|A^{-1}) = (I|A^{-1})$$

By Theorem 3.5.6, A^{-1} is a product of elementary matrices,

$$A^{-1} = E_1 E_2 \cdots E_p$$

Therefore,

$$E_1 E_2 \cdots E_p(A|I) = (I|A^{-1})$$

By successive applications of Corollary 3.5.4, we conclude that $(I|A^{-1})$ can be obtained from $(A|I)$ by means of a finite number of elementary row operations.

(b) Suppose that $(A|I)$ is transformed into $(I|B)$ by means of a finite sequence of elementary row operations. By Theorem 3.5.3, each such row operation can be associated with an elementary matrix E_i, such that

$$E_1 E_2 \cdots E_p(A|I) = (I|B)$$

Let $C = E_1 E_2 \cdots E_p$. Then

$$(I|B) = C(A|I)$$
$$= (CA|C)$$

Therefore, $CA = I$ and $B = C$. It follows that $B = C = A^{-1}$.   ■

Example 4   Use the method of Theorem 3.5.7 to compute the inverse of

$$A = \begin{bmatrix} 1 & 3 \\ 2 & 4 \end{bmatrix}$$

First form the 2×4 matrix

$$M = \begin{bmatrix} 1 & 3 & 1 & 0 \\ 2 & 4 & 0 & 1 \end{bmatrix}$$

Next, perform elementary row operations on M to transform the first two columns into I:

$$\begin{bmatrix} 1 & 3 & 1 & 0 \\ 2 & 4 & 0 & 1 \end{bmatrix} \to \begin{bmatrix} 1 & 3 & 1 & 0 \\ 0 & -2 & -2 & 1 \end{bmatrix} \to \begin{bmatrix} 1 & 3 & 1 & 0 \\ 0 & 1 & 1 & -0.5 \end{bmatrix}$$

$$\to \begin{bmatrix} 1 & 0 & -2 & 1.5 \\ 0 & 1 & 1 & -0.5 \end{bmatrix}$$

Therefore,

$$A^{-1} = \begin{bmatrix} -2 & 1.5 \\ 1 & -0.5 \end{bmatrix}$$

Compare this method of computing the inverse with the formula of Theorem 3.4.6.

Example 5   Use the method of Theorem 3.5.7 to compute the inverse of

$$A = \begin{bmatrix} 2 & -2 & 5 \\ 1 & -1 & 2 \\ 1 & 1 & 3 \end{bmatrix}$$

Form the 3×6 matrix

$$M = \begin{bmatrix} 2 & -2 & 5 & 1 & 0 & 0 \\ 1 & -1 & 2 & 0 & 1 & 0 \\ 1 & 1 & 3 & 0 & 0 & 1 \end{bmatrix}$$

Then perform elementary row operations on M to transform the first three columns into I:

$$\begin{bmatrix} 2 & -2 & 5 & 1 & 0 & 0 \\ 1 & -1 & 2 & 0 & 1 & 0 \\ 1 & 1 & 3 & 0 & 0 & 1 \end{bmatrix} \rightarrow \begin{bmatrix} 1 & -1 & 2 & 0 & 1 & 0 \\ 2 & -2 & 5 & 1 & 0 & 0 \\ 1 & 1 & 3 & 0 & 0 & 1 \end{bmatrix}$$

$$\rightarrow \begin{bmatrix} 1 & -1 & 2 & 0 & 1 & 0 \\ 0 & 0 & 1 & 1 & -2 & 0 \\ 0 & 2 & 1 & 0 & -1 & 1 \end{bmatrix}$$

$$\rightarrow \begin{bmatrix} 1 & -1 & 2 & 0 & 1 & 0 \\ 0 & 2 & 1 & 0 & -1 & 1 \\ 0 & 0 & 1 & 1 & -2 & 0 \end{bmatrix}$$

$$\rightarrow \begin{bmatrix} 1 & -1 & 2 & 0 & 1 & 0 \\ 0 & 1 & \frac{1}{2} & 0 & -\frac{1}{2} & \frac{1}{2} \\ 0 & 0 & 1 & 1 & -2 & 0 \end{bmatrix}$$

$$\rightarrow \begin{bmatrix} 1 & 0 & \frac{5}{2} & 0 & \frac{1}{2} & \frac{1}{2} \\ 0 & 1 & \frac{1}{2} & 0 & -\frac{1}{2} & \frac{1}{2} \\ 0 & 0 & 1 & 1 & -2 & 0 \end{bmatrix}$$

$$\rightarrow \begin{bmatrix} 1 & 0 & 0 & -\frac{5}{2} & \frac{11}{2} & \frac{1}{2} \\ 0 & 1 & 0 & -\frac{1}{2} & \frac{1}{2} & \frac{1}{2} \\ 0 & 0 & 1 & 1 & -2 & 0 \end{bmatrix}$$

Therefore,

$$A^{-1} = \begin{bmatrix} -\frac{5}{2} & \frac{11}{2} & \frac{1}{2} \\ -\frac{1}{2} & \frac{1}{2} & \frac{1}{2} \\ 1 & -2 & 0 \end{bmatrix}$$

Example 6   For the linear transformation $T: R^3 \to R^3$ defined by $T(x, y, z) = (2x - 2y + 5z, x - y + 2z, x + y + 3z)$, find $T^{-1}(x, y, z)$. The standard matrix of T is

$$A = \begin{bmatrix} 2 & -2 & 5 \\ 1 & -1 & 2 \\ 1 & 1 & 3 \end{bmatrix}$$

Therefore, by Example 5,

$$A^{-1} = \begin{bmatrix} -\frac{5}{2} & \frac{11}{2} & \frac{1}{2} \\ -\frac{1}{2} & \frac{1}{2} & \frac{1}{2} \\ 1 & -2 & 0 \end{bmatrix}$$

Since

$$A^{-1} \begin{bmatrix} x \\ y \\ z \end{bmatrix} = \begin{bmatrix} -\frac{5}{2} & \frac{11}{2} & \frac{1}{2} \\ -\frac{1}{2} & \frac{1}{2} & \frac{1}{2} \\ 1 & -2 & 0 \end{bmatrix} \begin{bmatrix} x \\ y \\ z \end{bmatrix}$$

$$= \begin{bmatrix} -\frac{5}{2}x + \frac{11}{2}y + \frac{1}{2}z \\ -\frac{1}{2}x + \frac{1}{2}y + \frac{1}{2}z \\ x - 2y \end{bmatrix}$$

we have by Theorem 3.4.12 that

$$T^{-1}(x, y, z) = \left(-\frac{5}{2}x + \frac{11}{2}y + \frac{1}{2}z, \; -\frac{1}{2}x + \frac{1}{2}y + \frac{1}{2}z, \; x - 2y \right)$$

Application to Economics: The Leontief Model

In 1973, Wassily Leontief won the Nobel Prize in Economics for his work in developing a mathematical model to describe various economic phenomena. To illustrate some of the ideas from his work we will consider two simplified models.

1. *The Closed Model.* Suppose that we have a simple society consisting of three people (industries): a farmer who grows all the food, a tailor who makes all the clothing, and a carpenter who builds all the housing. We make the assumption that

TABLE 3.5.1

	Food	Clothing	Housing
Farmer	0.40	0.20	0.20
Tailor	0.10	0.70	0.20
Carpenter	0.50	0.10	0.60

each person sells to and buys from a central pool and that everything produced is consumed. Because no commodities either enter or leave the system, this model is called the **closed model**.

The proportion of each commodity that is consumed by each person is listed in Table 3.5.1.

Let p_1, p_2, and p_3 represent the incomes of the farmer, tailor, and carpenter, respectively. For example, since the farmer's income is derived completely from his sale of food, p_1 also represents the value of the food he produces. To assure that the society survives, we shall require that the consumption of each person equals his income. (We will say more about this requirement later.) This condition translates into the following system of equations:

$$\text{Consumption} \qquad = \text{Income}$$

Farmer: $\quad 0.40p_1 + 0.20p_2 + 0.20p_3 = \quad p_1$

Tailor: $\quad 0.10p_1 + 0.70p_2 + 0.20p_3 = \quad p_2$

Carpenter: $0.50p_1 + 0.10p_2 + 0.60p_3 = \quad p_3$

This assumption is called the **equilibrium condition**. Writing this system in matrix notation, we have $A\mathbf{p} = \mathbf{p}$, where

$$A = \begin{bmatrix} 0.40 & 0.20 & 0.20 \\ 0.10 & 0.70 & 0.20 \\ 0.50 & 0.10 & 0.60 \end{bmatrix} \quad \text{and} \quad \mathbf{p} = \begin{bmatrix} p_1 \\ p_2 \\ p_3 \end{bmatrix}$$

A is called the **input–output (consumption) matrix**.

The reader may be tempted to replace the equilibrium condition by $A\mathbf{p} \le \mathbf{p}$; that is, each component of the vector $A\mathbf{p}$ is less than or equal to the corresponding component of the vector $\mathbf{p}$. This condition would require that consumption be less than or equal to income. However, it will be left as an exercise to show that the condition $A\mathbf{p} \le \mathbf{p}$ is equivalent to the condition $A\mathbf{p} = \mathbf{p}$.

We proceed to find all vectors $\mathbf{p}$ such that $A\mathbf{p} = \mathbf{p}$. Let $\mathbf{x}$ denote the unknown value of $\mathbf{p}$. Thus, we seek the solutions for the equation $A\mathbf{x} = \mathbf{x}$. Since $\mathbf{x} = I\mathbf{x}$, we have that $A\mathbf{x} = I\mathbf{x}$. Thus, $I\mathbf{x} - A\mathbf{x} = \mathbf{0}$, or

$$(I - A)\mathbf{x} = \mathbf{0}$$

This is a homogeneous system of linear equations. In solving this system, the augmented matrix row reduces to

$$\begin{bmatrix} 0.1 & -0.3 & 0.2 & 0 \\ 0 & 1.6 & -1.4 & 0 \\ 0 & 0 & 0 & 0 \end{bmatrix}$$

The solution space is spanned by $\mathbf{p} = (5, 7, 8)$. We may interpret this result to mean: The society will survive if the farmer, tailor, and carpenter have incomes in the proportions $5:7:8$, respectively.

We should state that for a problem such as this, we are interested only in nontrivial solutions $\mathbf{p}$ to the system $(I - A)\mathbf{x} = \mathbf{0}$ whose coordinates are nonnegative. It can be shown that such a solution will necessarily exist because A is a nonnegative matrix all of whose columns sum to 1. To obtain a solution $\mathbf{p}$ all of whose entries are positive, it is sufficient to require that all the entries of some power of A be positive. In Section 6.3 we shall refer to a matrix with this property as a *regular matrix*. For another condition, see (1) in the References.

2. *The Open Model.* In the **open model** we assume that there is also an outside demand for the products produced in our society. We let x_1, x_2, and x_3 be the amounts of food, clothing, and housing produced with outside demands d_1, d_2, and d_3, respectively. As in the closed model, we define the 3×3 input–output matrix A by the condition that A_{ij} equals the proportion of product i consumed in producing product j. We can express the *surplus* of food in this society as

$$x_1 - (A_{11}x_1 + A_{12}x_2 + A_{13}x_3)$$

This quantity represents the amount of food produced less the amount of food consumed to produce the three quantities. As in the closed model, we require an equilibrium condition, namely, that the surplus of the three quantities must equal the corresponding demand for the three quantities. This condition can be translated into the following system of linear equations:

$$x_1 - (A_{11}x_1 + A_{12}x_2 + A_{13}x_3) = d_1$$
$$x_2 - (A_{21}x_1 + A_{22}x_2 + A_{23}x_3) = d_2$$
$$x_3 - (A_{31}x_1 + A_{32}x_2 + A_{33}x_3) = d_3$$

This system is equivalent to the matrix equation $(I - A)\mathbf{x} = \mathbf{d}$, where $\mathbf{d}$ is the 3×1 column vector whose entries are d_1, d_2, and d_3. If $(I - A)$ is invertible, the solution is

$$\mathbf{x} = (I - A)^{-1}\mathbf{d}$$

As in the closed model, for this solution to be meaningful, we require that the entries of $\mathbf{x}$ be nonnegative.

For example, suppose that

$$A = \begin{bmatrix} 0.30 & 0.20 & 0.30 \\ 0.10 & 0.40 & 0.10 \\ 0.30 & 0.20 & 0.30 \end{bmatrix} \quad \text{and} \quad \mathbf{d} = \begin{bmatrix} 30 \\ 20 \\ 10 \end{bmatrix}$$

Then

$$I - A = \begin{bmatrix} 0.70 & -0.20 & -0.30 \\ -0.10 & 0.60 & -0.10 \\ -0.30 & -0.20 & 0.70 \end{bmatrix} \quad \text{and} \quad (I - A)^{-1} = \begin{bmatrix} 2.0 & 1.0 & 1.0 \\ 0.5 & 2.0 & 0.5 \\ 1.0 & 1.0 & 2.0 \end{bmatrix}$$

Thus,

$$\mathbf{x} = (I - A)^{-1}\mathbf{d} = \begin{bmatrix} 90 \\ 60 \\ 70 \end{bmatrix}$$

We conclude that the society should produce 90 units of food, 60 units of clothing, and 70 units of housing to meet the corresponding demands of 30, 20, and 10. Because $(I - A)^{-1}$ has positive entries, it follows that there will always be a (unique) solution with positive entries for any nonzero demand vector $\mathbf{d}$.

Exercises

1. For matrices A and B, find an elementary matrix E such that $EA = B$.

   (a) $A = \begin{bmatrix} 1 & 3 \\ 4 & 7 \end{bmatrix}$

   $B = \begin{bmatrix} 4 & 7 \\ 1 & 3 \end{bmatrix}$

   (b) $A = \begin{bmatrix} 1 & 2 & 3 \\ 1 & 0 & 1 \\ 1 & 1 & -1 \end{bmatrix}$

   $B = \begin{bmatrix} 1 & 2 & 3 \\ 1 & 0 & 1 \\ 4 & 7 & 8 \end{bmatrix}$

   (c) $A = \begin{bmatrix} 2 & 1 & -1 \\ 2 & 4 & 6 \\ 3 & 0 & 1 \end{bmatrix}$

   $B = \begin{bmatrix} 2 & 1 & -1 \\ 1 & 2 & 3 \\ 3 & 0 & 1 \end{bmatrix}$

   (d) $A = \begin{bmatrix} 1 & 0 \\ 1 & 1 \end{bmatrix}$

   $B = \begin{bmatrix} 1 & 0 \\ 0 & 1 \end{bmatrix}$

2. For matrices A and B, find an elementary matrix E such that $AE = B$.

   (a) $A = \begin{bmatrix} 1 & 0 \\ 0 & 1 \end{bmatrix}$

   $B = \begin{bmatrix} 1 & 2 \\ 0 & 1 \end{bmatrix}$

   (b) $A = \begin{bmatrix} 1 & 2 & -1 \\ 2 & 1 & 1 \\ 1 & 0 & 3 \end{bmatrix}$

   $B = \begin{bmatrix} 2 & 1 & -1 \\ 1 & 2 & 1 \\ 0 & 1 & 3 \end{bmatrix}$

(c) $A = \begin{bmatrix} 1 & 0 & 2 \\ 2 & 1 & -1 \\ 3 & 0 & 1 \end{bmatrix}$

$B = \begin{bmatrix} 1 & 0 & 2 \\ 0 & 1 & -1 \\ 3 & 0 & 1 \end{bmatrix}$

3. Find the inverse of each of the following elementary matrices.

(a) $\begin{bmatrix} 1 & 0 \\ 3 & 1 \end{bmatrix}$ (b) $\begin{bmatrix} 0 & 1 \\ 1 & 0 \end{bmatrix}$

(c) $\begin{bmatrix} 2 & 0 \\ 0 & 1 \end{bmatrix}$ (d) $\begin{bmatrix} 0 & 0 & 1 \\ 0 & 1 & 0 \\ 1 & 0 & 0 \end{bmatrix}$

(e) $\begin{bmatrix} 1 & 2 & 0 \\ 0 & 1 & 0 \\ 0 & 0 & 1 \end{bmatrix}$ (f) $\begin{bmatrix} 1 & 0 & 0 \\ 0 & -1 & 0 \\ 0 & 0 & 1 \end{bmatrix}$

4. For each matrix A, find A^{-1} provided that it exists.

(a) $\begin{bmatrix} 1 & 2 \\ 1 & 3 \end{bmatrix}$ (b) $\begin{bmatrix} 1 & 2 & 1 \\ 2 & 5 & 2 \\ 1 & 3 & 2 \end{bmatrix}$

(c) $\begin{bmatrix} 1 & 2 & 3 \\ 1 & 3 & 5 \\ 2 & 3 & 6 \end{bmatrix}$ (d) $\begin{bmatrix} 1 & 2 & 1 \\ 2 & 1 & 3 \\ 3 & 0 & 5 \end{bmatrix}$

(e) $\begin{bmatrix} 2 & 5 & 4 \\ 1 & 3 & 1 \\ 1 & 3 & 2 \end{bmatrix}$ (f) $\begin{bmatrix} 1 & 1 & 2 \\ 1 & 2 & 4 \\ 3 & 4 & 7 \end{bmatrix}$

(g) $\begin{bmatrix} 1 & 2 & 3 & 4 \\ 1 & 3 & 4 & 6 \\ 2 & 4 & 7 & 9 \\ 1 & 2 & 1 & 3 \end{bmatrix}$ (h) $\begin{bmatrix} 2 & 5 & 1 & 0 \\ 0 & 1 & -1 & 2 \\ 1 & 3 & 0 & 1 \\ 1 & 2 & 1 & 0 \end{bmatrix}$

5. Apply the technique used in the proof of Theorem 3.5.6 to represent each of the following matrices as a product of elementary matrices.

(a) $\begin{bmatrix} 1 & 1 \\ 1 & 2 \end{bmatrix}$ (b) $\begin{bmatrix} 1 & 3 & 1 \\ 0 & 1 & 1 \\ 1 & 3 & 4 \end{bmatrix}$ (c) $\begin{bmatrix} 1 & 2 & 1 \\ 1 & 4 & 4 \\ 0 & 2 & 4 \end{bmatrix}$

6. For each linear transformation $T: R^n \to R^n$, find $T^{-1}(x_1, x_2, \ldots, x_n)$.
   (a) $T(x, y) = (x + 2y, x - y)$
   (b) $T(x, y, z) = (x, x + y, y + z)$
   (c) $T(x, y, z) = (x + y + 2z, -y + 4z, x + y + 3z)$
   (d) $T(x, y, z, w) = (x + y + 2z + w, 2x + 3y + 4z + 2w, 3x + 4y + 7z + w, 4x + 6y + 9z + 3w)$
   (e) $T(x_1, x_2, \ldots, x_n) = (x_n, x_1, \ldots, x_{n-1})$
   (f) $T(x_1, x_2, \ldots, x_n) = (x_n, x_{n-1}, \ldots, x_2, x_1)$

7. Prove that if A is invertible, and n is a positive integer, then A^n is invertible, and $(A^n)^{-1} = (A^{-1})^n$.

8. Suppose that A is an $n \times n$ invertible matrix and B is an $n \times m$ matrix. Prove that if the augmented matrix $(A|B)$ is transformed into the matrix $(I_n|C)$ by means of elementary row operations, then $C = A^{-1}B$.

9. Suppose that A is an $n \times n$ matrix and B is an $n \times m$ matrix such that the augmented matrix $(A|B)$ is in triangular form. Prove that A is in triangular form.

10. Let A be a square matrix. Prove that if $A^k = I$ for some positive integer k, then A is invertible.

11. A square matrix A is called **nilpotent** if $A^k = O$ for some positive integer k. Prove that nilpotent matrices are not invertible.

†12. Let A be an $m \times n$ matrix. Use Theorems 3.5.6 and 3.3.10 and Corollaries 3.3.8 and 3.5.4 to prove that for any invertible matrices P and Q, where P has order m and Q has order n,

$$\text{rank } PAQ = \text{rank } A$$

13. Prove Theorem 3.5.5 for the case that E is obtained from I be means of an elementary row operation of type 2.

Exercises 14–16 deal with the Leontief economic models.

† This exercise will be used in a subsequent section.

14. In the Leontief closed model, suppose that the input–output matrix is given by

$$\begin{bmatrix} \frac{7}{16} & \frac{1}{2} & \frac{3}{16} \\ \frac{5}{16} & \frac{1}{6} & \frac{5}{16} \\ \frac{1}{4} & \frac{1}{3} & \frac{1}{2} \end{bmatrix}$$

At what ratio must the farmer, tailor, and carpenter produce in order for the equilibrium condition to be satisfied?

15. Prove that the condition $A\mathbf{p} \leq \mathbf{p}$ in the Leontief closed model is equivalent to the condition $A\mathbf{p} = \mathbf{p}$. (*Hint*: Use the fact that each column of A must sum to 1.)

16. In the Leontief open model, suppose that the input–output matrix and demand vector are given by

$$\begin{bmatrix} 0.20 & 0.20 & 0.30 \\ 0.50 & 0.50 & 0 \\ 0 & 0 & 0.20 \end{bmatrix} \quad \text{and} \quad \begin{bmatrix} 10 \\ 14 \\ 18 \end{bmatrix}$$

respectively. How much of each quantity must be produced to satisfy the demand?

*17. Use the program MATRIX to find the inverse of the following matrices.

(a) $\begin{bmatrix} 1 & 2 & 3 & 4 \\ 5 & 6 & 7 & 8 \\ 1 & 3 & 2 & 4 \\ 0 & 1 & 2 & 6 \end{bmatrix}$
(b) $\begin{bmatrix} 1 & 2 & 3 & 5 & 6 \\ 1 & 3 & 5 & 6 & 9 \\ 1 & 3 & 6 & 9 & 9 \\ 1 & 2 & 3 & 6 & 8 \\ 0 & 1 & 3 & 5 & 6 \end{bmatrix}$

*18. Use the program MATRIX to compute the inverse used to solve Exercise 16. Compare your computation to the original one. Can you explain the discrepancy?

*19. Use the program MATRIX to find the inverse of each of the linear transformations T given below.
  (a) $T(x, y, z) = (x + 3y + 2z, 2x + 8y + 5z, 3x + 7y + 4z)$
  (b) $T(x, y, z, w) = (x + y + 3z + 2w, -y + 2z + 3w, x + 2y + 2z + 4w, x + y + 4z + 6w)$

* This problem should be solved by using one of the programs noted in Appendix B.

3.6 THEORY OF SYSTEMS OF LINEAR EQUATIONS

Much of the activity of this chapter has involved the gathering of information about matrices. Since a system of linear equations can be represented as a single matrix equation, it seems reasonable to expect that we can apply some of this information to deduce properties about systems. We have already done this to some extent. For example, Theorem 3.4.4 gives us a method for solving a system when the coefficient matrix is invertible. In this section we obtain additional information about systems using results about matrices.

We begin our study with homogeneous systems. Recall that a system $A\mathbf{x} = \mathbf{b}$ is homogeneous if $\mathbf{b} = \mathbf{0}$. By Corollary 2.5.2 the set of solutions to a homogeneous system of linear equations in n unknowns is a subspace of R^n. The rank of the coefficient matrix can be used to compute the dimension of this subspace.

Theorem 3.6.1 *Let $A\mathbf{x} = \mathbf{0}$ be a homogeneous system of linear equations in n unknowns. Then the dimension of the solution space is $n -$ rank A.*

Proof

It is clear that the solution space to the system is the null space of L_A. Therefore, by the dimension theorem,

$$\text{dim (solution space)} = \text{nullity } A = n - \text{rank } A \qquad \blacksquare$$

Example 1   Find the dimension of the solution space of the system

$$x + y + z - w = 0$$
$$x - 2y + z + 2w = 0$$
$$x + 4y + z + 3w = 0$$

The coefficient matrix for this system is

$$A = \begin{bmatrix} 1 & 1 & 1 & -1 \\ 1 & -2 & 1 & 2 \\ 1 & 4 & 1 & 3 \end{bmatrix}$$

which can easily be shown to have rank equal to 3. Since the system has $n = 4$ unknowns, the dimension of the solution space is

$$n - \text{rank } A = 4 - 3$$
$$= 1$$

Since the solution space is not zero-dimensional, there are infinitely many solutions to the system.

Example 1 is a special case of the following result.

Corollary 3.6.2   *If a homogeneous system of linear equations has more unknowns than equations, then the system has infinitely many solutions.*

Proof

Suppose that $A\mathbf{x} = \mathbf{0}$ is a homogeneous system of m equations in n unknowns, and suppose that $m < n$. Then, by Corollary 3.3.5, L_A is not one-to-one. Therefore, the solution space of $A\mathbf{x} = \mathbf{0}$, which is the same as $N(L_A)$, is not the zero subspace. Thus, the solution space contains infinitely many vectors. $\blacksquare$

Every homogeneous system of linear equations has at least one solution, namely, the zero vector. However, as we noted in Chapter 1, there are linear systems with no solutions, namely, the inconsistent systems. There is a test based on the rank of a matrix for determining whether or not a system of linear equations is consistent.

Example 2   Consider the system of linear equations

$$x + 3y + z = 3$$
$$x - 5y + 3z = 1$$
$$x - y + 2z = 4$$

By means of elementary row operations, the augmented matrix of this system can be transformed into the matrix in triangular form

$$\begin{bmatrix} 1 & 3 & 1 & 3 \\ 0 & 4 & -1 & 1 \\ 0 & 0 & 0 & 2 \end{bmatrix}$$

Therefore, by Theorem 1.5.3 we have that the system is inconsistent. This matrix has rank equal to 3, although the rank of the coefficient matrix of the system is 2.

Theorem 3.6.3   *The system of linear equations* $A\mathbf{x} = \mathbf{b}$ *is consistent if and only if*

$$\text{rank } A = \text{rank } (A|\mathbf{b})$$

Proof
First use elementary row operations to transform the augmented matrix $(A|\mathbf{b})$ into a matrix $(C|\mathbf{d})$ in triangular form. Notice that the matrix C is the result of transforming A by the same elementary row operations, and C is also in triangular form. Since elementary row operations preserve rank, we have that rank $(A|\mathbf{b}) = $ rank $(C|\mathbf{d})$ and rank $A = $ rank C. By Theorem 1.5.3 the system $A\mathbf{x} = \mathbf{b}$ is consistent if and only if the last nonzero row of $(C|\mathbf{d})$ has a nonzero entry before the last column. Thus, the system $A\mathbf{x} = \mathbf{b}$ is consistent if and only if C and $(C|\mathbf{d})$ have the same number of nonzero rows. By Corollary 3.3.9, we have that the system is consistent if and only if C and $(C|\mathbf{d})$ have the same rank. Therefore, the system is consistent if and only if A and $(A|\mathbf{b})$ have the same rank.   ■

Example 3   Determine whether or not the systems below are consistent.

(a) $2x + 3y + z = 4$ (b) $x - 3y + z + w = 4$

$\quad\quad x - y + z = 2$ $\quad\quad x + y + z - w = 2$

$\quad\quad x + 4y \quad\quad = 2$ $\quad\quad x + 5y + z - 3w = 1$

First consider the system (a). The augmented matrix of the system is

$$(A|\mathbf{b}) = \begin{bmatrix} 2 & 3 & 1 & 4 \\ 1 & -1 & 1 & 2 \\ 1 & 4 & 0 & 2 \end{bmatrix}$$

By means of elementary row operations, this matrix can be transformed into a matrix in triangular form

$$(C/\mathbf{d}) = \begin{bmatrix} 1 & -1 & 1 & 2 \\ 0 & 5 & -1 & 0 \\ 0 & 0 & 0 & 0 \end{bmatrix}$$

It is clear that rank $(A|\mathbf{b}) = 2$. The same row operations which transform $(A|\mathbf{b})$ into $(C/\mathbf{d})$ will transform A into the matrix C,

$$\begin{bmatrix} 1 & -1 & 1 \\ 0 & 5 & -1 \\ 0 & 0 & 0 \end{bmatrix}$$

This matrix is also in triangular form and has rank equal to 2. Therefore, rank $A = 2$. We conclude that the system is consistent.

Next, consider the system (b). The augmented matrix can be transformed by means of elementary row operations to the matrix in triangular form

$$\begin{bmatrix} 1 & -3 & 1 & 1 & 4 \\ 0 & 4 & 0 & -2 & -2 \\ 0 & 0 & 0 & 0 & 1 \end{bmatrix}$$

This matrix has rank equal to 3, and hence rank $(A|\mathbf{b}) = 3$. As above, applying the same elementary row operations to A yields

$$\begin{bmatrix} 1 & -3 & 1 & 1 \\ 0 & 4 & 0 & -2 \\ 0 & 0 & 0 & 0 \end{bmatrix}$$

This matrix has rank equal to 2. Therefore, rank $A = 2$. We conclude that the system (b) is inconsistent.

Since the solution space of a system of homogeneous linear equations in n unknowns is a subspace of R^n, it is easily described. For example, it has a basis. Solution sets of nonhomogeneous systems are not subspaces. However, they can be described very nicely in terms of subspaces.

Definition   For a system of linear equations $A\mathbf{x} = \mathbf{b}$, the system $A\mathbf{x} = \mathbf{0}$ is called the **associated homogeneous system**.

Theorem 3.6.4 *Let K be the solution set for the system of linear equations, $A\mathbf{x} = \mathbf{b}$. Let H be the solution space for the associated homogeneous system. Then, for any solution $\mathbf{z}$ of $A\mathbf{x} = \mathbf{b}$,*

$$K = \{\mathbf{z} + \mathbf{x} : \mathbf{x} \text{ is in } H\}$$

Proof
Consider any solution $\mathbf{y}$ in K. Let $\mathbf{x} = \mathbf{y} - \mathbf{z}$. Then

$$A\mathbf{x} = A(\mathbf{y} - \mathbf{z})$$
$$= A\mathbf{y} - A\mathbf{z}$$
$$= \mathbf{b} - \mathbf{b}$$
$$= \mathbf{0}$$

Therefore, $\mathbf{x}$ lies in H. Furthermore, since $\mathbf{x} = \mathbf{y} - \mathbf{z}$, we have that $\mathbf{y} = \mathbf{z} + \mathbf{x}$. Therefore, $\mathbf{y}$ is of the form $\mathbf{z} + \mathbf{x}$ for some $\mathbf{x}$ in H.

Now suppose that $\mathbf{y}$ is any vector of the form $\mathbf{z} + \mathbf{x}$ for some $\mathbf{x}$ in H. Then

$$A\mathbf{y} = A(\mathbf{z} + \mathbf{x})$$
$$= A\mathbf{z} + A\mathbf{x}$$
$$= \mathbf{b} + \mathbf{0}$$
$$= \mathbf{b}$$

Thus, $\mathbf{y}$ lies in K. Hence, $K = \{\mathbf{z} + \mathbf{x} : \mathbf{x} \text{ is in } H\}$.   ∎

Geometrically, Theorem 3.6.4 states that the solution set of $A\mathbf{x} = \mathbf{b}$ is a *translate* of the solution space of $A\mathbf{x} = \mathbf{0}$.

Example 4 We apply Theorem 3.6.4 to describe all solutions to the system of one linear equation in three unknowns:

$$x + 2y - z = 4$$

First we find any solution to the system, for example, $x = 4$, $y = z = 0$. Next we find a basis for the associated homogeneous system

$$x + 2y - z = 0$$

The coefficient matrix $A = (1, 2, -1)$ has rank 1, and $n = 3$. Therefore, the solution space for the associated homogeneous system has dimension 2. So any linearly independent set consisting of two vectors is a basis for this solution space. Notice that $(1, 0, 1)$ and $(-2, 1, 0)$ are solutions. Furthermore, they are linearly independent. Therefore, they constitute a basis for the solution space. Thus, any solution to the associated homogeneous system is a unique linear combination $s(1, 0, 1) + t(-2, 1, 0)$. By Theorem 3.6.4, any solution to the original system can be represented by the sum

$$(4, 0, 0) + s(1, 0, 1) + t(-2, 1, 0)$$

and conversely, any sum of this form is a solution. Notice that the solution space for the associated homogeneous system is a plane through the origin, and that the solution set is a translate of this plane.

Of course, in practice one does not solve systems of linear equations by the method of Example 4 since Gaussian elimination is more practical. The purpose of Theorem 3.6.4 and Example 4 is to provide us with a means of understanding the structure of solution sets. In particular, they show how a solution set can be represented in terms of a subspace of R^n. Thus, the letters s and t in the final equation serve as parameters. In fact, if one were to solve the original system by means of back-substitution, one would obtain equivalent results. In this case, the number of parameters is 2. This number is equal to the dimension of the solution space of the associated homogeneous system. This solution space is also the null space of the coefficient matrix of the system. These remarks suggest the following result.

Theorem 3.6.5 *Let $A\mathbf{x} = \mathbf{b}$ be a consistent system of linear equations in n unknowns. If the system is solved by the method of Gaussian elimination, then the number of parameters used is equal to $n - \text{rank } A$.*

Proof

Let $r = \text{rank } A$. By Theorem 3.6.3, $r = \text{rank } (A|\mathbf{b})$. Suppose that $(A|\mathbf{b})$ is transformed by means of elementary row operations into a matrix B in triangular form. Then rank $B = r$, and by Corollary 3.3.9 it follows that B has exactly r nonzero rows. The first nonzero entry of any row of B is the coefficient for an initial unknown for the system. Therefore, there are r initial unknowns. The other $n - r$ unknowns are assigned parameters according to the technique of back-substitution. We conclude that there are $n - r$ parameters. ∎

This result allows us to conclude (see Exercise 5) that the generating set obtained by applying Gaussian elimination to a system of homogeneous linear equations is a basis for the solution space. For example, we can conclude that the set $\{(1, 1, 0), (-3, 0, 1)\}$ of Example 6 of Section 1.4 is a basis for the solution space of the system of linear equations of that example.

Exercises

1. Without solving the system, find the dimension of the solution space for each of the following homogeneous systems of linear equations.
   (a) $2x + y + z = 0$
       $x + y - z = 0$
   (b) $2x + 3y + z - w = 0$
       $x + y + z + w = 0$
       $x \quad\quad + 2z + 4w = 0$
   (c) $x + y + z = 0$
       $3x + y - z = 0$
       $x + y + 2z = 0$
   (d) $x - y + z + w = 0$
       $2x - 2y + 2z + 2w = 0$
   (e) $x_1 + x_2 + \cdots + x_n = 0$
   (f) $x_1 + x_2 + \cdots + x_n = 0$
       $x_2 + \cdots + x_n = 0$
       $\vdots \quad\quad \vdots$
       $x_n = 0$

2. For each of the following systems, use Theorem 3.6.3 to determine whether or not the system is consistent.
   (a) $x + y = 1$
       $x + y = 2$
   (b) $x + y + 2z = 3$
       $x - y + z = 2$
       $x + 3y + 3z = 4$
   (c) $x + y - z = 3$
       $x + y = 2$
   (d) $x + y + z + w = 1$
       $x - y + z + w = 2$
       $x \quad\quad + z + w = 3$

3. Prove that if $\mathbf{z}_1$ is a solution to the system of linear equations $A\mathbf{x} = \mathbf{b}_1$, and if $\mathbf{z}_2$ is a solution to the system of linear equations $A\mathbf{x} = \mathbf{b}_2$, then
   (a) $\mathbf{z}_1 + \mathbf{z}_2$ is a solution to the system $A\mathbf{x} = \mathbf{b}_1 + \mathbf{b}_2$.

   (b) For any scalar s, $s\mathbf{z}_1$ is a solution to the system $A\mathbf{x} = s\mathbf{b}_1$.

4. For any system of m linear equations $A\mathbf{x} = \mathbf{b}$, and for any $m \times m$ invertible matrix C, prove that the system $(CA)\mathbf{x} = CB$ is equivalent to the original system.

5. Suppose that in solving the system of linear equations $A\mathbf{x} = \mathbf{b}$ by the method of Gaussian elimination, the vector form of the general solution is

   $$\mathbf{y} = \mathbf{z} + t_1\mathbf{x}_1 + t_2\mathbf{x}_2 + \cdots + t_k\mathbf{x}_k$$

   where $t_1, t_2, \ldots, t_k$ are parameters.
   (a) Prove that the set $(\mathbf{x}_1, \mathbf{x}_2, \ldots, \mathbf{x}_k)$ is a basis for the solution space of the corresponding homogeneous system of linear equations.
   (b) Prove that for any particular solution $\mathbf{y}$, the values of $t_1, \ldots, t_k$ are unique.

6. Prove this generalization of Corollary 3.6.2: If the *consistent* system of linear equations $A\mathbf{x} = \mathbf{b}$ has more unknowns than equations, then the system has infinitely many solutions.

*7. Use the program LINSYST to solve the following system of linear equations.

   $$2x + y - z + 3w = 2$$
   $$3x - y + 7z + 3w = 3$$

   Use your results to find a basis for the solution space of the corresponding homogeneous system of linear equations.

* This problem should be solved by using one of the programs noted in Appendix B.

Key Words

4

Coordinate Vectors and Matrix Representations

The components of a vector in R^2 or R^3 depend on how the coordinate axes are drawn. For example, consider the vector $\mathbf{z}$ of length 2 drawn in Figure 4.1.1(a). If we draw an xy-coordinate system so that $\mathbf{z}$ lies in the positive direction of the x-axis, the components of $\mathbf{z}$ are $(2, 0)$ [see Figure 4.1.1(b)]. However, if we draw the coordinate system so that the x-axis makes a $45°$ angle with $\mathbf{z}$, placing $\mathbf{z}$ in the fourth quadrant, then $\mathbf{z}$ has components $(\sqrt{2}, -\sqrt{2})$ [see Figure 4.1.1(c)]. As we shall see, the way that

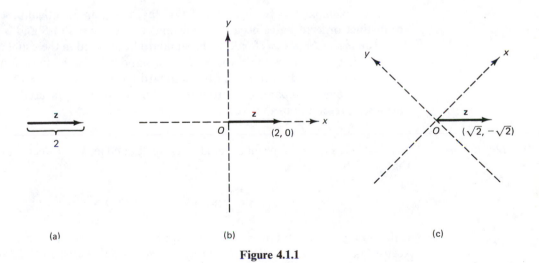

Figure 4.1.1

197

a matrix represents a linear transformation also depends on how the coordinate axes are drawn.

In this chapter we learn how to compare these various representations. This is important because for a given problem, there is often a favored coordinate system whose choice simplifies computations, making the problem easier to solve. For example, if the equation of a conic section (an ellipse, parabola, or hyperbola) has an xy-term, we can perform a rotation so that in the rotated coordinate system this mixed product is eliminated.

4.1 COORDINATE VECTORS AND CHANGE OF COORDINATE MATRICES

We begin by considering a precise formulation of a vector representation. The coordinates of a vector $\mathbf{x} = (x_1, x_2, \ldots, x_n)$ in R^n are identical to the coefficients used to represent $\mathbf{x}$ as a linear combination of the standard vectors:

$$\mathbf{x} = x_1\mathbf{e}_1 + x_2\mathbf{e}_2 + \cdots + x_n\mathbf{e}_n$$

If we represent $\mathbf{x}$ as a linear combination of the vectors of a different basis, we obtain different coefficients. We may think of these new coefficients as a new set of "coordinates" for $\mathbf{x}$. However, since the order of the coefficients depends on the order of the vectors in the basis, we must take care to note the order in which the vectors of a basis are listed. This consideration leads to the definition of "ordered basis."

Definition   An *ordered basis* is a basis which is listed in a specific order. That is, the first, second, and so on, vectors of the basis are specified.

For example, $S = \{\mathbf{e}_1, \mathbf{e}_2, \mathbf{e}_3\}$ and $S' = \{\mathbf{e}_2, \mathbf{e}_1, \mathbf{e}_3\}$ are identical bases for R^3, but are distinct ordered bases because the listings of the vectors in S and S' are distinct. The *standard ordered basis* for R^n is the standard basis listed in the usual order, that is, $\{\mathbf{e}_1, \mathbf{e}_2, \ldots, \mathbf{e}_n\}$. In the example above, S is the standard ordered basis for R^3, but S' is not even though S' does represent the standard basis for R^3.

We are now prepared to introduce a definition that can be used to compare the various representations of a vector.

Definition   Let $S = \{\mathbf{x}_1, \mathbf{x}_2, \ldots, \mathbf{x}_n\}$ be an ordered basis for R^n, and let $\mathbf{x}$ be a vector in R^n. Suppose that

$$\mathbf{x} = a_1\mathbf{x}_1 + a_2\mathbf{x}_2 + \cdots + a_n\mathbf{x}_n$$

is the (unique) representation of $\mathbf{x}$ as a linear combination of the vectors in S. Then the vector $(a_1, a_2, \ldots, a_n)$ is called the *coordinate vector of x relative to S*. The coefficient a_i

is called the *ith coordinate of x relative to S*. We denote the coordinate vector of x relative to S by $[x]_S$. That is,

$$[x]_S = (a_1, a_2, \ldots, a_n)$$

Recall that Theorem 3.1.4 allows us to conclude that the coefficients above are indeed unique.

We now relate this definition to the vector z in Figure 4.1.1. For $S = \{e_1, e_2\}$ we have that $[z]_S = (2, 0)$. For $S' = \{(1/\sqrt{2}, 1/\sqrt{2}), (-1/\sqrt{2}, 1/\sqrt{2})\}$, the result of rotating the vectors in S by $45°$ in the counterclockwise direction, $[z]_{S'} = (\sqrt{2}, -\sqrt{2})$.

From the discussion at the beginning of this section, it is clear that if S is the standard ordered basis for R^n, then $[x]_S = x$.

Example 1   Let $S = \{(1, 1, 1), (1, -1, 1), (1, 2, 2)\}$. It can be verified that S is an (ordered) basis for R^3. Suppose that for some vector x in R^3, $[x]_S = (1, 3, -2)$. We shall find x. By the definition of coordinate vector,

$$x = 1(1, 1, 1) + 3(1, -1, 1) - 2(1, 2, 2)$$
$$= (2, -6, 0)$$

Example 2   Let $S = \{(2, 3), (1, -1)\}$ and $x = (8, 7)$. We wish to find $[x]_S$. Suppose that $[x]_S = (x, y)$. Then

$$(8, 7) = x(2, 3) + y(1, -1)$$

Therefore,

$$2x + y = 8$$
$$3x - y = 7$$

This system has the solution $x = 3$ and $y = 2$. Hence,

$$(8, 7) = 3(2, 3) + 2(1, -1)$$

We conclude that $[x]_S = (3, 2)$.

If we interchange the vectors in S to form the ordered basis $S' = \{(1, -1), (2, 3)\}$, then $[x]_{S'} = (2, 3)$.

There is a simpler method for computing coordinate vectors than the one given in Example 2. It is based on matrix products. Recall from Example 4 of Section 2.2

that for an $m \times n$ matrix A with columns $^1A, \,^2A, \dots, ^nA$ and for an $n \times 1$ column vector $\mathbf{b}$,

$$\mathbf{b} = \begin{bmatrix} b_1 \\ b_2 \\ \vdots \\ b_n \end{bmatrix}$$

$$Ab = (^1A, \,^2A, \cdots, ^nA) \begin{bmatrix} b_1 \\ b_2 \\ \vdots \\ b_n \end{bmatrix}$$

$$= b_1 \,^1A + b_2 \,^2A + \cdots + b_n \,^nA$$

Theorem 4.1.1 *Let S be an ordered basis for R^n. Let P be the $n \times n$ matrix whose columns are given by the vectors in S. Then, for any column vector $\mathbf{x}$ in R^n, if $[\mathbf{x}]_S$ is represented as a column vector, we have*

$$[\mathbf{x}]_S = P^{-1}\mathbf{x}$$

Proof

First notice that P is invertible. This follows from the fact that the columns of P are linearly independent, and therefore, rank $P = n$. Next, let $\mathbf{x}$ be any vector in R^n represented as an $n \times 1$ column vector. Then there exist (unique) scalars $a_1, a_2, \dots, a_n$ such that

$$[\mathbf{x}]_S = \begin{bmatrix} a_1 \\ a_2 \\ \vdots \\ a_n \end{bmatrix}$$

Thus, if $S = \{\mathbf{x}_1, \mathbf{x}_2, \dots, \mathbf{x}_n\}$, we have that

$$\mathbf{x} = a_1\mathbf{x}_1 + a_2\mathbf{x}_2 + \cdots + a_n\mathbf{x}_n$$

$$= (\mathbf{x}_1, \mathbf{x}_2, \dots, \mathbf{x}_n) \begin{bmatrix} a_1 \\ a_2 \\ \vdots \\ a_n \end{bmatrix}$$

$$= P[\mathbf{x}]_S$$

Therefore, $P[\mathbf{x}]_S = \mathbf{x}$ and hence $[\mathbf{x}]_S = P^{-1}\mathbf{x}$. ∎

Example 3   We repeat the problem of Example 2 making use of Theorem 4.1.1. For $S = \{(2, 3),$ $(1, -1)\}$,

$$P = \begin{bmatrix} 2 & 1 \\ 3 & -1 \end{bmatrix} \quad \text{and} \quad P^{-1} = \begin{bmatrix} \frac{1}{5} & \frac{1}{5} \\ \frac{3}{5} & -\frac{2}{5} \end{bmatrix}$$

Thus, if we represent $\mathbf{x} = (8, 7)$ as a column vector, we have

$$[\mathbf{x}]_S = P^{-1}\mathbf{x}$$

$$= \begin{bmatrix} \frac{1}{5} & \frac{1}{5} \\ \frac{3}{5} & -\frac{2}{5} \end{bmatrix} \begin{bmatrix} 8 \\ 7 \end{bmatrix}$$

$$= \begin{bmatrix} 3 \\ 2 \end{bmatrix}$$

In the next example we use the fact from analytic geometry that the equation of an ellipse whose center is at the origin and whose major axis is horizontal is of the form

$$\frac{x^2}{a^2} + \frac{y^2}{b^2} = 1$$

where a and b are the lengths of the semimajor and semiminor axes, respectively (see Figure 4.1.2).

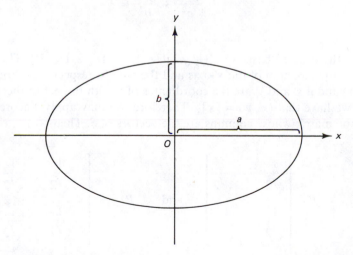

Figure 4.1.2

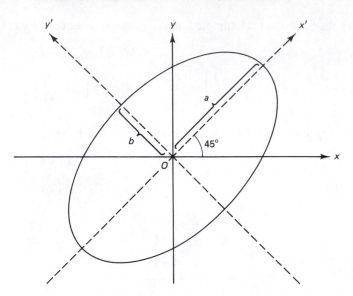

Figure 4.1.3

Example 4 Apply Theorem 4.1.1 to find the equation of an ellipse with center at the origin, whose major axis makes a 45° angle with the x-axis, whose semimajor axis has length a, and whose semiminor axis has length b (see Figure 4.1.3).

First draw an $x'y'$-coordinate system whose x'-axis makes a 45° angle with the x-axis as in Figure 4.1.3. The y'-axis is determined since it is perpendicular to the x'-axis. Then, with respect to this coordinate system, the equation of the ellipse drawn in Figure 4.1.3 is

$$\frac{(x')^2}{a^2} + \frac{(y')^2}{b^2} = 1$$

Consider the ordered basis $S = \{(1/\sqrt{2}, 1/\sqrt{2}), (-1/\sqrt{2}, 1/\sqrt{2})\}$. These are the unit vectors in the direction of the x'-axis and the y'-axis, respectively. For this reason, if $\mathbf{x} = (x, y)$ and if x' and y' are the coordinates of $\mathbf{x}$ with respect to the $x'y'$-coordinate system, we have that $(x', y') = [\mathbf{x}]_S$. Therefore, we may apply Theorem 4.1.1. Let P denote the matrix whose columns are the vectors of S. Then

$$P = \begin{bmatrix} \dfrac{1}{\sqrt{2}} & -\dfrac{1}{\sqrt{2}} \\[2mm] \dfrac{1}{\sqrt{2}} & \dfrac{1}{\sqrt{2}} \end{bmatrix} \quad \text{and} \quad P^{-1} = \begin{bmatrix} \dfrac{1}{\sqrt{2}} & \dfrac{1}{\sqrt{2}} \\[2mm] -\dfrac{1}{\sqrt{2}} & \dfrac{1}{\sqrt{2}} \end{bmatrix}$$

Thus, we have that

$$\begin{bmatrix} x' \\ y' \end{bmatrix} = P^{-1} \begin{bmatrix} x \\ y \end{bmatrix}$$

$$= \begin{bmatrix} \dfrac{1}{\sqrt{2}} & \dfrac{1}{\sqrt{2}} \\ -\dfrac{1}{\sqrt{2}} & \dfrac{1}{\sqrt{2}} \end{bmatrix} \begin{bmatrix} x \\ y \end{bmatrix}$$

$$= \frac{1}{\sqrt{2}} \begin{bmatrix} x + y \\ y - x \end{bmatrix}$$

Hence,

$$x' = \frac{1}{\sqrt{2}} (x + y) \qquad \text{and} \qquad y' = \frac{1}{\sqrt{2}} (y - x)$$

If we substitute these expressions for x' and y' into the equation of the ellipse, we have the equation of the ellipse in terms of x and y,

$$\frac{(x + y)^2}{2a^2} + \frac{(y - x)^2}{2b^2} = 1$$

Theorem 4.1.1 provides a method of converting the coordinates of $\mathbf{x}$ relative to the standard ordered basis into the coordinates of $\mathbf{x}$ relative to another ordered basis for R^n. The following theorem generalizes this process.

Theorem 4.1.2 *Let S and S' be ordered bases for R^n. Let P and Q be the square matrices whose columns are given by the vectors of S and S', respectively. Then, for any vector $\mathbf{x}$ in R^n*

$$[\mathbf{x}]_{S'} = Q^{-1} P [\mathbf{x}]_S$$

Proof
By Theorem 4.1.1, $[\mathbf{x}]_S = P^{-1} \mathbf{x}$ and $[\mathbf{x}]_{S'} = Q^{-1} \mathbf{x}$. Therefore, $\mathbf{x} = P[\mathbf{x}]_S$. Hence,

$$[\mathbf{x}]_{S'} = Q^{-1} \mathbf{x} = Q^{-1} P [\mathbf{x}]_S \qquad \blacksquare$$

Definition The matrix $Q^{-1} P$ in Theorem 4.1.2 is called the ***change of coordinate matrix from S to S'***.

Example 5   For the ordered bases

$$S = \{(1, 1, 1), (2, 3, 1), (1, 2, 1)\} \quad \text{and} \quad S' = \{(1, 0, 1), (0, 1, 1), (1, 0, 0)\}$$

(a) Find the change of coordinate matrix from S to S'.
(b) Given that $[\mathbf{x}]_S = (1, -2, 2)$, find $[\mathbf{x}]_{S'}$.

(a) Let

$$P = \begin{bmatrix} 1 & 2 & 1 \\ 1 & 3 & 2 \\ 1 & 1 & 1 \end{bmatrix} \quad \text{and} \quad Q = \begin{bmatrix} 1 & 0 & 1 \\ 0 & 1 & 0 \\ 1 & 1 & 0 \end{bmatrix}$$

Then

$$Q^{-1} = \begin{bmatrix} 0 & -1 & 1 \\ 0 & 1 & 0 \\ 1 & 1 & -1 \end{bmatrix}$$

and

$$Q^{-1}P = \begin{bmatrix} 0 & -1 & 1 \\ 0 & 1 & 0 \\ 1 & 1 & -1 \end{bmatrix}\begin{bmatrix} 1 & 2 & 1 \\ 1 & 3 & 2 \\ 1 & 1 & 1 \end{bmatrix}$$

$$= \begin{bmatrix} 0 & -2 & -1 \\ 1 & 3 & 2 \\ 1 & 4 & 2 \end{bmatrix}$$

(b) By Theorem 4.1.2,

$$[\mathbf{x}]_{S'} = Q^{-1}P[\mathbf{x}]_S$$

$$= \begin{bmatrix} 0 & -2 & -1 \\ 1 & 3 & 2 \\ 1 & 4 & 2 \end{bmatrix}\begin{bmatrix} 1 \\ -2 \\ 2 \end{bmatrix}$$

$$= \begin{bmatrix} 2 \\ -1 \\ -3 \end{bmatrix}$$

It is easy to verify that $(1, -2, 2)$ and $(2, -1, -3)$ are actually coordinate vectors for the same vector in $\mathbf{R}^3$. From $[\mathbf{x}]_S = (1, -2, 2)$, we have that

$$\mathbf{x} = 1(1, 1, 1) - 2(2, 3, 1) + 2(1, 2, 1)$$

$$= (-1, -1, 1)$$

From $[\mathbf{x}]_{S'} = (2, -1, -3)$, we have that

$$\mathbf{x} = 2(1, 0, 1) - (0, 1, 1) - 3(1, 0, 0)$$
$$= (-1, -1, 1)$$

Exercises

1. Given the ordered basis S and the coordinate vector $[\mathbf{x}]_S$, find the vector $\mathbf{x}$.
   (a) $S = \{(1, 2), (3, -1)\}$ and $[\mathbf{x}]_S = (2, 3)$
   (b) $S = \{(1, 2, 3), (1, -1, 1), (1, 0, 1)\}$ and $[\mathbf{x}]_S = (3, 1, -4)$
   (c) $S = \{(1, -1, 1, 1), (1, 1, -1, 1), (0, 1, 1, 1), (1, 1, 1, 1)\}$ and $[\mathbf{x}]_S = (1, -1, 2, 4)$

2. Given the ordered basis S and the vector $\mathbf{x}$, find the coordinate vector $[\mathbf{x}]_S$.
   (a) $S = \{(1, 2), (3, 5)\}$ and $\mathbf{x} = (2, 6)$
   (b) $S = \{(1, 1), (3, 5)\}$ and $\mathbf{x} = (1, -2)$
   (c) $S = \{(1, 1, 1), (2, 3, 2), (3, 2, 4)\}$ and $\mathbf{x} = (2, 1, -1)$
   (d) $S = \{(1, 1, 1), (2, 3, 2), (3, 2, 4)\}$ and $\mathbf{x} = (1, 0, 3)$

3. Use the method of Example 4 to find the equation of the ellipse with center at the origin, whose major axis makes a $30°$ angle with the x-axis, and whose semimajor axis has length 3 while its semiminor axis has length 2.

4. Use the method of Example 4 to find the equation of the hyperbola $xy = 1$ in terms of the $x'y'$-coordinate system, where the x'-axis makes an angle of $45°$ with the x-axis.

5. For each of the following, find the change of coordinate matrix from S to S'.
   (a) $S = \{(1, 2), (3, -1)\}$ and $S' = \{(2, 3), (3, 4)\}$
   (b) $S = \{(2, 1, 1), (1, -1, 1), (3, 1, 2)\}$ and $S' = \{(1, -1, 2), (2, -1, 7), (1, -1, 1)\}$

6. Using the change of coordinate matrices of Exercise 5 and the coordinate vector $[\mathbf{x}]_S$, find $[\mathbf{x}]_{S'}$.
   (a) $[\mathbf{x}]_S = (3, 7)$, where S and S' are the bases of Exercise 5(a)
   (b) $[\mathbf{x}]_S = (3, -1, 2)$, where S and S' are the bases of Exercise 5(b)

7. (a) Referring to Exercise 6(a), use the bases S and S' of Exercise 5(a) to compute the vector $\mathbf{x}$ from both $[\mathbf{x}]_S$ and $[\mathbf{x}]_{S'}$. You should get the same result in each case.
   (b) Referring to Exercise 6(b), use the bases S and S' of Exercise 5(b) to compute the vector $\mathbf{x}$ from both $[\mathbf{x}]_S$ and $[\mathbf{x}]_{S'}$. You should get the same result in each case.

8. If M is the change of coordinate matrix from S to S', show that M^{-1} is the change of coordinate matrix from S' to S.

9. Prove: If M_1 is the change of coordinate matrix from S to S', and if M_2 is the change of coordinate matrix from S' to S'', then $M_2 M_1$ is the change of coordinate matrix from S to S''.

10. Let S be an ordered basis for R^n. Prove that for any vectors $\mathbf{x}$ and $\mathbf{y}$ in R^n, and for any scalar c, we have:
    (a) $[\mathbf{x} + \mathbf{y}]_S = [\mathbf{x}]_S + [\mathbf{y}]_S$.
    (b) $[c\mathbf{x}]_S = c[\mathbf{x}]_S$.

11. Let S be an ordered basis for R^n. Let T be the mapping from R^n to R^n defined by $T(\mathbf{x}) = [\mathbf{x}]_S$ for $\mathbf{x}$ in R^n.
    (a) Use Exercise 10 to prove that $\mathbf{T}$ is an invertible linear transformation.
    (b) Describe the standard matrix of T.

*12. For the ordered bases

$$S = \{(1, 3, 1), (1, 4, 3), (2, 5, 1)\}$$

and

$$S' = \{(1, 2, 1, -1), (1, 3, 0, 2), (2, 3, 4, -6), (1, 3, 1, 5)\}$$

use the program MATRIX to compute:
(a) $(1, 6, 2)_S$ (d) $(1, 2, 3, 4)_{S'}$
(b) $(-1, 1, 4)_S$ (e) $(2, 3, -3, 1)_{S'}$
(c) $(2, 0, 8)_S$ (f) $(1, 0, -5, 2)_{S'}$

* This problem should be solved by using one of the programs noted in Appendix B.

4.2 MATRIX REPRESENTATIONS OF A LINEAR TRANSFORMATION

In the preceding section we considered the problem of representing a vector according to a choice of an ordered basis. In this section we consider the analogous problem for linear transformations. We shall restrict our study to those transformations which map R^n into itself.

We are familiar with the relationship between a linear transformation on R^n and its standard $n \times n$ matrix. We have used the term "standard" to indicate that this representation has something to do with the standard ordered basis for R^n. In fact, if $T: R^n \to R^n$ is linear and A is the standard matrix of T, then the jth column of A is determined from

$$T(\mathbf{e}_j) = a_{1j}\mathbf{e}_1 + a_{2j}\mathbf{e}_2 + \cdots + a_{nj}\mathbf{e}_j$$

That is,

$$^jA = [T(\mathbf{e}_j)]_S$$

where S is the standard ordered basis of R^n.

If we apply this approach to another ordered basis for R^n, we usually obtain a matrix which differs from the standard matrix of T. This leads us to the following definition.

Definition   Let $T: R^n \to R^n$ be linear, and let $S = \{\mathbf{x}_1, \mathbf{x}_2, \ldots, \mathbf{x}_n\}$ be an ordered basis for R^n. Let $[T]_S$ denote the $n \times n$ matrix whose jth column is the coordinate vector for $T(\mathbf{x}_j)$ relative to S. Then $[T]_S$ is called the ***matrix representation of T relative to the ordered basis S***.

We shall see that it is often possible to choose an ordered basis S so that the matrix representation $[T]_S$ has a simpler form than the standard matrix of T.

Example 1   Consider the linear transformation $T: R^3 \to R^3$ defined by

$$T(x, y, z) = (x + y, z, -x)$$

and the ordered basis $S = \{(1, 1, 1), (1, 2, 1), (0, -1, 1)\}$ for R^3. Find the matrix representation $[T]_S$ of T relative to S.

The first column of $[T]_S$ is the coordinate vector for $T(1, 1, 1)$ relative to S. To obtain this vector, we first compute $T(1, 1, 1) = (2, 1, -1)$. Next, applying Theorem 4.1.1, we form the matrix

$$P = \begin{bmatrix} 1 & 1 & 0 \\ 1 & 2 & -1 \\ 1 & 1 & 1 \end{bmatrix}$$

find its inverse,

$$P^{-1} = \begin{bmatrix} 3 & -1 & -1 \\ -2 & 1 & 1 \\ -1 & 0 & 1 \end{bmatrix}$$

and compute the product

$$P^{-1}(T(1, 1, 1)) = \begin{bmatrix} 3 & -1 & -1 \\ -2 & 1 & 1 \\ -1 & 0 & 1 \end{bmatrix} \begin{bmatrix} 2 \\ 1 \\ -1 \end{bmatrix}$$

$$= \begin{bmatrix} 6 \\ -4 \\ -3 \end{bmatrix}$$

Similarly, to find the second column of $[T]_S$, we compute

$$P^{-1}(T(1, 2, 1)) = \begin{bmatrix} 3 & -1 & -1 \\ -2 & 1 & 1 \\ -1 & 0 & 1 \end{bmatrix} \begin{bmatrix} 3 \\ 1 \\ -1 \end{bmatrix}$$

$$= \begin{bmatrix} 9 \\ -6 \\ -4 \end{bmatrix}$$

Finally, for the third column of $[T]_S$, we have

$$P^{-1}(T(0, -1, 1)) = \begin{bmatrix} 3 & -1 & -1 \\ -2 & 1 & 1 \\ -1 & 0 & 1 \end{bmatrix} \begin{bmatrix} -1 \\ 1 \\ 0 \end{bmatrix}$$

$$= \begin{bmatrix} -4 \\ 3 \\ 1 \end{bmatrix}$$

Putting these columns together, we obtain the matrix

$$[T]_S = \begin{bmatrix} 6 & 9 & -4 \\ -4 & -6 & 3 \\ -3 & -4 & 1 \end{bmatrix}$$

A careful study of the procedure used in Example 1 suggests a general method for computing $[T]_S$ from the standard matrix of T. We state this method as Theorem

4.2.1. When reading the proof of the theorem, keep in mind the procedure used in Example 1.

Theorem 4.2.1   *Let $T: R^n \to R^n$ be linear, and let $S = \{\mathbf{x}_1, \mathbf{x}_2, \ldots, \mathbf{x}_n\}$ be an ordered basis for R^n. If P is the $n \times n$ matrix whose jth column is $\mathbf{x}_j$, and if A is the standard matrix of T, we have*

$$[T]_S = P^{-1}AP \quad \text{or} \quad A = P[T]_S P^{-1}$$

Proof

Let $B = [T]_S$. Then as in Example 1, the jth column ${}^j B$ of B is computed as

$${}^j B = P^{-1}T(\mathbf{x}_j)$$

Since A is the standard matrix of T, we have that $T(\mathbf{x}_j) = A\mathbf{x}_j$. Therefore, ${}^j B = P^{-1}A\mathbf{x}_j$. Hence,

$$\begin{aligned}
B &= ({}^1 B, {}^2 B, \cdots, {}^n B) \\
&= (P^{-1}A\mathbf{x}_1, P^{-1}A\mathbf{x}_2, \cdots, P^{-1}A\mathbf{x}_n) \\
&= P^{-1}A(\mathbf{x}_1, \mathbf{x}_2, \cdots, \mathbf{x}_n) \\
&= P^{-1}AP
\end{aligned}$$

Now solve this equation for A to obtain that $A = PBP^{-1}$. ∎

Example 2   Apply Theorem 4.2.1 to recompute the matrix $[T]_S$ of Example 1.

Let A be the standard matrix of the transformation $T(x, y, z) = (x + y, z, -x)$ of Example 1. Then

$$A = \begin{bmatrix} 1 & 1 & 0 \\ 0 & 0 & 1 \\ -1 & 0 & 0 \end{bmatrix}$$

Therefore, by Theorem 4.2.1 we have

$$\begin{aligned}
[T]_S &= P^{-1}AP \\
&= \begin{bmatrix} 3 & -1 & -1 \\ -2 & 1 & 1 \\ -1 & 0 & 1 \end{bmatrix} \begin{bmatrix} 1 & 1 & 0 \\ 0 & 0 & 1 \\ -1 & 0 & 0 \end{bmatrix} \begin{bmatrix} 1 & 1 & 0 \\ 1 & 2 & -1 \\ 1 & 1 & 1 \end{bmatrix} \\
&= \begin{bmatrix} 6 & 9 & -4 \\ -4 & -6 & 3 \\ -3 & -4 & 1 \end{bmatrix}
\end{aligned}$$

Reflections in the *xy*-Plane

Let L be a line in R^2 which passes through the origin. The transformation $T: R^2 \to R^2$ defined according to Figure 4.2.1(a) is called the **reflection** about L. It can be shown that T is a linear transformation. However, it is not obvious how one would proceed to obtain the standard matrix of T. In the next two examples we shall see how matrix representations can be applied to simplify this problem. The trick is to choose a basis for R^2 so that $[T]_S$ is a diagonal matrix.

Example 3   For the reflection T defined above, describe how to find an ordered basis S for R^2 so that $[T]_S$ is a diagonal matrix.

Let $\mathbf{x}_1$ be a nonzero vector parallel to L, and let $\mathbf{x}_2$ be a nonzero vector orthogonal to L [see Figure 4.2.1(b)]. Let $S = \{\mathbf{x}_1, \mathbf{x}_2\}$. Since $\mathbf{x}_1$ and $\mathbf{x}_2$ are not collinear, S is linearly independent and hence is an ordered basis for R^2. Clearly,

$$T(\mathbf{x}_1) = \mathbf{x}_1$$
$$= (1)\mathbf{x}_1 + (0)\mathbf{x}_2$$

and

$$T(\mathbf{x}_2) = -\mathbf{x}_2$$
$$= (0)\mathbf{x}_1 + (-1)\mathbf{x}_2$$

Thus,

$$[T(\mathbf{x}_1)]_S = \begin{bmatrix} 1 \\ 0 \end{bmatrix} \quad \text{and} \quad [T(\mathbf{x}_2)]_S = \begin{bmatrix} 0 \\ -1 \end{bmatrix}$$

We conclude that

$$[T]_S = \begin{bmatrix} 1 & 0 \\ 0 & -1 \end{bmatrix}$$

Example 4   Apply Theorem 4.2.1 and the method in Example 3 to compute $T(x, y)$, where T is the reflection of R^2 about the line L with equation $y = 2x$.

Notice that the vector $\mathbf{x}_1 = (1, 2)$ lies in the direction of L, and that the vector $\mathbf{x}_2 = (-2, 1)$ lies in a direction orthogonal to L. Let $S = \{\mathbf{x}_1, \mathbf{x}_2\}$. Then, as in Example 3, S is an ordered basis for R^2, and

$$[T]_S = \begin{bmatrix} 1 & 0 \\ 0 & -1 \end{bmatrix}$$

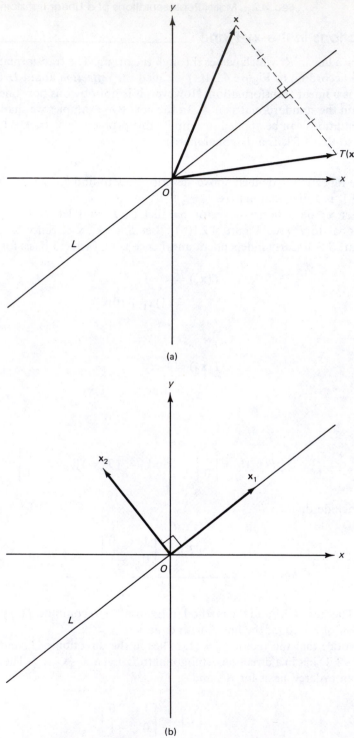

(a)

(b)

Figure 4.2.1

Let A be the standard matrix of this reflection, and let P be the matrix whose jth column is the jth vector of S. Then

$$P = \begin{bmatrix} 1 & -2 \\ 2 & 1 \end{bmatrix} \quad \text{and} \quad P^{-1} = \begin{bmatrix} \frac{1}{5} & \frac{2}{5} \\ -\frac{2}{5} & \frac{1}{5} \end{bmatrix}$$

By Theorem 4.2.1,

$$A = P[T]_S P^{-1}$$

$$= \begin{bmatrix} 1 & -2 \\ 2 & 1 \end{bmatrix} \begin{bmatrix} 1 & 0 \\ 0 & -1 \end{bmatrix} \begin{bmatrix} \frac{1}{5} & \frac{2}{5} \\ -\frac{2}{5} & \frac{1}{5} \end{bmatrix}$$

$$= \begin{bmatrix} -\frac{3}{5} & \frac{4}{5} \\ \frac{4}{5} & \frac{3}{5} \end{bmatrix}$$

Since

$$A \begin{bmatrix} x \\ y \end{bmatrix} = \begin{bmatrix} -\frac{3}{5} & \frac{4}{5} \\ \frac{4}{5} & \frac{3}{5} \end{bmatrix} \begin{bmatrix} x \\ y \end{bmatrix}$$

$$= \begin{bmatrix} -\frac{3}{5}x + \frac{4}{5}y \\ \frac{4}{5}x + \frac{3}{5}y \end{bmatrix}$$

we have that $T(x, y) = (-\frac{3}{5}x + \frac{4}{5}y, \frac{4}{5}x + \frac{3}{5}y)$.

Theorem 4.2.1 provides us with a method for converting the standard matrix of a linear transformation T into the matrix representation of T relative to an ordered basis S. A more general result would be one that describes a method for converting the matrix representation of T relative to an ordered basis S into the matrix representation of T relative to another ordered basis S'.

Corollary 4.2.2

Let $T: R^n \to R^n$ be linear, and let S and S' be ordered bases for R^n. Let M be the change of coordinate matrix from S' to S. Then

$$[T]_{S'} = M^{-1}[T]_S M$$

Proof

Let P be the matrix whose jth column is the jth vector of S, and let Q be the matrix whose jth column is the jth vector of S'. Then, by the results of Section 4.1, $M = P^{-1}Q$. Furthermore, by Theorem 4.2.1, we have

$$A = P[T]_S P^{-1} \quad \text{and} \quad A = Q[T]_{S'} Q^{-1}$$

Combining these two equations, we have that

$$Q[T]_{S'}Q^{-1} = P[T]_S P^{-1}$$

Therefore, solving for $[T]_{S'}$, we have

$$[T]_{S'} = Q^{-1}P[T]_S P^{-1}Q$$
$$= (P^{-1}Q)^{-1}[T]_S P^{-1}Q$$
$$= M^{-1}[T]_S M$$

$\blacksquare$

Example 5   Let $S = \{(1, 2), (1, 3)\}$ and $S' = \{(1, -1), (1, 0)\}$. Suppose that for a particular linear transformation $T: R^2 \to R^2$,

$$[T]_S = \begin{bmatrix} 3 & 2 \\ -1 & 1 \end{bmatrix}$$

We shall find $[T]_{S'}$. Let

$$P = \begin{bmatrix} 1 & 1 \\ 2 & 3 \end{bmatrix} \quad \text{and} \quad Q = \begin{bmatrix} 1 & 1 \\ -1 & 0 \end{bmatrix}$$

Then

$$M = P^{-1}Q$$
$$= \begin{bmatrix} 3 & -1 \\ -2 & 1 \end{bmatrix}\begin{bmatrix} 1 & 1 \\ -1 & 0 \end{bmatrix}$$
$$= \begin{bmatrix} 4 & 3 \\ -3 & -2 \end{bmatrix}$$

So

$$M^{-1} = \begin{bmatrix} -2 & -3 \\ 3 & 4 \end{bmatrix}$$

Thus,

$$[T]_{S'} = M^{-1}[T]_S M$$
$$= \begin{bmatrix} -2 & -3 \\ 3 & 4 \end{bmatrix}\begin{bmatrix} 3 & 2 \\ -1 & 1 \end{bmatrix}\begin{bmatrix} 4 & 3 \\ -3 & -2 \end{bmatrix}$$
$$= \begin{bmatrix} 9 & 5 \\ -10 & -5 \end{bmatrix}$$

We are familiar with the use of the standard matrix of a linear transformation to compute the image of a vector. The matrix representation of a linear transformation is also used to compute the image of a vector, but the computation is in terms of coordinate vectors.

Theorem 4.2.3 *Let $T: R^n \to R^n$ be linear and let S be an ordered basis for R^n. Then, for any vector $\mathbf{x}$ in R^n,*

$$[T(\mathbf{x})]_S = [T]_S[\mathbf{x}]_S$$

Proof

Let A be the standard matrix of T, and let P be the matrix whose jth column is the jth vector of S. Then

$$[T(\mathbf{x})]_S = [A\mathbf{x}]_S$$
$$= P^{-1}A\mathbf{x}$$
$$= P^{-1}APP^{-1}\mathbf{x}$$
$$= (P^{-1}AP)(P^{-1}\mathbf{x})$$
$$= [T]_S[\mathbf{x}]_S \qquad \blacksquare$$

Example 6   Let $S = \{(1, 2, 3), (2, -1, 1), (1, 1, 1)\}$. Then S is an ordered basis for R^3. Suppose that for a particular linear transformation $T: R^3 \to R^3$, we have

$$[T]_S = \begin{bmatrix} 1 & 2 & 1 \\ -1 & 1 & 0 \\ 0 & 1 & 2 \end{bmatrix}$$

and that for a particular vector $\mathbf{x}$ in R^3 we have $[\mathbf{x}]_S = (2, 1, -3)$. Then

$$[T(\mathbf{x})]_S = [T]_S[\mathbf{x}]_S$$
$$= \begin{bmatrix} 1 & 2 & 1 \\ -1 & 1 & 0 \\ 0 & 1 & 2 \end{bmatrix} \begin{bmatrix} 2 \\ 1 \\ -3 \end{bmatrix}$$
$$= \begin{bmatrix} 1 \\ -1 \\ -5 \end{bmatrix}$$

Notice that the ordered basis S was not used directly in the computation.

Example 7   Recall the reflection T and the basis S in Example 4. Given a vector $\mathbf{x}$ in R^2 such that

$$[\mathbf{x}]_S = \begin{bmatrix} 2 \\ 3 \end{bmatrix}$$

Use Theorem 4.2.3 to compute $T(\mathbf{x})$.

By Theorem 4.2.3 we have

$$[T(\mathbf{x})]_S = \begin{bmatrix} 1 & 0 \\ 0 & -1 \end{bmatrix} \begin{bmatrix} 2 \\ 3 \end{bmatrix} = \begin{bmatrix} 2 \\ -3 \end{bmatrix}$$

Therefore,

$$T(\mathbf{x}) = 2(1, 2) - 3(-2, 1) = (8, 1)$$

Notice that $T(\mathbf{x})$ can be computed directly using the expression derived in Example 4. Since

$$\mathbf{x} = 2(1, 2) + 3(-2, 1) = (-4, 7)$$

we have that

$$T(-4, 7) = \left(-\frac{3}{5}(-4) + \frac{4}{5}(7), \frac{4}{5}(-4) + \frac{3}{5}(7) \right)$$

$$= (8, 1)$$

A brief examination of Corollary 4.2.2 shows that if A and B are matrices that represent the same linear transformation $T: R^n \to R^n$, then $B = M^{-1}AM$ for some invertible $n \times n$ matrix M. This suggests the following definition.

Definition   If the square matrices A and B are related so that $B = M^{-1}AM$ for some invertible matrix M, we say that A is ***similar*** to B, and write $A \sim B$.

It now follows by Corollary 4.2.2 that if two square matrices represent the same linear transformation, one is similar to the other. This "similarity" relation has certain formal properties which are worth noting.

Theorem 4.2.4   *Let A, B, and C be square matrices. Then:*
  (a) $A \sim A$.
  (b) If $A \sim B$, then $B \sim A$.
  (c) If $A \sim B$, and $B \sim C$, then $A \sim C$.

Proof

(a) Since I is invertible and $A = I^{-1}AI$, part (a) follows.

(b) Suppose that $A \sim B$. Then there exists an invertible matrix M such that $B = M^{-1}AM$. Since M^{-1} is invertible and $(M^{-1})^{-1} = M$, we have that $A = MBM^{-1} = (M^{-1})^{-1}B(M^{-1})$. Therefore, $B \sim A$.

(c) Now suppose that $A \sim B$ and that $B \sim C$. Then there exist invertible matrices M and N such that $B = M^{-1}AM$ and $C = N^{-1}BN$. Thus,

$$
\begin{aligned}
C &= N^{-1}BN \\
&= N^{-1}(M^{-1}AM)N \\
&= (N^{-1}M^{-1})A(MN) \\
&= (MN)^{-1}A(MN)
\end{aligned}
$$

Therefore, $A \sim C$.   ∎

In view of Theorem 4.2.4(b) we can disregard the order in which the matrices are listed. We often say that matrices A and B are ***similar*** rather than that A is similar to B. As we noted earlier, two matrices that represent the same linear transformation are similar. It is of some interest that the converse is also true. That is, if two matrices are similar, they may represent the same linear transformation relative to different ordered bases. This result is aesthetically pleasing, but it will not be very useful for our purposes. Its proof is left as Exercise 13.

A computational method for determining when two matrices are similar is beyond the scope of this book. See, for example, page 332 of (1) in the References.

Similar matrices share many properties. Among these is rank. This fact can be used to prove that the rank of any matrix representation of a linear transformation is equal to the rank of the transformation (see Corollary 4.2.6).

Theorem 4.2.5   *Similar matrices have the same rank.*

Proof

Suppose that A and B are similar. Then there exists an invertible matrix M such that $B = M^{-1}AM$. By Exercise 12 of Section 3.5,

$$\text{rank } B = \text{rank } (M^{-1}AM) = \text{rank } A \qquad ∎$$

Example 8   Apply Theorem 4.2.5 to show that the matrices

$$
\begin{bmatrix} 1 & 2 & 1 \\ 2 & 1 & 2 \\ 1 & 0 & 2 \end{bmatrix}
\quad \text{and} \quad
\begin{bmatrix} 1 & 3 & 1 \\ -1 & 2 & 0 \\ 1 & 8 & 2 \end{bmatrix}
$$

are not similar.

Transforming these matrices into triangular form, we have

$$\begin{bmatrix} 1 & 2 & 1 \\ 0 & -3 & 0 \\ 0 & 0 & 1 \end{bmatrix} \quad \text{and} \quad \begin{bmatrix} 1 & 3 & 1 \\ 0 & 5 & 1 \\ 0 & 0 & 0 \end{bmatrix}$$

It follows that the first matrix has rank equal to 3, and the second has rank equal to 2. Therefore, the matrices cannot be similar.

Corollary 4.2.6

If B is a matrix representation for the linear transformation $T: R^n \to R^n$, then rank $B =$ rank T.

Proof
Suppose that B is a matrix representation of the linear transformation $T: R^n \to R^n$. By Theorem 4.2.1, $B = P^{-1}AP$, where A is the standard matrix of T and P is an invertible matrix. Therefore, A and B are similar. By Theorem 4.2.5, rank $B =$ rank A. But A and T have the same rank. Therefore, rank $B =$ rank T.   ∎

Example 9

Let $T: R^3 \to R^3$ be linear. Suppose that for some basis $S = \{\mathbf{x}, \mathbf{y}, \mathbf{z}\}$ for R^3, $T(\mathbf{x}) = \mathbf{x}$, $T(\mathbf{y}) = \mathbf{x} + 2\mathbf{y}$, and $T(\mathbf{z}) = \mathbf{x} + 2\mathbf{y} + 3\mathbf{z}$. Find the rank of T.
First note that

$$[T]_S = \begin{bmatrix} 1 & 1 & 1 \\ 0 & 2 & 2 \\ 0 & 0 & 3 \end{bmatrix}$$

Since the rank of this matrix is 3, it follows that rank $T = 3$.

Example 10

Let $T: R^2 \to R^2$ be a linear transformation. Suppose that for some ordered basis $S = \{\mathbf{x}, \mathbf{y}\}$ of R^2 we have that $T(\mathbf{x}) = \mathbf{x} + 2\mathbf{y}$ and $T(\mathbf{y}) = 2\mathbf{x} - \mathbf{y}$. Show that T is invertible. First note that

$$[T]_S = \begin{bmatrix} 1 & 2 \\ 2 & -1 \end{bmatrix}$$

Since the rank of this matrix is 2, we have that rank $T = 2$. Therefore, by Corollary 3.4.12, T is invertible.

This last example suggests the following corollary.

Corollary 4.2.7

Let $T: R^n \to R^n$ be linear. Let B be a matrix representation of T. Then T is invertible if and only if B is invertible.

We omit the proof (see Exercise 9).

Exercises

1. For the linear transformation T and the ordered basis S, find $[T]_S$.
   (a) $T(x, y) = (2x + y, x - y)$ and $S = \{(2, 1), (1, 0)\}$
   (b) $T(x, y) = (x - y, y)$ and $S = \{(2, 3), (1, 1)\}$
   (c) $T(x, y) = (x + 2y, x + y)$ and $S = \{(1, 1), (2, 1)\}$
   (d) $T(x, y, z) = (x + y, y - 2z, 2x - y + 3z)$ and $S = \{(1, 1, 1), (2, 3, 2), (1, 2, 2)\}$
   (e) $T(x, y, z, w) = (x + y, y - z, x + 2w, y - z + 3w)$ and $S = \{(1, -1, 2, 3), (1, -2, 1, 4), (1, -2, 0, 3), (0, 1, 1, -2)\}$

2. For the given ordered basis S and the description of the linear transformation T:
   (i)   Find $[T]_S$.
   (ii) Use the result of (i) to find the standard matrix for T.
   (iii) Use the result of (ii) to find a general expression for $T(\mathbf{x})$.
   (a) $S = \{(1, 1), (1, 2)\}$, $T(1, 1) = (1, 2)$, and $T(1, 2) = (1, 1)$
   (b) $S = \{(1, 3), (1, 0)\}$, $T(1, 3) = (1, 3) + (1, 0)$, and $T(1, 0) = (1, 3) - (1, 0)$
   (c) $S = \{(1, 0, 1), (0, 1, 0), (1, 1, 0)\}$, $T(1, 0, 1) = (0, -1, 0)$, $T(0, 1, 0) = (1, 1, 0)$, and $T(1, 1, 0) = (1, 1, 1)$

3. For the given ordered basis S for R^n, find $[T]_S$, where $T: R^n \to R^n$ is the linear transformation which satisfies the given equations.
   (a) $S = \{\mathbf{x}, \mathbf{y}\}$, $T(\mathbf{x}) = \mathbf{x} + 2\mathbf{y}$, and $T(\mathbf{y}) = -\mathbf{x} - \mathbf{y}$
   (b) $S = \{\mathbf{x}, \mathbf{y}, \mathbf{z}\}$, $T(\mathbf{x}) = \mathbf{x} + \mathbf{y}$, $T(\mathbf{y}) = 2\mathbf{x} - \mathbf{z}$, and $T(\mathbf{z}) = \mathbf{x} - 3\mathbf{y} + 2\mathbf{z}$
   (c) $S = \{\mathbf{x}_1, \mathbf{x}_2, \ldots, \mathbf{x}_n\}$, $T(\mathbf{x}_i) = \mathbf{x}_{i+1}$ if $i < n$, and $T(\mathbf{x}_n) = \mathbf{x}_1$

4. Consider the reflection $T: R^2 \to R^2$ about the line with equation $x + y = 0$.
   (a) Find a basis S for R^2 such that

   $$[T]_S = \begin{bmatrix} 1 & 0 \\ 0 & -1 \end{bmatrix}$$

   (b) Find the standard matrix for T.

   (c) Use the result from part (b) to find an expression for $T(x, y)$.

5. Let $T: R^2 \to R^2$ be the reflection about the line L with equation $y = mx$. Find an expression for $T(x, y)$.

6. Let L be a line in R^2 which passes through the origin. The transformation $T: R^2 \to R^2$ defined according to Figure 4.2.2 is called the *projection* on L. It can be shown that T is a linear transformation. Now suppose that L is

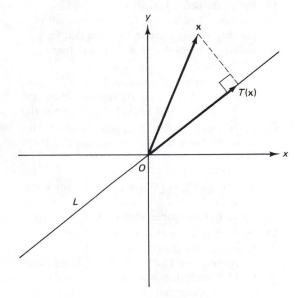

Figure 4.2.2

the line with equation $y = 2x$. Let T be the projection on L and let $S = \{(1, 2), (-2, 1)\}$.
   (a) Find $[T]_S$.
   (b) Use part (a) to find the standard matrix for T.
   (c) Use part (b) to find an expression for $T(x, y)$.

7. Let $T: R^2 \to R^2$ be the projection on the line L with equation $y = mx$ (see Exercise 6). Find an expression for $T(x, y)$.

8. For $n \times n$ matrices A and B, prove:
   (a) If $A \sim B$, then for any scalar c, we have that $cA \sim cB$.
   (b) For any scalar c, we have that $A \sim cI$ if and only if $A = cI$.
   (c) For any scalars c and d, $cI \sim dI$ if and only if $c = d$.
   (d) If $A \sim B$, then $A^n \sim B^n$ for any positive integer n.
   (e) If $A \sim B$ and A is invertible, then B is invertible and $A^{-1} \sim B^{-1}$.
   (f) If $A \sim B$, then $A^t \sim B^t$.

9. Prove Corollary 4.2.7.

10. Recall the definition of "trace" given in Exercise 15 of Section 2.1. Use the result of part (d) of that exercise to prove that for square matrices A and B, if $A \sim B$, then $\text{tr}(A) = \text{tr}(B)$.

11. For a linear transformation $T: R^n \to R^n$, prove that if A and B are matrix representations for T, then $\text{tr}(A) = \text{tr}(B)$. (*Hint*: Use Exercise 10.)

12. Let T and U be linear transformations from R^n to R^n. Let S be an ordered basis for R^n. Prove:
   (a) $[UT]_S = [U]_S[T]_S$.
   (b) $[U + cT]_S = [U]_S + c[T]_S$ for any scalar c.
   (c) If T is invertible, then $[T^{-1}]_S = ([T]_S)^{-1}$.

13. Prove that if A and B are similar $n \times n$ matrices, then there exists a linear transformation $T: R^n \to R^n$ and ordered bases S and S' such that $A = [T]_S$ and $B = [T]_{S'}$. (*Hint*: Suppose that $B = M^{-1}AM$. Let $T = L_A$. Choose S to be the standard ordered basis for R^n, and S' to be the ordered basis whose jth vector is the jth column of M.)

14. Recall right multiplication R_A by a matrix A as defined in Exercise 28 of Section 2.4. Prove that for any $n \times n$ matrix A, if S is the standard ordered basis for R^n, then:
   (a) $[L_A]_S = A$.
   (b) $[R_A]_S = A^t$.

The following definitions pertain to Exercises 15 and 16.

Definitions. Let V be a plane in R^3 which passes through the origin, and let $\mathbf{x}$ be a vector in R^3. The vector $\mathbf{x}'$ in V determined by the perpendicular from the tip of $\mathbf{x}$ to V is called the *projection* of $\mathbf{x}$ on V (see Figure 4.2.3). The extension of this perpendicular an equal distance on the other side of V determines a vector $\mathbf{x}''$ called the *reflection* of X about V (see Figure 4.2.3). The transformation

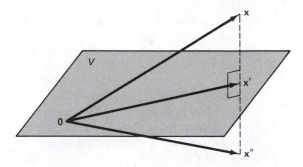

Figure 4.2.3

$T: R^3 \to R^3$ which takes any vector to its projection on V is called the *projection* on V. Similarly, the transformation which takes any vector of R^3 to its reflection about V is called the *reflection* about V. It can be shown that projections and reflections in R^3 are linear transformations.

15. Let V be the plane in R^3 with equation $x + 2y - 3z = 0$. Let $T: R^3 \to R^3$ be the projection of R^3 on V. Let $S = \{(3, 0, 1), (-2, 1, 0), (1, 2, -3)\}$.
   (a) Show that the first two vectors of S lie in V and that the third vector is perpendicular to V.
   (b) Find $[T]_S$.
   (c) Use part (b) to find the standard matrix for T.
   (d) Use part (c) to find an expression for $T(x, y, z)$.

16. Let T be the reflection about the plane V of Exercise 15.
   (a) Find the standard matrix for T.
   (b) Find an expression for $T(x, y, z)$.

*17. For the ordered bases

$$S = \{(1, 1, 1, 2), (2, 3, 3, 3), (1, 3, 4, 1),$$

$$(4, 5, 8, 8)\}$$

and

$S' = \{(1, 1, 2, 0, 1), (3, 2, 5, 1, 2),$

$(1, 3, 3, -2, 2), (0, 1, 2, -1, 3), (4, 1, 7, 9, 1)\}$

and the linear transformations T and U defined by

$T(x, y, z, w) = (x - y + z + 2w, 2x - 3w,$

$x + y + z, 3z - w)$

and

$U(x, y, z, u, v) = (2x - y + z + 3u,$

$x - z + u + 2v, y - v, 2z + u - v, x - y + 3u)$

use the program MATRIX to compute:
(a) $[T]_S$ (b) $[U]_{S'}$

*18. Do Exercises 15(c) and 16(a) with the assistance of the program MATRIX.

* This problem should be solved by using one of the programs noted in Appendix B.

Key Words

Change of coordinate matrix   203
Coordinate vector   198
Matrix representation of a linear transformation   206

Ordered basis   198
Similar matrices   214
Standard ordered basis   198

5

Determinants

In this chapter we explore the concept of a "determinant" of a square matrix. Although determinants were originally developed in the eighteenth century and have played an important role in many areas of mathematics, particularly linear algebra, their use as a computational tool has greatly diminished in recent years. One reason for this is related to the fact that the introduction of high-speed computers has drawn attention to the efficiency of various methods or algorithms in the computations of solutions of large systems of (linear) equations. In this regard, determinants do not provide a very practical tool in solving such systems. We will use determinants primarily in the development of "eigenvalues" in Chapter 6.

5.1 DETERMINANTS OF 2 × 2 MATRICES

In the first three chapters we have seen that many of the well-known arithmetic properties of scalars have extended to matrices. However, some properties, for example, the commutativity of multiplication does not extend to matrices. Another important property that does not generalize to matrices is the one dealing with the existence of a (multiplicative) inverse. For a scalar, the test for an inverse is easy. The scalar possesses an inverse if and only if it does not equal zero. On the other hand, we have seen many nonzero matrices that have no inverses. Not all is lost, however. From Theorem 3.4.6 we have that if

$$A = \begin{bmatrix} a & b \\ c & d \end{bmatrix}$$

then A is invertible if and only if $ad - bc \neq 0$. We use this fact to motivate the definition of determinant.

Definition   Let A be the 2×2 matrix given above. We define the **determinant** of A written **det** A or $|A|$ to be the scalar $ad - bc$, where A has the form given above.

    With this definition and the remark above, we have that A is invertible if and only if $|A| \neq 0$.

Example 1   If

$$A = \begin{bmatrix} 6 & 5 \\ 2 & -1 \end{bmatrix} \quad \text{and} \quad B = \begin{bmatrix} 2 & 4 \\ 4 & 8 \end{bmatrix}$$

then det $A = (6)(-1) - (5)(2) = -16$ and det $B = (2)(8) - (4)(4) = 0$. So, by our earlier remark, we have that A is invertible and B is not invertible.

    In other texts [see (1) in the References] the determinant is defined as a function, denoted *det*, on the set of $n \times n$ matrices obeying certain abstract properties. From this definition a number of properties are derived as well as a formula for evaluating the determinant of an arbitrary $n \times n$ matrix. Although this approach possesses a number of desirable characteristics, we have chosen a more concrete development.

    One of the original motivations for studying the determinant was the recognition that the solutions to particular systems of equations could be simply expressed in terms of the determinants of certain matrices.

    Consider the system

$$ax + by = e$$
$$cx + dy = e'$$

We compute the solution as

$$x = \frac{ed - be'}{ad - bc} \qquad y = \frac{ae' - ce}{ad - bc}$$

or

$$x = \frac{\begin{vmatrix} e & b \\ e' & d \end{vmatrix}}{\begin{vmatrix} a & b \\ c & d \end{vmatrix}} \qquad y = \frac{\begin{vmatrix} a & e \\ c & e' \end{vmatrix}}{\begin{vmatrix} a & b \\ c & d \end{vmatrix}}$$

Thus, the solution may be expressed in terms of quotients of determinants provided, of course, that the determinant of the coefficient matrix is not equal to zero, that is, provided that the coefficient matrix is invertible.

The fact that x and y may be expressed as quotients of determinants is a special case of *Cramer's rule*, which was discovered in the latter half of the eighteenth century by G. Cramer of Switzerland. In the next section we shall see that this result extends to larger systems.

Example 2   We will solve the system below by Cramer's rule.

$$3x - y = 3$$
$$2x + y = 7$$

By Cramer's rule, we have

$$x = \frac{\begin{vmatrix} 3 & -1 \\ 7 & 1 \end{vmatrix}}{\begin{vmatrix} 3 & -1 \\ 2 & 1 \end{vmatrix}} = \frac{10}{5} = 2 \qquad y = \frac{\begin{vmatrix} 3 & 3 \\ 2 & 7 \end{vmatrix}}{\begin{vmatrix} 3 & -1 \\ 2 & 1 \end{vmatrix}} = \frac{15}{5} = 3$$

Although the use of Cramer's rule in Example 2 required little computation, in the next section we will see that Cramer's rule is quite impractical for large systems.

Elementary Row Operations and the Determinant

In Section 1.4 we discovered the importance of the elementary row operations in Gaussian elimination. In what follows, we explore the effect such operations have on the evaluation of the determinant of a matrix.

Theorem 5.1.1   *Let A be a 2 × 2 matrix.*

(a) *If B is the matrix formed from A by interchanging two rows of A, then*

$$\det B = -\det A$$

(b) *If B is the matrix formed from A by multiplying a row of A by a scalar k, then*

$$\det B = k \det A$$

(c) *If B is the matrix formed from A by adding a multiple of one row of A to another row of A, then*

$$\det B = \det A$$

(d) $\det I = 1.$

Proof

We will only prove (c). The other parts are left as exercises.

Suppose that we add k times the second row of A to the first row of A. Then if

$$A = \begin{bmatrix} a & b \\ c & d \end{bmatrix}$$

we have

$$B = \begin{bmatrix} a + kc & b + kd \\ c & d \end{bmatrix}$$

So

$$\det B = (a + kc)d - (b + kd)c$$
$$= ad + kcd - bc - kdc$$
$$= ad - bc$$
$$= \det A \qquad \blacksquare$$

In Example 3 we illustrate the interplay between the elementary row operations and the evaluation of a determinant. It should be pointed out that the use of elementary row operations to compute the determinant of a matrix will be more practical when we define determinants for matrices of order greater than 2.

Example 3 We leave it as an exercise to determine which parts of Theorem 5.1.1 are being used below.

$$\begin{vmatrix} 4 & 7 \\ 2 & 0 \end{vmatrix} = 2 \begin{vmatrix} 4 & 7 \\ 1 & 0 \end{vmatrix} = -2 \begin{vmatrix} 1 & 0 \\ 4 & 7 \end{vmatrix} = -2 \begin{vmatrix} 1 & 0 \\ 0 & 7 \end{vmatrix}$$

$$= (-2)(7) \begin{vmatrix} 1 & 0 \\ 0 & 1 \end{vmatrix} = -14|I| = -14(1) = -14$$

In Section 2.1 we introduced the rotation (by θ) matrix A_θ. As this matrix will be used in Section 5.3, we compute its determinant.

$$|A_\theta| = \begin{vmatrix} \cos \theta & -\sin \theta \\ \sin \theta & \cos \theta \end{vmatrix} = \cos^2 \theta + \sin^2 \theta = 1$$

for all values of θ.

Application to Geometry

Consider the parallelogram determined by the vectors $\mathbf{x} = (a, b)$ and $\mathbf{y} = (c, d)$ in Figure 5.1.1(a). We will show that the area of this parallelogram is given by $|\det A|$, where A is the matrix whose rows are given by the vectors $\mathbf{x}$ and $\mathbf{y}$. It is easy to see this in the special case where $\mathbf{y}$ lies along the x-axis [see Figure 5.1.1(b)]. In this case we may write $\mathbf{y} = (c, 0)$. Then, using the fact that the areas of regions I and III are equal, we have

$$\text{Area} = (\text{area of I}) + (\text{area of II}) + (\text{area of III})$$

$$= \tfrac{1}{2}ab + b(c - a) + \tfrac{1}{2}ab$$

$$= bc$$

$$= \left| \det \begin{bmatrix} a & b \\ c & 0 \end{bmatrix} \right|$$

$$= |\det A|$$

In Section 5.3 we prove that this result holds for an arbitrary parallelogram such as the one in Figure 5.1.1(a).

As a consequence of this result it is very easy to answer the following questions:

1. How does the area change if we triple the length of $\mathbf{x}$?
2. How does the area change if we triple the length of $\mathbf{x}$ and double the length of $\mathbf{y}$?
3. What can be said about the area if $\mathbf{y}$ is a multiple of $\mathbf{x}$, that is, if the parallelogram is degenerate?

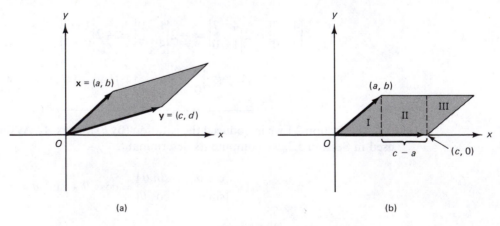

(a)   (b)

Figure 5.1.1

To answer question 1 we may apply Theorem 5.1.1(b) and obtain that the area of the new parallelogram is $3|\det A|$, that is, it is three times the original area. To answer question 2 we may apply Theorem 5.5.1(b) twice and obtain that the area of the new parallelogram is $(3)(2)|\det A| = 6|\det A|$. To answer question 3 we may conclude that because **x** is a multiple of **y**, we have that A is not invertible and so the determinant and hence the area is zero.

Example 4   Find the area of the parallelogram determined by the vectors $\mathbf{x} = (1, 5)$ and $\mathbf{y} = (-2, 3)$.

We need only compute

$$\begin{vmatrix} 1 & 5 \\ -2 & 3 \end{vmatrix} = 3 + 10 = 13$$

Exercises

In Exercises 1–3, compute the determinant for each 2×2 matrix.

1. $\begin{bmatrix} 6 & 3 \\ 4 & 5 \end{bmatrix}$ **2.** $\begin{bmatrix} 2 & -3 \\ -2 & 6 \end{bmatrix}$

3. $\begin{bmatrix} 2 & 9 \\ 0 & 7 \end{bmatrix}$

4. Let

$$A = \begin{bmatrix} 1 & 3 \\ 2 & 5 \end{bmatrix} \quad \text{and} \quad B = \begin{bmatrix} 3 & 4 \\ -1 & 6 \end{bmatrix}$$

Compute:
(a) $\det A + \det B$ (b) $\det (A + B)$
(c) $(\det A)(\det B)$ (d) $\det AB$
(e) $\det (A - 4I)$ (f) $\det A^t$

5. Prove parts (a), (b), and (d) of Theorem 5.1.1.

6. Determine which parts of Theorem 5.1.1 are being used in Example 3.

7. Provide examples to show that the following relations are *not* true.
(a) $\det A = 0$ if and only if $A = O$
(b) $\det (A + B) \le \det A + \det B$

8. Use Theorem 5.1.1 to compute the determinants of:

(a) $\begin{bmatrix} 3 & 7 \\ 4 & 8 \end{bmatrix}$ (b) $\begin{bmatrix} 5 & 1 \\ 2 & 1 \end{bmatrix}$ (c) $\begin{bmatrix} 2 & 4 \\ 4 & 8 \end{bmatrix}$

9. Suppose that

$$\begin{vmatrix} a & b \\ c & d \end{vmatrix} = 3$$

Use Theorem 5.1.1 to evaluate the following determinants by inspection.

(a) $\begin{vmatrix} 2a & 2b \\ c & d \end{vmatrix}$ (b) $\begin{vmatrix} c & d \\ a & b \end{vmatrix}$

(c) $\begin{vmatrix} a + 3c & b + 3d \\ c & d \end{vmatrix}$

10. Let $\mathbf{x} = (3, 7)$ and $\mathbf{y} = (4, 1)$.
(a) Find the area of the parallelogram with sides **x** and **y**.
(b) Use part (a) to find the area of the parallelogram with sides 2**x** and 5**y**.

11. Prove that $\det A = \det A^t$ for any 2×2 matrix A.

12. Prove that for any upper triangular 2×2 matrix A, $\det A$ equals the product of the diagonal entries.

13. Use the properties of determinants to find the values of x such that each of the following matrices is not invertible.

(a) $\begin{bmatrix} 2 + x & -1 \\ x & 3 \end{bmatrix}$ (b) $\begin{bmatrix} x - 2 & 3 \\ 3 & x - 2 \end{bmatrix}$

14. For any 2×2 matrix A define the function

$$f(x) = \det(xI - A)$$

   (a) Show that $f(x)$ is a polynomial of degree two.
   (b) Prove that if $f(c) = 0$, then there exists a nonzero vector $\mathbf{x}$ such that $A\mathbf{x} = c\mathbf{x}$.

15. Use the results of Exercise 14 to find a nonzero vector $\mathbf{x}$ and a scalar c for each of the following matrices such that $A\mathbf{x} = c\mathbf{x}$.

   (a) $A = \begin{bmatrix} 1 & 2 \\ -1 & 4 \end{bmatrix}$ (b) $A = \begin{bmatrix} 1 & 3 \\ 0 & 1 \end{bmatrix}$

16. Apply Cramer's rule to find solutions to the following systems.

   (a) $\begin{aligned} 2x - 3y &= 4 \\ 4x + 7y &= 0 \end{aligned}$ (b) $\begin{aligned} 5x + y &= 1 \\ x - 6y &= 8 \end{aligned}$

   (c) $\begin{aligned} x + y &= 2 \\ 3x - y &= 10 \end{aligned}$

17. Define the *classical adjoint* of a 2×2 matrix A, denoted by *adj* A as

$$\text{adj } A = \begin{bmatrix} d & -b \\ -c & a \end{bmatrix} \quad \text{where } A = \begin{bmatrix} a & b \\ c & d \end{bmatrix}$$

   Prove:
   (a) $(\text{adj } A)A = A(\text{adj } A) = (\det A)I$
   (b) $\det(\text{adj } A) = \det A$
   (c) $\text{adj } A^t = (\text{adj } A)^t$

18. Use Exercise 17(a) to prove that if a 2×2 matrix A is invertible, then $\det A \neq 0$ and

$$A^{-1} = \frac{1}{\det A} \text{ adj } A$$

5.2 DETERMINANTS OF $n \times n$ MATRICES

In Section 5.1 we developed several useful properties of determinants of 2×2 matrices. We used determinants to solve special systems of equations and to determine if a matrix is invertible. In this section the definition of determinant will be extended to arbitrary square matrices in such a way that these earlier properties are preserved.

We will need the following notation. For an $n \times n$ matrix A, we let M_{ij} denote the $(n - 1) \times (n - 1)$ submatrix of A formed by deleting the ith row and jth column of A, where $1 \leq i \leq n$ and $1 \leq j \leq n$.

Example 1   Let

$$A = \begin{bmatrix} 4 & 7 \\ 5 & 2 \end{bmatrix}$$

Then

$$M_{11} = (2), \quad M_{12} = (5), \quad M_{21} = (7), \quad M_{22} = (4)$$

Example 2   Let

$$A = \begin{bmatrix} 2 & -3 & 4 \\ 5 & 0 & 1 \\ 6 & 8 & 9 \end{bmatrix}$$

Then

$$M_{11} = \begin{bmatrix} 0 & 1 \\ 8 & 9 \end{bmatrix} \quad M_{22} = \begin{bmatrix} 2 & 4 \\ 6 & 9 \end{bmatrix} \quad M_{31} = \begin{bmatrix} -3 & 4 \\ 0 & 1 \end{bmatrix}$$

We begin by defining the ***determinant*** of a 1×1 matrix $A = (a)$ by *det* $A = |A| = a$. Notice that this definition preserves the property that A is invertible if and only if det A is not equal to zero. In fact, the definition preserves all the properties developed in Theorem 5.1.1.

To see how to extend the definition of a determinant to a 3×3 matrix, it will be useful to examine the definition for the 2×2 case more carefully. Let

$$A = \begin{bmatrix} A_{11} & A_{12} \\ A_{21} & A_{22} \end{bmatrix}$$

If we use our new notation, we have

$$|A| = A_{11}A_{22} - A_{12}A_{21} = A_{11}|M_{11}| - A_{12}|M_{12}|$$

Notice that the right side of the equation consists of products of the entries from the first row of A and the determinants of the corresponding submatrices. The signs between the terms are alternating.

Adopting a similar form for 3×3 matrices, we define the ***determinant*** of a 3×3 matrix as

$$\det A = |A| = A_{11}|M_{11}| - A_{12}|M_{12}| + A_{13}|M_{13}|$$

Example 3   Let

$$A = \begin{bmatrix} 3 & 7 & -2 \\ 2 & 1 & -3 \\ 4 & 0 & 6 \end{bmatrix}$$

Then

$$|A| = 3 \begin{vmatrix} 1 & -3 \\ 0 & 6 \end{vmatrix} - 7 \begin{vmatrix} 2 & -3 \\ 4 & 6 \end{vmatrix} + (-2) \begin{vmatrix} 2 & 1 \\ 4 & 0 \end{vmatrix}$$

$$= 3(6) - 7(24) - 2(-4)$$

$$= -142$$

Thus, the determinant of a 2×2 matrix may be defined in terms of the determinants of 1×1 matrices and, similarly, the determinant of a 3×3 matrix may be defined in terms of the determinants of 2×2 matrices. We will use this form of definition, called a *recursive* or *inductive* definition, to define the determinant of a higher-order matrix in terms of the determinants of lower-order matrices.

Definition   Suppose that A is an $n \times n$ matrix. We define the **determinant** of A denoted either **det** A or $|A|$ by

$$|A| = A_{11}|M_{11}| - A_{12}|M_{12}| + \cdots + (-1)^{1+n}A_{1n}|M_{1n}|$$

The term $(-1)^{i+j}|M_{ij}|$ is called the **ijth cofactor** of A and it is denoted by C_{ij}. The right side of the equation above is called the **cofactor expansion along the first row**.

Using the notation C_{ij}, we may write

$$\det A = A_{11}C_{11} + A_{12}C_{12} + \cdots + A_{1n}C_{1n}$$

Many authors call M_{ij} the **ijth minor** of A. We will not have use for this term.

Thus, if A is a 4×4 matrix, the right-hand side of the equation consists of the determinants of four 3×3 matrices.

Example 4   Let

$$A = \begin{bmatrix} 2 & 1 & -3 & 1 \\ -3 & -2 & 0 & 2 \\ 2 & 1 & 0 & -1 \\ 1 & 0 & 1 & 2 \end{bmatrix}$$

Then

$$\det A = (2)|M_{11}| - (1)|M_{12}| + (-3)|M_{13}| - (1)|M_{14}|$$

$$= (2)\begin{vmatrix} -2 & 0 & 2 \\ 1 & 0 & -1 \\ 0 & 1 & 2 \end{vmatrix} - (1)\begin{vmatrix} -3 & 0 & 2 \\ 2 & 0 & -1 \\ 1 & 1 & 2 \end{vmatrix}$$

$$+ (-3)\begin{vmatrix} -3 & -2 & 2 \\ 2 & 1 & -1 \\ 1 & 0 & 2 \end{vmatrix} - (1)\begin{vmatrix} -3 & -2 & 0 \\ 2 & 1 & 0 \\ 1 & 0 & 1 \end{vmatrix}$$

We will leave it as an exercise for the reader to evaluate the four determinants above and verify that

$$\det A = (2)(0) - (1)(1) + (-3)(2) - (1)(1)$$

$$= -8$$

At this point the reader may wonder what would result if instead of using the first row of a matrix in the definition of a determinant, we instead used another row along with the corresponding cofactors, or for that matter, another column. It is rather surprising that the result remains unchanged. This fact, which we will not prove [see (1) in the References], may be used to our advantage if we choose a row or column that contains a large number of zero entries. We state the result below.

Theorem 5.2.1   Let A be an $n \times n$ matrix. Then, for $1 \le i \le n$ and for $1 \le j \le n$, we have

$$\det A = A_{i1}C_{i1} + A_{i2}C_{i2} + \cdots + A_{in}C_{in}$$

and

$$\det A = A_{1j}C_{1j} + A_{2j}C_{2j} + \cdots + A_{nj}C_{nj}$$

The right side of the first equation is called the *cofactor expansion along the ith row*. Similarly, the right side of the second equation is called the *cofactor expansion along the jth column*.

Example 5   We will evaluate the determinant of the matrix A of Example 3 by the cofactor expansion along the second column.

$$A = \begin{bmatrix} 3 & 7 & -2 \\ 2 & 1 & -3 \\ 4 & 0 & 6 \end{bmatrix}$$

$$|A| = -(7)\begin{vmatrix} 2 & -3 \\ 4 & 6 \end{vmatrix} + (1)\begin{vmatrix} 3 & -2 \\ 4 & 6 \end{vmatrix} - (0)\begin{vmatrix} 3 & -2 \\ 2 & -3 \end{vmatrix}$$

$$= (-7)(24) + (26) - 0$$

$$= -142$$

Example 6   Evaluate the determinant of the matrix A given in the example above by the cofactor expansion along the third row.

$$|A| = (4)\begin{vmatrix} 7 & -2 \\ 1 & -3 \end{vmatrix} - (0)\begin{vmatrix} 3 & -2 \\ 2 & -3 \end{vmatrix} + (6)\begin{vmatrix} 3 & 7 \\ 2 & 1 \end{vmatrix}$$

$$= (4)(-19) - 0 + (6)(-11)$$

$$= -142$$

Example 7   Evaluate the determinant of the matrix A in Example 4 by the cofactor expansion along the third column.

We will omit the zero terms in the expansion.

$$A = \begin{bmatrix} 2 & 1 & -3 & 1 \\ -3 & -2 & 0 & 2 \\ 2 & 1 & 0 & -1 \\ 1 & 0 & 1 & 2 \end{bmatrix}$$

$$|A| = (-3) \begin{vmatrix} -3 & -2 & 2 \\ 2 & 1 & -1 \\ 1 & 0 & 2 \end{vmatrix} - (1) \begin{vmatrix} 2 & 1 & 1 \\ -3 & -2 & 2 \\ 2 & 1 & -1 \end{vmatrix}$$

$$= (-3)(2) - (2)$$

$$= -8$$

Numerical Considerations

As we mentioned earlier, determinants are impractical as computational tools. To illustrate this, we will derive an approximate count of the number of products needed to evaluate the determinant of an $n \times n$ matrix by cofactor expansion.

For $n = 2$ there are two products. If we consider cofactor expansion for $n = 3$, we see that there are three 2×2 determinants each of which is to be multiplied by the corresponding entry of some row (or column). Hence, there are $3(2) + 3 = 9$ products. In general, if we let $f(n)$ equal the number of products needed to evaluate the determinant of an $n \times n$ matrix by cofactor expansion, it can be shown that $f(n + 1) = (n + 1)(f(n) + 1)$. This result follows from the fact that for a matrix of order $n + 1$, each of the $n + 1$ cofactors contributes $f(n)$ products plus the product of the row entry and the corresponding determinant. To obtain a lower bound for $f(n)$, we note that $f(n + 1) > nf(n)$ for every n. So using the fact that $f(2) = 2$,

$$f(n + 1) > nf(n) > n(n - 1)f(n - 1) > \cdots > n!$$

In fact, it can be shown that if we let $g(n)$ equal the number of additions (or subtractions) for such a computation, then

$$f(n) = n! \sum_{k=1}^{n-1} \frac{1}{k!} \quad \text{and} \quad g(n) = n! - 1$$

A good estimate of $f(n)$ is $n!(e - 1)$. So the evaluation of the determinant of an $n \times n$ matrix by cofactor expansion requires approximately $n!(e - 1) + n! = en!$ arithmetic operations. To obtain some perspective concerning the time it would take a computer to evaluate the determinant of various-sized matrices, we will assume that such a computer performs 1 million arithmetic operations per second. The results listed in Table 5.2.1 should surprise most readers. We see that it would take over 200,000 years to evaluate the determinant of a 20×20 matrix!

TABLE 5.2.1

Order of the Matrix	Number of Operations	Time Needed to Evaluate the Determinant
5	326	3.26×10^{-4} seconds
10	9,864,101	9.86 seconds
15	3.55×10^{12}	41 days
20	6.61×10^{18}	210,000 years
50	8.27×10^{64}	2.62×10^{51} years

The more efficient way to evaluate a determinant is to apply elementary row operations to reduce the matrix to an upper triangular matrix. It will then be very easy (Theorem 5.2.3) to evaluate the resulting determinant. This process will take less than 1 second on the same computer for a 15×15 matrix! We will see an application of this method in Example 11. First, we will need Theorem 5.2.2, which is a generalization of Theorem 5.1.1.

Theorem 5.2.2 *Let A be an $n \times n$ matrix.*

(a) *If B is the matrix formed from A by interchanging two rows of A, then*

$$\det B = -\det A$$

(b) *If B is the matrix formed from A by multiplying a row of A by the scalar k, then*

$$\det B = k \det A$$

(c) *If B is the matrix formed from A by adding a multiple of one row of A to different row of A, then*

$$\det B = \det A$$

(d) $\det I = 1$.

It should be noted that this theorem remains true if the word "row" is replaced throughout by "column." The reason for this follows from the result [Theorem 5.3.5(c)] that $\det A = \det A^t$.

Proof

We will prove part (c) and leave the rest as an exercise.

The proof will be by mathematical induction on the order of the matrix. That is, we will show that if the result is true for all matrices of order $n - 1$, the result must be true for matrices of order n. Because we have already proved the result for $n = 2$ [Theorem 5.1.1(c)], we will be done.

Let A be a matrix of order n and let A' denote the matrix formed from A by adding k times row r to row s. The key to the proof is to recognize that for $r \neq i \neq s$, the matrix formed from A' by deleting its ith row and jth column is the same matrix as the one formed from M_{ij} by adding k times its rth row to its sth row. This matrix is denoted by M'_{ij}.

We will expand A along some row other than r or s without loss of generality; let $i = 1$. We obtain

$$|A| = A_{11}|M_{11}| - A_{12}|M_{12}| + \cdots + (-1)^{1+n}A_{1n}|M_{1n}|$$

If we expand A' along row 1, we obtain

$$|A'| = A'_{11}|M'_{11}| - A'_{12}|M'_{12}| + \cdots + (-1)^{1+n}A'_{1n}|M'_{1n}|$$

Because we have assumed that the result is true for all matrices of order n, we have $|M_{1j}| = |M'_{1j}|$ for all j. Also, because $r \neq 1 \neq s$, we have that $A_{1j} = A'_{1j}$ for all j. Thus, $|A| = |A'|$. ∎

An advantageous use of this theorem is to apply elementary row operations in such a way as to create a large number of zeros in some row (column) and then to expand by cofactors along this row (column).

Example 8   Let

$$A = \begin{bmatrix} 3 & 5 & 8 \\ 1 & 1 & 2 \\ 1 & 2 & 1 \end{bmatrix}$$

We will indicate those parts of Theorem 5.2.2 which are used below.

$$\begin{align} |A| &= \overset{\text{(a)}}{-} \begin{vmatrix} 1 & 1 & 2 \\ 3 & 5 & 8 \\ 1 & 2 & 1 \end{vmatrix} \overset{\text{(c,c)}}{=} - \begin{vmatrix} 1 & 1 & 2 \\ 0 & 2 & 2 \\ 0 & 1 & -1 \end{vmatrix} \\ &\overset{\text{(c)}}{=} - \begin{vmatrix} 1 & 1 & 2 \\ 0 & 2 & 2 \\ 0 & 0 & -2 \end{vmatrix} = -(1)\begin{vmatrix} 2 & 2 \\ 0 & -2 \end{vmatrix} \\ &= -(1)(-4) = 4 \end{align}$$

Theorem 5.2.2 is particularly useful for hand computations if the matrix contains many fractions. The next example illustrates this and uses the same notation as in Example 8.

Example 9

$$\begin{vmatrix} \frac{1}{3} & -\frac{1}{4} & \frac{1}{6} \\ \frac{1}{5} & 1 & \frac{2}{3} \\ \frac{1}{8} & \frac{1}{2} & \frac{1}{4} \end{vmatrix} \overset{(b,b,b)}{=} \frac{1}{(12)(15)(8)} \begin{vmatrix} 4 & -3 & 2 \\ 3 & 15 & 10 \\ 1 & 4 & 2 \end{vmatrix}$$

$$\overset{(a)}{=} -\frac{1}{1440} \begin{vmatrix} 1 & 4 & 2 \\ 3 & 15 & 10 \\ 4 & -3 & 2 \end{vmatrix}$$

$$\overset{(c,c)}{=} -\frac{1}{1440} \begin{vmatrix} 1 & 4 & 2 \\ 0 & 3 & 4 \\ 0 & -19 & -6 \end{vmatrix}$$

$$= -\frac{1}{1440} \begin{vmatrix} 3 & 4 \\ -19 & -6 \end{vmatrix}$$

$$= -\frac{58}{1440}$$

Example 10   Let A be the matrix

$$A = \begin{bmatrix} 3 & 1 & 2 \\ 5 & 3 & 4 \\ 7 & 1 & 1 \end{bmatrix}$$

To compute $|A|$ it is convenient to begin with the column operation of interchanging columns 1 and 2, and then to proceed with row operations.

$$|A| = -\begin{vmatrix} 1 & 3 & 2 \\ 3 & 5 & 4 \\ 1 & 7 & 1 \end{vmatrix} \overset{(c,c)}{=} -\begin{vmatrix} 1 & 3 & 2 \\ 0 & -4 & -2 \\ 0 & 4 & -1 \end{vmatrix} \overset{(c)}{=} -\begin{vmatrix} 1 & 3 & 2 \\ 0 & -4 & -2 \\ 0 & 0 & -3 \end{vmatrix}$$

$$= -(1)(12 - 0) = -12$$

Example 11   Let A be the 4×4 matrix defined below.

$$A = \begin{bmatrix} 1 & -1 & 2 & 1 \\ 2 & -1 & -1 & 4 \\ -4 & 5 & -10 & -6 \\ 3 & -2 & 10 & -1 \end{bmatrix}$$

By adding multiples of the first row to the other rows, we are able to create three zeros in the first column. By Theorem 5.2.2(c), det A remains unchanged. So

$$|A| = \begin{vmatrix} 1 & -1 & 2 & 1 \\ 0 & 1 & -5 & 2 \\ 0 & 1 & -2 & -2 \\ 0 & 1 & 4 & -4 \end{vmatrix}$$

Now we use the second row to create additional zeros in the second column.

$$|A| = \begin{vmatrix} 1 & -1 & 2 & 1 \\ 0 & 1 & -5 & 2 \\ 0 & 0 & 3 & -4 \\ 0 & 0 & 9 & -6 \end{vmatrix}$$

Finally, we use the third row to produce an upper triangular matrix.

$$|A| = \begin{vmatrix} 1 & -1 & 2 & 1 \\ 0 & 1 & -5 & 2 \\ 0 & 0 & 3 & -4 \\ 0 & 0 & 0 & 6 \end{vmatrix}$$

We may now easily compute the determinant of A by cofactor expansion along the first column.

$$|A| = (1)\begin{vmatrix} 1 & -5 & 2 \\ 0 & 3 & -4 \\ 0 & 0 & 6 \end{vmatrix} = (1)(1)\begin{vmatrix} 3 & -4 \\ 0 & 6 \end{vmatrix}$$

$$= (1)(1)(3)(6) = 18$$

Notice that Example 11 illustrates a very easy method to compute the determinant of a matrix which does not contain many zero entries.

Use elementary row or column operations to reduce A to an upper triangular matrix. The determinant of this matrix is the product of the diagonal entries.

This last fact is so useful that we state it as a theorem. The proof is straightforward and it is left as an exercise.

Theorem 5.2.3 *Let A be an upper triangular matrix, then the determinant of A is the product of its diagonal entries.*

Corollary 5.2.4

If A is a diagonal matrix, then the determinant of A is the product of its diagonal entries.

The Determinant of a 3 × 3 Matrix

There is a trick for computing the determinant of a 3×3 matrix. The reader is warned that this method works *only* for 3×3 matrices.

Let A be the matrix given in Example 5.

$$A = \begin{bmatrix} 3 & 7 & -2 \\ 2 & 1 & -3 \\ 4 & 0 & 6 \end{bmatrix}$$

Adjoin the first and second columns as follows:

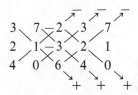

Now sum the products along the bottom arrows and subtract the products along the top. We obtain

$$[18 + (-84) + 0] - [(-8) + 0 + 84] = -142$$

Cramer's Rule

Although we have only illustrated Cramer's rule in the preceding section for linear systems with two equations and two unknowns, the rule also extends to larger systems, provided that the coefficient matrix is invertible. This method is highly impractical for systems with more than five or six equations because of the number of determinants which must be computed. The appropriate method is Gaussian elimination.

For the system $A\mathbf{x} = \mathbf{b}$ we will denote by M_i the matrix obtained from A by replacing the *i*th column of A by $\mathbf{b}$. If $\det A \neq 0$, then the (unique) solution $\mathbf{x} = (x_1, \ldots, x_n)$ is given by

$$x_i = \frac{|M_i|}{|A|} \qquad \text{for } i = 1, \ldots, n$$

Readers should convince themselves that this formula does indeed generalize the 2×2 case. The proof of Cramer's rule is included in the next section.

Example 12   Apply Cramer's rule to solve the following system:

$$x + 2y + 3z = 2$$
$$x \qquad + z = 3$$
$$x + y - z = 1$$

We will compute x and y by Cramer's rule and leave the computation of z as an exercise.

$$|M_1| = \begin{vmatrix} 2 & 2 & 3 \\ 3 & 0 & 1 \\ 1 & 1 & -1 \end{vmatrix} = 15 \qquad |M_2| = \begin{vmatrix} 1 & 2 & 3 \\ 1 & 3 & 1 \\ 1 & 1 & -1 \end{vmatrix} = -6$$

and

$$|A| = \begin{vmatrix} 1 & 2 & 3 \\ 1 & 0 & 1 \\ 1 & 1 & -1 \end{vmatrix} = 6$$

Thus,

$$x = \frac{15}{6} = \frac{5}{2} \quad \text{and} \quad y = -\frac{6}{6} = -1$$

Geometric Applications

In the preceding section, we saw that the determinant of a 2×2 matrix could be used to compute the area of a parallelogram. Below we list several other geometric uses of determinants. The proofs of these results are contained in elementary books on vector analysis. Additional applications to geometry are contained in the exercises.

1. The ***cross product*** of two vectors $\mathbf{x}$ and $\mathbf{y}$ in R^3, denoted $\mathbf{x} \times \mathbf{y}$, is defined *symbolically* as

$$\mathbf{x} \times \mathbf{y} = \begin{vmatrix} \mathbf{i} & \mathbf{j} & \mathbf{k} \\ x_1 & x_2 & x_3 \\ y_1 & y_2 & y_3 \end{vmatrix}$$

where $\mathbf{i} = (1, 0, 0)$, $\mathbf{j} = (0, 1, 0)$, $\mathbf{k} = (0, 0, 1)$, $\mathbf{x} = (x_1, x_2, x_3)$,   and $\mathbf{y} = (y_1, y_2, y_3)$. The $\mathbf{i}, \mathbf{j}, \mathbf{k}$ notation is the standard one used in the context of cross products.

To see how this definition works, let $\mathbf{x} = (1, 2, 1)$ and $\mathbf{y} = (3, 0, 4)$. Then

$$\mathbf{x} \times \mathbf{y} = \begin{vmatrix} \mathbf{i} & \mathbf{j} & \mathbf{k} \\ 1 & 2 & 1 \\ 3 & 0 & 4 \end{vmatrix}$$

$$= \begin{vmatrix} 2 & 1 \\ 0 & 4 \end{vmatrix} \mathbf{i} - \begin{vmatrix} 1 & 1 \\ 3 & 4 \end{vmatrix} \mathbf{j} + \begin{vmatrix} 1 & 2 \\ 3 & 0 \end{vmatrix} \mathbf{k}$$

$$= 8\mathbf{i} - \mathbf{j} - 6\mathbf{k}$$

$$= (8, -1, -6)$$

The main purpose of a cross product is to construct a vector that is perpendicular to the original two vectors.

In the example above, we compute dot products and obtain

$$(1, 2, 1) \cdot (8, -1, -6) = 8 - 2 - 6 \quad = 0$$

$$(3, 0, 4) \cdot (8, -1, -6) = 24 - 0 - 24 = 0$$

2. It can be shown that the volume of a *parallelepiped* (see Figure 5.2.1) with sides determined by $\mathbf{x}$, $\mathbf{y}$, and $\mathbf{z}$ in R^3 is given by the formula

$$\text{Volume} = \left| \det \begin{bmatrix} x_1 & x_2 & x_3 \\ y_1 & y_2 & y_3 \\ z_1 & z_2 & z_3 \end{bmatrix} \right|$$

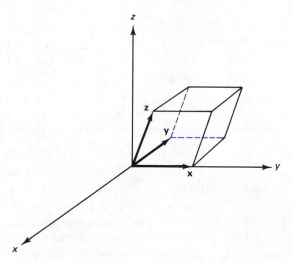

Figure 5.2.1

Exercises

In Exercises 1–6, compute the determinants of the matrices by the method of cofactor expansion.

1. $\begin{bmatrix} 4 & 1 & 0 \\ 2 & 3 & -2 \\ 3 & 5 & 2 \end{bmatrix}$

2. $\begin{bmatrix} 7 & 4 & 5 \\ 3 & 0 & 1 \\ 2 & 0 & 4 \end{bmatrix}$

3. $\begin{bmatrix} 3 & 1 & 4 & 1 \\ 2 & -1 & 0 & 1 \\ 2 & 3 & 0 & -3 \\ 0 & 1 & 2 & 5 \end{bmatrix}$

4. $\begin{bmatrix} 1 & 2 & 1 & -1 \\ 0 & 3 & 0 & 2 \\ 4 & -1 & 2 & 3 \\ 2 & 5 & -1 & 0 \end{bmatrix}$

5. $\begin{bmatrix} 2 & 3 & -1 & 4 \\ 0 & 0 & 5 & 6 \\ -2 & 3 & 1 & 0 \\ 4 & 2 & 2 & 1 \end{bmatrix}$

6. $\begin{bmatrix} 3 & -1 & 2 & 2 & 0 \\ 1 & 0 & 5 & 0 & -1 \\ 6 & -2 & 0 & 1 & 0 \\ 0 & 0 & 0 & 3 & 6 \\ 0 & 3 & 4 & 0 & 1 \end{bmatrix}$

In Exercises 7–12, compute the determinants of the matrices by first creating zero entries using elementary row operations.

7. $\begin{bmatrix} 1 & 3 & 2 \\ 4 & 14 & 6 \\ 2 & 5 & 3 \end{bmatrix}$

8. $\begin{bmatrix} 1 & -1 & 2 & -1 \\ -3 & 4 & 1 & -1 \\ 2 & -5 & -3 & 8 \\ -2 & 6 & -4 & 1 \end{bmatrix}$

9. $\begin{bmatrix} 1 & 4 & 2 \\ -2 & -11 & -5 \\ -2 & -8 & -10 \end{bmatrix}$

10. $\begin{bmatrix} 2 & 1 & 4 & 2 \\ -3 & -4 & -2 & 5 \\ 5 & 3 & 4 & 2 \\ 5 & 3 & 6 & -4 \end{bmatrix}$

11. $\begin{bmatrix} \frac{2}{3} & \frac{1}{6} & \frac{1}{3} \\ \frac{1}{2} & \frac{1}{4} & \frac{1}{2} \\ 0 & 2 & 1 \end{bmatrix}$

12. $\begin{bmatrix} \frac{1}{5} & \frac{2}{5} & \frac{3}{5} \\ \frac{1}{8} & 0 & \frac{1}{4} \\ \frac{3}{8} & \frac{3}{4} & 2 \end{bmatrix}$

In Exercises 13–18, compute the determinants of the matrices by the following method: Use elementary row operations to reduce each matrix to an upper triangular matrix and then apply Theorem 5.2.3.

13. $\begin{bmatrix} 3 & 2 & 1 \\ 6 & 2 & 1 \\ -6 & -2 & 3 \end{bmatrix}$

14. $\begin{bmatrix} -2 & 1 & 4 \\ -2 & 7 & 14 \\ 4 & -2 & -8 \end{bmatrix}$

15. $\begin{bmatrix} 2 & 1 & 5 & 2 \\ 2 & -1 & 5 & 3 \\ 2 & -1 & 8 & 1 \\ 4 & -2 & 10 & 8 \end{bmatrix}$

16. $\begin{bmatrix} 6 & 5 & 12 & 1 \\ 6 & 4 & 14 & 4 \\ 3 & 2 & 14 & 3 \\ 3 & 3 & 5 & -1 \end{bmatrix}$

17. $\begin{bmatrix} 1 & -1 & 2 & 1 \\ 2 & -1 & -1 & 4 \\ -4 & 5 & -10 & -6 \\ 3 & -2 & 10 & -1 \end{bmatrix}$

18. $\begin{bmatrix} 1 & 0 & -2 & 3 \\ -3 & 1 & 1 & 2 \\ 0 & 4 & -1 & 1 \\ 2 & 3 & 0 & 1 \end{bmatrix}$

19. For the matrices given in Exercises 1 and 2, verify that the determinant of their product equals the product of their determinants.

20. Compute the determinants of the 3×3 matrices given in Exercises 1, 2, 7, and 9 by augmenting each matrix by its first two columns and then using the method developed in this section.

21. Verify the computations given in Example 4.

22. Verify the computations given in Example 7.

23. Suppose that

$$\begin{vmatrix} a & b & c \\ a' & b' & c' \\ a'' & b'' & c'' \end{vmatrix} = 4$$

Evaluate the following determinants by inspection.

(a) $\begin{vmatrix} 2a & 2b & 2c \\ a' & b' & c' \\ a'' & b'' & c'' \end{vmatrix}$

(b) $\begin{vmatrix} a' & b' & c' \\ a & b & c \\ a'' & b'' & c'' \end{vmatrix}$

(c) $\begin{vmatrix} -a & -b & -c \\ 5a' & 5b' & 5c' \\ 3a'' & 3b'' & 3c'' \end{vmatrix}$

(d) $\begin{vmatrix} a + a'' & b + b'' & c + c'' \\ a - a' & b - b' & c - c' \\ a'' & b'' & c'' \end{vmatrix}$

24. (a) Use elementary row operations to prove that the determinant of the *Vandermonde matrix* below satisfies the following equation:

$$\begin{vmatrix} 1 & 1 & 1 \\ a & b & c \\ a^2 & b^2 & c^2 \end{vmatrix} = (b - a)(c - a)(c - b)$$

(b) Suppose that a, b, and c are distinct scalars and that a', b', c' are arbitrary scalars. Use part (a) to show that there exists a polynomial f of degree 2 such that $f(a) = a'$, $f(b) = b'$, and $f(c) = c'$.

25. Use Cramer's rule to solve for z in Example 12 and check your result by substitution.

26. Apply Cramer's rule to solve the following systems of equations.

(a) $\begin{aligned} 2x + y + z &= 0 \\ x - y + 5z &= 0 \\ y - z &= 4 \end{aligned}$

(b) $\begin{aligned} x - y + z &= 6 \\ 2x + 3z &= 1 \\ 2y - z &= 0 \end{aligned}$

(c) $\begin{aligned} x - 2y &= 2 \\ x + y + z &= 1 \\ x - y + 2z &= 5 \end{aligned}$

27. Solve the systems of equations in Exercise 26 by Gaussian elimination and decide if this method is an improvement over Cramer's rule.

28. Verify that if $g(n)$ represents the total number of additions and subtractions required to evaluate the determinant of an $n \times n$ matrix by cofactor expansion, then $g(n + 1) = (n + 1)g(n) + n$.

29. For each part below use cross products to find a vector perpendicular to each of the two given vectors.

(a) $(1, 2, -1)$ and $(3, 0, 2)$

(b) $(2, 2, 1)$ and $(1, 4, 5)$

(c) $(0, 1, 2)$ and $(3, 3, 1)$

30. Using the $\mathbf{i}, \mathbf{j}, \mathbf{k}$ vectors defined with cross products, verify the following relations.

(a) $\mathbf{i} \times \mathbf{i} = \mathbf{j} \times \mathbf{j} = \mathbf{k} \times \mathbf{k} = \mathbf{0}$

(b) $\mathbf{i} \times \mathbf{j} = \mathbf{k}, \mathbf{j} \times \mathbf{k} = \mathbf{i}, \mathbf{k} \times \mathbf{i} = \mathbf{j}$

31. Use the definition of cross product to verify the following relations involving arbitrary vectors $\mathbf{x}$, $\mathbf{y}$, and $\mathbf{z}$.

(a) $\mathbf{x} \times \mathbf{y} = -(\mathbf{y} \times \mathbf{x})$

(b) $\mathbf{x} \times \mathbf{x} = \mathbf{0}$

(c) $c\mathbf{x} \times \mathbf{y} = c(\mathbf{x} \times \mathbf{y})$ for any scalar c

32. Find the volume of each of the parallelepipeds whose sides are determined by the following vectors.

(a) $(2, 0, 0)$, $(0, 3, 0)$, and $(0, 0, 4)$

(b) $(1, 2, 1)$, $(1, 1, 3)$, and $(2, 3, 0)$

33. Prove that the equation of the line through the points (a, b) and (c, d) is given by

$$\begin{vmatrix} 1 & 1 & 1 \\ x & a & c \\ y & b & d \end{vmatrix} = 0$$

34. Prove that the area of a triangle with vertices (x_1, y_1), (x_2, y_2), (x_3, y_3) is given by

$$\frac{1}{2}\det\begin{bmatrix} x_1 & y_1 & 1 \\ x_2 & y_2 & 1 \\ x_3 & y_3 & 1 \end{bmatrix}$$

35. Use Theorem 5.2.2 to provide a direct proof that the determinant of a diagonal matrix is the product of the diagonal entries.

36. Let A be an $n \times n$ matrix. Prove that $|kA| = k^n|A|$ for any scalar k.

37. Prove parts (a), (b), and (d) of Theorem 5.2.2.

*38. Use the program ROW REDUCTION to evaluate the determinants of the matrices in Exercises 3–6 by reducing them to upper triangular form.

*39. Use the program MATRIX to evaluate the determinants in Exercises 3–6.

*40. Use the program MATRIX to evaluate the determinants in Exercises 13–16.

* This problem should be solved by using one of the programs noted in Appendix B.

5.3 ADDITIONAL PROPERTIES OF DETERMINANTS, THE CLASSICAL ADJOINT, AND CODING

In the preceding two sections, we developed several important properties of determinants which were very useful in their evaluation. In addition, we have seen applications of determinants to geometry and to the solution of systems of equations. In this section we study their relationship to various matrix operations, such as matrix multiplication, inversion, and transposition. We will use the elementary matrices as our "building blocks."

Lemma 5.3.1   *For each part below, E will denote an $n \times n$ elementary matrix.*

    (a) *Let E be the elementary matrix that corresponds to interchanging two rows. Then $\det E = -1$.*

    (b) *Let E be the elementary matrix that corresponds to multiplying a row by a scalar k. Then $\det E = k$.*

    (c) *Let E be the elementary matrix that corresponds to adding a multiple of one row to another row. Then $\det E = 1$.*

    (d) *For any $n \times n$ matrix A, we have $|EA| = |E| \cdot |A|$.*

Proof

In each of the parts (a), (b), and (c), E is the matrix formed by applying the corresponding row operation to the identity matrix I. Hence, we may apply Theorem 5.2.2 to obtain the results in parts (a), (b), and (c) immediately. For example, to prove part (a) we may apply Theorem 5.2.2(a) and part (d) to obtain

$$\det E = -\det I = -1$$

To prove part (d) we again appeal to Theorem 5.2.2. For example, if E corresponds to multiplying a row by the scalar k, then by part (b) of Theorem 5.2.2 and by part (b) above, we have

$$|EA| = k|A| = |E||A|$$

The other cases of part (d) are handled similarly. ∎

Observe that because matrix multiplication is an associative operation, part (d) allows us to conclude that for any matrix A and for any s elementary matrices, we have

$$|E_1 E_2 \cdots E_s A| = |E_1| \cdot |E_2| \cdots |E_s| \cdot |A|$$

If we let $A = I$, we obtain

$$|E_1 E_2 \cdots E_s| = |E_1| \cdot |E_2| \cdots |E_s| \qquad (1)$$

Theorem 5.3.2 *Let A be an $n \times n$ matrix. Then A is invertible if and only if $\det A \neq 0$.*

Proof

If A is invertible, then by Theorem 3.5.6, A is a product of elementary matrices, say $A = E_1 E_2 \cdots E_s$. By equation (1) we have

$$|A| = |E_1| \cdot |E_2| \cdots |E_s|$$

By the lemma, each of the factors on the right side of the equation above is nonzero. Hence, $|A| \neq 0$.

If A is not invertible, it may be reduced to an upper triangular matrix B which contains a row of zeros. Thus, there exists elementary matrices $E_1, E_2, \ldots, E_s$ such that

$$E_1 E_2 \cdots E_s A = B$$

So

$$|E_1| \cdot |E_2| \cdots |E_s| \cdot |A| = |B|$$

If we evaluate $|B|$ by cofactor expansion along the zero row, we obtain that $|B| = 0$. Because all the factors on the left side involving elementary matrices are nonzero, we must conclude that $|A| = 0$. ∎

Corollary 5.3.3 *Let A be an $n \times n$ matrix. If A has two equal rows (columns), then $|A| = 0$.*

Proof

If A has two equal rows (columns), then A has rank less than n, so A is not invertible. ∎

Example 1   Theorem 5.3.2 may sometimes be used to save time in computing a determinant if we can recognize that the matrix is not invertible. Suppose that

$$A = \begin{bmatrix} 2 & -1 & 3 \\ 1 & 1 & 2 \\ 3 & 0 & 5 \end{bmatrix}$$

For this matrix it is easy to see that the third row is the sum of the first two rows. This means that the rank of A is less than three, so A is not invertible. Thus, $|A| = 0$.

Theorem 5.3.4   *Let A and B be $n \times n$ matrices. Then*

$$|AB| = |A| \cdot |B|$$

Proof
Case 1:   Suppose that either A or B is not invertible. It then follows that AB is not invertible by Corollary 3.4.10. So by Theorem 5.3.2 it follows that both sides are zero.
Case 2:   Suppose that both A and B are invertible. Then each matrix is a product of elementary matrices. By equation (1) the determinant of a product of elementary matrices is the product of the determinants. From this result, the theorem follows immediately.   ■

Suppose that we know that $A^3 = O$ for some square matrix A. Although we may apply earlier results to conclude that A is not invertible, a quick application of the preceding two theorems yields

$$0 = |O| = |A^3| = |A|^3$$

But this implies that $|A| = 0$, so A is not invertible. Theorem 5.3.4 is a very useful tool for proving additional properties of determinants. We will see such an application in the proof of the next theorem.

Theorem 5.3.5   *Let A be an $n \times n$ matrix.*
   (a) *If A is invertible, then*

$$|A^{-1}| = |A|^{-1}$$

   (b) *If A is similar to B, then $|A| = |B|$.*
   (c) *$|A^t| = |A|$.*

Proof
(a) If A is invertible, we may apply Theorem 5.3.4 and write

$$1 = |I| = |AA^{-1}| = |A| \cdot |A^{-1}|$$

The result follows immediately.

(b) If A is similar to B, there exists an invertible matrix P such that

$$A = PBP^{-1}$$

So

$$|A| = |PBP^{-1}| = |P| \cdot |B| \cdot |P^{-1}| = |P| \cdot |B| \cdot |P|^{-1} = |B|$$

(c) We leave it as an exercise to show that if E is an elementary matrix, then $|E| = |E^t|$. Now suppose that A is not invertible. Then A^t is also not invertible; hence both $|A|$ and $|A^t|$ are zero by Theorem 5.3.2.

If A is invertible, we may write A as a product of elementary matrices, say $A = E_1 E_2 \cdots E_s$. So

$$
\begin{aligned}
|A^t| &= |(E_1 E_2 \cdots E_s)^t| \\
&= |E_s^t E_{s-1}^t \cdots E_1^t| \\
&= |E_s^t| \cdot |E_{s-1}^t| \cdots |E_1^t| \\
&= |E_s| \cdot |E_{s-1}| \cdots |E_1| \\
&= |E_1| \cdot |E_2| \cdots |E_s| \\
&= |E_1 E_2 \cdots E_s| \\
&= |A| \qquad \blacksquare
\end{aligned}
$$

Determinants of Block Matrices

In Section 2.2 we studied block matrices and saw that matrix multiplication and addition treated blocks as if they were scalars. Unfortunately, we will see in Example 2 that the determinant function does not treat blocks exactly as it treats scalars. In particular, the next example demonstrates that in general,

$$
\begin{vmatrix} A & B \\ C & D \end{vmatrix} \neq |A||D| - |B||C|
$$

Example 2   Let

$$
A = \begin{bmatrix} 1 & 0 & 1 & 0 \\ 0 & 1 & 0 & 1 \\ 1 & 0 & 2 & 0 \\ 0 & 1 & 0 & 1 \end{bmatrix} = \begin{bmatrix} I & I \\ I & D \end{bmatrix}
$$

Then $|A| = 0$ by Corollary 5.3.3 because the second and fourth rows are equal. On the other hand,

$$|I||D| - |I||I| = (1)(2) - (1)(1) = 1 \neq 0$$

All is not lost, however, for we do have the following theorem.

Theorem 5.3.6 *Let H be the block matrix*

$$H = \begin{bmatrix} A & B \\ O & D \end{bmatrix}$$

where the blocks A and D are square matrices. Then

$$|H| = |A| \cdot |D|$$

Before we prove this theorem, we should note the following result, which is a slightly stronger version of Theorem 1.5.2:

(∗)   *A square matrix may be transformed into an upper triangular matrix by repeated applications of elementary row (column) operations of type 3.*

Thus, by Theorem 5.2.2(c), we have

$$\begin{vmatrix} 0 & 2 & 3 \\ 2 & 1 & 4 \\ 6 & 1 & 5 \end{vmatrix} = \begin{vmatrix} 2 & 3 & 7 \\ 2 & 1 & 4 \\ 6 & 1 & 5 \end{vmatrix} \quad \text{added row 2 to row 1}$$

$$= \begin{vmatrix} 2 & 3 & 7 \\ 0 & -2 & -3 \\ 0 & -8 & -16 \end{vmatrix} \quad \begin{matrix} \text{added } -(\text{row 1}) \text{ to row 2} \\ \text{added } -3 \times (\text{row 1}) \text{ to row 3} \end{matrix}$$

$$= \begin{vmatrix} 2 & 3 & 7 \\ 0 & -2 & -3 \\ 0 & 0 & -4 \end{vmatrix} \quad \text{added } -4 \times (\text{row 2}) \text{ to row 3}$$

$$= 2(-2)(-4) = 16$$

We are now ready for the proof of the theorem.

Proof

Because of (∗) above, we may transform A into an upper triangular matrix A' where $|A'| = |A|$. These row operations transform B into some matrix, say B'. Thus, H is transformed into

$$\begin{bmatrix} A' & B' \\ O & D \end{bmatrix}$$

We similarly transform D into an upper triangular matrix D' where $|D'| = |D|$. We obtain the upper triangular matrix

$$\begin{bmatrix} A' & B' \\ O & D' \end{bmatrix}$$

If we let p and q denote the products of the diagonal entries of the matrices A' and D', respectively, we have

$$\begin{vmatrix} A & B \\ O & D \end{vmatrix} = \begin{vmatrix} A' & B' \\ O & D' \end{vmatrix} = pq = |A'| \cdot |D'| = |A| \cdot |D| \qquad \blacksquare$$

Example 3

$$\begin{vmatrix} 2 & 1 & -3 & 1 \\ 6 & 2 & 4 & 4 \\ 0 & 0 & 3 & -1 \\ 0 & 0 & 5 & 6 \end{vmatrix} = \begin{vmatrix} 2 & 1 \\ 6 & 2 \end{vmatrix} \begin{vmatrix} 3 & -1 \\ 5 & 6 \end{vmatrix} = (-2)(23)$$

$$= -46$$

To apply the properties derived in this section, we return to the formula for the area of a parallelogram discussed in Section 5.1. Recall that if the sides of the parallelogram are determined by the vectors $\mathbf{x}$ and $\mathbf{y}$, then the area is given by $|\det A|$, where the rows of A are given by $\mathbf{x}$ and $\mathbf{y}$. Notice that because $\det A = \det A^t$ by Theorem 5.3.5(c), we may assume that the columns of A are given by $\mathbf{x}$ and $\mathbf{y}$. If we use block notation, we may write the area as $|\det (\mathbf{x}, \mathbf{y})|$.

Our first proof assumed that $\mathbf{y}$ is parallel to the x-axis. Suppose that we want to extend this formula to the parallelogram in Figure 5.3.1(a). The area of this

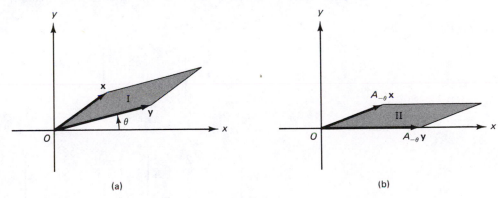

(a)   (b)

Figure 5.3.1

parallelogram is the same as the one obtained in Figure 5.3.1(b) by rotating (counterclockwise) the parallelogram by $-\theta$. This parallelogram has sides determined by $A_{-\theta}\mathbf{x}$ and $A_{-\theta}\mathbf{y}$. Using Theorem 5.3.4 and the properties of the rotation matrix A_θ, we have

$$
\begin{aligned}
\text{Area(I)} &= \text{area(II)} \\
&= |\det(A_{-\theta}\mathbf{x}, A_{-\theta}\mathbf{y})| \\
&= |\det(A_{-\theta}(\mathbf{x}, \mathbf{y}))| \\
&= |\det A_{-\theta}|\,|\det(\mathbf{x}, \mathbf{y})| \\
&= (1)|\det(\mathbf{x}, \mathbf{y})| \\
&= |\det(\mathbf{x}, \mathbf{y})|
\end{aligned}
$$

The Classical Adjoint and Proof of Cramer's Rule (Optional)

A final topic is one that is interesting in its own right but has little application in the remainder of the book. We assume throughout that A is an $n \times n$ matrix with ijth cofactor C_{ij}, that is,

$$
C_{ij} = (-1)^{i+j}|M_{ij}|
$$

Recall that

$$
|A| = A_{i1}C_{i1} + \cdots + A_{in}C_{in} \tag{2}
$$

represents the cofactor expansion of det A along the ith row.

Definition   The **classical adjoint** of A, denoted by **adj** A, is defined to be the matrix whose ijth entry is C_{ji}.

In other words, if we let C be the matrix of cofactors, then adj $A = C^t$.

Example 4   Let

$$
A = \begin{bmatrix} 1 & 2 & 3 \\ 1 & 0 & 1 \\ 1 & 1 & -1 \end{bmatrix}
$$

Then

$$
\text{adj } A = \begin{bmatrix} -1 & 2 & 1 \\ 5 & -4 & 1 \\ 2 & 2 & -2 \end{bmatrix}^t = \begin{bmatrix} -1 & 5 & 2 \\ 2 & -4 & 2 \\ 1 & 1 & -2 \end{bmatrix}
$$

Lemma 5.3.7   *If $i \neq j$, then*

$$A_{i1}C_{j1} + \cdots + A_{in}C_{jn} = 0 \tag{3}$$

Notice that in formula (3) we have replaced the cofactors of row i with the cofactors of row j.

Proof
Let B denote the matrix formed from A by replacing the jth row of A by the ith row of A. Because B has two equal rows, we may conclude from Corollary 5.3.3 that $\det B = 0$. However, it is easy to see that the left side of (3) is precisely the cofactor expansion of $\det B$ along the jth row. ∎

Combining equations (2) and (3), we obtain

$$A_{i1}C_{j1} + \cdots + A_{in}C_{jn} = \begin{cases} \det A & \text{if } i = j \\ 0 & \text{if } i \neq j \end{cases}$$

The left side of the equation above is by definition the ijth entry of the matrix product $A \text{ adj } A$, whereas the right side of the equation is the ijth entry of the matrix $(\det A) I$. As a consequence of these observations, we have the following theorem.

Theorem 5.3.8   *For an $n \times n$ matrix A we have*

$$A \text{ adj } A = (\det A)I$$

Corollary 5.3.9   *If A is invertible, then*

$$A^{-1} = \frac{1}{|A|} \text{ adj } A$$

Proof
The formula for the inverse follows from Theorem 5.3.8, the fact that $|A| \neq 0$, and the result that the inverse is unique. ∎

Example 5   To illustrate Theorem 5.3.8, we use the matrix A of Example 4. By cofactor expansion along the second row of A, we obtain

$$\det A = (-1)(1)(-5) + (-1)(1)(-1) = 6$$

So

$$A \text{ adj } A = \begin{bmatrix} 1 & 2 & 3 \\ 1 & 0 & 1 \\ 1 & 1 & -1 \end{bmatrix} \begin{bmatrix} -1 & 5 & 2 \\ 2 & -4 & 2 \\ 1 & 1 & -2 \end{bmatrix} = \begin{bmatrix} 6 & 0 & 0 \\ 0 & 6 & 0 \\ 0 & 0 & 6 \end{bmatrix}$$

$$= (\det A)I$$

Although the formula above provides a theoretical representation for the inverse, the reader is warned that the computation of adj A is prohibitive for $n \geq 5$.

We are now ready for the proof of Cramer's rule. Recall that for the system $A\mathbf{x} = \mathbf{b}$, we denote by M_i the matrix formed from A by replacing the ith column of A by $\mathbf{b}$.

Theorem 5.3.10

(Cramer's Rule)   Let A be an $n \times n$ matrix which is invertible. Then the (unique) solution to $A\mathbf{x} = \mathbf{b}$ is given by the vector $\mathbf{x} = (x_1, \ldots, x_n)$ whose coordinates are defined by

$$x_i = \frac{\det M_i}{\det A} \quad for \quad i = 1, \ldots, n$$

Proof
Because A is invertible, we may write

$$\mathbf{x} = A^{-1}\mathbf{b}$$

$$= \frac{(\text{adj } A)\mathbf{b}}{\det A}$$

So if the components of $\mathbf{b}$ are $b_1, \ldots, b_n$, then

$$x_i = \frac{b_1 C_{1i} + \cdots + b_n C_{ni}}{\det A}$$

The numerator of the fraction above is precisely the determinant of M_i expanded along the ith column. ∎

Application to Coding

The use of mathematics in cryptography or coding has become particularly significant in recent years. A simple example occurs when the sender encodes a message using a square matrix A. The receiver decodes the message with the inverse of A. Unless one knows the matrix A, the code may be difficult to break. It becomes yet more difficult to break the code if the matrix A is changed periodically. Hence, the availability of a large number of matrices with *nice* inverses becomes very important. Before we show how to create such matrices, we illustrate how the coding actually works.

We first assign a number to each letter of the alphabet. To keep the numbers relatively small, we use negative ones as well as positive ones. Say that $A = 1, B = -1,$ $C = 2, D = -2, E = 3,$ and so on. To allow us to separate the words, we let 0 correspond to a blank or space. The following message illustrates the method:

$$
\begin{array}{ccccccccccc}
\text{R} & \text{E} & \text{M} & \text{A} & \text{I} & \text{N} & & \text{H} & \text{O} & \text{M} & \text{E} \\
-9 & 3 & 7 & 1 & 5 & -7 & 0 & -4 & 8 & 7 & 3
\end{array}
$$

To encode the message, we use the matrix

$$
A = \begin{bmatrix} 1 & 2 & -2 \\ 2 & 3 & 0 \\ 1 & 1 & 1 \end{bmatrix}
$$

Because A is a 3×3 matrix, we divide the numbers above into four 3×1 column vectors as follows:

$$
\begin{bmatrix} -9 \\ 3 \\ 7 \end{bmatrix} \quad \begin{bmatrix} 1 \\ 5 \\ -7 \end{bmatrix} \quad \begin{bmatrix} 0 \\ -4 \\ 8 \end{bmatrix} \quad \begin{bmatrix} 7 \\ 3 \\ 0 \end{bmatrix}
$$

Notice that the last vector is filled with as many zeros as needed. We now let B be the matrix whose columns are the vectors above and encode the message by computing the product AB. The columns of AB are simply the products of A and each of the column vectors. Thus,

$$
AB = \begin{bmatrix} 1 & 2 & -2 \\ 2 & 3 & 0 \\ 1 & 1 & 1 \end{bmatrix} \begin{bmatrix} -9 & 1 & 0 & 7 \\ 3 & 5 & -4 & 3 \\ 7 & -7 & 8 & 0 \end{bmatrix} = \begin{bmatrix} -17 & 25 & -24 & 13 \\ -9 & 17 & -12 & 23 \\ 1 & -1 & 4 & 10 \end{bmatrix}
$$

We send the message:

$$
-17, -9, 1, 25, 17, \ldots, 10
$$

The receiver decodes the message by computing $A^{-1}(AB) = B$. In this case,

$$
A^{-1} = \begin{bmatrix} 3 & -4 & 6 \\ -2 & 3 & -4 \\ -1 & 1 & -1 \end{bmatrix}
$$

For this process to work smoothly, it is important to have matrices such as A whose entries are integers and whose inverses also have entries that are integers. After computing many inverses in Chapter 3, the reader may doubt that many such matrices exist. We will use the ideas developed in this chapter to show how such matrices may be produced quite simply.

The idea is to begin with a matrix A whose determinant equals ± 1 and whose entries are integers. Because A has integral entries, its cofactors are integers and hence the entries of adj A are integers. Because of Corollary 5.3.9, we have $A^{-1} = (1/|A|)$ adj A, so A^{-1} has integral entries.

A practical approach is to begin with an upper triangular matrix with integral entries and such that each of its diagonal entries is ± 1. This latter condition guarantees that the determinant equals ± 1 by Theorem 5.2.3. By Theorem 5.2.2(c) the determinant remains unchanged if we add a multiple of one row to another row.

For example, we might begin with

$$A = \begin{bmatrix} 1 & 5 & -4 \\ 0 & -1 & 2 \\ 0 & 0 & 1 \end{bmatrix}$$

Now det $A = -1$. We first add $-2 \times$ row 1 to row 2 and then add $3 \times$ the new row 2 to row 3. We obtain

$$\begin{bmatrix} 1 & 5 & -4 \\ -2 & -11 & 10 \\ -6 & -33 & 31 \end{bmatrix}$$

By our previous comments, this matrix has an inverse consisting of integral entries. In fact, its inverse is

$$A^{-1} = \begin{bmatrix} 11 & 23 & -6 \\ -2 & -7 & 2 \\ 0 & -3 & 1 \end{bmatrix}$$

Exercises

1. Determine by inspection which of the following matrices have determinants equal to zero.

   (a) $\begin{bmatrix} 1 & 2 & -5 \\ -2 & -4 & 10 \\ 2 & 3 & 1 \end{bmatrix}$

   (b) $\begin{bmatrix} 2 & 3 & 4 & 2 \\ 1 & -1 & 5 & 6 \\ 0 & 0 & 2 & 3 \\ 0 & 0 & 0 & 7 \end{bmatrix}$

   (c) $\begin{bmatrix} 1 & 3 & 1 \\ 2 & 4 & 3 \\ 3 & 7 & 4 \end{bmatrix}$

2. For each of the following matrices A, first compute det A and adj A. Then use Corollary 5.3.9 to compute A^{-1}. Check your answer by verifying that $AA^{-1} = I$.

   (a) $\begin{bmatrix} 1 & 3 & -1 \\ -1 & -4 & 2 \\ 2 & 0 & 1 \end{bmatrix}$

   (b) $\begin{bmatrix} 1 & 1 & 2 \\ 0 & -1 & 3 \\ 2 & 2 & 6 \end{bmatrix}$

(c) $\begin{bmatrix} 1 & -2 & 1 \\ 0 & -1 & 0 \\ 1 & 1 & 2 \end{bmatrix}$ (d) $\begin{bmatrix} 0 & 0 & 1 \\ 2 & -1 & 3 \\ 0 & 2 & 1 \end{bmatrix}$

3. For the matrices in parts (a) and (b) of Exercise 2, verify that the determinant of the product is the product of the determinants.

4. Suppose that A is a square matrix and $|A| = 3$. Use the appropriate theorem(s) to determine the determinants of:
   (a) A^2 (b) A^{-1} (c) A^t

5. Prove that if A is a square matrix, then $|A^m| = |A|^m$ for any positive integer m. Is the result true for any integer m if A is invertible?

6. Let A and B be square matrices of the same size. Prove $|AB| = |BA|$.

7. Let A be an $n \times n$ matrix. Prove that $|A| = 0$ if and only if the rank of A is less than n.

8. Use elementary row operations and Theorem 5.3.2 to prove that the following matrix is not invertible for any values of x, y, and z.

$$\begin{bmatrix} x & y & z \\ x+1 & y+1 & z+1 \\ x+2 & y+2 & z+2 \end{bmatrix}$$

9. A square matrix A is said to be an **orthogonal matrix** if $AA^t = I$. Prove that if A is orthogonal, then $|A| = \pm 1$.

10. An $n \times n$ matrix A is said to be **skew symmetric** if $A^t = -A$. Suppose that n is odd. Prove that if A is skew symmetric, then $|A| = 0$.

11. Prove that for any elementary matrix E, we have $|E^t| = |E|$.

12. Prove that the set of $n \times n$ matrices that have determinant equal to one is *closed* under multiplication, inversion, and transposition.

13. Evaluate the determinant of each of the following block matrices using Theorem 5.3.6.

(a) $\begin{bmatrix} 2 & 3 & 1 & 8 \\ 8 & -3 & 2 & 5 \\ 0 & 0 & 4 & 6 \\ 0 & 0 & 3 & 6 \end{bmatrix}$

(b) $\begin{bmatrix} 4 & 2 & -1 \\ 0 & 7 & 3 \\ 0 & 5 & 4 \end{bmatrix}$

(c) $\begin{bmatrix} 3 & 0 & 4 & -1 \\ 0 & 2 & 1 & 2 \\ 0 & -1 & 0 & 4 \\ 0 & 3 & 3 & 1 \end{bmatrix}$

(d) $\begin{bmatrix} 2 & 1 & -5 & 2 \\ 3 & 2 & -2 & 4 \\ 0 & 0 & 2 & 1 \\ 0 & 0 & 4 & 4 \end{bmatrix}$

†14. For any $n \times n$ matrix A, define the function

$$f(x) = \det(xI - A)$$

(a) This function is called the **characteristic polynomial** of A. Prove that f is a polynomial of degree n. (Use mathematical induction.)

(b) Prove that if $f(c) = 0$, there exists a nonzero vector $\mathbf{x}$ such that $A\mathbf{x} = c\mathbf{x}$. The scalar c is called an **eigenvalue** of A and the vector $\mathbf{x}$ is called an **eigenvector** of A.

(c) Prove that if c is an eigenvalue of A, then c is an eigenvalue of A^t.

(d) For the matrix A given below, find two eigenvalues and their corresponding eigenvectors.

$$A = \begin{bmatrix} 1 & 3 \\ 4 & 2 \end{bmatrix}$$

15. A matrix A is said to be **idempotent** if $A^2 = A$. Prove that if A is idempotent, then $\det A = 0$ or 1.

16. A matrix A is said to be **nilpotent** if $A^k = O$ for some positive integer k. Prove that if A is nilpotent, then $|A| = 0$.

17. Use the method developed in this section to construct two invertible matrices whose entries are nonzero integers and whose inverses also have entries which are integers.

† This exercise will be used in a subsequent section.

18. Use your answers to Exercise 17 to produce two encodings of the message

    ALL IS WELL

*19. Use the program MATRIX to verify Theorems 5.3.4 and 5.3.5(a) for the matrices A and B given below.

$$A = \begin{bmatrix} 1 & 2 & -1 \\ 2 & 5 & 4 \\ 3 & 7 & 4 \end{bmatrix} \quad \text{and} \quad B = \begin{bmatrix} 1 & 0 & 1 \\ 3 & 3 & 4 \\ 2 & 2 & 3 \end{bmatrix}$$

* This problem should be solved by using one of the programs noted in Appendix B.

Key Words _____

6

Eigenvalues, Eigenvectors, and Their Applications

The increased use of linear algebra in such diverse areas as physics, sociology, biology, economics, and statistics has focused considerable attention on "eigenvalues" and "eigenvectors"—their applications and their computation. In this chapter we introduce these very important tools together with a variety of applications. In Chapter 9 we turn our attention to the numerical techniques related to the computation of eigenvalues and eigenvectors.

6.1 EIGENVECTORS AND EIGENVALUES

One of the areas in which linear algebra has provided both clarity and simplification is geometry. Many geometric results that in the past required tedious proofs have yielded to easily verifiable statements about linear transformations. Three important linear transformations in geometry which we have already encountered are projections, reflections, and rotations. In Section 4.2 we found a simple matrix representation of a reflection by choosing a basis of vectors, each of which was carried into a multiple of itself under the reflection. We shall find vectors with the same property for the transformations given in the next several examples.

Example 1 Let $T: R^2 \rightarrow R^2$ be the linear transformation defined by $T(x, y) = (x, 0)$. Recall that T is the projection on the x-axis (see Figure 6.1.1). There are two linearly independent vectors which are of interest to us, namely, $\mathbf{e}_1 = (1, 0)$ and $\mathbf{e}_2 = (0, 1)$. Notice that both of these vectors are carried by T into multiples of themselves. That is,

$$T(\mathbf{e}_1) = (1)\mathbf{e}_1 \qquad \text{and} \qquad T(\mathbf{e}_2) = (0)\mathbf{e}_2$$

253

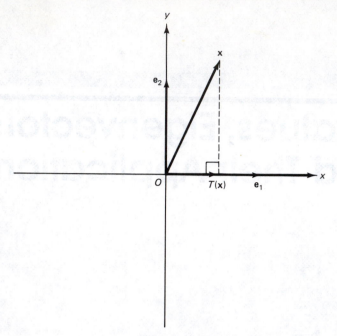

Figure 6.1.1

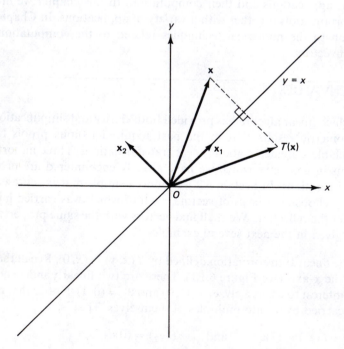

Figure 6.1.2

Example 2   Let T denote the linear transformation which reflects a vector in R^2 about the line $y = x$ (see Figure 6.1.2). If we choose $\mathbf{x}_1 = (1, 1)$ and $\mathbf{x}_2 = (-1, 1)$, then it is easy to see that

$$T(\mathbf{x}_1) = \mathbf{x}_1 \text{ and } T(\mathbf{x}_2) = -\mathbf{x}_2$$

That is, $\mathbf{x}_1$ and $\mathbf{x}_2$ are carried by T into multiples of themselves.

From Examples 1 and 2, we see that vectors that are carried into multiples of themselves are in some sense "natural" for the given transformation.

Definition   Let T be a linear transformation from $R^n \rightarrow R^n$. A nonzero vector $\mathbf{x}$ in R^n is called an ***eigenvector*** of T if $T(\mathbf{x}) = \lambda\mathbf{x}$ for some scalar λ. The scalar λ is called the ***eigenvalue of T corresponding to x***.

It is traditional to use the Greek letter λ (lambda) to denote an eigenvalue. Other texts on linear algebra use a variety of terms for the prefix "eigen-" such as "characteristic," "proper," "latent," and others.

For the transformation in Example 1, we see that $\mathbf{e}_1$ and $\mathbf{e}_2$ are eigenvectors with corresponding eigenvalues 1 and 0, respectively. It is easy to show that all nonzero multiples of these vectors are also eigenvectors with the same corresponding eigenvalues.

It should be pointed out that by this definition, eigenvalues are automatically real numbers. However, there are many practical applications where complex eigenvalues are needed, for example, in the solution of differential equations that describe harmonic motion. We shall see such an application in Section 6.3. In the complex case, the transformation T is defined on the set C^n of n-tuples of complex numbers instead of R^n. Many of the theorems that we will encounter in Chapters 6 and 7 dealing with eigenvalues and eigenvectors will also hold for complex eigenvalues and eigenvectors with complex coordinates. Although *eigenvalue* will mean *real eigenvalue* for us in this chapter, we will provide some optional material for those interested in the complex case. Additional coverage of the situation where scalars can be either complex or real will occur in Chapters 8 and 9.

Because we will be concerned in this chapter only with linear transformations from R^n into itself, we introduce the term "linear operator."

Definition   A linear transformation from R^n to R^n is called a ***linear operator*** on R^n.

Example 3   Let T represent the linear operator on R^2 which rotates a vector by an angle θ (counterclockwise) where $0 \le \theta < 2\pi$ (see Figure 6.1.3). Three interesting cases arise:

1. If $\theta = 0$, then T is the identity transformation, in which case every vector is carried into itself [see Figure 6.1.3(a)]. Thus *every* nonzero vector is an eigenvector of T with eigenvalue equal to 1.

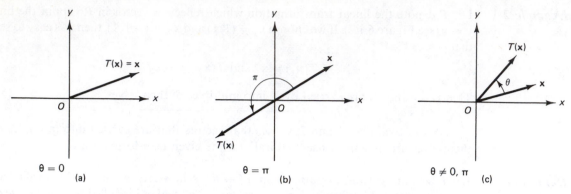

Figure 6.1.3

2. If $\theta = \pi$, then $T(\mathbf{x}) = -\mathbf{x}$ for every vector $\mathbf{x}$ [see Figure 6.1.3(b)]. Thus, *every nonzero vector is an eigenvector of T with eigenvalue equal to -1.* In this case, T is the *reflection about the origin.*

3. If θ is not an integral multiple of π, then T has no eigenvectors and hence no eigenvalues [see Figure 6.1.3(c)].

Another geometric transformation different from the ones in Example 3 is called a ***shear transformation.*** A shear transformation in the x-direction is an operator on R^2 defined by

$$T(x, y) = (x + ky, y) \qquad \text{for some nonzero scalar } k$$

In Figure 6.1.4 notice that vectors along the x-axis are not affected by T because in this case $y = 0$; however, vectors whose heads lie far from the x-axis, that is, a large y-coordinate, are changed the most. It is clear geometrically, as well as by the algebraic definition of T, that for $k \neq 0$, the only eigenvectors of T corresponding to eigenvalue 1 are nonzero multiples of $\mathbf{e}_1$.

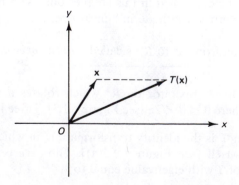

Figure 6.1.4

Thus far, we have considered only relatively simple geometric transformations where the eigenvectors and eigenvalues are easy to identify. For other types of transformations, it will be more practical to examine the standard matrix of the transformation. To do this, we must first introduce the following definition.

Definition

Let A be an $n \times n$ matrix. A nonzero $n \times 1$ vector $\mathbf{x}$ is an ***eigenvector*** of A if $A\mathbf{x} = \lambda \mathbf{x}$ for some scalar λ. The scalar λ is called the ***eigenvalue of A corresponding to $\mathbf{x}$***.

If A is the standard matrix of T, then $A\mathbf{x} = \lambda \mathbf{x}$ if and only if $T(\mathbf{x}) = \lambda \mathbf{x}$. So both A and T have exactly the same eigenvectors and corresponding eigenvalues. To find the eigenvectors of A, we have to solve the system of equations represented by $A\mathbf{x} = \lambda \mathbf{x}$, or $(\lambda I - A)\mathbf{x} = \mathbf{0}$. Unfortunately, the scalar λ is unknown. The next theorem tells us how to find λ.

Theorem 6.1.1

Let A be an $n \times n$ matrix. Then a scalar λ is an eigenvalue of A if and only if λ is a (real) root of the equation

$$\det (xI - A) = 0$$

Proof

A scalar λ is an eigenvalue of A if and only if there exists a nonzero vector $\mathbf{x}$ such that $A\mathbf{x} = \lambda \mathbf{x}$, or $(\lambda I - A)\mathbf{x} = \mathbf{0}$. By Corollary 3.4.8, this is true if and only if $(\lambda I - A)$ is not invertible. However, by Theorem 5.3.2 this result is equivalent to the statement that $\det (\lambda I - A) = 0$. ∎

As we noted earlier, complex eigenvalues are useful in a number of applications. Just as in the case of real eigenvalues, they may be found by solving the equation given in Theorem 6.1.1.

To see how this theorem is applied, let A be the standard matrix of the shear transformation described earlier. Then

$$A = \begin{bmatrix} 1 & k \\ 0 & 1 \end{bmatrix}$$

We must solve the equation

$$\det (xI - A) = \left| x \begin{bmatrix} 1 & 0 \\ 0 & 1 \end{bmatrix} - \begin{bmatrix} 1 & k \\ 0 & 1 \end{bmatrix} \right|$$

$$= \begin{vmatrix} x - 1 & -k \\ 0 & x - 1 \end{vmatrix}$$

$$= (x - 1)^2$$

$$= 0$$

The only solution is $\lambda = 1$. This agrees with our earlier observation. To find the eigenvectors corresponding to $\lambda = 1$, we must solve the system $A\mathbf{x} = 1\mathbf{x}$, or $(1I - A)\mathbf{x} = \mathbf{0}$. That is,

$$\begin{bmatrix} 0 & -k \\ 0 & 0 \end{bmatrix}\begin{bmatrix} x \\ y \end{bmatrix} = \begin{bmatrix} 0 \\ 0 \end{bmatrix}$$

or

$$0x - ky = 0$$

$$0x + 0y = 0$$

Because x is not an initial unknown, we assign it the parameter s. Since $k \neq 0$, it follows that $y = 0$. We obtain the solution $\mathbf{y} = (s, 0) = s(1, 0)$. That is, the solution space is spanned by the vector $\mathbf{e}_1$. Thus, the nonzero multiples of $\mathbf{e}_1$ are the eigenvectors of A corresponding to the eigenvalue 1.

Notice that to find the eigenvalues of the shear transformation, we had to find the zeros of the polynomial $f(x) = (x - 1)^2$, which has degree two. In general, if A is a matrix of order n, then $f(x) = \det(xI - A)$ is a polynomial of degree n (see Exercise 14 of Section 5.3).

Definition   Let A be an $n \times n$ matrix. The nth degree polynomial

$$f(x) = \det(xI - A)$$

is called the **characteristic polynomial** of A. The equation

$$\det(xI - A) = 0$$

is called the **characteristic equation** of A. If A is the standard matrix of an operator T, we say that $f(x)$ is the **characteristic polynomial** of T and $f(x) = 0$ is the **characteristic equation** of T.

_____

From Theorem 6.1.1 it follows that the (real) roots of the characteristic equation, that is, the (real) zeros of the characteristic polynomial, are the eigenvalues of A. In the box below we summarize the method for finding the eigenvalues and eigenvectors of a matrix A. For a linear operator, we apply the method to its standard matrix.

Computation of Eigenvalues and Eigenvectors of a Matrix *A*

1. Compute the roots of the characteristic equation of *A*.
2. For each root λ, the nonzero solutions of the system $(\lambda I - A)\mathbf{x} = \mathbf{0}$ are the eigenvectors corresponding to the eigenvalue λ.

At this time, we will not worry about the problem of finding the zeros of those characteristic polynomials that are not easy to factor. In fact, the problem of finding the characteristic polynomial as well as its zeros for large matrices can be quite challenging. In Chapter 9 we develop numerical techniques for approximating eigenvalues and eigenvectors of large matrices without first computing the characteristic polynomial.

Example 4   Find the eigenvalues and eigenvectors of the matrix *A*.

$$A = \begin{bmatrix} 1 & 3 \\ 4 & 2 \end{bmatrix}$$

The characteristic polynomial of *A* is

$$f(x) = \left| x \begin{bmatrix} 1 & 0 \\ 0 & 1 \end{bmatrix} - \begin{bmatrix} 1 & 3 \\ 4 & 2 \end{bmatrix} \right| = \begin{vmatrix} x - 1 & -3 \\ -4 & x - 2 \end{vmatrix}$$

$$= (x - 1)(x - 2) - 12$$

$$= x^2 - 3x - 10$$

$$= (x - 5)(x + 2)$$

So the eigenvalues are 5 and -2.

To find the eigenvectors corresponding to 5, we must solve the system $(5I - A)\mathbf{x} = \mathbf{0}$. By row reduction of the augmented matrix, we have

$$\begin{bmatrix} 4 & -3 & 0 \\ -4 & 3 & 0 \end{bmatrix} \rightarrow \begin{bmatrix} 4 & -3 & 0 \\ 0 & 0 & 0 \end{bmatrix}$$

Thus, we must solve the equivalent system,

$$4x - 3y = 0$$

The solution space is spanned by the vector $\mathbf{x}_1 = (3, 4)$. Therefore, the eigenvectors corresponding to the eigenvalue 5 are the nonzero multiples of $(3, 4)$.

To find the eigenvectors corresponding to -2, we must solve the system $(-2I - A)\mathbf{x} = \mathbf{0}$. By row reduction of the augmented matrix, we have

$$\begin{bmatrix} -3 & -3 & 0 \\ -4 & -4 & 0 \end{bmatrix} \rightarrow \begin{bmatrix} -3 & -3 & 0 \\ 0 & 0 & 0 \end{bmatrix}$$

Thus, we must solve the equivalent system,

$$-3x - 3y = 0$$

The solution space is spanned by the vector $\mathbf{x}_2 = (1, -1)$. Therefore, the nonzero multiples of $(1, -1)$ provide all the eigenvectors corresponding to -2. We conclude that A has two linearly independent eigenvectors, $\mathbf{x}_1 = (3, 4)$ and $\mathbf{x}_2 = (1, -1)$, corresponding to the eigenvalues 5 and -2, respectively.

Example 5   Let

$$A = \begin{bmatrix} 0 & -1 \\ 1 & 0 \end{bmatrix}$$

Then the characteristic polynomial of A is

$$f(x) = \left| x\begin{bmatrix} 1 & 0 \\ 0 & 1 \end{bmatrix} - \begin{bmatrix} 0 & -1 \\ 1 & 0 \end{bmatrix} \right| = \begin{vmatrix} x & 1 \\ -1 & x \end{vmatrix}$$
$$= x^2 + 1$$

Because this polynomial has no (real) zeros, the matrix A has no eigenvalues and hence no eigenvectors. This is not surprising since A represents the rotation by $\pi/2$, considered in Example 3.

   Complex case (optional): Suppose that we want to find the complex eigenvalues and eigenvectors of A. The complex solutions of the characteristic equation are $\lambda = \pm i$. We must therefore solve the systems

$$(iI - A)\mathbf{x} = \mathbf{0} \qquad \text{and} \qquad (-iI - A)\mathbf{x} = \mathbf{0}$$

For the first system, we row reduce the augmented matrix as follows:

$$\begin{bmatrix} i & 1 & 0 \\ -1 & i & 0 \end{bmatrix} \rightarrow \begin{bmatrix} -1 & i & 0 \\ i & 1 & 0 \end{bmatrix} \rightarrow \begin{bmatrix} -1 & i & 0 \\ 0 & 0 & 0 \end{bmatrix}$$

This yields the system

$$-x + iy = 0$$

One of the solutions to this system is $\mathbf{x} = (i, 1)$; that is, $\mathbf{x}$ is an eigenvector of A corresponding to the eigenvalue i. Because $\mathbf{x}$ is an element of C^2, the set of ordered pairs of complex numbers, there is no simple geometric interpretation. Similarly, it can be shown that an eigenvector corresponding to $-i$ is $(-i, 1)$.

Definition

Let A be an $n \times n$ matrix and let λ be an eigenvalue of A. The subset $E(\lambda)$ of vectors $\mathbf{x}$ in R^n such that $A\mathbf{x} = \lambda\mathbf{x}$ is called the ***eigenspace of A corresponding to the eigenvalue λ***.

Let T be a linear operator on R^n and let λ be an eigenvalue of T. The subset $E(\lambda)$ of vectors $\mathbf{x}$ in R^n such that $T(\mathbf{x}) = \lambda\mathbf{x}$ is called the ***eigenspace of T corresponding to the eigenvalue λ***.

For the matrix A in Example 4, the eigenspace $E(5)$ is the subspace of R^2 spanned by the vector $\mathbf{x}_1 = (3, 4)$, and the eigenspace $E(-2)$ is the subspace of R^2 spanned by the vector $\mathbf{x}_2 = (1, -1)$.

It is easy to see that an eigenspace $E(\lambda)$ consists of the zero vector and all eigenvectors corresponding to the eigenvalue λ. In Example 4, the eigenspaces were both subspaces of R^2. This is true in general because the eigenspace is always the solution space of the homogeneous system: $(\lambda I - A)\mathbf{x} = \mathbf{0}$. For a linear operator T on R^n, we have that $E(\lambda) = N(\lambda I - T)$, which is also a subspace R^n.

Example 6

Let T be the linear operator on R^3 defined by

$$T(x, y, z) = (4x + z, 2x + 3y + 2z, x + 4z)$$

If A is the standard matrix of T, then

$$A = \begin{bmatrix} 4 & 0 & 1 \\ 2 & 3 & 2 \\ 1 & 0 & 4 \end{bmatrix}$$

The characteristic polynomial of A is

$$f(x) = \begin{vmatrix} x - 4 & 0 & -1 \\ -2 & x - 3 & -2 \\ -1 & 0 & x - 4 \end{vmatrix}$$

$$= (x - 4) \begin{vmatrix} x - 3 & -2 \\ 0 & x - 4 \end{vmatrix} + (-1) \begin{vmatrix} -2 & x - 3 \\ -1 & 0 \end{vmatrix}$$

$$= (x - 4)(x - 3)(x - 4) - (x - 3)$$

$$= (x - 3)((x - 4)^2 - 1)$$

$$= (x - 3)(x^2 - 8x + 15)$$

$$= (x - 3)^2(x - 5)$$

So the eigenvalues of A (and hence of T) are 5 and 3. We will now find the eigenspaces, $E(5)$ and $E(3)$.

First, we solve the system $(5I - A)\mathbf{x} = \mathbf{0}$. We apply the elementary row operations to the augmented matrix as follows:

$$\begin{bmatrix} 1 & 0 & -1 & 0 \\ -2 & 2 & -2 & 0 \\ -1 & 0 & 1 & 0 \end{bmatrix} \rightarrow \begin{bmatrix} 1 & 0 & -1 & 0 \\ 0 & 2 & -4 & 0 \\ 0 & 0 & 0 & 0 \end{bmatrix}$$

From the reduced matrix, we see that $x_1 = x_3$ and $x_2 = 2x_3$. So the solution space or the eigenspace $E(5)$ is the subspace spanned by the eigenvector $\mathbf{x}_1 = (1, 2, 1)$.

To find the eigenspace $E(3)$, we must solve the system $(3I - A)\mathbf{x} = \mathbf{0}$. Again, we apply the elementary row operations:

$$\begin{bmatrix} -1 & 0 & -1 & 0 \\ -2 & 0 & -2 & 0 \\ -1 & 0 & -1 & 0 \end{bmatrix} \rightarrow \begin{bmatrix} 1 & 0 & 1 & 0 \\ 0 & 0 & 0 & 0 \\ 0 & 0 & 0 & 0 \end{bmatrix}$$

Notice that the second column of the matrix above consists of zeros. This means that the unknown x_2 can take on any value. Also, $x_1 = -x_3$. So the eigenspace $E(3)$ is the subspace spanned by the vectors $\mathbf{x}_2 = (1, 0, -1)$ and $\mathbf{x}_3 = (0, 1, 0)$. If we let $S = \{\mathbf{x}_1, \mathbf{x}_2, \mathbf{x}_3\}$, then S is a (ordered) basis of R^3 and

$$[T]_S = \begin{bmatrix} 5 & 0 & 0 \\ 0 & 3 & 0 \\ 0 & 0 & 3 \end{bmatrix}$$

For example, the entries of the first column are determined by the equation

$$T(\mathbf{x}_1) = 5\mathbf{x}_1 = 5\mathbf{x}_1 + 0\mathbf{x}_2 + 0\mathbf{x}_3$$

The fact, illustrated in Example 6, that certain linear transformations may be represented as diagonal matrices will play a significant role in the remainder of this text. From the same example the reader may get the impression that the dimension of the eigenspace corresponding to the eigenvalue λ is equal to the number of times the factor $(x - \lambda)$ occurs in the factored form of the characteristic polynomial. Unfortunately, this is not always the case. For example, if

$$A = \begin{bmatrix} 1 & 1 \\ 0 & 1 \end{bmatrix}$$

then the characteristic polynomial of A is $f(x) = (x - 1)^2$. However, the eigenspace corresponding to the eigenvalue 1 is generated by $(1, 0)$, and thus has dimension one. We will deal with these considerations in more detail later.

Example 7   Let T be the linear operator on R^3 which projects a vector on the plane $2x - y + 5z = 0$. Clearly, any nonzero vector in the plane is an eigenvector with corresponding eigenvalue 1. So the eigenspace $E(1)$ may be represented geometrically by the plane $2x - y + 5z = 0$. The vector $(2, -1, 5)$ is a normal to the plane, and therefore the image of this vector under T is the zero vector. So the eigenspace $E(0)$ may be represented geometrically as the line through the origin which is parallel to $(2, -1, 5)$. Geometrically, it is clear that T has no other eigenvalues.

Exercises

For each of the linear operators in Exercises 1–8, determine the eigenvalues and a basis for each of the eigenspaces *without* computing the characteristic polynomial.

1. T is the zero operator on R^n.

2. T is the identity operator on R^n.

3. T is the operator on R^3 which projects a vector on the xy-plane.

4. T is the operator on R^3 which projects a vector on the plane $5x - 6y + 3z = 0$.

5. T is the operator on R^2 which projects a vector on the line $y = x$.

6. T is the operator on R^2 which reflects a vector about the x-axis.

7. T is the operator on R^2 which rotates a vector by $60°$.

8. T is the operator on R^3 which has the form $T(\mathbf{x}) = \mathbf{x} \times \mathbf{y}$, where $\mathbf{y}$ is a fixed nonzero vector in R^3 and $\times$ represents cross product (see Section 5.2).

For each of the matrices or linear operators in Exercises 9–16, compute the characteristic polynomial, the eigenvalues (if there are any), and the bases for the eigenspaces.

9. $\begin{bmatrix} 1 & 1 \\ 4 & 1 \end{bmatrix}$ 10. $\begin{bmatrix} 1 & 2 \\ -1 & 4 \end{bmatrix}$ 11. $\begin{bmatrix} 3 & 2 \\ 3 & -2 \end{bmatrix}$

12. $\begin{bmatrix} 3 & -2 \\ 2 & -1 \end{bmatrix}$ 13. $\begin{bmatrix} 1 & -2 \\ 1 & -1 \end{bmatrix}$ 14. $\begin{bmatrix} 2 & 2 \\ -1 & 5 \end{bmatrix}$

15. $T(x, y, z) = (x + 2y - z, x + z, 4x - 4y + 5z)$

16. $T(x, y, z) = (8x + 2y - 2z, 3x + 3y - z, 24x + 8y - 6z)$

17. Prove that the eigenvalues of an upper triangular matrix are precisely the diagonal entries.

18. Suppose that H is a square matrix which has the block form

$$\begin{bmatrix} A & B \\ O & C \end{bmatrix}$$

where A and C are square matrices. Prove that the characteristic polynomial of H is the product of the characteristic polynomials of A and C.

19. Apply Exercise 18 to find the characteristic polynomial of

$$\begin{bmatrix} 3 & 4 & 6 & -3 \\ 2 & 1 & 5 & 4 \\ 0 & 0 & 1 & 4 \\ 0 & 0 & -3 & 2 \end{bmatrix}$$

20. Prove that a matrix is invertible if and only if 0 is not an eigenvalue.

21. Show that if λ is an eigenvalue of an invertible matrix A, then $\lambda \neq 0$ and $1/\lambda$ is an eigenvalue of A^{-1}.

22. Suppose that λ is an eigenvalue of a linear operator T.
    (a) Show that λ^2 is an eigenvalue of T^2.
    (b) Show that $\lambda + 1$ is an eigenvalue of $T + I$.
    (c) If $g(x)$ is any polynomial, define $g(T)$ to be the linear operator formed by substituting T for x. Prove that $g(\lambda)$ is an eigenvalue of $g(T)$.

23. Show that if $f(x)$ is the characteristic polynomial of an $n \times n$ matrix A, then $f(0) = (-1)^n \det A$.

24. Prove that similar matrices have the same characteristic polynomial.

25. Prove that a matrix and its transpose have the same eigenvalues. (Use properties of determinants.)

26. Recall (Exercise 16 of Section 5.3) that a matrix is nilpotent if there exists a nonnegative integer k such that $A^k = O$. Prove that 0 is the only eigenvalue of a nilpotent matrix.

27. Recall (Exercise 15 of Section 5.3) that a matrix A is idempotent if $A^2 = A$. Suppose that A is an idempotent matrix and that λ is an eigenvalue. Show that λ must equal 1 or 0.

28. Recall (Exercise 15 of Section 2.1) that for a matrix A, the trace of A, denoted tr A, is defined to be the sum of the diagonal entries. If A is a 2×2 matrix, prove that its characteristic polynomial has the form:

$$f(x) = x^2 - (\text{tr } A)x + \det A$$

29. Suppose that T is a linear operator on R^n. A subspace W of R^n is said to be ***invariant under***

T if $T(\mathbf{x})$ is in W for every vector $\mathbf{x}$ in W. Prove that every eigenspace of T is invariant under T.

*30. (i) Use the program CHAR POLYN to compute the characteristic polynomials and the eigenvalues of the matrices below.

   (ii) Use the program LINSYST to find bases for the eigenspaces corresponding to the eigenvalues found in (i).

(a) $\begin{bmatrix} 6 & -8 & 6 \\ 6 & -15 & 14 \\ 6 & -19 & 18 \end{bmatrix}$

(b) $\begin{bmatrix} 2 & -1.5 & 3 \\ 0 & 5 & -6 \\ 0 & 3 & -4 \end{bmatrix}$

(c) $\begin{bmatrix} 1 & 4 & -7 \\ 5 & 0 & -7 \\ 5 & 4 & -11 \end{bmatrix}$

(d) $\begin{bmatrix} -1 & 9 & -9 & 12 \\ -5 & 16 & -14 & 19 \\ -9 & 22 & -20 & 31 \\ -4 & 9 & -9 & 15 \end{bmatrix}$

* This problem should be solved by using one of the programs noted in Appendix B.

6.2 DIAGONALIZATION

In Section 2.1 we encountered the transition matrix

$$A = \text{To} \begin{array}{c} \\ \text{City} \\ \text{Suburbs} \end{array} \begin{array}{c} \text{From} \\ \text{City} \quad \text{Suburbs} \\ \begin{bmatrix} 0.85 & 0.03 \\ 0.15 & 0.97 \end{bmatrix} \end{array}$$

The entries represented the proportions (or probabilities) of people moving from one area to another in a given year. For example, 15% of the people living in the city in a given year will move to the suburbs. We saw that if $\mathbf{p}$ were the vector whose coordinates represented the number of people living in the city and suburbs in a given year, then $A\mathbf{p}$ represents the respective populations in the following year. To find the populations in 5 years, we have to compute $A^5\mathbf{p}$. If we want to project far into the future, we have to compute $A^m\mathbf{p}$ for large values of m. This is not a pleasant task even for $m = 5$. Not only is time a factor, but repeated round-off errors might make any result suspect.

In this section we will see that for many matrices, the computation of matrix powers may be made relatively simple. Later, we will look at a rather diverse collection of examples where the computation of matrix powers is essential. For the remainder of this chapter, the word "basis" will be understood to mean "ordered basis."

To make our notation simpler, we shall denote the diagonal matrix

$$D = \begin{bmatrix} c_1 & 0 & 0 & \cdots & 0 \\ 0 & c_2 & 0 & \cdots & 0 \\ 0 & 0 & c_3 & \cdots & 0 \\ \cdot & \cdot & \cdot & \cdots & \cdot \\ \cdot & \cdot & \cdot & \cdots & \cdot \\ \cdot & \cdot & \cdot & \cdots & \cdot \\ 0 & 0 & 0 & \cdots & c_n \end{bmatrix}$$

by

$$D = \text{diag}\,(c_1, \ldots, c_n)$$

Before we state the principal theorem of this section, we review a computation similar to the one we made at the end of Example 6 of the preceding section. Suppose that T is a linear operator on R^n and S is a basis of eigenvectors of T such that $T(\mathbf{x}_i) = \lambda_i \mathbf{x}_i$ for each i. For the sake of concreteness, assume that $n = 3$. Then

$$T(\mathbf{x}_1) = \lambda_1 \mathbf{x}_1 + 0\mathbf{x}_2 + 0\mathbf{x}_3$$

$$T(\mathbf{x}_2) = 0\mathbf{x}_1 + \lambda_2 \mathbf{x}_2 + 0\mathbf{x}_3$$

$$T(\mathbf{x}_3) = 0\mathbf{x}_1 + 0\mathbf{x}_2 + \lambda_3 \mathbf{x}_3$$

So

$$[T]_S = \begin{bmatrix} \lambda_1 & 0 & 0 \\ 0 & \lambda_2 & 0 \\ 0 & 0 & \lambda_3 \end{bmatrix} = \text{diag}\,(\lambda_1, \lambda_2, \lambda_3)$$

Theorem 6.2.1 *Let A be an $n \times n$ matrix which has n linearly independent eigenvectors with the corresponding eigenvalues $\lambda_1, \ldots, \lambda_n$. If P denotes the (invertible) matrix whose columns are the eigenvectors of A and $D = \text{diag}\,(\lambda_1, \ldots, \lambda_n)$, then*

$$P^{-1}AP = D$$

Conversely, suppose that A is an n × n matrix and P is an invertible matrix such that $P^{-1}AP = D = \text{diag}(\lambda_1, \ldots, \lambda_n)$. Then the columns of P are (linearly independent) eigenvectors of A with the corresponding eigenvalues $\lambda_1, \ldots, \lambda_n$.

Proof

Let $T = L_A$ and let S be a basis of eigenvectors corresponding to the eigenvalues $\lambda_1, \ldots, \lambda_n$. By the definition of the matrix $[T]_S$, we have that

$$[T]_S = \text{diag}(\lambda_1, \ldots, \lambda_n) = D$$

By Theorem 4.2.1 we may conclude immediately that $P^{-1}AP = D$.

To prove the converse, suppose that $P^{-1}AP = D = \text{diag}(\lambda_1, \ldots, \lambda_n)$, or $AP = PD$. Using block multiplication, we have that the ith column of AP is

$$^i(AP) = A(^iP)$$

On the other hand, by block multiplication of a $1 \times n$ matrix and $n \times n$ matrix, we have

$$PD = (^1P, \ldots, {}^nP)\begin{bmatrix} \lambda_1 & 0 & \cdots & 0 \\ 0 & \lambda_2 & \cdots & 0 \\ \cdot & \cdot & \cdots & \cdot \\ \cdot & \cdot & \cdots & \cdot \\ \cdot & \cdot & \cdots & \cdot \\ 0 & 0 & \cdots & \lambda_n \end{bmatrix}$$

$$= (\lambda_1\, {}^1P, \ldots, \lambda_n\, {}^nP)$$

So if we equate the ith columns of AP and PD, we obtain

$$A(^iP) = \lambda_i\, {}^iP$$

This equation is equivalent to the statement that the ith column of P is an eigenvector of A with eigenvalue λ_i. ∎

The reader may wonder how this theorem relates to our original problem of computing powers of a given matrix A. We attend to this now. Suppose that A satisfies the conditions of Theorem 6.2.1 with the diagonal matrix D. Write $A = PDP^{-1}$. Then

$$A^2 = (PDP^{-1})(PDP^{-1}) = PD(P^{-1}P)DP^{-1} = PDIDP^{-1} = PD^2P^{-1}$$

Similarly,

$$A^3 = A^2A = (PD^2P^{-1})(PDP^{-1}) = PD^3P^{-1}$$

and, in general,

$$A^m = PD^mP^{-1} \qquad \text{for any positive integer } m$$

This result allows us to compute A^m with only two matrix multiplications. It is easy to show that $D^m = \text{diag}\,(\lambda_1^m, \ldots, \lambda_n^m)$.

To see how this computation is applied, let us return to the transition matrix A introduced at the beginning of the section. It can be shown that A has eigenvectors $(1, 5)$ and $(1, -1)$ with corresponding (approximate) eigenvalues 1 and 0.82, respectively. So

$$P = \begin{bmatrix} 1 & 1 \\ 5 & -1 \end{bmatrix} \qquad P^{-1} = \frac{1}{6}\begin{bmatrix} 1 & 1 \\ 5 & -1 \end{bmatrix} \qquad D = \begin{bmatrix} 1 & 0 \\ 0 & 0.82 \end{bmatrix}$$

Using the fact that $A^m = PD^mP^{-1}$, we have

$$A^m = \begin{bmatrix} 1 & 1 \\ 5 & -1 \end{bmatrix}\begin{bmatrix} 1 & 0 \\ 0 & (0.82)^m \end{bmatrix}\left(\frac{1}{6}\right)\begin{bmatrix} 1 & 1 \\ 5 & -1 \end{bmatrix}$$

$$= \frac{1}{6}\begin{bmatrix} 1 + 5(0.82)^m & 1 - (0.82)^m \\ 5 - 5(0.82)^m & 5 + (0.82)^m \end{bmatrix}$$

In particular, with two-place accuracy we have

$$A^5 = \begin{bmatrix} 0.48 & 0.10 \\ 0.52 & 0.90 \end{bmatrix} \qquad \text{and} \qquad A^8 = \begin{bmatrix} 0.34 & 0.13 \\ 0.66 & 0.87 \end{bmatrix}$$

If we begin with an initial population vector $Q = (50, 75)$ where the units are in thousands of people, then by computing

$$A^5Q = \begin{bmatrix} 31.65 \\ 93.35 \end{bmatrix} \qquad \text{and} \qquad A^8Q = \begin{bmatrix} 26.80 \\ 98.20 \end{bmatrix}$$

we know the populations in 5 and 8 years. In particular, if the present population of the city is 50 thousand people, then in 5 years there will be 31,650 people living in the city and in 8 years there will be 26,800 people living in the city. Does this mean that at some point in the future, there will be virtually no one living in the city? Although we will take up this question in more generality when we study Markov chains, we can answer this question easily for our matrix A.

Because $(0.82)^m$ approaches 0 as m approaches infinity, we say that "A^m approaches" the matrix

$$\begin{bmatrix} \frac{1}{6} & \frac{1}{6} \\ \frac{5}{6} & \frac{5}{6} \end{bmatrix}$$

So for *any* initial population vector $\mathbf{p} = (a, b)$ whose coordinates add up to the total population, say $a + b = 125$ thousand people, we have that $A^m\mathbf{p}$ approaches

$$\begin{bmatrix} \frac{1}{6} & \frac{1}{6} \\ \frac{5}{6} & \frac{5}{6} \end{bmatrix}\begin{bmatrix} a \\ b \end{bmatrix} = \begin{bmatrix} \frac{1}{6}(a + b) \\ \frac{5}{6}(a + b) \end{bmatrix} = \begin{bmatrix} 20.83 \\ 104.17 \end{bmatrix}$$

We conclude that no matter how the population is initially distributed, there will eventually be about 21 thousand people living in the city!

Example 1   In Example 4 of Section 6.1, we showed that if

$$A = \begin{bmatrix} 1 & 3 \\ 4 & 2 \end{bmatrix}$$

then $(3, 4)$ and $(1, -1)$ are linearly independent eigenvectors with corresponding eigenvalues 5 and -2, respectively. So let

$$P = \begin{bmatrix} 3 & 1 \\ 4 & -1 \end{bmatrix} \quad \text{and} \quad P^{-1} = \frac{1}{7}\begin{bmatrix} 1 & 1 \\ 4 & -3 \end{bmatrix}$$

Then

$$A = \begin{bmatrix} 3 & 1 \\ 4 & -1 \end{bmatrix}\begin{bmatrix} 5 & 0 \\ 0 & -2 \end{bmatrix}\left(\frac{1}{7}\right)\begin{bmatrix} 1 & 1 \\ 4 & -3 \end{bmatrix}$$

Therefore,

$$\begin{aligned} A^m &= \begin{bmatrix} 3 & 1 \\ 4 & -1 \end{bmatrix}\begin{bmatrix} 5^m & 0 \\ 0 & (-2)^m \end{bmatrix}\left(\frac{1}{7}\right)\begin{bmatrix} 1 & 1 \\ 4 & -3 \end{bmatrix} \\ &= \frac{1}{7}\begin{bmatrix} 3(5^m) + 4(-2)^m & 3(5^m) - 3(-2)^m \\ 4(5^m) - 4(-2)^m & 4(5^m) + 3(-2)^m \end{bmatrix} \end{aligned}$$

For example, if $m = 4$, we obtain

$$A^4 = \begin{bmatrix} 277 & 261 \\ 348 & 364 \end{bmatrix}$$

Example 2   Let A be the matrix of Example 6 of Section 6.1. This matrix has linearly independent eigenvectors $(1, 2, 1)$, $(1, 0, -1)$, and $(0, 1, 0)$ with corresponding eigenvalues 5, 3, and 3. From Theorem 6.2.1 it follows that if

$$P = \begin{bmatrix} 1 & 1 & 0 \\ 2 & 0 & 1 \\ 1 & -1 & 0 \end{bmatrix}$$

then $P^{-1}AP = D$, where $D = \text{diag}(5, 3, 3)$. It is easier to verify this result if we write it as $AP = PD$. That is,

$$
\begin{bmatrix} 4 & 0 & 1 \\ 2 & 3 & 2 \\ 1 & 0 & 4 \end{bmatrix}\begin{bmatrix} 1 & 1 & 0 \\ 2 & 0 & 1 \\ 1 & -1 & 0 \end{bmatrix} = \begin{bmatrix} 5 & 3 & 0 \\ 10 & 0 & 3 \\ 5 & -3 & 0 \end{bmatrix}
$$

$$
= \begin{bmatrix} 1 & 1 & 0 \\ 2 & 0 & 1 \\ 1 & -1 & 0 \end{bmatrix}\begin{bmatrix} 5 & 0 & 0 \\ 0 & 3 & 0 \\ 0 & 0 & 3 \end{bmatrix}
$$

Definition   A matrix is said to be ***diagonalizable*** if it is similar to a diagonal matrix.

From Theorem 6.2.1 we see that a matrix is diagonalizable if and only if the matrix possesses a set of n linearly independent eigenvectors. Thus, the matrices of Examples 1 and 2 are diagonalizable.

An example of a matrix that is not diagonalizable is the matrix

$$
A = \begin{bmatrix} 0 & 1 \\ 0 & 0 \end{bmatrix}
$$

To see why A is *not* diagonalizable, suppose that $A = PDP^{-1}$. Because the characteristic polynomial of A is $f(x) = x^2$, the only eigenvalue of A is 0. So, if A were diagonalizable, then $D = O$. But this would mean that $A = POP^{-1} = O$, a contradiction.

As we have seen in this section, diagonalizable matrices have desirable properties. The question of concern to us now is how to recognize if a given matrix is diagonalizable. Our next theorem is a first step in this direction.

Theorem 6.2.2 *Let A and B be similar matrices. Then A is diagonalizable if and only if B is diagonalizable.*

Proof
Since A and B are similar, we may write $A \sim B$. B is diagonalizable if and only if $B \sim D$ for some diagonal matrix D. But by Theorem 4.2.4 $B \sim D$ if and only if $A \sim D$. The theorem follows immediately.   ∎

Notice that the proof demonstrates the fact that similar matrices have the same eigenvalues if they are diagonalizable. However, this fact follows from the more general result given in Theorem 6.2.3.

Theorem 6.2.3 *Let A and B be similar matrices. Then A and B have the same characteristic polynomial.*

Proof

If A is similar to B, then there exists an invertible matrix Q such that $Q^{-1}BQ = A$. Hence

$$xI - A = xI - Q^{-1}BQ = Q^{-1}(xI - B)Q$$

So, by Theorems 5.3.4 and 5.3.5(a), we have

$$|xI - A| = |Q^{-1}| \cdot |xI - B| \cdot |Q| = |xI - B| \qquad \blacksquare$$

Corollary 6.2.4

If A and B are matrix representations of a linear operator T, then A and B have the same characteristic polynomial as T.

Proof

The result follows from Corollary 4.2.2. $\qquad \blacksquare$

We are now ready for the definition of a diagonalizable operator.

Definition

A linear operator T on R^n is **diagonalizable** if there exists a basis of R^n which consists of eigenvectors of T.

It is easy to see that T is diagonalizable if and only if its standard matrix is diagonalizable. However, by Theorem 6.2.2 and Corollary 4.2.2, we can say more. *T is diagonalizable if and only if any matrix representation of T is diagonalizable.* From this observation it follows that statements concerning diagonalization about matrices can be restated for linear operators, and vice versa.

Diagonalizable Matrices and Linear Operators

In the remainder of this section, we are concerned with the problem of finding conditions which ensure that a matrix (or linear operator) is diagonalizable.

Theorem 6.2.5   *For a linear operator T on R^n (or an $n \times n$ matrix), any set of eigenvectors corresponding to distinct eigenvalues is linearly independent.*

Proof

The proof is by induction on the number r of distinct eigenvalues. Because an eigenvector must be a nonzero vector, it is clear that the theorem is true for $r = 1$. Now suppose that the theorem is true for any $r - 1$ distinct eigenvalues. Let $\lambda_1, \ldots, \lambda_r$ be distinct eigenvalues with corresponding eigenvectors $\mathbf{x}_1, \ldots, \mathbf{x}_r$. We must show that $\mathbf{x}_1, \ldots, \mathbf{x}_r$ are linearly independent. To accomplish this, assume that

$$a_1\mathbf{x}_1 + \cdots + a_r\mathbf{x}_r = \mathbf{0} \qquad (1)$$

Apply the linear operator $(\lambda_r I - T)$ to both sides of the equation. Using the fact that $(\lambda_r I - T)\mathbf{x}_i = (\lambda_r - \lambda_i)\mathbf{x}_i$ for $i = 1, \ldots, r$, we obtain

$$a_1(\lambda_r - \lambda_1)\mathbf{x}_1 + \cdots + a_r(\lambda_r - \lambda_{r-1})\mathbf{x}_{r-1} + \mathbf{0} = \mathbf{0}$$

By the induction assumption, we have that $\mathbf{x}_1, \ldots, \mathbf{x}_{r-1}$ are linearly independent. Hence, the coefficients $a_1(\lambda_r - \lambda_1), \ldots, a_{r-1}(\lambda_r - \lambda_{r-1})$ are all zero, and therefore, $a_1, \ldots, a_{r-1}$ are all zero. So, from (1), we obtain that $a_r \mathbf{x}_r = \mathbf{0}$. Because $\mathbf{x}_r$ is a nonzero vector, we have that $a_r = 0$. Thus, all of the a_i's are zero, so $\mathbf{x}_1, \ldots, \mathbf{x}_r$ are linearly independent. ∎

Theorem 6.2.6 *If the characteristic polynomial of a linear operator on R^n (or of an $n \times n$ matrix) has n distinct zeros, then the linear operator (matrix) is diagonalizable.*

Proof
Suppose that the characteristic polynomial of a linear operator T on R^n has n distinct zeros, $\lambda_1, \ldots, \lambda_n$. For each eigenvalue λ_j there exists an eigenvector $\mathbf{x}_j$, for $j = 1, \ldots, n$. By Theorem 6.2.5 the vectors $\mathbf{x}_1, \ldots, \mathbf{x}_n$ are linearly independent, and thus must form a basis of R^n. Hence, T is diagonalizable. ∎

The matrix of Example 2 is diagonalizable but it does not have three distinct eigenvalues; thus the converse to the theorem is not true. On the other hand, something can be said for the case when the eigenvalues are not all distinct. We need the following definition first.

Definition Let λ be a zero of the polynomial $f(x)$. The (***algebraic***) ***multiplicity*** of λ is the largest positive integer m such that $(x - \lambda)^m$ is a factor of $f(x)$.

When speaking of the multiplicity of an eigenvalue of a linear operator T, the polynomial $f(x)$ is the characteristic polynomial of T.

   If $f(x) = (x - 6)^3(x - 4)$, then the multiplicities of 6 and 4 are 3 and 1, respectively.

Theorem 6.2.7 *Let T be a linear operator on R^n (or an $n \times n$ matrix). If λ is an eigenvalue of T, then the dimension of the eigenspace $E(\lambda)$ is less than or equal to the multiplicity of λ.*

Proof
Let $r = \dim E(\lambda)$ and let $\mathbf{x}_1, \ldots, \mathbf{x}_r$ be a basis of $E(\lambda)$. We may extend this basis to a basis $\mathbf{x}_1, \ldots, \mathbf{x}_n$ of all of R^n. The matrix representation of T in this basis may be written in the block form

$$\begin{bmatrix} \lambda I_r & B \\ O & C \end{bmatrix}$$

Appealing to Corollary 5.2.4 and Theorem 5.3.6, the characteristic polynomial of T is

$$f(x) = \begin{vmatrix} (x - \lambda)I_r & -B \\ O & xI_{n-r} - C \end{vmatrix}$$

$$= |(x - \lambda)I_r| \cdot |(xI_{n-r} - C)|$$

$$= (x - \lambda)^r |(xI_{n-r} - C)|$$

Because $(x - \lambda)^r$ is a factor of $f(x)$, it must be the case that r cannot exceed the multiplicity of λ. ∎

Some authors use the term *geometric multiplicity* for the dimension of $E(\lambda)$.

Notice that if the multiplicity of an eigenvalue is 1, the dimension of the corresponding eigenspace must also be 1. So the multiplicity in this case is guaranteed to equal the dimension of the eigenspace. This fact will be very useful to us when we attempt to determine if a matrix is diagonalizable.

For the next theorem we assume that the characteristic polynomial factors into (not necessarily distinct) linear factors, that is,

$$f(x) = (x - \lambda_1)(x - \lambda_2) \cdots (x - \lambda_n)$$

As the reader may be aware, there are polynomials such as $f(x) = x^2 + 1$ which do not factor into linear factors (unless complex numbers are allowed). On the other hand, if a linear operator T is diagonalizable, then its characteristic polynomial does factor into linear factors. This follows easily if we let B be a diagonal matrix that represents T. Then it is clear that $f(x) = \det(xI - B)$ has the form above. Equivalently, we have that if the characteristic polynomial of a linear operator does not have such a factorization, it is not diagonalizable. Therefore, we need only consider the case when such a factorization takes place.

Theorem 6.2.8 *Let T be a linear operator on R^n (or an $n \times n$ matrix) whose characteristic polynomial factors into linear factors. Then T is diagonalizable if and only if for each eigenvalue λ, the dimension of $E(\lambda)$ equals the multiplicity of λ.*

Proof

Assume that T is diagonalizable. Let $\lambda_1, \ldots, \lambda_r$ be the distinct eigenvalues of T. For $1 \le i \le r$, let m_i be the multiplicity of λ_i and let $d_i = \dim E(\lambda_i)$. Because the degree of the characteristic polynomial is n, we have that

$$m_1 + \cdots + m_r = n \tag{2}$$

Because T is diagonalizable, there is a basis of n linearly independent eigenvectors. The fact that each eigenvector belongs to some eigenspace implies that

$$n \le d_1 + \cdots + d_r \tag{3}$$

By (2), (3), and Theorem 6.2.7, we have

$$n \le d_1 + \cdots + d_r \le m_1 + \cdots + m_r \le n \tag{4}$$

But (4) implies that

$$d_1 + \cdots + d_r = m_1 + \cdots + m_r$$

or

$$(m_1 - d_1) + \cdots + (m_r - d_r) = 0$$

Because each term in the sum is nonnegative by Theorem 6.2.7, we have that each term is 0; that is, $m_i = d_i$ for $i = 1, \ldots, r$. This completes the proof of the first half of the theorem.

To prove the second half of the theorem we need to assume that the dimension of each $E(\lambda)$ equals the multiplicity of λ. The proof that T is diagonalizable will follow easily if it can be shown that the union S of bases of the eigenspaces is a linearly independent set. For in this case S will be a basis of R^n since the number of vectors in S equals the sum of the multiplicities, which is equal to n. The proof that S is linearly independent uses the technique developed in the proof of Theorem 6.2.5. The details are left to the reader. ∎

As an example, suppose that the characteristic polynomial of T factors as

$$f(x) = (x - 3)^2(x - 4)(x - 5)$$

By our comment following Theorem 6.2.7, we need not worry about the eigenvalues of multiplicity equal to 1. Hence, if we can show that dim $E(3) = 2$, we know that T is diagonalizable by Theorem 6.2.8.

From the dimension theorem we have that for any eigenvalue λ of a matrix A,

$$\text{dim } E(\lambda) = \text{dim (solution space of } (\lambda I - A)\mathbf{x} = \mathbf{0})$$

$$= n - \text{rank } (\lambda I - A)$$

So, using Theorem 6.2.8, we have that A is diagonalizable if and only if the multiplicity of $\lambda = n - \text{rank } (\lambda I - A)$ for every eigenvalue λ. The test is given below.

TEST FOR DIAGONALIZATION

Let A be an $n \times n$ matrix (or T a linear operator on R^n whose standard matrix is A). Then A is diagonalizable if and only if the following two conditions hold:

1. The characteristic polynomial of A factors into a product of linear factors.

2. The multiplicity of each eigenvalue λ equals $n - \text{rank } (\lambda I - A)$.

Notice that the second condition need not be checked for eigenvalues of multiplicity 1.

Example 3   Let T be the linear operator defined by

$$T(x, y, z) = (9x + y, 9y, 7z)$$

Determine if T is diagonalizable.

Let A be the standard matrix of T, then

$$A = \begin{bmatrix} 9 & 1 & 0 \\ 0 & 9 & 0 \\ 0 & 0 & 7 \end{bmatrix}$$

The characteristic polynomial of A is $f(x) = (x - 9)^2(x - 7)$. So condition 1 is met. Because 7 has multiplicity 1, we need only examine the rank of $(9I - A)$.

$$9I - A = \begin{bmatrix} 0 & -1 & 0 \\ 0 & 0 & 0 \\ 0 & 0 & 2 \end{bmatrix}$$

Clearly, rank $(9I - A) = 2$. Therefore, A and hence T is not diagonalizable because $3 - \text{rank}\,(9I - A) = 3 - 2 = 1 \neq 2 = $ multiplicity of 9.

Example 4   For the matrix

$$A = \begin{bmatrix} 3 & -2 & 0 \\ -2 & 3 & 0 \\ 0 & 0 & 5 \end{bmatrix}$$

determine if A is diagonalizable.

The characteristic polynomial of A is $f(x) = (x - 5)^2(x - 1)$. So condition 1 is satisfied. As in Example 3, we need only check the rank of $(5I - A)$.

$$5I - A = \begin{bmatrix} 2 & 2 & 0 \\ 2 & 2 & 0 \\ 0 & 0 & 0 \end{bmatrix}$$

Clearly, rank $(5I - A) = 1$. Therefore, A is diagonalizable because $3 - \text{rank}\,(5I - A) = 3 - 1 = 2 = $ multiplicity of 5.

Virtually all the preceding theorems and proofs hold if the eigenvalues are complex and if the eigenvectors have complex coordinates. In one sense it is "easier" for a matrix to be diagonalizable in the complex case than in the real case because it can be shown that *all* polynomials may be factored into a product of complex linear factors. The next example provides an illustration.

Example 5   (*Complex eigenvalues*—optional)   Let

$$A = \begin{bmatrix} 0 & -1 \\ 1 & 0 \end{bmatrix}$$

The characteristic polynomial of A is $f(x) = x^2 + 1$. Clearly, condition 1 is not satisfied, so A is not diagonalizable if we allow only real eigenvalues. However, we may write

$$f(x) = (x + i)(x - i)$$

Because both of the eigenvalues have multiplicity 1, A is diagonalizable. In fact, from Example 5 of Section 6.1 we have that $(i, 1)$ and $(-i, 1)$ are eigenvectors with corresponding eigenvalues i and $-i$. Therefore, if we let P be the matrix whose columns are $(i, 1)$ and $(-i, 1)$, respectively, and $D = \text{diag}\,(i, -i)$, then $P^{-1}AP = D$, or

$$\frac{1}{2i} \begin{bmatrix} 1 & i \\ -1 & i \end{bmatrix} \begin{bmatrix} 0 & -1 \\ 1 & 0 \end{bmatrix} \begin{bmatrix} i & -i \\ 1 & 1 \end{bmatrix} = \begin{bmatrix} i & 0 \\ 0 & -i \end{bmatrix}$$

It should be pointed out that in Exercises 17 and 18, there are two partial checks given to determine if the eigenvalues of a given matrix have been computed correctly.

In summary, to diagonalize an $n \times n$ matrix A (or a linear operator), carry out the following steps:

DIAGONALIZATION OF A MATRIX A

1. Check that the characteristic polynomial of A has the form

$$f(x) = (x - \lambda_1)^{m_1} \cdots (x - \lambda_r)^{m_r}$$

where $\lambda_1, \ldots, \lambda_r$ are distinct. If not, then A is not diagonalizable.

2. Verify that $n - \text{rank}\,(\lambda_i I - A) = m_i$ for the multiplicities that are greater than 1. If equality fails for any m_i, then A is not diagonalizable.

3. Find a basis for the solution space of $(\lambda_i I - A)\mathbf{x} = \mathbf{0}$ for each i.

4. Let P be the matrix whose columns are the basis vectors found in (3) (in same order as the λ_i's). Then

$$P^{-1}AP = D$$

where

$$D = \text{diag}\,(\underbrace{\lambda_1, \ldots, \lambda_1}_{m_1}, \ldots, \underbrace{\lambda_r, \ldots, \lambda_r}_{m_r})$$

Exercises

In Exercises 1–6, a matrix or a linear operator and its characteristic polynomial are given. Use the Test for Diagonalization to determine if the given matrix or linear operator is diagonalizable.

1. $\begin{bmatrix} 1 & 1 & 0 \\ 0 & 2 & 2 \\ 0 & 0 & 3 \end{bmatrix}$; $f(x) = (x-1)(x-2)(x-3)$

2. $\begin{bmatrix} 3 & -1 & -2 \\ 2 & 0 & -2 \\ 2 & -1 & -1 \end{bmatrix}$; $f(x) = x(x-1)^2$

3. $\begin{bmatrix} -1 & 1 & 0 \\ 0 & 5 & 0 \\ 4 & -2 & 5 \end{bmatrix}$; $f(x) = (x+1)(x-5)^2$

4. $\begin{bmatrix} -1 & -3 & -9 \\ 0 & 5 & 18 \\ 0 & -2 & -7 \end{bmatrix}$; $f(x) = (x+1)^3$

5. $T(x, y, z) = (-y, x, 3z)$;
$f(x) = (x^2 + 1)(x - 3)$

6. $T(x, y, z) = (8x + 2y - 2z, 3x + 3y - z, 24x + 8y - 6z)$; $f(x) = (x-2)^2(x-1)$

7. For the matrix in Exercise 1, find an invertible matrix P and a diagonal matrix D such that $P^{-1}AP = D$.

8. For the standard matrix A of the linear operator T in Exercise 6:
   (a) Find an invertible matrix P and a diagonal matrix D such that $P^{-1}AP = D$.
   (b) Find a basis S of R^3 such that $[T]_S = D$.

In Exercises 9–13, an $n \times n$ matrix, its eigenvalues, and a basis of R^n of eigenvectors (in the same order) are given. Use the fact that the matrix is diagonalizable to compute A^7.

9. $\begin{bmatrix} 2 & 2 \\ -1 & 5 \end{bmatrix}$; 4, 3; $\{(1, 1), (2, 1)\}$

10. $\begin{bmatrix} 1 & 1 \\ -2 & 4 \end{bmatrix}$; 3, 2; $\{(1, 2), (1, 1)\}$

11. $\begin{bmatrix} 3 & 2 \\ 3 & -2 \end{bmatrix}$; 4, −3; $\{(2, 1), (-1, 3)\}$

12. $\begin{bmatrix} 3 & -2 & 0 \\ -2 & 3 & 0 \\ 0 & 0 & 5 \end{bmatrix}$; 5, 5, 1; $\{(-1, 1, 0), (0, 0, 1), (1, 1, 0)\}$

13. $\begin{bmatrix} 1 & 2 & -1 \\ 1 & 0 & 1 \\ 4 & -4 & 5 \end{bmatrix}$; 1, 2, 3; $\{(-1, 1, 2), (-2, 1, 4), (-1, 1, 4)\}$

14. In reference to the transition matrix given in this section, determine how many people will be living in the city in 10 years.

15. Determine the 3×3 matrix A that has eigenvalues 3, 2, and 2 with corresponding eigenvectors $(2, 1, 1)$, $(1, 0, 1)$, and $(0, 0, 4)$.

16. Prove that if A is diagonalizable, then A^m is diagonalizable for any positive integer m.

Exercises 17 and 18 provide a partial check to see if the correct eigenvalues of a diagonalizable matrix have been computed.

17. Recall that the trace of a matrix A is defined to be the sum of its diagonal entries. Prove that if a matrix is diagonalizable, then its trace is the sum of its eigenvalues (counting multiplicities). In the notation of this section, show that

$$\text{tr } A = \sum_{i=1}^{r} m_i \lambda_i$$

(Use Exercise 10 of Section 4.2.)

18. By evaluating the characteristic polynomial at 0, show that if a matrix A is diagonalizable, then its determinant is the product of its eigenvalues (counting multiplicities). In the notation of this section, show that

$$\det A = \lambda_1^{m_1} \cdots \lambda_r^{m_r}$$

19. Suppose that you are told that the eigenvalues of the matrix A given below are 3, −1, 2, and 5. Use Exercises 17 and 18 to give two reasons why these values cannot be correct.

$$A = \begin{bmatrix} 3 & 1 & 0 & 0 \\ -1 & 0 & 0 & 0 \\ 0 & 0 & 2 & 1 \\ 0 & 0 & 3 & 7 \end{bmatrix}$$

20. A matrix B is said to be a ***square root*** of a matrix A if $B^2 = A$. Show that if A is diagonalizable and has only nonnegative eigenvalues, then A has a square root. (*Hint:* First write $A = PDP^{-1}$ as in this section.)

21. Use the hint of Exercise 20 to prove that if A is diagonalizable, then A has a ***cube root*** B, that is, $A = B^3$.

22. Suppose that T and U are diagonalizable linear operators on R^n which have the same eigenvectors. Prove that T and U commute, that is, $TU = UT$.

23. Suppose that A is an $n \times n$ matrix.
(a) Show that if A is diagonalizable and λ is an eigenvalue of A with multiplicity n, then $A = \lambda I$.
(b) Use part (a) to show that the matrix A below is not diagonalizable.

$$A = \begin{bmatrix} 1 & 1 \\ 0 & 1 \end{bmatrix}$$

24. Prove that if a matrix is *nilpotent* (see Exercise 26 of Section 6.1) and diagonalizable, the matrix must be the zero matrix.

25. The *Cayley–Hamilton* theorem states that for any matrix A with characteristic polynomial $f(x)$, it is the case that $f(A) = O$. [$f(A)$ is the matrix formed by substituting A for x.]
(a) Verify the Cayley–Hamilton theorem for

$$A = \begin{bmatrix} 1 & 2 \\ -1 & 4 \end{bmatrix}$$

(b) Prove the Cayley–Hamilton theorem in the special case in which A is diagonalizable. [*Hint:* Write $A = PDP^{-1}$ as in this section, and show that $f(A) = Pf(D)P^{-1}$.]

*26. Use the program MATRIX to find a matrix A with the given eigenvalues and corresponding eigenvectors.
(a) $2, -1, 3; (2, 6, 4), (3, 8, 4), (-2, -3, -3)$
(b) $2, -2, 3; (1, 3, 6), (-1, -2, -4),$
    $(2, 2, 5)$

* This problem should be solved by using one of the programs noted in Appendix B.

6.3 APPLICATIONS: DIFFERENCE EQUATIONS, MARKOV CHAINS, AND SYSTEMS OF DIFFERENTIAL EQUATIONS

The number of applications of the theory of eigenvalues and eigenvectors is so extensive that we are forced to be highly selective in our choices for this section. As a result we have chosen three areas which are rich in examples: difference equations, Markov chains, and systems of differential equations. The applications are all independent of one another and so can be covered or omitted at the discretion of the reader.

Difference Equations

To introduce difference equations, we begin with a counting problem. This problem is typical of the type that occurs in the study of combinatorial analysis. This field has gained considerable attention in recent years because of its applications to computer science and operations research.

TABLE 6.3.1

n	Arrangements	$r(n)$
1	Y	1
2	YY, R, G	3
3	YYY, RY, YR, GY, YG	5

Suppose that we have a large number of blocks. The blocks are of three colors: red, yellow, and green. The red and green blocks each take up two spaces and the yellow blocks each take up one space. Question: How many ways are there to arrange the blocks in a line so as to fill up n spaces? Denote the answer to the problem by $r(n)$ and by R, Y, and G, the red, yellow, and green blocks, respectively. Let's first answer the question for several small values of n (see Table 6.3.1).

The reader is correct if he guesses that $r(n)$ grows very quickly as n increases. To solve the problem, we look at three cases:

Case 1: The first block is red: In this case, there are $n - 2$ spaces left to fill. The number of ways of doing this is $r(n - 2)$.

Case 2: The first block is yellow: In this case, there are $n - 1$ spaces left to fill. The number of ways of doing this is $r(n - 1)$.

Case 3: The first block is green: In this case, there are $n - 2$ spaces left to fill. The number of ways of doing this is $r(n - 2)$.

So all together we have

$$r(n) = r(n - 2) + r(n - 1) + r(n - 2)$$
$$= r(n - 1) + 2r(n - 2) \tag{1}$$

This is our first example of a ***difference equation***. The name may seem more appropriate if we write equation (1) in the form $r(n) - r(n - 1) = 2r(n - 2)$. Here the left side represents the difference in the process at two stages. Equation (1) is also described as a *recurrence relation*.

But how do we find a formula for $r(n)$ which is a simple function of n? The trick is to rewrite (1) as a matrix equation. First, write

$$r(n + 1) = r(n) + 2r(n - 1)$$
$$r(n) = r(n)$$

The first equation is formed by replacing n by $n + 1$ in equation (1). Although the second equation does not seem to be very useful, the system may be written in matrix form:

$$\begin{bmatrix} r(n + 1) \\ r(n) \end{bmatrix} = \begin{bmatrix} 1 & 2 \\ 1 & 0 \end{bmatrix} \begin{bmatrix} r(n) \\ r(n - 1) \end{bmatrix}$$

Or $\mathbf{s}(n) = A\mathbf{s}(n-1)$, where

$$\mathbf{s}(n) = \begin{bmatrix} r(n+1) \\ r(n) \end{bmatrix} \quad \text{and} \quad A = \begin{bmatrix} 1 & 2 \\ 1 & 0 \end{bmatrix}$$

From Table 6.3.1 we have that

$$\mathbf{s}(1) = \begin{bmatrix} r(2) \\ r(1) \end{bmatrix} = \begin{bmatrix} 3 \\ 1 \end{bmatrix}$$

Thus, $\mathbf{s}(n+1) = A\mathbf{s}(n) = A^2\mathbf{s}(n-1) = \cdots = A^n\mathbf{s}(1)$. Now, to determine $\mathbf{s}(n+1)$ we are back to the problem of the last section, that is, the computation of the powers of a matrix. We saw that if the matrix is diagonalizable, the computation is made considerably easier. To discover if A is diagonalizable, we first compute its characteristic polynomial.

$$f(x) = \begin{vmatrix} x-1 & -2 \\ -1 & x \end{vmatrix} = (x-2)(x+1)$$

Since the characteristic polynomial factors into distinct linear factors, we know from Theorem 6.2.6 that A is diagonalizable. It is easy to show that the eigenvectors corresponding to the eigenvalues 2 and -1 are multiples of $(2,1)$ and $(1,-1)$, respectively. So define

$$P = \begin{bmatrix} 2 & 1 \\ 1 & -1 \end{bmatrix} \quad \text{and} \quad D = \begin{bmatrix} 2 & 0 \\ 0 & -1 \end{bmatrix}$$

Then, by Theorem 6.2.1 we have that $A = PDP^{-1}$, and by the same reasoning as in Section 6.2, we have $A^n = PD^nP^{-1}$. Thus,

$$\mathbf{s}(n+1) = PD^nP^{-1}\mathbf{s}(1)$$

or

$$\begin{bmatrix} r(n+2) \\ r(n+1) \end{bmatrix} = \begin{bmatrix} 2 & 1 \\ 1 & -1 \end{bmatrix} \begin{bmatrix} 2^n & 0 \\ 0 & (-1)^n \end{bmatrix} \begin{bmatrix} \frac{1}{3} & \frac{1}{3} \\ \frac{1}{3} & -\frac{2}{3} \end{bmatrix} \begin{bmatrix} 3 \\ 1 \end{bmatrix}$$

$$= \frac{1}{3} \begin{bmatrix} 2^{n+3} + (-1)^n \\ 2^{n+2} + (-1)^{n+1} \end{bmatrix}$$

Thus, $r(n+1) = (2^{n+2} + (-1)^{n+1})/3$, or

$$r(n) = \frac{2^{n+1} + (-1)^n}{3}$$

As a check, we compute $r(1) = 1$, $r(2) = 3$, and $r(3) = 5$. These results agree with those in Table 6.3.1. Of course, it is now easy to compute $r(n)$ for larger values of n. For example, $r(10) = 683$, $r(20) = 699,051$, and $r(32)$ is almost 3 billion! Needless to say, a complete listing to find out the number of ways these three blocks may be arranged in 32 spaces is rather impractical.

In general, a **kth-order (homogeneous) difference equation** (or **recurrence relation**) is a relation of the form $r(n) = a_{n-1}r(n-1) + a_{n-2}r(n-2) + \cdots + a_{n-k}r(n-k)$ for $n = k + 1, k + 2, \ldots$ and $a_{n-k} \neq 0$.

Our block problem was represented by a second-order difference equation. The **initial conditions** were $r(1) = 1$ and $r(2) = 3$. Notice that the number of initial conditions equals the order of the difference equation. From the previous example, it is not surprising that if we represent a kth-order difference equation by the matrix equation $\mathbf{s}(n + 1) = A\mathbf{s}(n)$, where the $k \times k$ matrix A has k distinct eigenvalues $\lambda_1, \ldots, \lambda_k$ (see Exercise 8), then the general form of the solution may be obtained as before from $\mathbf{s}(n + 1) = PD^nP^{-1}\mathbf{s}(1)$. It can be shown that when the matrices are multiplied together, $r(n)$ will have the form

$$r(n) = b_1\lambda_1^n + \cdots + b_k\lambda_k^n \tag{2}$$

The b_i's are determined by the initial conditions, which are given by the vector $\mathbf{s}(1)$. In particular, if $b_1 = 1$ and the remaining b_i's are 0, then $r(n) = \lambda_1^n$ is a solution. The case in which the eigenvalues are not distinct will not be discussed.

Equation (2) offers us an alternative method for finding $r(n)$ *without* computing the eigenvectors. We illustrate this method with another example.

It is known that rabbits reproduce at a very rapid rate. For the sake of simplicity, we suppose that a pair of rabbits will not produce any offspring during the first month of their lives but that they will produce exactly one pair (male and female) each month thereafter. Assume also that initially we have one pair of newborn rabbits and that no rabbits die. How many rabbits will there be after n months? Let's try to answer this question for $n = 0, 1, 2, 3$. Let $r(n) =$ the number of pairs of rabbits after n months.

After zero months, we have only the initial pair. Similarly, after 1 month, we still have only the initial pair. So $r(1) = r(0) = 1$. After 2 months, we have the initial pair and their offspring, that is, $r(2) = 2$. After 3 months, we have what we had before and, in addition, the offspring of the pair that we had over a month ago; that is, $r(3) = r(2) + r(1) = 2 + 1 = 3$. In general, after n months we have the rabbits we had last month and the offspring of those rabbits which are over 1 month old. Thus,

$$r(n) = r(n - 1) + r(n - 2) \tag{3}$$

The numbers generated by equation (3) are 1, 1, 2, 3, 5, 8, 13, 21, 34, ... Each number is the sum of the preceding two numbers. Such a sequence is quite famous and is called a **Fibonacci sequence**. It occurs in a variety of contexts, including the number of spirals of various plants. We will soon see another application in which this sequence arises. First, however, we will find the solution to our rabbit problem.

By our earlier comments, we know that the general form of the solution is a linear combination of terms of the form λ^n. To find the values of λ_1 and λ_2, we can either construct a matrix A and compute the roots of its characteristic equation as before, or substitute λ^n for $r(n)$ in (3). We will not prove that the two methods are equivalent. This substitution produces

$$\lambda^n = \lambda^{n-1} + \lambda^{n-2}$$

or

$$\lambda^2 = \lambda + 1$$

The roots of this equation are $(1/2)(1 \pm \sqrt{5})$. Using these roots, we may write the general solution in the form

$$r(n) = b_1\left(\frac{1}{2} + \frac{\sqrt{5}}{2}\right)^n + b_2\left(\frac{1}{2} - \frac{\sqrt{5}}{2}\right)^n$$

To find b_1 and b_2, we use the initial conditions:

$$1 = r(0) = \qquad (1)b_1 + \qquad (1)b_2$$

$$1 = r(1) = \left(\frac{1}{2} + \frac{\sqrt{5}}{2}\right)b_1 + \left(\frac{1}{2} - \frac{\sqrt{5}}{2}\right)b_2$$

This system has the solution $b_1 = (1/\sqrt{5})(\frac{1}{2} + \sqrt{5}/2)$ and $b_2 = -(1/\sqrt{5})(\frac{1}{2} - \sqrt{5}/2)$. Thus, the general solution is

$$r(n) = \frac{1}{\sqrt{5}}\left(\frac{1}{2} + \frac{\sqrt{5}}{2}\right)^{n+1} - \frac{1}{\sqrt{5}}\left(\frac{1}{2} - \frac{\sqrt{5}}{2}\right)^{n+1}$$

This complicated result should surprise most readers because $r(n)$ is a positive integer for every value of n. To find the fiftieth Fibonacci number, we compute $r(50)$ to be over 20 billion!

Our final example involves a code, such as Morse code, which involves two signals. In Morse code each signal is either a dot (short) or a dash (long). Each signal takes a certain amount of time to transmit, say s and t units, respectively. The question is: How many messages (a message is a sequence of signals) may be sent in time w? Let $r(w) = $ the number of such messages. We consider two cases:

Case 1: The message begins with signal 1: Because signal 1 takes up s units, the number of such messages that may be sent is $r(w - s)$.

Case 2: The message begins with signal 2: Because signal 2 takes up t units, the number of such messages that may be sent is $r(w - t)$.

So the total number of messages sent in time w is

$$r(w) = r(w - s) + r(w - t)$$

In the special case that $s = 1$ and $t = 2$, we let $n = w$ and once again obtain the Fibonacci sequence

$$r(n) = r(n - 1) + r(n - 2)$$

Notice that in this case the initial conditions are different than in the rabbit problem.

Markov Chains

The subject of Markov chains deserves considerably more attention than we are able to give it here. However, because of its dependence on the material that we have developed so far, we include it as an interesting application. The subject is named after the Russian mathematician Andrei Markov, who developed the fundamentals of the theory at the beginning of this century.

A *Markov chain* is a process that consists of a finite number of *states* and known probabilities p_{ij}, where p_{ij} represents the probability of moving from state j to state i. Our example dealing with movement between the city and suburbs, reintroduced in Section 6.2, has two states: living in the city and living in the suburbs. The matrix

$$A = \begin{bmatrix} 0.85 & 0.03 \\ 0.15 & 0.97 \end{bmatrix}$$

gives us the probabilities of moving from one state to another. In general, a matrix A in which $A_{ij} = p_{ij}$ for all i and j is called the *transition* or *stochastic matrix* for the Markov chain. In fact, any matrix all of whose entries are nonnegative and each of whose columns sum to 1 is called a transition matrix.

Other examples might include: political affiliation as three states, Democrat, Republican, and Independent, and p_{ij} represents the probability of a son belonging to party i if his father belonged to party j; cholesterol levels as the states, say, high, normal, and low, and p_{ij} represents the probability of moving from one level to another in a given length of time; the class populations of a particular high school as three states, sophomore, junior, and senior, and p_{ij} represents the probability of moving from one class to another in 1 year.

A vector whose coordinates are nonnegative and sum to 1 is called a *probability vector*. By definition, the columns of a transition matrix are probability vectors. In the "city–suburb" example, we used an initial population vector of $(50, 75)$. This vector could have easily been replaced by the probability vector $(\frac{50}{125}, \frac{75}{125}) = (\frac{2}{5}, \frac{3}{5})$. The information that this vector provides is that during the initial time period, $\frac{2}{5}$ of the population lived in the city and $\frac{3}{5}$ of the population lived in the suburbs.

Of particular interest is a probability vector **p** such that $A\mathbf{p} = \mathbf{p}$, that is, an eigenvector of A that corresponds to the eigenvalue 1. Such a vector is called a *steady-*

state vector. In our population example, it represents those proportions (of people living in the city and suburbs) which do not change from year to year. For example,

$$\begin{bmatrix} 0.85 & 0.03 \\ 0.15 & 0.97 \end{bmatrix} \begin{bmatrix} \frac{1}{6} \\ \frac{5}{6} \end{bmatrix} = \begin{bmatrix} \frac{1}{6} \\ \frac{5}{6} \end{bmatrix}$$

So if $\mathbf{p} = (\frac{1}{6}, \frac{5}{6})$, we have that $A\mathbf{p} = \mathbf{p}$. That is, although individual people may move between the city and suburbs, the proportions remain the same. This result could have a significant effect on city planning.

The reader may recall that in Section 6.2 we showed that A^k approaches a matrix each of whose columns is equal to the steady-state vector $\mathbf{p}$. This result is no coincidence, as is indicated in the theorem below [the proof of this theorem may be found on page 268 of (1) in the References]. Before we state this theorem, we need one more definition.

Definition   A transition matrix is a ***regular matrix*** if some power of the matrix has only positive entries.

Our matrix A is clearly a regular matrix. Suppose that we consider the matrices B and B^2 given below.

$$B = \begin{bmatrix} 0.5 & 0 & 0.2 \\ 0 & 0.6 & 0.8 \\ 0.5 & 0.4 & 0 \end{bmatrix} \quad \text{and} \quad B^2 = \begin{bmatrix} 0.35 & 0.08 & 0.10 \\ 0.40 & 0.68 & 0.48 \\ 0.25 & 0.24 & 0.42 \end{bmatrix}$$

Although B has several zero entries, we see that B^2 has only positive entries. Thus, B is a regular matrix.

On the other hand, if

$$C = \begin{bmatrix} \frac{1}{2} & 0 \\ \frac{1}{2} & 1 \end{bmatrix}$$

then

$$C^k = \begin{bmatrix} \dfrac{1}{2^k} & 0 \\ c(k) & 1 \end{bmatrix}$$

where $c(k) = \frac{1}{2} + 1/2^2 + \cdots + 1/2^k = 1 - 1/2^k$. So C is not a regular matrix. (A less "exotic" example of a matrix that is not regular is the identity matrix.)

The property of being regular has an interesting meaning in terms of the Markov chain. We noted in Section 2.2 that the entries of A^k have a probabilistic interpretation of their own. Recall that $(A^k)_{ij}$ represents the probability of moving from state j to state i in k steps (years). So "a matrix is regular" means that there is a positive probability that every state may be reached from every other state within a finite number of steps. In the matrix C above, state 1 cannot be reached from state 2 in a finite number of steps.

We have used the term "approaches" in regards to matrices rather loosely. To say that A^k **approaches the matrix B as k approaches infinity** will mean that $(A^k)_{ij}$ approaches B_{ij} for all i,j as k approaches infinity.

Theorem 6.3.1   *Let A be a regular transition matrix. Then*

(a) *A^k approaches a matrix all of whose columns equal the unique steady-state vector $\mathbf{p}$ as k approaches infinity.*

(b) *For any probability vector $\mathbf{x}$, $A^k\mathbf{x}$ approaches $\mathbf{p}$ as k approaches infinity.*

In Section 6.2 we saw the interesting effect that part (b) has on our city–suburb example. We were able to conclude that no matter what the initial population proportions are, the eventual proportions are $\frac{1}{6}$ and $\frac{5}{6}$.

If we consider the regular matrix B given earlier, we have (approximately)

$$B^5 = \begin{bmatrix} 0.147 & 0.110 & 0.121 \\ 0.550 & 0.593 & 0.594 \\ 0.302 & 0.297 & 0.285 \end{bmatrix} \qquad B^{10} = \begin{bmatrix} 0.119 & 0.117 & 0.117 \\ 0.587 & 0.589 & 0.588 \\ 0.294 & 0.294 & 0.294 \end{bmatrix}$$

and

$$B^{20} = \begin{bmatrix} 0.118 & 0.118 & 0.118 \\ 0.588 & 0.588 & 0.588 \\ 0.294 & 0.294 & 0.294 \end{bmatrix}$$

As Theorem 6.3.1(a) indicates, the columns of the powers of B approach a common vector $\mathbf{p}$. We leave it as an exercise to verify that $B\mathbf{p} = \mathbf{p}$.

As a final example, suppose that domestic automobile manufacturers are trying to estimate their eventual share of the market. Suppose also that the consumer is in one of three states: he owns an American car, he owns a European car, or he owns a Japanese car. Assume that the probabilities that in the next year his car is one of these types, is given by transition matrix A below.

From

	Amer.	Eur.	Jap.
Amer.	0.60	0.30	0.20
To Eur.	0.15	0.40	0.30
Jap.	0.25	0.30	0.50

TABLE 6.3.2 Market Shares after n Years

	0	1	2	3	4	5
			n			
Amer.	0.650	0.485	0.417	0.392	0.383	0.380
Eur.	0.250	0.228	0.250	0.262	0.267	0.269
Jap.	0.100	0.288	0.333	0.346	0.350	0.351

Since A is a regular transition matrix, we know that powers of A approach a matrix whose columns are the steady-state vector. With three significant digits, we have

$$A^{14} = \begin{bmatrix} 0.378 & 0.378 & 0.378 \\ 0.270 & 0.270 & 0.270 \\ 0.351 & 0.351 & 0.351 \end{bmatrix}$$

Our conclusion is that no matter what the present shares of the market are today, eventually the American share will be approximately 38%, the European share will be 27%, and the Japanese share will be 35%. For example, suppose that initially the American share is 65%, the European share is 25%, and the Japanese share is 10%. The vector $\mathbf{p}_0 = (0.65, 0.25, 0.10)$ represents the initial state. After n years, the vector $A^n\mathbf{p}_0$ gives the respective shares. These shares are given in Table 6.3.2 (to three significant digits). Notice that the vectors $A^n\mathbf{p}_0$ which give the columns of Table 6.3.2 are approaching the steady-state vector $\mathbf{p} = (0.378, 0.270, 0.351)$ as Theorem 6.3.1 indicates. It can also be shown that $A\mathbf{p} = \mathbf{p}$. Of course, $\mathbf{p}$ can be computed directly as the probability vector that is a solution to the system $(I - A)\mathbf{p} = \mathbf{0}$.

Systems of Differential Equations

This section assumes that the reader is familiar with the differential equation

$$y' = ky \tag{4}$$

Included in the typical calculus course are examples of substances whose quantities change at rates that are proportional to the amount present at time t. Examples include the decay of radioactive material and the unrestricted growth of bacteria and other organisms. If $y = f(t)$ represents the amount present at time t, and k represents the constant of proportionality, then we obtain the differential equation $f'(t) = kf(t)$. In the conventional notation of differential equations, this equation is represented as in (4). The *general solution* to the differential equation (4) is given by

$$y = ae^{kt}$$

where a is an arbitrary constant. That is, if we insert ae^{kt} and ake^{kt} for y and y', respectively, into (4), we obtain an identity. To find the value of a, we need an **initial condition**. For example, we need to know how much of the quantity $f(t)$ is present at some time t, say $t = 0$. Suppose that 3 units were present at this time. This fact is represented by the equation $y(0) = 3$. From this condition we have

$$3 = y(0) = ae^{k(0)} = a(1) = a$$

So $a = 3$ and the **particular solution** to (4) is $y = 3e^{kt}$.

Now suppose that we have a system of three differential equations

$$y' = 3y$$

$$u' = 4u$$

$$v' = 5v$$

This system is just as easy to solve as (4). The general solution is given by

$$y = ae^{3t}$$

$$u = be^{4t}$$

$$v = ce^{5t}$$

If we have the initial conditions $y(0) = 10$, $u(0) = 12$, and $v(0) = 15$, then the particular solution is given by

$$y = 10e^{3t}$$

$$u = 12e^{4t}$$

$$v = 15e^{5t}$$

This system may be represented by the matrix equation

$$\begin{bmatrix} y' \\ u' \\ v' \end{bmatrix} = \begin{bmatrix} a & 0 & 0 \\ 0 & b & 0 \\ 0 & 0 & c \end{bmatrix} \begin{bmatrix} y \\ u \\ v \end{bmatrix}$$

If we let $\mathbf{y}'$ denote the vector of derivatives of the coordinate functions of $\mathbf{y} = (y, u, v)$ and let D denote the diagonal matrix above, we may represent the system as

$$\mathbf{y}' = D\mathbf{y}$$

with the initial condition

$$\mathbf{y}(0) = \begin{bmatrix} 10 \\ 12 \\ 15 \end{bmatrix}$$

The more general system, however, is given by

$$\mathbf{y}' = A\mathbf{y} \tag{5}$$

where A is not necessarily a diagonal matrix. That is,

$$y_1' = A_{11}y_1 + A_{12}y_2 + \cdots + A_{1n}y_n$$
$$y_2' = A_{21}y_1 + A_{22}y_2 + \cdots + A_{2n}y_n$$
$$\vdots \qquad \vdots \qquad \vdots \qquad \vdots$$
$$y_n' = A_{n1}y_1 + A_{n2}y_2 + \cdots + A_{nn}y_n$$

This type of system might arise from a continuous version of a Markov chain, where the rate of change of each population is given by a linear combination of the other populations. For example, we might have three species of animals that interact with each other. In this case it is reasonable to assume that the growth rate of each species is dependent (as a function of time) on the numbers present in each of the species.

The solution of (5) will depend on the appropriate substitution for $\mathbf{y}$. Suppose that we introduce a *change of variable*; that is, we define $\mathbf{z} = P^{-1}\mathbf{y}$ or $\mathbf{y} = P\mathbf{z}$ for some invertible matrix P. It is not hard to prove that $\mathbf{y}' = P\mathbf{z}'$ (see Exercise 23). If we substitute $P\mathbf{z}$ for $\mathbf{y}$ and $P\mathbf{z}'$ for $\mathbf{y}'$ in (5), we obtain

$$P\mathbf{z}' = AP\mathbf{z}$$

or

$$\mathbf{z}' = P^{-1}AP\mathbf{z}$$

Thus, if we are able to choose P such that $P^{-1}AP$ is a diagonal matrix D, we will have the system

$$\mathbf{z}' = D\mathbf{z}$$

which is of the same form as the one we solved earlier. Of course, once we know $\mathbf{z}$, the solution to our original system (5) is $\mathbf{y} = P\mathbf{z}$. From our knowledge of diagonalization, we know how to pick P; that is, if A is diagonalizable, choose P to be the matrix whose columns are linearly independent eigenvectors of A. In this case the diagonal entries of

D are the eigenvalues of A. The method for an $n \times n$ system whose coefficient matrix is diagonalizable is summarized below.

SOLUTION OF $y' = Ay$

1. Find the eigenvalues $\lambda_1, \ldots, \lambda_n$ of A and a set of corresponding linearly independent eigenvectors.
2. Let $D = \text{diag}(\lambda_1, \ldots, \lambda_n)$ and let P be the matrix whose columns consist of the eigenvectors of A found in (1).
3. Solve the diagonal system $\mathbf{z}' = D\mathbf{z}$.
4. The solution to the original system is $\mathbf{y} = P\mathbf{z}$.

For example, consider the system

$$u' = 4u + v$$
$$v' = 3u + 2v$$

This system may be put into the form $\mathbf{y}' = A\mathbf{y}$, where

$$\mathbf{y} = \begin{bmatrix} u \\ v \end{bmatrix} \quad \text{and} \quad A = \begin{bmatrix} 4 & 1 \\ 3 & 2 \end{bmatrix}$$

By the methods of Sections 6.1 and 6.2 it can be shown that A has linearly independent eigenvectors $(1, -3)$ and $(1, 1)$ with corresponding eigenvalues 1 and 5, respectively. So first let

$$P = \begin{bmatrix} 1 & 1 \\ -3 & 1 \end{bmatrix} \quad \text{and} \quad D = \begin{bmatrix} 1 & 0 \\ 0 & 5 \end{bmatrix}$$

Now solve the system $\mathbf{z}' = D\mathbf{z}$,

$$z_1' = z_1$$
$$z_2' = 5z_2$$

This system has the solution

$$\mathbf{z} = \begin{bmatrix} ae^t \\ be^{5t} \end{bmatrix}$$

So the general solution to our original system is

$$\mathbf{y} = P\mathbf{z} = \begin{bmatrix} 1 & 1 \\ -3 & 1 \end{bmatrix} \begin{bmatrix} ae^t \\ be^{5t} \end{bmatrix} = \begin{bmatrix} ae^t + be^{5t} \\ -3ae^t + be^{5t} \end{bmatrix}$$

or

$$u = \quad ae^t + be^{5t}$$

$$v = -3ae^t + be^{5t}$$

Notice that it is not necessary to compute P^{-1}.

Suppose that the variables u and v represented the numbers of two species present at a particular time, and suppose that $u(0) = 120$ and $v(0) = 40$. Then, to find a particular solution to the system, we must solve

$$120 = u(0) = \quad a + b$$

$$40 = v(0) = -3a + b$$

We obtain $a = 20$ and $b = 100$. So a particular solution is

$$u = \quad 20e^t + 100e^{5t}$$

$$v = -60e^t + 100e^{5t}$$

Systems of differential equations similar to the one above arise in the context of *prey-predator models*. For example, u and v might represent the numbers of rabbits and foxes, respectively (see Exercise 21), or of food fish and sharks.

The procedure we have described for finding the general solution to $\mathbf{y}' = A\mathbf{y}$ does *not* cover the situation when the matrix A is not diagonalizable. In this case the *Jordan canonical form* must be used. This technique is beyond the scope of this text and will not be pursued.

Finally, it should be noted that only a slight modification is needed for the nonhomogeneous system $\mathbf{y}' = A\mathbf{y} + \mathbf{b}$, where $\mathbf{b} \neq \mathbf{0}$.

It is interesting to observe that a system of first-order differential equations may sometimes be used to solve a higher-order differential equation. For example, consider the third-order differential equation

$$y''' - 6y'' + 11y' - 6y = 0$$

We will make the substitutions $u = y'$ and $v = y'' = u'$. Using these substitutions in the differential equation above, we have that $v' = y''' = 6y'' - 11y' + 6y = 6v - 11u + 6y$. These equations may be rewritten as the system

$$y' = \qquad u$$

$$u' = \qquad\qquad v$$

$$v' = 6y - 11u + 6v$$

or, in matrix form,

$$\begin{bmatrix} y' \\ u' \\ v' \end{bmatrix} = \begin{bmatrix} 0 & 1 & 0 \\ 0 & 0 & 1 \\ 6 & -11 & 6 \end{bmatrix} \begin{bmatrix} y \\ u \\ v \end{bmatrix}$$

The 3×3 matrix has linearly independent eigenvectors $(1, 1, 1)$, $(1, 2, 4)$, and $(1, 3, 9)$ with corresponding eigenvalues 1, 2, and 3, respectively. Using the previous method, we may solve (see Exercise 24) for y, u, and v. Of course, in this case we are only interested in y. The general solution to the third-order differential equation is

$$y = ae^t + 2be^{2t} + 3ce^{3t}$$

The Complex Case (Optional)

The situation where the eigenvalues are complex occurs quite naturally when dealing with *harmonic motion*. Suppose that a body of weight w is suspended from a spring (see Figure 6.3.1). The body is now moved from its resting position and set in motion. The distance of the body (measured positive downward and negative upward) from its resting point at time t is given by $y(t)$. If k is the spring constant, g is the force of gravity, and $by'(t)$ represents a *damping force* which reflects the viscosity of the medium in which the motion takes place, then

$$\frac{w}{g} y''(t) + by'(t) + ky(t) = 0$$

For example, suppose that the body weighs 8 pounds, the spring constant is 2.4 pounds per foot, and $b = 0.8$. Then the differential equation above may be simplified to

$$y'' + 3.2y' + 9.6y = 0$$

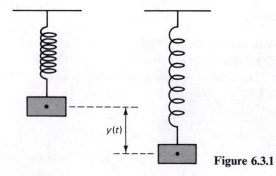

$y(t)$

Figure 6.3.1

If we let $u = y'$, the differential equation is equivalent to the system

$$u' = -3.2u - 9.6y$$
$$y' = \quad u$$

or

$$\begin{bmatrix} u' \\ y' \end{bmatrix} = \begin{bmatrix} -3.2 & -9.6 \\ 1 & 0 \end{bmatrix} \begin{bmatrix} u \\ y \end{bmatrix}$$

The characteristic equation of the matrix above is

$$x^2 + 3.2x + 9.6 = 0$$

Notice that the coefficients in this equation are the same constants that occur in the differential equation. This is not a coincidence. The (complex) roots of the equation are approximately $-1.6 \pm 2.65i$, so the general solution can be shown to be

$$y = ae^{(-1.6 + 2.65i)t} + be^{(-1.6 - 2.65i)t}$$

Referring to Appendix A, we have

$$e^{s+it} = e^s(\cos t + i \sin t)$$

We may rewrite the solution in the form

$$y = e^{-1.6t}(c \cos 2.65t + d \sin 2.65t)$$

The values c and d may be determined from the initial conditions, such as the initial displacement and the initial velocity of the body. It can be shown from this solution that the body oscillates with smaller and smaller amplitudes.

Exercises

Difference Equations

Solve the difference equations given in each of Exercises 1–4 by either of the methods developed in this section. Then use your result to find $r(6)$.

1. $r(n) = r(n - 1) + 2r(n - 2)$; $r(1) = 2$ and $r(2) = 16$
2. $r(n) = 3r(n - 1) - 2r(n - 2)$; $r(1) = 3$ and $r(2) = 7$
3. $r(n) = 3r(n - 1) + 4r(n - 2)$; $r(0) = 1$ and $r(1) = 1$

4. $r(n) = 2r(n - 1)$; $r(1) = 5$

5. Suppose that we have a large number of colored blocks. The blocks are of five colors: red, yellow, green, orange, and blue. Each of the red and yellow blocks takes up one space and each of the green, orange, and blue blocks takes up two spaces. Let $r(n)$ be the number of ways the blocks may be arranged in n spaces.
   (a) Determine $r(1)$, $r(2)$, and $r(3)$ by listing the possibilities.

(b) Write the difference equation involving $r(n)$.

(c) Solve the difference equation in part (b) by either of the methods developed in this section.

(d) Use your answer to part (c) to check your answers in part (a).

6. Suppose that a bank pays 8% annual interest on savings. Use the appropriate difference equation to determine how much money would be in a savings account after n years if initially there was $1000. What is the answer for the following values of n: $n = 5, 10, 50$?

7. Write the third-order difference equation

$$r(n) = 4r(n - 1) - 2r(n - 2) + 5r(n - 3)$$

in matrix notation $\mathbf{s}(n + 1) = A\mathbf{s}(n)$ as we did in this section.

8. Extend Exercise 7 by writing the general kth-order difference equation in matrix form.

Markov Chains

9. Which of the following transition matrices are regular?

(a) $\begin{bmatrix} 0.5 & 0.5 \\ 0.5 & 0.5 \end{bmatrix}$ (b) $\begin{bmatrix} 0.25 & 0 \\ 0.75 & 1 \end{bmatrix}$

(c) $\begin{bmatrix} 0.9 & 0.5 & 0 \\ 0 & 0.5 & 0.4 \\ 0.1 & 0 & 0.6 \end{bmatrix}$ (d) $\begin{bmatrix} 0.5 & 0 & 1 \\ 0.5 & 0 & 0 \\ 0 & 1 & 0 \end{bmatrix}$

10. Show that the matrix B given following the definition of "regular matrix" has the vector $\mathbf{p} = (0.117, 0.588, 0.294)$ as an approximate steady-state vector.

11. Prove that a transition matrix A is regular if and only if A^2 is regular.

12. Suppose that the probability that the child of a college-educated parent also becomes college educated is 0.75, and that the probability that the child of a non-college-educated parent becomes college educated is 0.35.

(a) Assuming that the information above describes a Markov process, determine the states.

(b) Write down a transition matrix that describes the information above.

(c) If 30% of the parents are college educated, what (approximate) proportion of the population in one, two, and three generations will be college educated?

(d) Without any knowledge of the present proportion of college-educated parents, determine the eventual proportion of college-educated people.

13. In reference to the example concerning the shares of American, European, and Japanese cars given in this section, what are the respective shares in 1, 2, and 3 years if the initial shares are 80%, 10%, and 10%, respectively?

14. Suppose that a particular region with a fixed population is divided into three areas: the city, suburbs, and country. The probability that a person living in the city moves to the suburbs (in one year) is 0.10 and moves to the country is 0.50. The probability that a person living in the suburbs moves to the city is 0.20 and moves to the country is 0.10. The probability that a person living in the country moves to the city is 0.20 and moves to the suburbs is 0.20. Suppose initially that 50% of the people live in the city, 30% live in suburbs, and 20% live in the country.

(a) Determine the transition matrix for the three states.

(b) Determine the percentage of people living in each area after 1, 2, and 3 years.

*(c) Use the program MATRIX to find the percentage of people living in each area after 5 and 8 years.

(d) Determine the eventual percentages of people in each area.

*15. Use the program MATRIX and Theorem 6.3.1 to estimate the steady-state vector of the matrix

$$\begin{bmatrix} 0.2 & 0.4 & 0.1 \\ 0.3 & 0.2 & 0.3 \\ 0.5 & 0.4 & 0.6 \end{bmatrix}$$

* This problem should be solved by using one of the programs noted in Appendix B.

Systems of Differential Equations

In each of the following systems of differential equations, find the general solution by the matrix method developed in this section. If initial conditions are given, also find a particular solution.

16. $u' = 1u + 2v$
    $v' = -1u + 4v$

17. $u' = 3u + 2v$
    $v' = 3u - 2v$

18. $u' = u + v$
    $v' = 4u + v$
    $u(0) = 15, \quad v(0) = -10$

19. $u' = 2u + 2v$
    $v' = -1u + 5v$
    $u(0) = 7, \quad v(0) = 5$

20. $y' = y + 2u - v$
    $u' = y + v$
    $v' = 4y - 4u + 5v$

21. Let the number of rabbits in a certain area at time t be given by $u(t)$, and the number of foxes be given by $v(t)$. Suppose that at time $t = 0$, there are 900 rabbits and 300 foxes. Assume that the following system of

differential equations expresses the rate of change for the two species.

$$u' = 2u - 4v$$
$$v' = u - 3v$$

(a) Find the particular solution for this system.
(b) How many (approximately) of each species will be present at times $t = 1, 2,$ and 3? For each of these times compute the ratio of foxes to rabbits.
(c) What is the eventual (approximate) ratio of foxes to rabbits? Does this ratio depend on the initial numbers present?

22. Convert the following third-order differential equation to a system of differential equations, and then find the general solution.

$$y''' - 2y'' - 8y' = 0$$

23. Let $\mathbf{z}$ be a 3×1 column vector of (differentiable) functions and let P be a 3×3 matrix. Prove that if $\mathbf{y} = P\mathbf{z}$, then $\mathbf{y}' = P\mathbf{z}'$.

24. Complete the details for solving the third order differential equation in the section.

Key Words

Orthonormal Sets and Orthogonal Matrices

In this chapter we return to the concept of orthogonality introduced in Section 1.6. At this point, however, we have a number of sophisticated techniques at our disposal which were not available to us at that time, such as the construction of bases of subspaces and eigenspaces of linear transformations and matrices. We shall see that if we choose the appropriate basis of orthogonal vectors, we will be able to provide a number of interesting applications of linear algebra to such diverse areas as statistics and geometry. Throughout this chapter the word "basis" will be understood to mean "ordered" basis.

7.1 ORTHONORMAL SETS

We have already seen that an appropriate choice of basis can greatly simplify computations. In this section we examine one of the most useful properties of a set of vectors, namely, the property that the vectors are mutually orthogonal. An example of such a set is the standard basis. This set also has the property that every vector is a unit vector. In fact, many of the properties of the standard basis carry over to "orthonormal sets."

Definition   A set S of vectors in R^n is **orthogonal** if every pair of distinct vectors in S is orthogonal. S is **orthonormal** if S is orthogonal, and every vector in S is a unit vector.

Example 1   Let $S = \{(3, 4), (4, -3)\}$. Clearly, S is an orthogonal subset of R^2. If we divide each vector in S by its length, we obtain the orthonormal set $\{(\frac{3}{5}, \frac{4}{5}), (\frac{4}{5}, -\frac{3}{5})\}$.

Notice that if $S = \{\mathbf{x}_1, \mathbf{x}_2, \ldots, \mathbf{x}_k\}$ is an orthonormal set of vectors, then

$$\mathbf{x}_i \cdot \mathbf{x}_j = \begin{cases} 1 & \text{if } i = j \\ 0 & \text{if } i \neq j \end{cases}$$

This result will be used frequently in many of our computations involving orthonormal sets.

It is easy to see that any orthogonal set of nonzero vectors in R^2 or R^3 is linearly independent. In fact, the next theorem tells us that this result holds generally.

Theorem 7.1.1 *Suppose that S is an orthogonal set of nonzero vectors in R^n. Then S is linearly independent.*

Proof
Suppose that $S = \{\mathbf{x}_1, \mathbf{x}_2, \ldots, \mathbf{x}_k\}$. Assume that

$$a_1 \mathbf{x}_1 + \cdots + a_k \mathbf{x}_k = \mathbf{0}$$

If we take the dot product of both sides of this equation with the vector $\mathbf{x}_1$ and use the fact that the dot product distributes over vector addition, we obtain

$$a_1 \mathbf{x}_1 \cdot \mathbf{x}_1 + a_2 \mathbf{x}_2 \cdot \mathbf{x}_1 + \cdots + a_k \mathbf{x}_k \cdot \mathbf{x}_1 = 0$$

Because S is orthogonal, this equation reduces to

$$a_1 \|\mathbf{x}_1\|^2 = 0$$

We conclude that $a_1 = 0$ since $\|\mathbf{x}_1\| \neq 0$. Similarly, $a_i = 0$ for the other a_i's. Thus, S is linearly independent. ∎

Corollary 7.1.2   *If S is an orthonormal subset of R^n, then S is linearly independent.*

One reason that the standard ordered basis is particularly useful is that it allows us to express a vector very easily as a linear combination of its members. For example, if $\mathbf{x} = (a, b, c)$, we may write

$$\mathbf{x} = a\mathbf{e}_1 + b\mathbf{e}_2 + c\mathbf{e}_3$$

Notice that we may also write $\mathbf{x}$ in the following way:

$$\mathbf{x} = (\mathbf{x} \cdot \mathbf{e}_1)\mathbf{e}_1 + (\mathbf{x} \cdot \mathbf{e}_2)\mathbf{e}_2 + (\mathbf{x} \cdot \mathbf{e}_3)\mathbf{e}_3$$

This ease in determining the coefficients of a vector represented as a linear combination of the standard vectors extends to all other orthonormal bases.

Theorem 7.1.3 *Let $S = \{x_1, x_2, \ldots, x_k\}$ be an orthonormal basis of a subspace W of R^n. If x is a vector in W, then we may write*

$$x = \sum_{i=1}^{k} (x \cdot x_i) x_i$$

Proof

Let x be a vector in W. Because S spans W, we may write x as a linear combination of the vectors in S, say

$$x = a_1 x_1 + \cdots + a_k x_k$$

Using the same technique as in the proof of Theorem 7.1.1, we take the dot product of both sides of the equation above with x_1 and obtain

$$x \cdot x_1 = a_1 x_1 \cdot x_1 + a_2 x_2 \cdot x_1 + \cdots + a_k x_k \cdot x_1$$

$$= a_1(1) + a_2(0) + \cdots + a_k(0)$$

$$= a_1$$

A similar argument will show that $a_i = x \cdot x_i$ for every i.   ■

Example 2   Let $S = \{(1/\sqrt{2})(1, 1, 0), (1/\sqrt{3})(1, -1, 1), (1/\sqrt{6})(-1, 1, 2)\}$. It is easy to show that S is an orthonormal subset of R^3. By Corollary 7.1.2, it follows that S is linearly independent and so it is a basis of R^3. We shall verify Theorem 7.1.3 for $x = (2, 1, 3)$. If we take the dot product of x with each of the three vectors in S, we obtain the three coefficients $3/\sqrt{2}$, $4/\sqrt{3}$, and $5/\sqrt{6}$. It is now an easy matter to check that

$$(2, 1, 3) = \frac{3}{\sqrt{2}} \frac{1}{\sqrt{2}} (1, 1, 0) + \frac{4}{\sqrt{3}} \frac{1}{\sqrt{3}} (1, -1, 1)$$

$$+ \frac{5}{\sqrt{6}} \frac{1}{\sqrt{6}} (-1, 1, 2)$$

One of the reasons that orthonormal sets are desirable is the ease in which computations with dot products are performed. In this regard the theorem below will prove to be very useful.

Theorem 7.1.4 *Let $\{x_1, x_2, \ldots, x_k\}$ be an orthonormal set. If*

$$x = \sum_{i=1}^{k} a_i x_i \quad and \quad y = \sum_{i=1}^{k} b_i x_i$$

then

$$\mathbf{x} \cdot \mathbf{y} = \sum_{i=1}^{k} a_i b_i$$

In particular,

$$\|\mathbf{x}\|^2 = \sum_{i=1}^{k} a_i^2$$

Proof

The reader should observe the use of the distributive properties of the dot product given below.

$$\mathbf{x} \cdot \mathbf{y} = \left(\sum_{i=1}^{k} a_i \mathbf{x}_i \right) \cdot \left(\sum_{j=1}^{k} b_j \mathbf{x}_j \right)$$

$$= \sum_{i=1}^{k} \sum_{j=1}^{k} a_i b_j (\mathbf{x}_i \cdot \mathbf{x}_j)$$

$$= \sum_{i=1}^{k} a_i b_i \qquad \blacksquare$$

Example 3   Apply the last statement of Theorem 7.1.4 to the vector $\mathbf{x} = (2, 1, 3)$ in Example 2. It is easy to see that $\|\mathbf{x}\|^2 = 14$. On the other hand, we have

$$\left(\frac{3}{\sqrt{2}} \right)^2 + \left(\frac{4}{\sqrt{3}} \right)^2 + \left(\frac{5}{\sqrt{6}} \right)^2 = \frac{9}{2} + \frac{16}{3} + \frac{25}{6} = 14$$

Many of the results we have proven thus far would be of little use if orthonormal sets were not readily available. The next theorem allows us to "transform" any linearly independent set into an orthonormal set which spans the same subspace spanned by the original set. The method of construction which is used is called the *Gram–Schmidt (orthogonalization) process.*

Theorem 7.1.5   *(Gram–Schmidt)   Let $S = \{\mathbf{y}_1, \ldots, \mathbf{y}_k\}$ be a linearly independent subset of R^n. Suppose that we define the set $S' = \{\mathbf{x}_1, \mathbf{x}_2, \ldots, \mathbf{x}_k\}$ by*

$$\mathbf{x}_1 = \mathbf{y}_1$$

$$\mathbf{x}_2 = \mathbf{y}_2 - \frac{\mathbf{y}_2 \cdot \mathbf{x}_1}{\|\mathbf{x}_1\|^2} \mathbf{x}_1$$

$$\mathbf{x}_3 = \mathbf{y}_3 - \frac{\mathbf{y}_3 \cdot \mathbf{x}_1}{\|\mathbf{x}_1\|^2} \mathbf{x}_1 - \frac{\mathbf{y}_3 \cdot \mathbf{x}_2}{\|\mathbf{x}_2\|^2} \mathbf{x}_2$$

and, in general, for $1 < i \leq k,$

$$\mathbf{x}_i = \mathbf{y}_i - \sum_{j=1}^{i-1} \frac{\mathbf{y}_i \cdot \mathbf{x}_j}{\|\mathbf{x}_j\|^2} \mathbf{x}_j$$

Then S' is an orthogonal set of nonzero vectors which spans the same subspace as S.

Notice that if the original set S is orthogonal, then $\mathbf{x}_i = \mathbf{y}_i$ for each i.

Before we proceed with the proof, we shall consider the result from a geometric point of view (see Figure 7.1.1). To construct a vector $\mathbf{x}_2$ which is orthogonal to $\mathbf{y}_1$ and

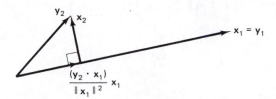

Figure 7.1.1

which is in the subspace spanned by $\mathbf{y}_1$ and $\mathbf{y}_2$, we need only let $\mathbf{x}_2$ be the vector difference of $\mathbf{y}_2$ and the orthogonal projection of $\mathbf{y}_2$ on $\mathbf{y}_1$. The Gram–Schmidt process allows us to extend this construction to any set of k independent vectors.

Proof

First observe that once it is known that S' is an orthogonal set of nonzero vectors, it then follows from Theorem 7.1.1 that S' is linearly independent. Thus, S and S' both span subspaces of equal dimension. By definition the vectors of S' are linear combinations of the vectors of S, and so the span of S' is contained in the span of S. It now follows from Theorem 3.2.7 that S and S' span the same subspace.

We shall only show that $\mathbf{x}_1$ and $\mathbf{x}_2$ are orthogonal. The proof that S' is orthogonal follows easily by mathematical induction on the number of elements k in S'.

$$\mathbf{x}_1 \cdot \mathbf{x}_2 = \mathbf{x}_1 \cdot \left(\mathbf{y}_2 - \frac{\mathbf{y}_2 \cdot \mathbf{x}_1}{\|\mathbf{x}_1\|^2} \mathbf{x}_1 \right)$$

$$= \mathbf{x}_1 \cdot \mathbf{y}_2 - \frac{(\mathbf{y}_2 \cdot \mathbf{x}_1)(\mathbf{x}_1 \cdot \mathbf{x}_1)}{\|\mathbf{x}_1\|^2}$$

$$= \mathbf{x}_1 \cdot \mathbf{y}_2 - \mathbf{y}_2 \cdot \mathbf{x}_1$$

$$= 0$$

∎

Example 4 Let $S = \{(1, 1, 0), (2, 0, 1), (2, 2, 1)\}$. We shall construct the orthogonal set S' defined above. First, let $\mathbf{x}_1 = (1, 1, 0)$. Then $\|\mathbf{x}_1\|^2 = 2$. So

$$\mathbf{x}_2 = \mathbf{y}_2 - \frac{\mathbf{y}_2 \cdot \mathbf{x}_1}{\|\mathbf{x}_1\|^2} \mathbf{x}_1$$

$$= (2, 0, 1) - \frac{2}{2}(1, 1, 0)$$

$$= (1, -1, 1)$$

Finally,

$$\mathbf{x}_3 = (2, 2, 1) - \left(\frac{4}{2}\right)(1, 1, 0) - \left(\frac{1}{3}\right)(1, -1, 1)$$

$$= \left(-\frac{1}{3}, \frac{1}{3}, \frac{2}{3}\right)$$

It is now an easy matter to check that the set S' is indeed orthogonal.

One of the most important consequences of the Gram–Schmidt process is stated below as a theorem.

Theorem 7.1.6 *Every (nonzero) subspace of R^n has an orthonormal basis.*

Proof
Let W be a subspace of R^n and let S be a basis of W. By the Gram–Schmidt process we may construct an orthogonal set S' of nonzero vectors which also spans W. By dividing each vector in S' by its length, we obtain an orthonormal set S'' of vectors which is linearly independent because of Theorem 7.1.1. The set S'' is the desired basis of W. ∎

Orthogonal Complements and Orthogonal Projections

In Section 1.6 we defined the orthogonal projection of a vector on a line. We shall now consider orthogonal projections on arbitrary subspaces.

Definition Let S be a (nonempty) subset of R^n. Define the ***orthogonal complement***, denoted $S^\perp$ (read "S perp"), to be the set of those vectors in R^n which are orthogonal to every vector in S.

Notice that the only vector which lies in both S and $S^\perp$ is the zero vector. This follows from the fact that the zero vector is the only vector that is orthogonal to itself.

Example 5   We shall determine the orthogonal complements of the subsets given below.
(a) Let S denote the set $\{(1, 1)\}$ in R^2. Then $S^\perp$ is the set of vectors that is represented by the line $y = -x$.
(b) Let W be the subspace of R^3 which corresponds to the plane $3x - 2y + 4z = 0$. If we let $\mathbf{x} = (3, -2, 4)$, then $\mathbf{x}$ is a normal to the plane. So $W^\perp$ is represented by the line through the origin with direction vector $\mathbf{x}$.
(c) Let W be the zero subspace of R^n. Then $W^\perp$ is all of R^n.
(d) Let W be all of R^n. Then $W^\perp$ is the zero subspace of R^n.

Theorem 7.1.7   *Let S be a subset of R^n. Then $S^\perp$ is a subspace of R^n.*

Proof
First observe that $S^\perp$ is not empty because it contains the zero vector. Now let $\mathbf{y}_1$ and $\mathbf{y}_2$ be vectors in $S^\perp$ and let c be a scalar. Suppose that $\mathbf{x}$ is any vector in S. Then

$$(\mathbf{y}_1 + \mathbf{y}_2) \cdot \mathbf{x} = \mathbf{y}_1 \cdot \mathbf{x} + \mathbf{y}_2 \cdot \mathbf{x}$$
$$= 0 + 0$$
$$= 0$$

and

$$(c\mathbf{y}_1) \cdot \mathbf{x} = c(\mathbf{y}_1 \cdot \mathbf{x}) = c0 = 0$$

So $S^\perp$ is closed under addition and scalar multiplication, and thus is a subspace. ∎

Theorem 7.1.8   *Let W be a subspace of R^n and let $\mathbf{y}$ be a vector in R^n. Then $\mathbf{y}$ may be written uniquely as*

$$\mathbf{y} = \mathbf{y}_1 + \mathbf{y}_2$$

where $\mathbf{y}_1$ is in W and $\mathbf{y}_2$ is in $W^\perp$.

Proof
Let $S = \{\mathbf{x}_1, \ldots, \mathbf{x}_k\}$ be an orthonormal basis for W. Now extend S to a basis for R^n. If we apply the Gram–Schmidt process to this basis, we obtain an orthonormal basis S' of R^n. Because the vectors in S are already orthogonal, we may write $S' = \{\mathbf{x}_1, \ldots, \mathbf{x}_k, \mathbf{x}_{k+1}, \ldots, \mathbf{x}_n\}$. For $\mathbf{y}$ in R^n, we have

$$\mathbf{y} = \sum_{i=1}^{n} a_i \mathbf{x}_i$$
$$= \sum_{i=1}^{k} a_i \mathbf{x}_i + \sum_{i=k+1}^{n} a_i \mathbf{x}_i$$
$$= \mathbf{y}_1 + \mathbf{y}_2$$

Because of the way $\mathbf{y}_1$ and $\mathbf{y}_2$ are defined above, it is clear that $\mathbf{y}_1$ is in W and $\mathbf{y}_2$ is in $W^\perp$.

To prove uniqueness, assume that

$$\mathbf{y}_1 + \mathbf{y}_2 = \mathbf{y} = \mathbf{y}_1' + \mathbf{y}_2'$$

where $y_1, y_1' \in W$ and $y_2, y_2' \in W^\perp$. Then $\mathbf{y}_1 - \mathbf{y}_1' = \mathbf{y}_2' - \mathbf{y}_2$ is a vector which is in both W and $W^\perp$, and hence must be the zero vector. Thus, $\mathbf{y}_1 = \mathbf{y}_1'$ and $\mathbf{y}_2 = \mathbf{y}_2'$. ∎

We are now in a position where we can easily define orthogonal projections on subspaces of R^n.

Definition

Let W be a subspace of R^n and let $\mathbf{y}$ be a vector in R^n. We call the (unique) vector $\mathbf{y}_1$ (defined in Theorem 7.1.8) the ***orthogonal projection of y on W***.

To use the proof of Theorem 7.1.8 to construct the orthogonal projection of a vector $\mathbf{y}$ on a subspace W of R^n, we must first obtain an orthonormal basis $\{\mathbf{x}_1, \mathbf{x}_2, \ldots, \mathbf{x}_k\}$ of W. Then the orthogonal projection $\mathbf{y}_1$ is given by

$$\mathbf{y}_1 = \sum_{i=1}^{k} (\mathbf{x} \cdot \mathbf{x}_i)\mathbf{x}_i$$

It is not difficult to see that this definition generalizes the two-dimensional case considered in Section 1.6, where we defined the orthogonal projection of a vector on a line. Notice that the same definition tells us that the orthogonal projection of $\mathbf{y}$ on $W^\perp$ is given by $\mathbf{y}_2$.

Example 6

Let W be the plane spanned by the vectors $(1, 1, 0)$ and $(2, 0, 1)$ and let $\mathbf{y} = (1, 2, -1)$.
(a) Find the orthogonal projection of $\mathbf{y}$ on W.
(b) Find the distance d from the point $P = (1, 2, -1)$ to the plane W (see Figure 7.1.2).

We use the results of Example 4 to obtain the orthonormal basis $S'' = \{(1/\sqrt{2})(1, 1, 0), (1/\sqrt{3})(1, -1, 1)\}$ of W. If we compute the dot products of $\mathbf{y}$ with the vectors in S'', we obtain the scalars $3/\sqrt{2}$ and $-2/\sqrt{3}$. So the orthogonal projection $\mathbf{y}_1$ of $\mathbf{y}$ on W is given by

$$\mathbf{y}_1 = \left(\frac{3}{\sqrt{2}}\right)\frac{1}{\sqrt{2}}(1, 1, 0) + \left(-\frac{2}{\sqrt{3}}\right)\frac{1}{\sqrt{3}}(1, -1, 1)$$

$$= \left(\frac{5}{6}, \frac{13}{6}, -\frac{4}{6}\right)$$

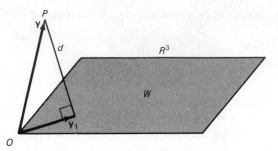

Figure 7.1.2

From Figure 7.1.2 we see that the distance d is given by

$$d = \|\mathbf{y} - \mathbf{y}_1\| = \left\|\left(\frac{1}{6}, -\frac{1}{6}, -\frac{2}{6}\right)\right\| = \frac{1}{\sqrt{6}}$$

We could also obtain d by noting that $\mathbf{y} - \mathbf{y}_1 = \mathbf{y}_2$, the orthogonal projection of $\mathbf{y}$ on $W^\perp$. From Example 4 we see that an orthonormal basis of S' is given by the vector $(1/\|\mathbf{x}_3\|)\mathbf{x}_3 = (1/\sqrt{6})(-1, 1, 2)$. The dot product of $\mathbf{y}$ with this vector is $-1/\sqrt{6}$. So

$$\mathbf{y}_2 = \left(-\frac{1}{\sqrt{6}}\right)\frac{1}{\sqrt{6}}(-1, 1, 2) = -\frac{1}{6}(-1, 1, 2)$$

Therefore,

$$d = \|\mathbf{y}_2\| = \frac{1}{\sqrt{6}}$$

Example 6 suggests that the orthogonal projection of a vector $\mathbf{y}$ on a subspace W has the property that it is the vector in W which is the *closest* to $\mathbf{y}$. This property of an orthogonal projection generalizes to R^n. The result is stated formally below.

Theorem 7.1.9 *Let $\mathbf{y}$ be a vector in R^n and let W be a subspace of R^n. Then the orthogonal projection $\mathbf{y}_1$ of $\mathbf{y}$ on W satisfies the inequality*

$$\|\mathbf{y} - \mathbf{y}_1\| \le \|\mathbf{y} - \mathbf{x}\|$$

for all $\mathbf{x}$ in W.

Proof
Let $S = \{\mathbf{x}_1, \mathbf{x}_2, \ldots, \mathbf{x}_k\}$ be an orthonormal basis of W. As in the proof of Theorem 7.1.8, we may extend S to an orthonormal basis $S' = \{\mathbf{x}_1, \ldots, \mathbf{x}_n\}$ of R^n. For any vector $\mathbf{y}$ in R^n and vector $\mathbf{x}$ in W we may write

$$\mathbf{y} = \sum_{i=1}^{n} a_i \mathbf{x}_i \quad \text{and} \quad \mathbf{x} = \sum_{i=1}^{k} b_i \mathbf{x}_i$$

So

$$\mathbf{y}_1 = \sum_{i=1}^{k} a_i \mathbf{x}_i$$

By Theorem 7.1.4, we have

$$\|\mathbf{y} - \mathbf{y}_1\|^2 = \left\| \sum_{i=1}^{n} a_i \mathbf{x}_i - \sum_{i=1}^{k} a_i \mathbf{x}_i \right\|^2$$

$$= \left\| \sum_{i=k+1}^{n} a_i \mathbf{x}_i \right\|^2$$

$$= \sum_{i=k+1}^{n} a_i^2$$

On the other hand,

$$\|\mathbf{y} - \mathbf{x}\|^2 = \left\| \sum_{i=1}^{n} a_i \mathbf{x}_i - \sum_{i=1}^{k} b_i \mathbf{x}_i \right\|^2$$

$$= \left\| \sum_{i=1}^{k} (a_i - b_i)\mathbf{x}_i + \sum_{i=k+1}^{n} a_i \mathbf{x}_i \right\|^2$$

$$= \sum_{i=1}^{k} (a_i - b_i)^2 + \sum_{i=k+1}^{n} a_i^2$$

$$\geq \sum_{i=k+1}^{n} a_i^2$$

$$= \|\mathbf{y} - \mathbf{y}_1\|^2$$

So the proof is complete. ∎

Least-Squares Approximation (Optional)

In almost all areas of empirical research there is an interest in finding (simple) mathematical relationships between variables. In economics the variables might be the gross national product, the unemployment rate, and the annual deficit. In the life sciences the variables of interest might be the incidence of smoking and heart disease. In sociology it might be birth order and frequency of juvenile delinquency.

Many relationships in science are *deterministic*, that is, information about one variable completely determines the value or measure of another variable. For example, the relationship between force f and acceleration a of an object of mass m is given by $f = ma$. f is completely specified by m and a. Another example might be the height of a (free) falling object and the time that it has been falling. On the other hand, the relationship between the height and weight of an individual is not deterministic. There are many people with the same weight but different heights. Yet there exist charts in hospitals which give the recommended height for a given weight. How are such charts determined? In this section we will use the tools of linear algebra to obtain reasonable mathematical relationships between variables that are not deterministic. Such relationships are often called *probabilistic* or *stochastic*.

Before we begin this topic, we first need to recall several results concerning dot product and rank developed in Section 3.3. For the remainder of this section we consider the vectors in R^n as column vectors. In this case the dot product between **x** and **y** may be written

$$\mathbf{x} \cdot \mathbf{y} = \mathbf{x}^t \mathbf{y}$$

It was shown in Theorem 3.3.6 that if A is an $m \times n$ matrix, then

1. $A\mathbf{x} \cdot \mathbf{y} = \mathbf{x} \cdot A^t \mathbf{y}$ for all vectors **x** in R^n and vectors **y** in R^m.

2. rank $A^t A$ = rank A.

To see how our results can be used, we shall assume that we are given a set of data $(x_1, y_1), \ldots, (x_n, y_n)$. For example, we might have a randomly selected sample of n people, where x_i represents the number of years of education of the ith person and y_i represents the annual income of the ith person. The data may be plotted as in Figure 7.1.3. Notice that there is an approximately straight line or linear relationship between x and y. To obtain this relationship, we would like to find the line that *best* fits the data. The usual criterion that statisticians use for finding this line is that the sum of the squared vertical distances of the data from the line be minimized. From Figure 7.1.3 we see that we must find a and b so that the quantity

$$E = \sum_{i=1}^{n} [y_i - (a + bx_i)]^2$$

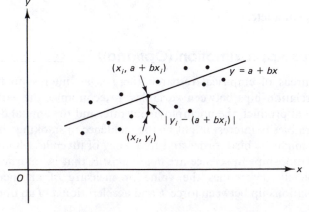

Figure 7.1.3

is minimized. The quantity E is called the **error sum of squares** and the line for which E is minimized is called the **least-squares line**. If we introduce the notation

$$\mathbf{y} = \begin{bmatrix} y_1 \\ \vdots \\ y_n \end{bmatrix} \qquad A = \begin{bmatrix} 1 & x_1 \\ \vdots & \vdots \\ 1 & x_n \end{bmatrix} \qquad \mathbf{x}_0 = \begin{bmatrix} a \\ b \end{bmatrix}$$

then we may represent (Exercise 27) the error sum of squares by

$$E = \|\mathbf{y} - A\mathbf{x}_0\|^2$$

If, in fact, all the points lie on the line $y = a + bx$, then the error sum of squares would equal zero, in which case $\mathbf{y} = A\mathbf{x}_0$. Of course, in stochastic relationships, it would be quite rare for this to happen.

To see what is involved geometrically in finding the vector $\mathbf{x}_0$, consider Figure 7.1.4 where we let W be the subspace $\{A\mathbf{x} : \mathbf{x} \in R^2\}$. We want the vector in W that is closest to $\mathbf{y}$. Notice that $\mathbf{y}$ is a vector in R^n. By Theorem 7.1.9, the desired vector is the orthogonal projection of $\mathbf{y}$ on W. Denote this vector by $A\mathbf{x}_0$. By Theorem 7.1.8 the vector $\mathbf{y} - A\mathbf{x}_0$ lies in $W^\perp$. So we have

$$A\mathbf{x} \cdot (\mathbf{y} - A\mathbf{x}_0) = 0$$

for all $\mathbf{x}$ in R^2. By Theorem 3.3.6(a), this equation is equivalent to

$$\mathbf{x} \cdot A^t(\mathbf{y} - A\mathbf{x}_0) = 0$$

Because this equation is valid for *all* $\mathbf{x}$ in R^2, it follows that

$$A^t(\mathbf{y} - A\mathbf{x}_0) = 0$$

or

$$A^t A\mathbf{x}_0 = A^t\mathbf{y}$$

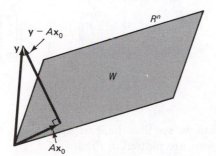

Figure 7.1.4

This system of equations is referred to as the **normal equations**. Notice that unless all the x_i's are equal, that is, the line is vertical, we have that rank $A = 2$. This means that the columns of A are linearly independent. In this case Theorem 3.3.6(b) allows us to conclude that rank $A^t A = 2$, so $A^t A$ is invertible. Thus, the solution we seek is given by

$$\mathbf{x}_0 = (A^t A)^{-1} A^t \mathbf{y}$$

The term *full rank* is applied to any matrix whose columns are linearly independent. We summarize our results in the next theorem.

Theorem 7.1.10

Let $(x_1, y_1), \ldots, (x_n, y_n)$ be a set of points in R^2 in which at least two of the x_i's are distinct, and let

$$\mathbf{y} = \begin{bmatrix} y_1 \\ \vdots \\ y_n \end{bmatrix} \qquad A = \begin{bmatrix} 1 & x_1 \\ \vdots & \vdots \\ 1 & x_n \end{bmatrix} \qquad \mathbf{x}_0 = \begin{bmatrix} a \\ b \end{bmatrix}$$

then the vector $\mathbf{x}_0$ which minimizes $E = \|\mathbf{y} - A\mathbf{x}_0\|^2$ is given by

$$\mathbf{x}_0 = (A^t A)^{-1} A^t \mathbf{y}$$

Example 7

Find the least-squares line for the points $(1, 2)$, $(3, 7)$, and $(5, 8)$. For these points, we let

$$A = \begin{bmatrix} 1 & 1 \\ 1 & 3 \\ 1 & 5 \end{bmatrix} \qquad \text{and} \qquad \mathbf{y} = \begin{bmatrix} 2 \\ 7 \\ 8 \end{bmatrix}$$

Then

$$A^t A = \begin{bmatrix} 1 & 1 & 1 \\ 1 & 3 & 5 \end{bmatrix} \begin{bmatrix} 1 & 1 \\ 1 & 3 \\ 1 & 5 \end{bmatrix} = \begin{bmatrix} 3 & 9 \\ 9 & 35 \end{bmatrix}$$

So

$$\mathbf{x}_0 = (A^t A)^{-1} A^t \mathbf{y}$$

$$= \begin{bmatrix} \frac{35}{24} & -\frac{3}{8} \\ -\frac{3}{8} & \frac{1}{8} \end{bmatrix} \begin{bmatrix} 1 & 1 & 1 \\ 1 & 3 & 5 \end{bmatrix} \begin{bmatrix} 2 \\ 7 \\ 8 \end{bmatrix}$$

$$= \begin{bmatrix} \frac{7}{6} \\ \frac{3}{2} \end{bmatrix}$$

From this computation we see that the least-squares line is given by $y = \frac{7}{6} + \frac{3}{2}x$. The line and the three points are plotted in Figure 7.1.5.

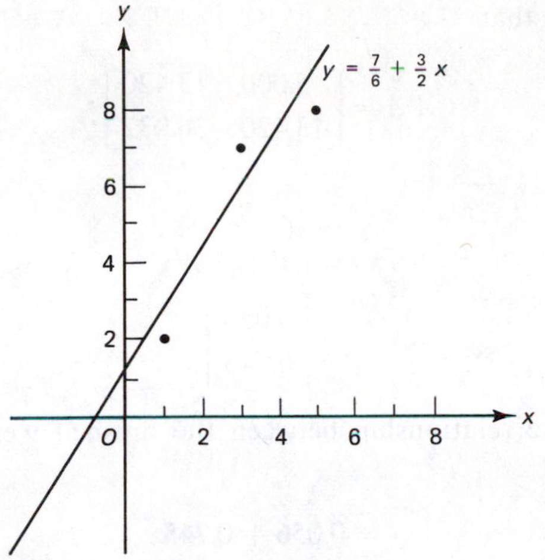

Figure 7.1.5

Example 8   In the manufacture of refrigerators, it is necessary to finish connecting rods. If the weight of the finished rod is above a certain amount, the rod must be discarded. As the finishing process is expensive, it would be of considerable value to the manufacturer to be able to estimate the relationship between the finished weight and the initial rough weight. In this way he could discard those rods whose rough weights were too high. From past experience, the manufacturer knows that this relationship is approximately linear.

From a sample of five rods, we let x_i be the rough weight and y_i the finished weight of the ith rod. The data are given below.

Rough weight, x_i	Finished weight, y_i
2.60	2.00
2.72	2.10
2.75	2.10
2.67	2.03
2.68	2.04

From this information we let

$$A = \begin{bmatrix} 1 & 2.60 \\ 1 & 2.72 \\ 1 & 2.75 \\ 1 & 2.67 \\ 1 & 2.68 \end{bmatrix} \quad \text{and} \quad \mathbf{y} = \begin{bmatrix} 2.00 \\ 2.10 \\ 2.10 \\ 2.03 \\ 2.04 \end{bmatrix}$$

Then it can be shown that

$$A^t A = \begin{bmatrix} 5.000 & 13.420 \\ 13.420 & 36.032 \end{bmatrix}$$

so

$$\mathbf{x}_0 = (A^t A)^{-1} A^t \mathbf{y}$$

$$= \begin{bmatrix} 0.056 \\ 0.745 \end{bmatrix}$$

Thus, the approximate relationship between the finished weight y and the rough weight x is given by

$$y = 0.056 + 0.745x$$

Therefore, if the rough weight of a rod is 2.65, then the (approximate) finished weight is $0.056 + 0.745(2.65) = 2.030$.

The method we have developed may also be applied to find the best quadratic fit, that is, $y = a + bx + cx^2$ for the n points $(x_1, y_1), \ldots, (x_n, y_n)$. The only modification to what we have discussed in the linear case is the *new* error sum of squares,

$$E = \sum_{i=1}^{n} [y_i - (a + bx_i + cx_i^2)]^2$$

E may be represented as $\|\mathbf{y} - A\mathbf{x}_0\|^2$, where

$$A = \begin{bmatrix} 1 & x_1 & x_1^2 \\ \vdots & \vdots & \vdots \\ 1 & x_n & x_n^2 \end{bmatrix} \qquad \mathbf{y} = \begin{bmatrix} y_1 \\ \vdots \\ y_n \end{bmatrix} \qquad \mathbf{x}_0 = \begin{bmatrix} a \\ b \\ c \end{bmatrix}$$

Example 9   It is known from physics that if a ball is thrown upward at a velocity of v_0 feet per second from a building of height s_0 feet, then the height of the ball after t seconds is given by $s = s_0 + v_0 t - \frac{1}{2}gt^2$, where g represents the force of gravity ($g = -32$ feet per second per second). To provide an empirical estimate of g, a ball is thrown upward from a building 100 feet high at a velocity of 30 feet per second. The height of the ball is observed at the times given below.

Time	Height
0	100
1	118
2	92
3	48
3.5	7

For these data we let

$$A = \begin{bmatrix} 1 & 0 & 0 \\ 1 & 1 & 1 \\ 1 & 2 & 4 \\ 1 & 3 & 9 \\ 1 & 3.5 & 12.25 \end{bmatrix} \quad \text{and} \quad \mathbf{y} = \begin{bmatrix} 100 \\ 118 \\ 92 \\ 48 \\ 7 \end{bmatrix}$$

Then, using

$$\mathbf{x}_0 = (A^t A)^{-1} A^t \mathbf{y}$$

it can be shown that

$$\mathbf{x}_0 = \begin{bmatrix} s_0 \\ v_0 \\ -\frac{1}{2}g \end{bmatrix} = \begin{bmatrix} 101.00 \\ 29.77 \\ -16.11 \end{bmatrix}$$

This yields the approximate relationship

$$s = 101.00 + 29.77t - 16.11t^2$$

The exact relationship is

$$s = 100 + 30t - \tfrac{1}{2}gt^2$$

Setting $-\frac{1}{2}g = -16.11$, we obtain the estimate 32.22 for g.

It should be pointed out that the same method may be used to discover the best-fitting polynomial of any desired degree. In fact, by using the appropriate change of variable, many more complicated relationships may be estimated by the same type of matrix computations.

Exercises

1. In each of the following parts a set S and a vector $\mathbf{x}$ are given. First show that S is orthonormal and then use Theorem 7.1.3 to write $\mathbf{x}$ as a linear combination of the vectors in S. Finally, check your results by performing vector addition.

   (a) $S = \{(1/\sqrt{2})(1, 1), (1/\sqrt{2})(1, -1)\};$
   $\mathbf{x} = (4, 1)$

   (b) $S = \{(1/5)(3, -4), (1/5)(4, -3)\};$
   $\mathbf{x} = (2, 3)$

   (c) $S = \{(1/\sqrt{2})(1, 1, 0), (1/\sqrt{6})(-1, 1, 2),$
   $(1/\sqrt{3})(1, -1, 1)\}; \mathbf{x} = (4, 1, 0)$

   (d) $S = \{(1/3)(2, 1, -2), (1/3)(1, 2, 2),$
   $(1/3)(2, -2, 1)\}; \mathbf{x} = (-2, 1, 3)$

2. For each of the vectors $\mathbf{x}$ in Exercise 1, compute $\|\mathbf{x}\|^2$. Then use Theorem 7.1.4 to verify your result.

3. In each of the following parts a set S is given. Use the Gram–Schmidt process to find an

orthogonal basis of the subspace spanned by
S. Also find an orthonormal basis.
(a) $S = \{(1, -1), (2, 3)\}$
(b) $S = \{(1, 2), (3, 4)\}$
(c) $S = \{(1, 2, 3), (2, 2, 1)\}$
(d) $S = \{(1, 0, 1), (0, 1, 1)\}$
(e) $S = \{(1, 0, 1), (1, 2, -2), (2, -1, 1)\}$
(f) $S = \{(1, 1, 0, 0), (0, 1, 1, 0), (1, 0, 1, 2)\}$
(g) $S = \{(1, 1, -1, 0), (0, 0, 1, 1), (1, 2, 0, 1)\}$

4. In the notation of Theorem 7.1.5 prove that if
S is orthogonal, then $\mathbf{x}_i = \mathbf{y}_i$ for $i = 1, \ldots, k$.

5. Let $S = \{(1/\sqrt{2})(1, 0, 1, 0), (1/\sqrt{10})(1, 2, -1, 2)\}$.
(a) Prove that S is an orthonormal subset of R^4.
(b) Extend S to an orthonormal basis of R^4.

6. In each of the following parts a subspace and
a vector are given. Find the orthogonal
projection of the vector on the subspace.
(a) The line $y = 2x$; $(1, 5)$
(b) The line $y = -6x$; $(2, -1)$
(c) The yz-plane; $(3, 5, 8)$
(d) The plane $x + 2y + z = 0$; $(1, 1, 1)$
(e) The plane $2x - y - 2z = 0$; $(1, 1, 0)$
(f) The line with direction vector $(3, 2, 1)$
which passes through the origin;
$(2, -1, 4)$
(g) The line with direction vector $(1, 3, 2)$
which passes through the origin; $(4, 0, 2)$

7. Let S be a basis of R^n and let P be the $n \times n$
matrix whose columns consist of the vectors
in S. Prove that S is orthonormal if and only
if $P^t P = I$.

8. Use the proof of Theorem 7.1.8 to show that
for any subspace W of R^n, we have

$$\dim W + \dim W^\perp = n$$

9. Let W be a subspace of R^n.
(a) Use Exercise 8 to prove that

$$\dim W = \dim W^{\perp\perp}$$

where $W^{\perp\perp}$ is the orthogonal complement
of $W^\perp$.

(b) Prove that $W = W^{\perp\perp}$. [*Hint:* First show
that W is contained in $W^{\perp\perp}$, then use part
(a).]

10. Suppose that W_1 and W_2 are subspaces of R^n.
Prove that if W_1 is contained in W_2, then $W_2^\perp$
is contained in $W_1^\perp$.

11. (*Parseval's identity*) Let $\{\mathbf{x}_1, \ldots, \mathbf{x}_k\}$ be an
orthonormal basis of a subspace W of R^n.
Prove that for any vectors $\mathbf{x}$ and $\mathbf{y}$ in W, we
have

$$\mathbf{x} \cdot \mathbf{y} = \sum_{i=1}^{k} (\mathbf{x} \cdot \mathbf{x}_i)(\mathbf{y} \cdot \mathbf{x}_i)$$

12. (*Bessel's inequality*) Let $\{\mathbf{x}_1, \ldots, \mathbf{x}_k\}$ be an
orthonormal subset of R^n. Then for any vector
$\mathbf{x}$ in R^n, prove that

$$\|\mathbf{x}\|^2 \geq \sum_{i=1}^{k} (\mathbf{x} \cdot \mathbf{x}_i)^2$$

Definition Let W be a nonzero subspace of
R^n. Let $T: R^n \to R^n$ be the function that
assigns to each vector $\mathbf{x}$ in R^n its orthogonal
projection on W. T is called the *orthogonal
projection operator on W*.

Exercises 13–20 will assume that T is the
orthogonal projection operator on a subspace W of
R^n.

13. Prove that T is a linear transformation.

14. Prove that W is T-*invariant*; that is, prove
that if $\mathbf{x}$ is in W, then $T(\mathbf{x})$ is in W.

15. Prove that $R(T) = W$ and $N(T) = W^\perp$.

16. Prove that $T^2 = T$.

17. Prove that the only eigenvalues of T are 0
and 1.

18. Use Exercises 15 and 17 to show that there
exists an orthonormal basis S of R^n such that

$$[T]_S = \begin{bmatrix} I_k & O \\ O & O \end{bmatrix}$$

where $k = \dim W$.

19. Let A be the standard matrix of T. Use Exercises 7 and 18 to show that there exists an invertible matrix P such that $P^t P = I$ and $P^t A P = D$, where D is a diagonal matrix consisting of 0's and 1's.

20. Let $T^\perp$ denote the orthogonal projection operator on the orthogonal complement $W^\perp$ of W. Prove
    (a) $TT^\perp = T^\perp T = T_0$
    (b) $T + T^\perp = I$

Least Squares

21. Find the least-squares line for the points $(0, 0)$, $(1, 3)$, and $(2, 4)$. Plot the points and the line.

22. Find the least-squares line for the points $(-2, -4)$, $(-1, 0)$, $(0, 2)$, $(1, 2)$, and $(2, 5)$. Plot the points and the line.

23. In physics *Hooke's law* states that (within certain limits) there is a linear relation between the length x of a spring and the force y applied to (or exerted by) the spring. That is, $y = a + bx$, where b is called the **spring constant**. Use the data below to estimate the spring constant. (The length is given in inches and the force is given in pounds.)

Force, x	Length, y
3.5	1.0
4.0	2.2
4.5	2.8
5.0	4.3

24. Find the best quadratic fit to the following points: $(-2, -2)$, $(-1, 0)$, $(0, 1)$, $(1, 4)$, and $(2, 4)$. Plot the points and your quadratic function.

25. A ball is thrown upward with a velocity of 40 feet per second from a building that is 50 feet high.
    (a) Determine the *exact* relationship between the height of the ball and time.

(b) Use the data below to determine the approximate relationship between the height of the ball and time.

Time	Height
0.0	50
1.0	72
2.0	68
2.5	52
3.0	22

26. Use the method of least squares to find the best *cubic fit* for the points: $(-2, -4)$, $(-1, 1)$, $(0, 1)$, $(2, 10)$, and $(3, 26)$.

27. Using the notation given in this section, verify that the error sum of squares for the least-squares line is given by $\|\mathbf{y} - A\mathbf{x}_0\|^2$.

*28. Use the program GRAM–SCHMIDT to find an orthogonal basis S' of the subspace spanned by the subset S given below.
    (a) $S = \{(1, 2, 1), (2, -3, 1)\}$
    (b) $S = \{(2, 2, 1), (1, 2, 2), (-1, -1, 1)\}$
    (c) $S = \{(1, 0, 1, 1), (-2, 1, 0, 1), (2, 0, 2, 1)\}$
    (d) $S = \{(1, 1, 1, 1), (2, -1, 1, 0), (3, 0, 2, 1)\}$

*29. For the data below, use the program MATRIX to find:
    (a) The least-squares line.
    (b) The best quadratic fit.

x	y
-2	15
-1	8
0	4
1	1
2	5
3	10

30. Plot the data in Exercise 29 and the linear and quadratic fits on the same set of axes.

* This problem should be solved by using one of the programs noted in Appendix B.

7.2 ORTHOGONAL AND SYMMETRIC MATRICES

Many of the concepts in linear algebra have their roots in geometry. The terms *angle*, *perpendicular*, and *distance* will play a strong role in this section. Just as linear transformations preserve the operations of vector addition and scalar multiplication, it will be seen that "orthogonal transformations" preserve angle, dot product, and distance. Recall the rotation matrix A_θ defined by

$$A_\theta = \begin{bmatrix} \cos\theta & -\sin\theta \\ \sin\theta & \cos\theta \end{bmatrix}$$

Because this matrix rotates a vector by an angle θ, we know that the associated transformation preserves the angle between two vectors; that is, the angle between two vectors $\mathbf{x}$ and $\mathbf{y}$ equals the angle between the vectors $A_\theta\mathbf{x}$ and $A_\theta\mathbf{y}$. It clearly preserves the length of a vector. Notice that the columns (or rows) of A_θ are orthonormal. It is this last property of rotation matrices that motivates the following definition.

Definition   A square matrix A is said to be an ***orthogonal matrix*** if the columns of A form an orthonormal set.

It can be shown that any 2×2 orthogonal matrix is either a rotation matrix or a matrix which is formed by interchanging two columns of a rotation matrix.

Example 1   The matrix A below is an orthogonal matrix.

$$A = \begin{bmatrix} \dfrac{1}{\sqrt{2}} & \dfrac{1}{\sqrt{3}} & -\dfrac{1}{\sqrt{6}} \\ \dfrac{1}{\sqrt{2}} & -\dfrac{1}{\sqrt{3}} & \dfrac{1}{\sqrt{6}} \\ 0 & \dfrac{1}{\sqrt{3}} & \dfrac{2}{\sqrt{6}} \end{bmatrix}$$

Theorem 7.2.1 *Let A be an $n \times n$ matrix. Then the following are equivalent.*
(a) *A is an orthogonal matrix.*
(b) *$A^t A = A A^t = I$.*
(c) *$A\mathbf{x} \cdot A\mathbf{y} = \mathbf{x} \cdot \mathbf{y}$ for all vectors $\mathbf{x}$ and $\mathbf{y}$ in R^n.*
(d) *$\|A\mathbf{x}\| = \|\mathbf{x}\|$ for all vectors $\mathbf{x}$ in R^n.*

Proof

We will prove this result by first showing that (a) implies (b), (b) implies (c), and (c) implies (a). This will demonstrate that (a), (b), and (c) are equivalent. Then we will show that (c) and (d) are equivalent.

(*a*) *implies* (*b*): The ijth entry of A^tA is found by computing the dot product of the ith row of A^t and the jth column of A, that is, the dot product of the ith column of A and the jth column of A. Because the columns of A form an orthonormal set, this dot product equals 1 if $i = j$ and equals 0 if $i \neq j$. On the other hand, the ijth entry of I is defined exactly the same way. So $A^tA = I$. By Theorem 3.4.9 we have that $A^t = A^{-1}$, and hence $AA^t = AA^{-1} = I$.

(*b*) *implies* (*c*): Using Theorem 3.3.6(a), we have for any vectors $\mathbf{x}$ and $\mathbf{y}$ in R^n,

$$A\mathbf{x} \cdot A\mathbf{y} = \mathbf{x} \cdot A^tA\mathbf{y} = \mathbf{x} \cdot I\mathbf{y} = \mathbf{x} \cdot \mathbf{y}$$

(*c*) *implies* (*a*): Let $\mathbf{x}$ and $\mathbf{y}$ be vectors in R^n. Then

$$\mathbf{x} \cdot \mathbf{y} = A\mathbf{x} \cdot A\mathbf{y} = \mathbf{x} \cdot A^tA\mathbf{y}$$

Using the properties of the dot product, we may rewrite this equation as

$$0 = \mathbf{x} \cdot A^tA\mathbf{y} - \mathbf{x} \cdot \mathbf{y} = \mathbf{x} \cdot (A^tA - I)\mathbf{y}$$

Because $\mathbf{x}$ is arbitrary, the equation above implies that $(A^tA - I)\mathbf{y} = \mathbf{0}$. Since $\mathbf{y}$ is arbitrary, it follows that $A^tA - I = O$ or $A^tA = I$. As we noted earlier, this result implies that the columns of A are orthogonal.

(*c*) *implies* (*d*): Let $\mathbf{x}$ and $\mathbf{y}$ be any vectors in R^n. Then

$$A\mathbf{x} \cdot A\mathbf{y} = \mathbf{x} \cdot \mathbf{y}$$

If we let $\mathbf{x} = \mathbf{y}$, we have $\|A\mathbf{x}\|^2 = \|\mathbf{x}\|^2$.

(*d*) *implies* (*c*): Let $\mathbf{x}$ and $\mathbf{y}$ be any vectors in R^n. Then

$$\|A(\mathbf{x} + \mathbf{y})\|^2 = \|\mathbf{x} + \mathbf{y}\|^2$$

So, applying the properties of dot product to both sides of the equation above, and using the fact that $\|\mathbf{z}\|^2 = \mathbf{z} \cdot \mathbf{z}$ for any vector $\mathbf{z}$, we have

$$\|A\mathbf{x}\|^2 + 2(A\mathbf{x} \cdot A\mathbf{y}) + \|A\mathbf{y}\|^2 = \|\mathbf{x}\|^2 + 2(\mathbf{x} \cdot \mathbf{y}) + \|\mathbf{y}\|^2$$

Now we may use our assumption that $\|A\mathbf{x}\| = \|\mathbf{x}\|$ and $\|A\mathbf{y}\| = \|\mathbf{y}\|$ to conclude that

$$A\mathbf{x} \cdot A\mathbf{y} = \mathbf{x} \cdot \mathbf{y} \qquad \blacksquare$$

There are a number of conclusions that we may draw from this theorem.

1. Part (b) tells us that the statement that the columns form an orthonormal set is equivalent to the statement that the rows form an orthonormal set. Thus, A is an orthogonal matrix if and only if A^t is orthogonal matrix. A direct proof of this result is challenging.

2. Because the angle between vectors may be defined in terms of dot product, we can conclude that orthogonal matrices also preserve the angles between vectors.

3. It follows immediately from part (b) that A is an orthogonal matrix if and only if $A^{-1} = A^t$.

Definition   A linear transformation $T: R^n \rightarrow R^n$ is said to be an **orthogonal transformation** if $\|T(\mathbf{x})\| = \|\mathbf{x}\|$ for all vectors $\mathbf{x}$ in R^n.

By Theorem 7.2.1(d) it is easy to see that T is an orthogonal transformation if and only if its standard matrix is an orthogonal matrix. Using part (c) of the same theorem, we may conclude that T is an orthogonal transformation if and only if T preserves dot products.

Example 2   Let T be the linear operator on R^3 that rotates a vector by an angle θ about the z-axis. Clearly, T preserves the length of a vector. Thus, T is an orthogonal transformation. To find the standard matrix of T, we note that for vectors in the xy-plane, T acts like a two-dimensional rotation. The z-axis remains fixed. Thus, the standard matrix of T is

$$\begin{bmatrix} \cos\theta & -\sin\theta & 0 \\ \sin\theta & \cos\theta & 0 \\ 0 & 0 & 1 \end{bmatrix}$$

We have seen how important the concept of similarity has been in the study of the diagonalization of matrices. We will now define the related notion of orthogonally equivalent.

Definition   Let A and B be $n \times n$ matrices. A is said to be **orthogonally equivalent** to B if there exists an orthogonal matrix P such that $A = P^t B P$.

The equation in the definition implies that

$$PAP^t = PP^t BPP^t = IBI = B$$

or

$$B = (P^t)^t A P^t$$

Because P^t is also an orthogonal matrix, we see that if A is orthogonally equivalent to B, then B is orthogonally equivalent to A. Thus, it makes sense to say instead that A and B are **orthogonally equivalent**. Because $P^t = P^{-1}$, it is clear that if A and B are orthogonally equivalent, then they are similar. We shall leave the next theorem as an exercise. Its proof is similar to the proof of Theorem 4.2.4.

Theorem 7.2.2   *Let A and B be $n \times n$ matrices. Then*

(a) *A is orthogonally equivalent to A.*

(b) *If A is orthogonally equivalent to B, then B is orthogonally equivalent to A.*

(c) *If A is orthogonally equivalent to B and B is orthogonally equivalent to C, then A is orthogonally equivalent to C.*

In Chapter 6 we derived the Test for Diagonalizability for a matrix to determine if a matrix is similar to a diagonal matrix. Now we ask the following question: *When is a matrix orthogonally equivalent to a diagonal matrix?* Notice that if a matrix A is orthogonally equivalent to a diagonal matrix D, that is,

$$A = P^t D P$$

for some orthogonal matrix P, then

$$A^t = (P^t D P)^t = P^t D^t P^{tt} = P^t D P = A$$

So A is symmetric. The next theorem is remarkable in its simplicity. It states that the converse also holds.

Theorem 7.2.3   *Let A be a symmetric matrix. Then A is orthogonally equivalent to a diagonal matrix.*

This theorem is sometimes called the *spectral theorem*, and it is used to decompose a symmetric matrix into a sum of special matrices (see Exercise 17). We will not prove this theorem [see (1) in the References]. A key result that is often used in the proof of the theorem is stated below:

The roots of the characteristic equation of a symmetric matrix are real.

We begin with some examples.

Example 3   Let

$$A = \begin{bmatrix} 3 & -4 \\ -4 & -3 \end{bmatrix}$$

It can be shown that the eigenvalues of A are 5 and -5 with associated eigenvectors

$(1/\sqrt{5})(-2, 1)$ and $(1/\sqrt{5})(1, 2)$, respectively. Notice that we have chosen unit vectors so that in this case the eigenvectors are orthonormal. If we let

$$P = \begin{bmatrix} -\dfrac{2}{\sqrt{5}} & \dfrac{1}{\sqrt{5}} \\ \dfrac{1}{\sqrt{5}} & \dfrac{2}{\sqrt{5}} \end{bmatrix} \quad \text{and} \quad D = \begin{bmatrix} 5 & 0 \\ 0 & -5 \end{bmatrix}$$

then it can be shown that $P^t AP = D$. Notice that P is an orthogonal matrix. This will always be the case for symmetric matrices if the eigenvalues are distinct and if we choose unit vectors as the eigenvectors.

Suppose that A is symmetric, in which case there exists an orthogonal matrix P and a diagonal matrix D such that $P^t AP = D$. Because the columns of P are the eigenvectors of A, we have the following corollary.

Corollary 7.2.4   *Let A be an $n \times n$ symmetric matrix. Then there exists an orthonormal basis of R^n consisting of eigenvectors of A.*

We now turn our attention to the problem of finding an orthonormal basis of eigenvectors of a symmetric matrix. Notice that the two eigenvectors of the matrix A in Example 3 are orthogonal and that their corresponding eigenvalues are distinct. The next theorem tells us that this result is not a coincidence.

Theorem 7.2.5   *Let A be a symmetric matrix with eigenvectors $\mathbf{x}_1$ and $\mathbf{x}_2$ with corresponding eigenvalues λ_1 and λ_2, respectively. If $\lambda_1 \neq \lambda_2$, then $\mathbf{x}_1$ and $\mathbf{x}_2$ are orthogonal.*

Proof

By Theorem 3.3.6(a) we have

$$\lambda_1(\mathbf{x}_1 \cdot \mathbf{x}_2) = \lambda_1 \mathbf{x}_1 \cdot \mathbf{x}_2$$
$$= A\mathbf{x}_1 \cdot \mathbf{x}_2$$
$$= \mathbf{x}_1 \cdot A^t \mathbf{x}_2$$
$$= \mathbf{x}_1 \cdot A\mathbf{x}_2$$
$$= \mathbf{x}_1 \cdot \lambda_2 \mathbf{x}_2$$
$$= \lambda_2(\mathbf{x}_1 \cdot \mathbf{x}_2)$$

So

$$(\lambda_1 - \lambda_2)(\mathbf{x}_1 \cdot \mathbf{x}_2) = 0$$

Because $\lambda_1 \neq \lambda_2$, we have that $\mathbf{x}_1 \cdot \mathbf{x}_2 = 0$, and hence $\mathbf{x}_1$ and $\mathbf{x}_2$ are orthogonal. ∎

From Theorem 7.2.5 we see that if all the eigenvalues of a symmetric matrix are distinct, then any basis of eigenvectors is orthogonal. Suppose, however, that there are eigenvalues of multiplicity greater than 1. In this case we can apply the Gram–Schmidt process to find an orthogonal basis of the corresponding eigenspace. The next example illustrates the technique.

Example 4   Find an orthonormal basis of eigenvectors of the matrix A given below. Then use this basis to construct an orthogonal matrix P such that $P^t A P$ is a diagonal matrix.

$$A = \begin{bmatrix} 4 & 2 & 2 \\ 2 & 4 & 2 \\ 2 & 2 & 4 \end{bmatrix}$$

Because A is a symmetric matrix, we know that A is diagonalizable. The characteristic polynomial of A is

$$f(x) = (x - 2)^2(x - 8)$$

It can be shown that the vectors $(-1, 1, 0)$ and $(-1, 0, 1)$ are eigenvectors corresponding to the eigenvalue 2. Because they are not orthogonal, we can use the Gram–Schmidt process to find an orthogonal basis of $E(2)$. Let $\mathbf{x}_1 = (-1, 1, 0)$ and $\mathbf{x}_2 = (-1, 0, 1) - (\frac{1}{2})(-1, 1, 0) = (-\frac{1}{2}, -\frac{1}{2}, 1) = (-\frac{1}{2})(1, 1, -2)$. We obtain the orthonormal basis of $E(2)$,

$$S = \left\{ \left(-\frac{1}{\sqrt{2}}, \frac{1}{\sqrt{2}}, 0 \right), \left(\frac{1}{\sqrt{6}}, \frac{1}{\sqrt{6}}, -\frac{2}{\sqrt{6}} \right) \right\}$$

By Theorem 7.2.5 any eigenvector corresponding to the eigenvalue 8 is orthogonal to the vectors in S. Because $(1, 1, 1)$ is such an eigenvector, we let

$$S' = \left(-\frac{1}{\sqrt{2}}, \frac{1}{\sqrt{2}}, 0 \right), \left(\frac{1}{\sqrt{6}}, \frac{1}{\sqrt{6}}, -\frac{2}{\sqrt{6}} \right), \left(\frac{1}{\sqrt{3}}, \frac{1}{\sqrt{3}}, \frac{1}{\sqrt{3}} \right)$$

Since S' is an orthonormal basis of eigenvectors, the matrix P whose columns are the vectors of S' is the desired orthogonal matrix. In fact, we have

$$\begin{bmatrix} -\dfrac{1}{\sqrt{2}} & \dfrac{1}{\sqrt{2}} & 0 \\ \dfrac{1}{\sqrt{6}} & \dfrac{1}{\sqrt{6}} & -\dfrac{2}{\sqrt{6}} \\ \dfrac{1}{\sqrt{3}} & \dfrac{1}{\sqrt{3}} & -\dfrac{1}{\sqrt{3}} \end{bmatrix} \begin{bmatrix} 4 & 2 & 2 \\ 2 & 4 & 2 \\ 2 & 2 & 4 \end{bmatrix} \begin{bmatrix} -\dfrac{1}{\sqrt{2}} & \dfrac{1}{\sqrt{6}} & \dfrac{1}{\sqrt{3}} \\ \dfrac{1}{\sqrt{2}} & \dfrac{1}{\sqrt{6}} & \dfrac{1}{\sqrt{3}} \\ 0 & -\dfrac{2}{\sqrt{6}} & \dfrac{1}{\sqrt{3}} \end{bmatrix}$$

$$= \begin{bmatrix} 2 & 0 & 0 \\ 0 & 2 & 0 \\ 0 & 0 & 8 \end{bmatrix}$$

For an arbitrary $n \times n$ symmetric matrix, it may be necessary to apply the Gram–Schmidt process as many times as there are eigenvalues of multiplicity greater than 1. By Theorem 7.2.5 the union of all the resulting orthogonal bases will be an orthogonal basis of R^n.

Application to Conic Sections

The conic sections suggest the following curves in R^2: the circle, ellipse, parabola, and the hyperbola. The algebraic form of all of these curves may be obtained from the equation

$$ax^2 + 2bxy + cy^2 + dx + ey + f = 0 \tag{1}$$

by making various choices for the coefficients. The coefficient $2b$ is used for computational convenience. For example, $a = c = 1$, $b = d = e = 0$, and $f = -9$ yields the equation

$$x^2 + y^2 = 9$$

which represents the circle of radius 3 with center at the origin. For a circle with center other than the origin, at least one of d or e is not equal to zero. For ellipses with the major axis not parallel to one of the coordinate axes, we would have b unequal to 0. This type of ellipse would be difficult to sketch from the equation without a change of variable. We will use our knowledge of orthogonal matrices and coordinate vectors to show how to draw graphs of these curves when b is unequal to zero.

We shall consider the *associated quadratic form* to equation (1) given below.

$$ax^2 + 2bxy + cy^2 \tag{2}$$

If we let

$$A = \begin{bmatrix} a & b \\ b & c \end{bmatrix} \quad \text{and} \quad \mathbf{x} = \begin{bmatrix} x \\ y \end{bmatrix}$$

then (2) may be rewritten as $\mathbf{x}^t A \mathbf{x}$. For example, the form $3x^2 + 4xy + 6y^2$ may be written as

$$(x, y) \begin{bmatrix} 3 & 2 \\ 2 & 6 \end{bmatrix} \begin{bmatrix} x \\ y \end{bmatrix}$$

Because A is symmetric, we may use Theorem 7.2.3 to obtain an orthogonal matrix P and a diagonal matrix $D = \text{diag}(\lambda_1, \lambda_2)$ such that $P^t A P = D$. Now let

$$\mathbf{x}' = \begin{bmatrix} x' \\ y' \end{bmatrix} = P^t \mathbf{x}$$

By Theorem 4.1.1 and the fact that $P^{-1} = P^t$, we have that $\mathbf{x}'$ is the coordinate vector of $\mathbf{x}$ relative to the basis $S = \{{}^1P, {}^2P\}$ of R^2. Furthermore, $\mathbf{x} = P\mathbf{x}'$, and hence

$$ax^2 + 2bxy + cy^2 = \mathbf{x}^t A \mathbf{x}$$
$$= (P\mathbf{x}')^t A (P\mathbf{x}')$$
$$= \mathbf{x}'^t P^t A P \mathbf{x}'$$
$$= \mathbf{x}'^t D \mathbf{x}'$$
$$= \lambda_1 x'^2 + \lambda_2 y'^2$$

Thus, the change of variables to $\mathbf{x}'$ allows us to rewrite the associated quadratic form without the xy-term.

To see how this works in practice, consider the equation

$$2x^2 - 4xy + 5y^2 - 36 = 0$$

The associated quadratic form is given by $2x^2 - 4xy + 5y^2$. We let

$$A = \begin{bmatrix} 2 & -2 \\ -2 & 5 \end{bmatrix}$$

The eigenvalues of A are 1 and 6 with corresponding eigenvectors $(1/\sqrt{5})(2, 1)$ and $(1/\sqrt{5})(-1, 2)$. We let

$$P = \begin{bmatrix} \dfrac{2}{\sqrt{5}} & -\dfrac{1}{\sqrt{5}} \\ \dfrac{1}{\sqrt{5}} & \dfrac{2}{\sqrt{5}} \end{bmatrix} \quad \text{and} \quad \mathbf{x}' = P^t \mathbf{x}$$

Then, as earlier, we have that

$$2x^2 - 4xy + 5y^2 = x'^2 + 6y'^2$$

Thus, the original equation becomes

$$x'^2 + 6y'^2 - 36 = 0$$

or

$$\frac{x'^2}{36} + \frac{y'^2}{6} = 1$$

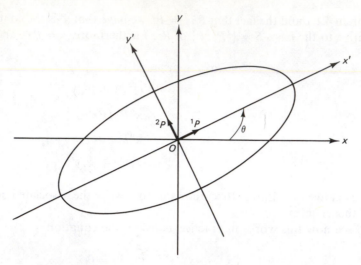

Figure 7.2.1

where x' and y' are the coordinates relative to

$$S = \left\{ \left(\frac{2}{\sqrt{5}}, \frac{1}{\sqrt{5}} \right), \left(-\frac{1}{\sqrt{5}}, \frac{2}{\sqrt{5}} \right) \right\}$$

We see that the equation represents an ellipse. To sketch the ellipse, we draw the x'-axis in the direction of $(2/\sqrt{5}, 1/\sqrt{5})$ and the y'-axis in the direction of $(-1/\sqrt{5}, 2/\sqrt{5})$ (see Figure 7.2.1).

A careful study of the matrix P indicates that it is a rotation matrix which rotates a vector by an angle $\theta = \cos^{-1}(2/\sqrt{5})$ (approximately 26.6°). Therefore, since $^1P = P\mathbf{e}_1$ and $^2P = P\mathbf{e}_2$, we see that the new x'- and y'-axes can be drawn by rotating the original x- and y-axes by θ.

Although not every orthogonal matrix is a rotation matrix, we can always choose the matrix P so that it is a rotation matrix, as in the example above. We state the following theorem without proof.

Theorem 7.2.6 *A 2×2 orthogonal matrix P is a rotation matrix if and only if* $\det P = 1$,

If the original choice of P does not have determinant equal to 1, then it has determinant equal to -1 (see Exercise 12). In the latter case, we need only multiply one of the columns of P by -1. The new matrix is an orthogonal matrix which is also a rotation matrix.

Positive Definite Matrices (Optional)

In many of the applications that have been introduced thus far, we have observed the significant role played by the eigenvalues of a matrix. A particularly important case occurs when the matrix is symmetric (and hence diagonalizable by Theorem 7.2.3) and the eigenvalues are all positive. This type of matrix arises in the discussion of numerical techniques of Chapter 9. We begin with a definition.

Definition Let A be a symmetric matrix. We say that A is ***positive definite*** if $A\mathbf{x} \cdot \mathbf{x} > 0$ for every nonzero vector $\mathbf{x}$.

The inequality in the definition may be rewritten as

$$\mathbf{x}^t A \mathbf{x} > 0$$

for every nonzero vector $\mathbf{x}$ (where $\mathbf{x}$ is represented as a column vector). An equivalent condition which is usually easier to work with is given in Theorem 7.2.7.

Theorem 7.2.7 *Let A be a symmetric matrix. Then A is positive definite if and only if all the eigenvalues of A are positive.*

Proof

First, suppose that A is positive definite and that λ is an eigenvalue of A with corresponding eigenvector $\mathbf{x}$. Then

$$0 < A\mathbf{x} \cdot \mathbf{x} = \lambda \mathbf{x} \cdot \mathbf{x} = \lambda \|\mathbf{x}\|^2$$

Because $\mathbf{x} \neq 0$, we can conclude that $\lambda > 0$.

Now suppose that A is an $n \times n$ matrix and that all of the eigenvalues of A are positive. By Theorem 7.2.3 we may write

$$A = PDP^t$$

where P is an orthogonal matrix and D is the diagonal matrix whose diagonal entries consist of the eigenvalues of A, say, $D = \text{diag}\,(\lambda_1, \ldots, \lambda_n)$. For $\mathbf{x} \neq \mathbf{0}$, we have

$$\begin{aligned}
A\mathbf{x} \cdot \mathbf{x} &= (A\mathbf{x})^t \mathbf{x} \\
&= \mathbf{x}^t A \mathbf{x} \\
&= \mathbf{x}^t (PDP^t)\mathbf{x} \\
&= (P^t\mathbf{x})^t D(P^t\mathbf{x})
\end{aligned}$$

Now let $\mathbf{y} = P^t\mathbf{x}$. Then $\mathbf{y} \neq \mathbf{0}$ because P^t is invertible and $\mathbf{x} \neq \mathbf{0}$. So

$$A\mathbf{x} \cdot \mathbf{x} = \mathbf{y}^t D\mathbf{y} = \sum_{i=1}^{n} \lambda_i y_i^2 > 0$$

where $\mathbf{y} = (y_1, \ldots, y_n)$. Thus, A is positive definite. ∎

Corollary 7.2.8

Let A be positive definite. Then A has a positive definite square root B; that is, $A = B^2$.

We leave the proof of the corollary as an exercise. An alternative proof of Theorem 7.2.7 uses the fact that the eigenvectors of A form an orthonormal basis of R^n. Manipulations with dot products permitted by Theorem 7.1.4 then allow us to conclude that the scalar $A\mathbf{x} \cdot \mathbf{x}$ is positive for any nonzero vector $\mathbf{x}$. The details of this proof are left as an exercise.

Exercises

For each of the matrices A in Exercises 1–8, find an orthogonal matrix P and a diagonal matrix D such that $P^t A P = D$.

1. $\begin{bmatrix} 3 & 1 \\ 1 & 3 \end{bmatrix}$

2. $\begin{bmatrix} 3 & 4 \\ 4 & -3 \end{bmatrix}$

3. $\begin{bmatrix} 1 & 2 \\ 2 & 1 \end{bmatrix}$

4. $\begin{bmatrix} 1 & -1 \\ -1 & 1 \end{bmatrix}$

5. $\begin{bmatrix} 3 & 2 & 2 \\ 2 & 2 & 0 \\ 2 & 0 & 4 \end{bmatrix}$

6. $\begin{bmatrix} 0 & 2 & 2 \\ 2 & 0 & 2 \\ 2 & 2 & 0 \end{bmatrix}$

7. $\begin{bmatrix} -1 & 0 & 0 \\ 0 & 0 & 2 \\ 0 & 2 & 3 \end{bmatrix}$

8. $\begin{bmatrix} -2 & 0 & -36 \\ 0 & -3 & 0 \\ -36 & 0 & -23 \end{bmatrix}$

9. Find an orthogonal matrix whose first row is:
   (a) $(1/\sqrt{5}, 2/\sqrt{5})$
   (b) $(2/\sqrt{29}, -5/\sqrt{29})$

10. Find an orthogonal matrix whose first row is:
   (a) $(-4/5, 0, 3/5)$
   (b) $(-2/3, 1/3, 2/3)$

11. Determine which of the following linear operators on R^2 are orthogonal transformations.
   (a) The reflection about the y-axis.
   (b) The reflection about the line $y = -4x$.
   (c) The projection on the line $y = 3x$.
   (d) The rotation by $45°$.
   (e) The shear transformation defined by $T(x, y) = (x, 3y)$.
   (f) The identity operator.
   (g) The operator defined by $T(\mathbf{x}) = 3\mathbf{x}$.
   (h) The operator defined by $T(x, y) = (2x, y/2)$.
   (i) The operator defined by $T(x, y) = (x + y, x - y)$.

12. Prove that if P is an orthogonal matrix, then $\det P = \pm 1$.

13. Suppose that P and Q are orthogonal matrices of the same size. Prove or give a counter-example that the following matrices are orthogonal.
   (a) PQ
   (b) P^{-1}
   (c) $P + Q$
   (d) P^2

14. Prove that if λ is an eigenvalue of an orthogonal transformation, then $\lambda = \pm 1$.

15. Prove Theorem 7.2.2.

16. Suppose that A is a symmetric matrix and that B is orthogonally equivalent to A. Prove that B is symmetric.

17. Let A be an $n \times n$ symmetric matrix. Suppose that $\mathbf{x}_1, \ldots, \mathbf{x}_n$ are orthonormal eigenvectors of A (considered as column vectors) and $\lambda_1, \ldots, \lambda_n$ are the corresponding eigenvalues. Prove that

$$A = \sum_{k=1}^{n} \lambda_k \mathbf{x}_k \mathbf{x}_k^t$$

   (*Hint:* Write $A = PDP^t$ and use Exercise 19 of Section 2.1).

18. Suppose that $T: R^n \to R^n$ is an orthogonal transformation. Prove that T is one-to-one and onto.

19. Suppose that $T: R^n \to R^n$ is an orthogonal transformation. Prove that if S is an orthonormal basis of R^n, then $T(S)$ is an orthonormal basis of R^n.

20. Let T be a linear operator on R^n. Suppose that for some orthonormal basis S of R^n it is true that $T(S)$ is also an orthonormal basis of R^n. Prove that T is an orthogonal transformation.

21. Prove that if a matrix is both orthogonal and upper triangular, then it must be a diagonal matrix.

22. Suppose that A is an $n \times n$ symmetric matrix and that $A\mathbf{x} \cdot \mathbf{x} = 0$ for all vectors $\mathbf{x}$ in R^n. Prove that A is the zero matrix.

Conic Sections

23. Consider the curve given by

$$2x^2 + 2xy + 2y^2 = 1$$

(a) Determine the symmetric matrix A such that the associated quadratic form may be written as $\mathbf{x}^t A \mathbf{x}$.
(b) Determine the orthogonal matrix P so that the change of variable $\mathbf{x}' = P^t \mathbf{x}$ allows us to rewrite the associated quadratic form without the xy-term.
(c) Use the previous parts to write the equation of the curve in new coordinates without the xy-term. Identify the curve.
(d) Draw the new axes and sketch the curve in the new coordinates.

24. Consider the curve given by

$$x^2 - 12xy - 4y^2 = 40$$

(a) Determine the symmetric matrix A such that the associated quadratic form may be written as $\mathbf{x}^t A \mathbf{x}$.
(b) Determine the orthogonal matrix P so that the change of variable $\mathbf{x}' = P^t \mathbf{x}$ allows us to rewrite the associated quadratic form without the xy-term.
(c) Use the previous parts to write the equation of the curve in new coordinates without the xy-term. Identify the curve.
(d) Draw the new axes and sketch the curve in the new coordinates.

Positive Definite Matrices

25. Determine which of the following matrices are positive definite.

(a) $\begin{bmatrix} 2 & 4 \\ 3 & 5 \end{bmatrix}$ (b) $\begin{bmatrix} 4 & 0 \\ 0 & -3 \end{bmatrix}$

(c) $\begin{bmatrix} 3 & 0 \\ 0 & 7 \end{bmatrix}$ (d) $\begin{bmatrix} 5 & 2 \\ 2 & 6 \end{bmatrix}$

(e) $\begin{bmatrix} -3 & 5 \\ 5 & 10 \end{bmatrix}$ (f) $\begin{bmatrix} 5 & 0 \\ 0 & 0 \end{bmatrix}$

26. Suppose that A is an $n \times n$ symmetric matrix. Prove that

$$\sum_{i,j} A_{ij} x_i x_j > 0$$

for all nonzero vectors $(x_1, \ldots, x_n)$ if and only if A is a positive definite matrix.

27. Prove that the determinant of a positive definite matrix is positive.

28. Suppose that A is a positive definite matrix and that B is orthogonally equivalent to A. Prove that B is positive definite.

29. Prove that if A is a positive definite matrix, then A^m is positive definite for every positive integer m.

30. Suppose that A is a positive definite matrix and that m is a positive integer. Prove that there exists a positive definite matrix B such that $B^m = A$.

31. Prove Corollary 7.2.8.

32. Prove that if a matrix A is positive definite, then all of its diagonal entries are positive. (*Hint:* Consider the product $\mathbf{e}_i^t A \mathbf{e}_i$ for each standard vector $\mathbf{e}_i$.)

33. Show that the converse of Exercise 32 is not true.

34. Suppose that A is a symmetric $n \times n$ matrix with orthonormal eigenvectors $\mathbf{x}_1, \ldots, \mathbf{x}_n$ and corresponding eigenvalues $\lambda_1, \ldots, \lambda_n$.
(a) Prove that if

$$\mathbf{x} = \sum_{i=1}^{n} a_i \mathbf{x}_i$$

then

$$A\mathbf{x} \cdot \mathbf{x} = \sum_{i=1}^{n} a_i^2 \lambda_i$$

(b) Use part (a) to show that if the λ_i's are positive, then A is positive definite.

7.3 ROTATION MATRICES IN R^3 AND COMPUTER GRAPHICS

In this section we consider rotations of R^3 about a line. In particular, we examine the rotations of R^3 about the x-, y-, and z-axes. We discuss how to compute these rotations and describe how these computations can be used for the graphical representation of three-dimensional objects.

We have seen that left multiplication by the orthogonal matrix of Example 2 of Section 7.2 corresponds to the linear operator which rotates a vector by an angle θ about the z-axis. We shall denote this matrix by P_θ. Therefore, we can write

$$P_\theta = \begin{bmatrix} \cos\theta & -\sin\theta & 0 \\ \sin\theta & \cos\theta & 0 \\ 0 & 0 & 1 \end{bmatrix}$$

It can be shown that the rotation of R^3 about any line which passes through the origin can be described in terms of left-multiplication by a 3×3 orthogonal matrix [see page 420 of (1) in the References]. Such matrices are called ***rotation matrices***.

By arguments similar to those in Example 2, we can produce these matrices for rotations about the x-axis and the y-axis in R^3. If Q_θ and R_θ are the matrices for rotations by an angle θ about the x-axis and the y-axis, respectively, then

$$Q_\theta = \begin{bmatrix} 1 & 0 & 0 \\ 0 & \cos\theta & -\sin\theta \\ 0 & \sin\theta & \cos\theta \end{bmatrix} \quad \text{and} \quad R_\theta = \begin{bmatrix} \cos\theta & 0 & \sin\theta \\ 0 & 1 & 0 \\ -\sin\theta & 0 & \cos\theta \end{bmatrix}$$

In each case we use the common convention that the positive direction of the rotation is counterclockwise as viewed from the positive direction of the axis of rotation (see Figure 7.3.1).

We can compose rotations by taking products of rotation matrices. For example, if a vector $\mathbf{x}$ is rotated about the z-axis by an angle θ and then the result is

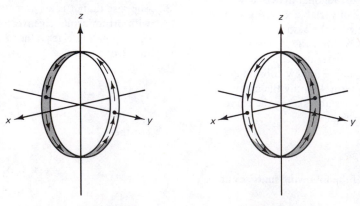

(a) Rotation about the x-axis   (b) Rotation about the y-axis   **Figure 7.3.1** rotated about the y-axis by an angle ϕ, the final position of the point is $R_\phi(P_\theta \mathbf{x}) = (R_\phi P_\theta)\mathbf{x}$. The effect of these rotations is depicted in Figure 7.3.2.

One should be careful to note that the order in which the rotations are made is important. That is, $R_\phi P_\theta$ is not usually equal to $P_\theta R_\phi$. To continue with this example, suppose that $\mathbf{x} = (1, 0, 0)$, $\theta = 30°$, and $\phi = 45°$. Then

$$R_\phi P_\theta \mathbf{x} = \begin{bmatrix} \dfrac{1}{\sqrt{2}} & 0 & \dfrac{1}{\sqrt{2}} \\[2ex] 0 & 1 & 0 \\[2ex] -\dfrac{1}{\sqrt{2}} & 0 & \dfrac{1}{\sqrt{2}} \end{bmatrix} \begin{bmatrix} \dfrac{\sqrt{3}}{2} & -\dfrac{1}{2} & 0 \\[2ex] \dfrac{1}{2} & \dfrac{\sqrt{3}}{2} & 0 \\[2ex] 0 & 0 & 1 \end{bmatrix} \begin{bmatrix} 1 \\[2ex] 0 \\[2ex] 0 \end{bmatrix}$$

$$= \begin{bmatrix} \dfrac{\sqrt{3}}{2\sqrt{2}} & -\dfrac{1}{2\sqrt{2}} & \dfrac{1}{\sqrt{2}} \\[2ex] \dfrac{1}{2} & \dfrac{\sqrt{3}}{2} & 0 \\[2ex] -\dfrac{\sqrt{3}}{2\sqrt{2}} & \dfrac{1}{2\sqrt{2}} & \dfrac{1}{\sqrt{2}} \end{bmatrix} \begin{bmatrix} 1 \\[2ex] 0 \\[2ex] 0 \end{bmatrix}$$

$$= \begin{bmatrix} \dfrac{\sqrt{3}}{2\sqrt{2}} \\[2ex] \dfrac{1}{2} \\[2ex] -\dfrac{\sqrt{3}}{2\sqrt{2}} \end{bmatrix}$$

Rotation matrices are important in computer graphics, because they can be used to compute various orientations of the same three-dimensional shape. Although computers can store the information necessary to construct many kinds of three-dimensional shapes, these shapes must be represented graphically on a two-dimensional surface such as a CRT (cathode ray tube) or a sheet of paper. From a mathematical viewpoint, an orientation of such a shape is projected on a plane. For example, the shape can be projected on the yz-plane by simply ignoring the first coordinates of points which constitute the shape, and plotting only second and third coordinates. To get different views, the shape is rotated in various ways before the projection is made. To illustrate the results of these procedures, a computer program

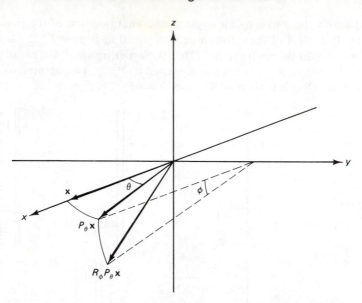

Figure 7.3.2

which creates three-dimensional shapes consisting of points connected by lines was written for a microcomputer. The coordinates of these points (vertices) and the information about which of the points are connected by lines (edges) are used as data in the program. The program plots the projection of the resulting shape on the yz-plane and represents the results as a printout on a dot matrix printer. Before making such a plot, the computer can rotate the shape about any one or combination of the three coordinate axes. For each such rotation, the computer multiplies the vertices of the shape by the appropriate rotation matrix. In the following illustration we use a crude rendering of a space shuttle. The first figure (Figure 7.3.3) is the projection of the shuttle onto the yz-plane without any rotations. For each of the subsequent figures, the shuttle is rotated about one or two axes before being projected. For Figure 7.3.4, the shuttle is rotated by $90°$ about the y-axis. For Figure 7.3.5 the shuttle is first rotated by $-30°$ about the z-axis, and then rotated by $20°$ about the y-axis. For Figure 7.3.6 the shuttle is first rotated by $45°$ about the x-axis, and then by $30°$ about the y-axis.

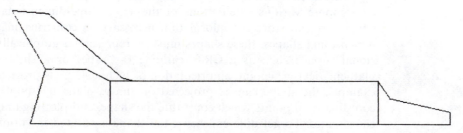

Figure 7.3.3

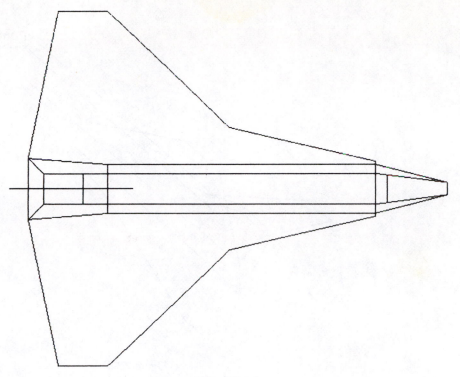

Figure 7.3.4

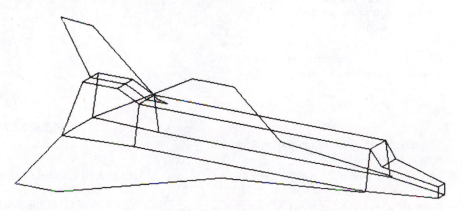

Figure 7.3.5

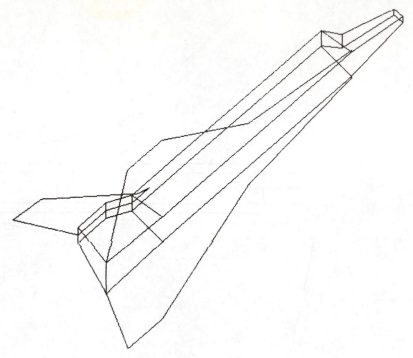

Figure 7.3.6

Exercises

1. Compute $P_\theta R_\phi$ for $\theta = 30°$ and $\phi = 45°$. Compare your result to the matrix $R_\phi P_\theta$ computed in this section.

2. (a) For any angles θ and ϕ, show that
   $P_{\theta + \phi} = P_\theta P_\phi$ and $P_{-\theta} = (P_\theta)^{-1}$.
   (b) State and prove a similar result for rotations about the x-axis and the y-axis.

3. Define the map $U: R^3 \to R^2$ by $U(x, y, z) = (y, z)$. This is the projection of R^3 on the yz-plane. Show that for any angle θ,

$$UP_\theta = A_\theta U$$

   where A_θ is the rotation matrix for R^2 which rotates by an angle θ.

4. Let L be any line in R^3 which passes through the origin. For any angle θ define the mapping $T = U_\theta: R^3 \to R^3$ as follows: For $\mathbf{x}$ in R^3, let $U_\theta\mathbf{x}$ be the result of rotating $\mathbf{x}$

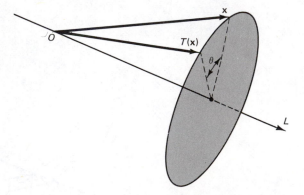

Figure 7.3.7

about L by an angle θ (see Figure 7.3.7). It can be shown that U_θ is a linear operator [see page 420 of (1) in the References]. Let $\mathbf{x}_1$ be any unit vector which is orthogonal to L. For

$\theta = 90°$, let $\mathbf{x}_2 = U_\theta \mathbf{x}_1$. Let $\mathbf{x}_3$ be a unit vector which is parallel to L.

(a) Use a geometric argument to show that $S = \{\mathbf{x}_1, \mathbf{x}_2, \mathbf{x}_3\}$ is an orthonormal basis for R^3.

(b) Show that $[U_\theta]_S$ is the same as the matrix P_θ.

5. Let U_θ, $\mathbf{x}_1$, $\mathbf{x}_2$, and $\mathbf{x}_3$ be as in Exercise 4.

(a) Show that $\mathbf{x}_3$ is an eigenvector of U_θ corresponding to the eigenvalue $\lambda = 1$.

(b) Show that L is the eigenspace for the eigenvalue $\lambda = 1$.

(c) Determine whether or not U_θ is diagonalizable.

*6. (a) Use the program MATRIX to approximate a 3×3 matrix A such that for any vector $\mathbf{x}$ in R^3, $A\mathbf{x}$ is the result of first rotating $\mathbf{x}$ by $45°$ about the x-axis, and then rotating the resulting vector by $30°$ about the y-axis. Notice that this is how the shuttle was reoriented in Figure 7.3.7.

(b) Apply A to the vector $(1, 1, 1)$, and each of the standard vectors in R^3.

* This problem should be solved by using one of the programs noted in Appendix B.

Key Words

Introduction to Abstract Vector Spaces and Linear Transformations

Up to now we have thought of linear algebra in terms of the study of vectors in R^n. However, there are other important mathematical systems which share many of the formal properties of R^n. Some years ago it became apparent to mathematicians that the methods and techniques used in the study of vectors could be adapted to a large variety of other mathematical systems. To accommodate these systems, the theory of vectors was extended or "abstracted." Many of the more important systems developed from the study of analysis. Consequently, a number of the examples in Chapter 8 require a background in calculus for a complete understanding.

In the abstract theory, a vector space is defined as any mathematical system that satisfies certain properties or axioms. The general theorems about vector spaces are then deduced from these axioms. Once it is shown that a system satisfies the axioms, it follows immediately that all of these theorems apply.

Most of the important properties of R^n can be deduced from Theorem 1.2.1. Because of this, the items that constitute this theorem are taken as the axioms for abstract vector spaces.

8.1 ELEMENTARY PROPERTIES OF VECTOR SPACES

We begin with the formal definition of vector space. Compare this definition with Theorem 1.2.1.

Definition A (*real*) *vector space* V is a set on which two operations called *vector addition* and *scalar multiplication* are defined so that for any elements $\mathbf{x}$ and $\mathbf{y}$ in V and any scalar (real number) a, the sum $\mathbf{x} + \mathbf{y}$ and the scalar product $a\mathbf{x}$ are uniquely defined elements of V, and such that the following axioms hold:

AXIOMS OF AN ABSTRACT VECTOR SPACE

1. $\mathbf{x} + \mathbf{y} = \mathbf{y} + \mathbf{x}$ for any $\mathbf{x}$ and $\mathbf{y}$ in V.
2. $(\mathbf{x} + \mathbf{y}) + \mathbf{z} = \mathbf{x} + (\mathbf{y} + \mathbf{z})$ for any $\mathbf{x}$, $\mathbf{y}$, and $\mathbf{z}$ in V.
3. There exists an element $\mathbf{0}$ in V such that $\mathbf{x} + \mathbf{0} = \mathbf{x}$ for any $\mathbf{x}$ in V.
4. For each element $\mathbf{x}$ in V there exists an element $-\mathbf{x}$ such that $\mathbf{x} + -\mathbf{x} = \mathbf{0}$.
5. $1\mathbf{x} = \mathbf{x}$ for any $\mathbf{x}$ in V.
6. $(ab)\mathbf{x} = a(b\mathbf{x})$ for any scalars a and b and any $\mathbf{x}$ in V.
7. $a(\mathbf{x} + \mathbf{y}) = a\mathbf{x} + a\mathbf{y}$ for any scalar a and any $\mathbf{x}$ and $\mathbf{y}$ in V.
8. $(a + b)\mathbf{x} = a\mathbf{x} + b\mathbf{x}$ for any scalars a and b and any $\mathbf{x}$ in V.

The elements of a vector space are called *vectors*. The vector $\mathbf{0}$ of axioms (3) and (4) is called the *zero vector*. We shall show that it is unique; that is, there cannot be two distinct vectors in a vector space which both satisfy axiom (3). For any vector $\mathbf{x}$ of a vector space V the vector $-\mathbf{x}$ of axiom (4) is called the *additive inverse* of $\mathbf{x}$. We shall see that every vector has a unique additive inverse.

For any positive integer n, R^n is a vector space under the operations of addition and scalar multiplication defined in Chapter 1 because of Theorem 1.2.1. By virtue of Theorem 1.3.1, any subspace of R^n is also a vector space. These are the examples with which we are already quite familiar.

Function Spaces

One of the most important examples of a vector space is the collection of real-valued functions defined on a set. Collections of functions frequently form vector spaces. Such vector spaces are called (appropriately) *function spaces*. The area of modern mathematics called "functional analysis" is devoted to the study of function spaces.

For any nonempty set S, we let $F(S)$ denote the set of all functions from S to R. For any function f in $F(S)$ and any element x in S, the image $f(x)$ of x under f is a real number. For functions f and g in $F(S)$, we define the *sum* $f + g$ to be the function from S to R which takes any element x in S into the number $f(x) + g(x)$. That is, $(f + g)(x) = f(x) + g(x)$ for all x in S. For any function f in $F(S)$ and any scalar a we define the *scalar product* af to be the function from S to R which takes any element x in S into the number $af(x)$. That is, $(af)(x) = af(x)$ for all x in S.

As a specific example of these definitions, suppose that $S = R$, $f(x) = x^2$, and $g(x) = 2x + 1$. Then $(f + g)(x) = x^2 + 2x + 1$, and $(3f)(x) = 3x^2$.

Before verifying that $F(S)$ is a vector space, we must define $\mathbf{0}$ and $-f$ for any function f. Let $\mathbf{0}$ denote the function which maps every element of S into the number 0. That is, $\mathbf{0}(x) = 0$ for all x in S. For any function f in $F(S)$, the function $-f$ is defined by $(-f)(x) = -f(x)$ for all x in S.

Theorem 8.1.1   *For any nonempty set S, $F(S)$ with the operations defined above forms a vector space.*

Proof

Because the sum of two functions or the product of a scalar and a function are functions, these two operations are defined on $F(S)$. To prove the theorem, we must show that axioms (1) through (8) are satisfied. We shall verify axioms (1), (3), and (7), leaving the rest as exercises. In the context of $F(S)$, each of the axioms is an equation involving functions. Two functions f and g are **equal** if $f(x) = g(x)$ for all x in S. We use this definition to help us verify that the equations are valid.

(1) Let f and g be functions in $F(S)$. Then for any x in S,

$$(f + g)(x) = f(x) + g(x) \quad \text{(definition of sum of functions)}$$

$$= g(x) + f(x) \quad \text{(commutative law for real numbers)}$$

$$= (g + f)(x) \quad \text{(definition of sum of functions)}$$

Therefore, $f + g = g + f$. Thus, axiom (1) is verified.

(3) Let f be any function in $F(S)$. Then for any x in S.

$$(f + \mathbf{0})(x) = f(x) + \mathbf{0}(x) \quad \text{(definition of sum of functions)}$$

$$= f(x) + 0 \quad \text{(definition of } \mathbf{0})$$

$$= f(x)$$

Therefore, $f + \mathbf{0} = f$. Thus, axiom (3) is verified.

(7) Let f and g be functions in $F(S)$, and let a be any scalar. Then for any x in S,

$$(a(f + g))(x) = a(f + g)(x) \quad \text{(definition of scalar product)}$$

$$= a(f(x) + g(x)) \quad \text{(definition of sum of functions)}$$

$$= af(x) + ag(x) \quad \text{(distributive law for real numbers)}$$

$$= (af)(x) + (ag)(x) \quad \text{(definition of scalar product)}$$

$$= (af + ag)(x) \quad \text{(definition of sum of functions)}$$

Therefore, $a(f + g) = af + ag$. This verifies axiom (7).   ■

Other Examples of Vector Spaces

Example 1 For fixed positive integers m and n, let $M_{m \times n}$ denote the set of all $m \times n$ matrices. It can be verified that $M_{m \times n}$ is a vector space under the operations of matrix addition and multiplication of a matrix by a scalar as defined in Section 2.1. The $m \times n$ zero matrix acts as the zero "vector," and for any $m \times n$ matrix A, we have that $(-1)A$ is the additive inverse $-A$. Axioms (1), (2), (3), and (7) follow from Theorem 2.1.1. The verifications of the other axioms are left as exercises.

Example 2 Let $L(R^n, R^m)$ denote the set of all linear transformations from R^n to R^m. It can be shown that $L(R^n, R^m)$ is a vector space under the operations of addition of linear transformations and the product of a linear transformation by a scalar as defined in Section 2.4. The zero transformation, T_0, plays the role of the zero vector, and for any transformation T, we have that $(-1)T$ is the additive inverse, $-T$.

The proof that $L(R^n, R^m)$ is a vector space is very similar to the proof that $F(S)$ is a vector space because in each case we are dealing with a set of functions. For this reason it is left as an exercise.

Example 3 Let P denote the set of all polynomials with real coefficients. For example, the polynomial $p(x) = x^2 + 3x - 2$ is a polynomial with real coefficients 1, 3, and -2. For any polynomials $p(x) = a_0 + a_1 x + \cdots + a_n x^n$ and $q(x) = b_0 + b_1 x + \cdots + b_n x^n$, and for any scalar c, define the sum $p(x) + q(x)$ the product $cp(x)$, and the additive inverse $-p(x)$ by

$$p(x) + q(x) = (a_0 + b_0) + (a_1 + b_1)x + \cdots + (a_n + b_n)x^n$$

$$cp(x) = ca_0 + ca_1 x + \cdots + ca_n x^n$$

and $-p(x) = (-1)p(x)$. It is a simple matter to verify that with respect to these operations, P is a vector space. The zero polynomial serves the role as the zero vector. We omit the details.

Elementary Properties of Vector Spaces

The following results are deduced entirely from the axioms given in the definition of vector space.

Theorem 8.1.2 (*Right cancellation law of addition*) *Let V be a vector space, and let $\mathbf{x}$, $\mathbf{y}$, and $\mathbf{z}$ be vectors in V such that $\mathbf{x} + \mathbf{y} = \mathbf{z} + \mathbf{y}$. Then $\mathbf{x} = \mathbf{z}$.*

Proof

Suppose that $\mathbf{x} + \mathbf{y} = \mathbf{z} + \mathbf{y}$. Then $\mathbf{x} = \mathbf{x} + \mathbf{0} = \mathbf{x} + (\mathbf{y} + -\mathbf{y}) = (\mathbf{x} + \mathbf{y}) + -\mathbf{y} = (\mathbf{z} + \mathbf{y}) + -\mathbf{y} = \mathbf{z} + (\mathbf{y} + -\mathbf{y}) = \mathbf{z} + \mathbf{0} = \mathbf{z}$. ∎

Applying axiom (1) to the right cancellation law of addition, we get the left cancellation law of addition. The details are left as an exercise.

Corollary 8.1.3   *(Left cancellation law of addition) Let V be a vector space, and let* $\mathbf{x}$, $\mathbf{y}$, *and* $\mathbf{z}$ *be vectors in V such that* $\mathbf{y} + \mathbf{x} = \mathbf{y} + \mathbf{z}$. *Then* $\mathbf{x} = \mathbf{z}$.

Corollary 8.1.4   *For any vector space V there is only one vector* $\mathbf{0}$ *which satisfies axiom (3).*

Proof
Suppose that there is another zero vector, $\mathbf{0}'$, for the vector space V. Then $\mathbf{0}' = \mathbf{0}' + \mathbf{0} = \mathbf{0}' + \mathbf{0}$. Therefore, by the left cancellation law of addition, $\mathbf{0}' = \mathbf{0}$. ∎

Corollary 8.1.5   *Let V be a vector space. For any vector* $\mathbf{x}$ *in V,* $\mathbf{x}$ *has a unique additive inverse.*

Proof
Suppose that a vector $\mathbf{x}$ has additive inverses $-\mathbf{x}$ and $\mathbf{x}'$. Then $\mathbf{x} + -\mathbf{x} = \mathbf{0}$ and $\mathbf{x} + \mathbf{x}' = \mathbf{0}$. Therefore,

$$\mathbf{x} + -\mathbf{x} = \mathbf{x} + \mathbf{x}'$$

Thus, by the left cancellation law, $-\mathbf{x} = \mathbf{x}'$. ∎

Theorem 8.1.6   *Let V be a vector space. Then for any vector* $\mathbf{x}$ *in V and any scalar a, we have:*
(a) $0\mathbf{x} = \mathbf{0}$
(b) $a\mathbf{0} = \mathbf{0}$
(c) $(-1)\mathbf{x} = -\mathbf{x}$
(d) $(-a)\mathbf{x} = a(-\mathbf{x}) = -(a\mathbf{x})$

Proof
We prove parts (a) and (c), leaving the proofs of parts (b) and (d) as exercises.
(a)

$$0\mathbf{x} + 0\mathbf{x} = (0 + 0)\mathbf{x} \quad \text{[axiom (8)]}$$
$$= 0\mathbf{x} \quad \text{(property of 0)}$$
$$= 0\mathbf{x} + \mathbf{0} \quad \text{[axiom (3)]}$$

Therefore, by the left cancellation law we have that $0\mathbf{x} = \mathbf{0}$.

(c)

$$
\begin{aligned}
\mathbf{x} + (-1)\mathbf{x} &= (1)\mathbf{x} + (-1)\mathbf{x} &&\text{[axiom (5)]} \\
&= (1 + -1)\mathbf{x} &&\text{[axiom (8)]} \\
&= 0\mathbf{x} \\
&= \mathbf{0} &&\text{by part (a)}
\end{aligned}
$$

Therefore, since additive inverses are unique (Corollary 8.1.5), $(-1)\mathbf{x} = -\mathbf{x}$. ■

Complex Vector Spaces

If we modify the definition of a real vector space so that the definition of "scalar" is extended to include all complex numbers, we obtain the definition for a **complex vector space**. For elementary properties of complex numbers, the reader should consult Appendix A.

For example, let C^n be the set of all ordered n-tuples of complex numbers. For $\mathbf{z} = (z_1, z_2, \ldots, z_n)$ and $\mathbf{w} = (w_1, w_2, \ldots, w_n)$ in C^n, and for any complex number c, we define the sum

$$
\mathbf{z} + \mathbf{w} = (z_1 + w_1, z_2 + w_2, \ldots, z_n + w_n)
$$

and the scalar product

$$
c\mathbf{z} = (cz_1, cz_2, \ldots, cz_n)
$$

It is an easy matter to verify that the set C^n with the operations defined above satisfies the eight axioms of a vector space allowing for complex scalars. Thus, C^n forms a complex vector space.

Example 4   Let S be any nonempty set. Let $F(S, C)$ denote the set of all functions from S to C, the complex numbers. Such functions are called **complex-valued functions**. The definitions of addition and scalar multiplication on $F(S, C)$ are identical to the corresponding definitions on $F(S)$ except that scalars for $F(S, C)$ are complex rather than real. It follows that $F(S, C)$ forms a complex vector space under these operations. The proof is identical to the one given for Theorem 8.1.1.

For example, let $S = R$, and let f and g be functions from R to C to be defined by

$$
f(x) = (2x - 1) + i(x + 3)
$$

and

$$
g(x) = x^2 + i(x + 1)
$$

Then

$$(f + g)(x) = (x^2 + 2x - 1) + i(2x + 4)$$

Furthermore, for the complex scalar $1 + i$,

$$((1 + i)f)(x) = (1 + i)((2x - 1) + i(x + 3))$$
$$= (x - 4) + i(3x + 2)$$

In subsequent sections the term "vector space" will apply to both real and complex vector spaces. We will have occasion, however, to distinguish between them.

Exercises

In Exercises 1–5, verify that the set V is a real vector space with respect to the indicated operations.

1. $V = P$ as defined in Example 3.

2. Fix a nonempty set S and a positive integer n. Let V denote the set of all functions from S to R^n. For any functions f and g in V and any scalar s define the sum $f + g$ by $(f + g)(x) = f(x) + g(x)$ for all x in S and the product sf by $(sf)(x) = sf(x)$ for all x in S.

3. Let V be the set of all positive real numbers. Define the operation of vector addition of two elements $\mathbf{x}$ and $\mathbf{y}$ of V (not to be confused with ordinary addition) by $\mathbf{x} + \mathbf{y} = \mathbf{xy}$, the usual product of these two elements as ordinary numbers. Define the product of the scalar s and the element $\mathbf{x}$ of V by $s\mathbf{x} = \mathbf{x}^s$, $\mathbf{x}$ raised to the power s.

4. Let V be the set of all 2×2 matrices of the form

$$\begin{bmatrix} a & 2a \\ b & -b \end{bmatrix}$$

where a and b are any real numbers. Addition and multiplication by scalars are defined in the usual way for matrices.

5. Let V be the set of all functions $f : R \to R$ for which $f(x) = 0$ whenever $x < 0$. Addition of functions and multiplication by scalars are defined as for function spaces.

6. Let $P(C)$ denote the set of all polynomials with complex coefficients with the operations as defined on P as in Example 3 except that the scalars are complex numbers. Show that $P(C)$ is a complex vector space.

7. Verify axioms (2), (4), (5), (6), and (8) for $F(S)$ to complete the proof of Theorem 8.1.1.

8. Verify axioms (4), (5), (6), and (8) for $M_{m \times n}$ in Example 1.

9. Verify the axioms of an abstract vector space for $L(R^n, R^m)$ in Example 2.

10. Prove the left cancellation law of vector addition, Corollary 8.1.3.

11. Prove Theorem 8.1.6(b).

12. Prove Theorem 8.1.6(d).

13. Prove that for any vector $\mathbf{x}$ in a vector space V, $-(-\mathbf{x}) = \mathbf{x}$.

14. Prove that for any vectors $\mathbf{x}$ and $\mathbf{y}$ in a vector space V, $-(\mathbf{x} + \mathbf{y}) = -\mathbf{x} + -\mathbf{y}$.

15. Let $\mathbf{x}$ and $\mathbf{y}$ be vectors in a vector space V, and suppose that $c\mathbf{x} = c\mathbf{y}$ for some scalar $c \neq 0$. Prove that $\mathbf{x} = \mathbf{y}$.

16. Prove that for any vector $\mathbf{x}$ in a vector space V, and any scalar c, if $c\mathbf{x} = \mathbf{0}$, then $c = 0$ or $\mathbf{x} = \mathbf{0}$.

17. Prove that for any vector $\mathbf{x}$ in a vector space V, and any scalar a, $(-a)(-\mathbf{x}) = a\mathbf{x}$.

18. Let V be any complex vector space. Show that if the scalars used in the scalar product are restricted to real numbers, then V becomes a real vector space.

19. Verify the axioms of an abstract vector space for the complex vector space C^n.

8.2 SUBSPACES, BASES, AND DIMENSION

In this section we reexamine the concepts of subspaces, linear combinations, linear dependence and independence, basis, and dimension in the more general context of abstract vector spaces.

A careful examination of Section 1.3 and Chapter 3 will reveal that the proofs of results dealing with these concepts for R^n can, with little or no modification, be adapted to establish the comparable results for abstract vector spaces, complex as well as real. This should come as no surprise because most of these proofs are based on or are consequences of the items listed in Theorem 1.2.1. Remember that these items are the axioms for abstract vector spaces. Consequently, we use some of these results in this chapter.

We begin our study with the concept of "subspace." Since the conventional definition makes use of abstract vector spaces, the definition of subspace given in Section 1.3 is necessarily different, although equivalent to the usual one for subspaces of R^n. We state the standard definition.

Definition   A subset W of a vector space V is called a ***subspace*** of V if W forms a vector space under the same operations that are defined on V.

Suppose that W is a subspace of a vector space V and $\mathbf{x}$ and $\mathbf{y}$ are vectors in W. Then $\mathbf{x} + \mathbf{y}$ is in W because W is a vector space under the operations of V. As in Section 1.3, we say that W is ***closed under vector addition*** defined on V. Similarly, for any scalar c and any vector $\mathbf{x}$ in W, $c\mathbf{x}$ is in W. Thus, we say that W is ***closed under scalar multiplication*** defined on V.

Example 1   Let S be a nonempty set. Pick an element x_0 in S. Let W be the set of all real-valued functions f on S such that $f(x_0) = 0$. Then W is a subset of $F(S)$. For functions f and g in S, $(f + g)(x_0) = f(x_0) + g(x_0) = 0 + 0 = 0$. Therefore, $f + g$ is in W. For f in W and any scalar a, we have that $(af)(x_0) = af(x_0) = a(0) = 0$. Thus, af is in W. We conclude that the operations of addition and scalar multiplication that are defined on V are also defined on W. That is, W is closed under these operations. In order to show that W is a vector space under these operations, we must show that the eight axioms (1) to (8) are satisfied. Notice that all of these but (3) and (4) are satisfied merely because the elements of W are also vectors of V. Axiom (3) is satisfied because the zero function lies in W. Finally, if f lies in W, then $(-1)f$ is in W. But by Theorem 8.1.6, $(-1)f = -f$. Therefore, W satisfies axiom (4). We conclude that W is a vector space, and hence a subspace of V.

The following theorem helps us to simplify the arguments used to establish that a subset of a vector space is a subspace.

Theorem 8.2.1   *Let V be a vector space. A subset W of V is a subspace of V if and only if the following three conditions are satisfied:*
  (a) *The zero vector of V is in W.*
  (b) *W is closed under vector addition.*
  (c) *W is closed under scalar multiplication.*

Proof

Suppose that W is a subspace of V. Since the operations of V are defined on W, it is clear that W is closed under these operations. Hence, we need only prove part (a). Because W is a vector space, it has an additive identity $\mathbf{0}'$. Thus, $\mathbf{0}' + \mathbf{0}' = \mathbf{0}'$. Since $\mathbf{0}'$ is in V and $\mathbf{0}$ is the identity of V, we have that $\mathbf{0}' + \mathbf{0} = \mathbf{0}'$. Therefore, $\mathbf{0}' + \mathbf{0}' = \mathbf{0}' + \mathbf{0}$. It follows from the left cancellation law (Corollary 8.1.3) that $\mathbf{0}' = \mathbf{0}$. This proves part (a).

Conversely, suppose that W satisfies the three conditions. We prove that W is a subspace of V. Because of the second and third conditions, the operations of V are also defined on W. Therefore, we proceed to the eight axioms of the definition of "vector space." As was argued in Example 1, we need only be concerned with axioms (3) and (4). Axiom (3) follows from condition (a). Hence, we need only consider axiom (4). Let $\mathbf{x}$ be any element of W. By Theorem 8.1.6, $-\mathbf{x} = (-1)\mathbf{x}$, which is in W. Therefore, axiom (4) holds. We conclude that W is a vector space, and hence it is a subspace of V. ∎

Example 2   Fix any positive integer n. Let W be the set of all $n \times n$ matrices with trace equal to zero (see Exercise 15 of Section 2.1). We use Theorem 8.2.1 to verify that W is a subspace of $M_{n \times n}$. Since the $n \times n$ zero matrix is in W, we have part (a) of the theorem. Suppose that A and B are matrices in W. Then by Exercise 15 of Section 2.1,

$$\mathrm{tr}(A + B) = \mathrm{tr}(A) + \mathrm{tr}(B)$$

$$= 0 + 0$$

$$= 0$$

Therefore, $A + B$ is in W. This proves part (b). We leave the verification of part (c) as an exercise. Thus, we have that W is a subspace of $M_{n \times n}$.

Example 3   Let W be the set of all continuous real-valued functions on R. Then W is a subset of $F(R)$, the vector space of Theorem 8.1.1. Since the zero function is continuous, the sum of continuous functions is continuous, and the product of a continuous function with a scalar is continuous, we have by Theorem 8.2.1 that W is a subspace of $F(R)$.

Example 4   Recall the vector space P of all polynomials considered in Example 3 of Section 8.1. Let n be a positive integer, and let P_n be the subset of P consisting of all polynomials of degree less than or equal to n. By convention, the zero polynomial lies in this set. It is clear that P_n is closed under the operations of addition and scalar multiplication as defined on P. We conclude that P_n is a subspace of P.

We now proceed to the concept of "linear combination."

Definition   Let $S = \{x_1, x_2, \ldots, x_k\}$ be a finite subset of a vector space V. A vector x in V is a **_linear combination_** of the vectors of S if there exist scalars $t_1, t_2, \ldots, t_k$ such that

$$x = t_1 x_1 + t_2 x_2 + \cdots + t_k x_k$$

Example 5   Let V be the vector space $M_{2 \times 2}$ of 2×2 matrices. Let

$$S = \left\{ \begin{bmatrix} 1 & 2 \\ 1 & 1 \end{bmatrix}, \begin{bmatrix} 1 & -1 \\ 1 & 2 \end{bmatrix} \right\}$$

and

$$A = \begin{bmatrix} 5 & 1 \\ 5 & 8 \end{bmatrix}$$

Then A is a linear combination of the vectors of S because

$$A = 2 \begin{bmatrix} 1 & 2 \\ 1 & 1 \end{bmatrix} + 3 \begin{bmatrix} 1 & -1 \\ 1 & 2 \end{bmatrix}$$

Example 6   Let $S = \{1, 2, 3\}$. For $i = 1, 2,$ and 3, define

$$f_i(k) = \begin{cases} 1 & \text{if } i = k \\ 0 & \text{if } i \neq k \end{cases}$$

For example, $f_2(1) = 0$, $f_2(2) = 1$, and $f_2(3) = 0$. Consider the function $f : S \to R$ defined by $f(1) = 8$, $f(2) = 6$, and $f(3) = 2$. Then f is a linear combination of the functions $f_1, f_2,$ and f_3. In fact,

$$f = 8f_1 + 6f_2 + 2f_3$$

To see this, let $g = 8f_1 + 6f_2 + 2f_3$. To show that $g = f$, we must show that $g(i) = f(i)$ for $i = 1, 2, 3$.

$$g(1) = (8f_1 + 6f_2 + 2f_3)(1)$$
$$= 8f_1(1) + 6f_2(1) + 2f_3(1)$$
$$= 8(1) + 6(0) + 2(0)$$
$$= 8$$
$$= f(1)$$

Similarly, $g(2) = f(2)$ and $g(3) = f(3)$. Therefore, $g = f$.

***

Example 7   Let S be the set of real-valued functions $S = \{\sin x, 1, \sin^2 x, \cos^2 x\}$. Then S is a subset of $F(R)$. Observe that the function $f(x) = \cos 2x$ is a linear combination of the vectors of S because

$$\cos 2x = \cos^2 x - \sin^2 x$$
$$= 0 \sin x + 0(1) + (-1)\sin^2 x + (1)\cos^2 x$$

***

Since any subspace of R^n has a finite basis, we have been able to restrict our study of linear combinations to finite sets. In general, however, it is useful to consider infinite sets. Since sums involve a finite number of vectors, we must formulate such a definition in terms of finite sets.

Definition   Let S be an infinite subset of a vector space V. A vector $\mathbf{x}$ in V is a **linear combination** of the vectors of S if $\mathbf{x}$ is a linear combination of the vectors of some finite subset of S.

***

For example, consider the vector space P of polynomial functions. Let S be the infinite subset

$$S = \{1, x, x^2, \ldots, x^n, \ldots\}$$

Let S' be a finite nonempty subset of S. Then S' is of the form

$$S' = \{x^{n_1}, x^{n_2}, \ldots, x^{n_k}\}$$

and an arbitrary linear combination of the vectors of S' is of the form

$$a_1 x^{n_1} + a_2 x^{n_2} + \cdots + a_k x^{n_k}$$

Since any polynomial is of this form, we have that any polynomial is a linear combination of the vectors of S. To be specific, consider the polynomial $f(x) = 1 + 2x - 3x^5$. Clearly $f(x)$ is a linear combination of the finite subset $S' = \{1, x, x^5\}$.

Having extended the definition of "linear combination" to include infinite sets, we can now define the *span* of an infinite set to be the collection of all linear combinations of the vectors in the set. Thus, in the example above, the span of S is all of P because any polynomial function can be represented as a linear combination of the vectors of S.

The concept of "linear combination" leads quite naturally to the concepts of linear dependence and linear independence. We are familiar with the definitions of these concepts for finite sets.

Definition   A subset S of a vector space V is *linearly dependent* if some finite subset of S is linearly dependent. A subset S is *linearly independent* if S is not linearly dependent, that is, if every finite subset of S is linearly independent.

Notice that if S is a finite set, these definitions reduce to the usual ones.

In the following three examples, we reexamine these concepts as applied to finite sets, but in the context of more general vector spaces.

Example 8   Let $Q = \{f_1, f_2, f_3\}$ be the subset of $F(S)$ as defined in Example 6. We show that Q is linearly independent. Consider any scalars a_1, a_2, and a_3 for which $a_1 f_1 + a_2 f_2 + a_3 f_3 = 0$. Then, in particular,

$$
\begin{aligned}
0 &= (a_1 f_1 + a_2 f_2 + a_3 f_3)(1) \\
&= a_1 f_1(1) + a_2 f_2(1) + a_3 f_3(1) \\
&= a_1(1) + a_2(0) + a_3(0) \\
&= a_1
\end{aligned}
$$

Thus, $a_1 = 0$. Similarly, $a_2 = a_3 = 0$. We conclude that Q is linearly independent.

Example 9   The subset $S = \{x^2, 1 + x^2, 3\}$ of P_2 (see Example 4) is linearly dependent because

$$
3x^2 + (-3)(1 + x^2) + (1)(3) = 0
$$

where 0 is the zero polynomial.

Example 10   Let $S = \{e^x, e^{2x}, e^{3x}\}$. We show that S is a linearly independent subset of $F(R)$. Suppose that a, b, and c are scalars such that

$$
ae^x + be^{2x} + ce^{3x} = 0
$$

If we differentiate both sides of this equation twice, we obtain the equations

$$ae^x + 2be^{2x} + 3ce^{3x} = \mathbf{0}$$

and

$$ae^x + 4be^{2x} + 9ce^{3x} = \mathbf{0}$$

Substituting $x = 0$ into these three equations, we obtain the homogeneous linear system

$$
\begin{aligned}
a + \ b + \ \ c &= 0 \\
a + 2b + 3c &= 0 \\
a + 4b + 9c &= 0
\end{aligned}
$$

It is easy to show that this system has only the trivial solution, $a = b = c = 0$. Therefore, S is linearly independent.

Example 11   Let S be the infinite subset

$$S = \{1, x, x^2, \ldots, x^n, \ldots\}$$

of the vector space P of polynomial functions. We have seen that any nontrivial linear combination of S is a polynomial function. Since the only polynomial that is zero is the one all of whose coefficients are zero, we have that S is linearly independent.

As was defined earlier, a subset S of a vector space V is a **basis** for V if S is linearly independent and spans V. If we apply this definition to the infinite set S of Example 11 above, we have that S is a basis for P, the vector space of polynomial functions. Thus, in contrast to the subspaces of R^n, P has an infinite basis. In Chapter 3 it was shown that any two bases for the same subspace of R^n contain the same number of vectors. These observations lead to three questions about bases of abstract vector spaces. We list these questions together with the answers. We shall justify the first two answers in this section. The third is beyond the scope of this book [see pages 52–55 of (1) in the References].

1. Is it possible for a vector space to have both an infinite and a finite basis? The answer is *no*.
2. If a vector space has a finite basis, is it the case that any two bases for the vector space have the same number of vectors (as it is the case for subspaces of R^n)? The answer is *yes*.

3. Does every vector space have a basis? If a certain mildly controversial axiom of set theory (called the "axiom of choice") is accepted, then the answer is *yes*.

To help us answer the first two of these questions, we introduce a special type of linear transformation called an "isomorphism."

Definition   Let V and W be vector spaces which are either both real or both complex. A mapping $T: V \rightarrow W$ is called an *isomorphism* if T is one-to-one, onto, and linear. Recall that T is *linear* if for any vectors $\mathbf{x}$ and $\mathbf{y}$ in V, and any scalar c, we have

$$T(\mathbf{x} + \mathbf{y}) = T(\mathbf{x}) + T(\mathbf{y})$$

and

$$T(c\mathbf{x}) = cT(\mathbf{x})$$

If $T: V \rightarrow W$ is an isomorphism, we say that V is *isomorphic* to W.

Any mapping that is both one-to-one and onto has an inverse. If T is an isomorphism, then its inverse T^{-1} is also linear and hence is an isomorphism (see Exercise 20). Therefore, if V is isomorphic to W, then W is isomorphic to V. For this reason we simply say that V and W are *isomorphic*.

Since an isomorphism preserves the essence of the structure of a vector space, we can expect that it will transfer any property that can be stated in terms of the operations of the vector spaces. The following restatement of Exercise 17 of Section 3.1 is an example of such a phenomenon. We state the result without proof.

Theorem 8.2.2 *Let V and W be vector spaces, and let $T: V \rightarrow W$ be an isomorphism. Then for any finite linearly independent subset $\{\mathbf{x}_1, \mathbf{x}_2, \ldots, \mathbf{x}_k\}$ of V, the set of images $\{T(\mathbf{x}_1), T(\mathbf{x}_2), \ldots, T(\mathbf{x}_k)\}$ is a linearly independent subset of W.*

Since we are familiar with the properties of R^n, it is useful to construct an isomorphism from a given vector space V to R^n. Naturally, this cannot always be done. The next result asserts that such an isomorphism exists provided that V has a finite basis. Before stating and proving the result, we examine the method of constructing such an isomorphism. Suppose that $S = \{\mathbf{x}_1, \mathbf{x}_2, \ldots, \mathbf{x}_n\}$ is a finite ordered basis for V. Then a vector $\mathbf{x}$ in V can be represented as a linear combination of the vectors in S in only one way. Hence, $\mathbf{x}$ determines a unique ordered n-tuple $(a_1, a_2, \ldots, a_n)$, where

$$\mathbf{x} = a_1\mathbf{x}_1 + a_2\mathbf{x}_2 + \cdots + a_n\mathbf{x}_n$$

The mapping T_S which takes $\mathbf{x}$ into this ordered n-tuple is the required isomorphism.

Theorem 8.2.3   *Let V be a real vector space with a finite basis $S = \{\mathbf{x}_1, \mathbf{x}_2, \ldots, \mathbf{x}_n\}$. Then the mapping $T_S\colon V \to R^n$ defined by*

$$T_S(a_1\mathbf{x}_1 + a_2\mathbf{x}_2 + \cdots + a_n\mathbf{x}_n) = (a_1, a_2, \ldots, a_n)$$

is an isomorphism between V and R^n.

Proof

Let $\mathbf{x} = a_1\mathbf{x}_1 + a_2\mathbf{x}_2 + \cdots + a_n\mathbf{x}_n$ and $\mathbf{y} = b_1\mathbf{x}_1 + b_2\mathbf{x}_2 + \cdots + b_n\mathbf{x}_n$ be vectors in V. Then

$$\mathbf{x} + \mathbf{y} = (a_1 + b_1)\mathbf{x}_1 + (a_2 + b_2)\mathbf{x}_2 + \cdots + (a_n + b_n)\mathbf{x}_n$$

Therefore,

$$
\begin{aligned}
T_S(\mathbf{x} + \mathbf{y}) &= T_S((a_1 + b_1)\mathbf{x}_1 + (a_2 + b_2)\mathbf{x}_2 + \cdots + (a_n + b_n)\mathbf{x}_n) \\
&= (a_1 + b_1, a_2 + b_2, \ldots, a_n + b_n) \\
&= (a_1, a_2, \ldots, a_n) + (b_1, b_2, \ldots, b_n) \\
&= T_S(\mathbf{x}) + T_S(\mathbf{y})
\end{aligned}
$$

Similarly, $T_S(c\mathbf{x}) = cT_S(\mathbf{x})$ for any scalar c and any vector $\mathbf{x}$ in V. Thus, T_S is linear.

To show that T_S is one-to-one, we show that the null space of T_S is trivial. Suppose that $T_S(\mathbf{x}) = \mathbf{0}$ for

$$\mathbf{x} = a_1\mathbf{x}_1 + a_2\mathbf{x}_2 + \cdots + a_n\mathbf{x}_n$$

Then by the definition of T_S, we have that $(a_1, a_2, \ldots, a_n) = \mathbf{0}$. Therefore, $a_i = 0$ for all i. We conclude that $\mathbf{x} = \mathbf{0}$. Therefore, T_S is one-to-one.

To show that T_S is onto, consider any vector $\mathbf{z} = (a_1, a_2, \ldots, a_n)$ in R^n. By the definition of T_S, we have that $\mathbf{z}$ is the image of the vector

$$\mathbf{x} = a_1\mathbf{x}_1 + a_2\mathbf{x}_2 + \cdots + a_n\mathbf{x}_n$$

Therefore, T_S is onto.   ∎

Corollary 8.2.4   *If a real vector space V has a finite basis consisting of n vectors, then V is isomorphic to R^n.*

We are now in a position to answer the first two questions.

Theorem 8.2.5   *Let V be a real vector space with a finite basis. Then*

    (a) Every basis of V is finite.

    (b) Every basis of V contains the same number of vectors.

Proof

Suppose that V is a vector space with a finite basis $S = \{\mathbf{x}_1, \mathbf{x}_2, \ldots, \mathbf{x}_n\}$.

(a) Let $T_S: V \to R^n$ be the isomorphism of Theorem 8.2.3. Let S' be any subset of V consisting of more than n vectors (S' could be infinite). Then we can find a subset $S'' = \{\mathbf{y}_1, \mathbf{y}_2, \ldots, \mathbf{y}_{n+1}\}$ of S' consisting of $n + 1$ distinct vectors. We assert that S'' is linearly dependent. For suppose that S'' is linearly independent. Then, by Theorem 8.2.2, the set of images, $\{T_S(\mathbf{y}_1), T_S(\mathbf{y}_2), \ldots, T_S(\mathbf{y}_{n+1})\}$ is linearly independent. Thus, we have a linearly independent subset of R^n consisting of $n + 1$ vectors, a contradiction. We conclude that S'' is linearly dependent. Since S'' is a subset of S', we have that S' is linearly dependent. Thus, any basis of V must contain at most n vectors. This establishes part (a).

(b) Let S_1 be any basis of V. By the proof of part (a) S_1 consists of m vectors for some $m \le n$. Reversing the roles of S and S_1, we have that $n \le m$. Therefore, $m = n$. We conclude that any two bases for V contain the same number of vectors. ∎

By Theorem 8.2.5, real vector spaces come in two varieties. There is the type that has a finite basis (and hence for which every basis is finite), and there is the type which has no finite basis. By suitable modifications of Theorems 8.2.3, 8.2.4, and 8.2.5, we can show that these results are valid for complex vector spaces (see the exercises). The first type of vector space is called *finite-dimensional*. By Theorem 8.2.5, any two bases of a finite-dimensional vector space have the same number of vectors. This number is called the *dimension* of the vector space. The other type of vector space is called *infinite-dimensional*. It can be shown that any infinite-dimensional vector space contains an infinite linearly independent set (see Exercise 22).

We conclude this section with three more examples.

Example 12   Recall the vector space P_n of polynomial functions of degree at most n of Example 4. The set $S = \{1, x, x^2, \ldots, x^n\}$ is a linearly independent subset of P_n. Furthermore, any polynomial $f(x)$ of degree at most n can be expressed as a linear combination of the polynomials of S,

$$f(x) = a_0 1 + a_1 x + a_2 x^2 + \cdots + a_n x^n$$

Therefore, S spans P_n. We conclude that S is a basis of P_n. Since S contains $n + 1$ polynomials, we have that P_n is finite-dimensional, with dimension $n + 1$.

Example 13   Recall the vector space of $m \times n$ matrices $M_{m \times n}$ of Example 1 of Section 8.1. For each i and j, $1 \le i \le m$, and $1 \le j \le n$, let $E(i, j)$ denote the $m \times n$ matrix with a 1 in the ijth position and zeros elsewhere. Let S be the set of all matrices of the form $E(i, j)$. Then S is a subset of $M_{m \times n}$ consisting of mn matrices. For any $m \times n$ matrix A,

$$A = \sum_{i=1}^{m} \sum_{j=1}^{n} A_{ij} E(i, j)$$

and furthermore, this representation is unique. Therefore, S is linearly independent and spans $M_{m \times n}$. Thus, S is a basis for $M_{m \times n}$. We conclude that $M_{m \times n}$ is finite-dimensional and has dimension mn.

Example 14   Recall the vector space $L(R^n, R^m)$ of linear transformations as defined in Example 2 of Section 8.1. Let $T: M_{m \times n} \to L(R^n, R^m)$ be defined by $T(A) = L_A$. We show that T is an isomorphism. For any $m \times n$ matrices A and B, we have

$$T(A + B) = L_{A+B}$$
$$= L_A + L_B$$
$$= T(A) + T(B)$$

Similarly, $T(cA) = cT(A)$ for any scalar c, and therefore T is linear. By Theorem 2.4.5, T is onto and by Corollary 2.1.3, T is one-to-one. We conclude that T is an isomorphism. By Exercise 16, isomorphisms preserve dimension. Thus, since $M_{m \times n}$ is isomorphic to $L(R^n, R^m)$, we have that $L(R^n, R^m)$ is finite-dimensional and has dimension mn.

Exercises

In Exercises 1–4, determine whether or not the set V is a subspace of the vector space $M_{n \times n}$.

1. V is the set of all $n \times n$ matrices with determinant equal to 0.

2. V is the set of all $n \times n$ symmetric matrices.

3. V is the set of all $n \times n$ matrices A such that $A^2 = A$.

4. Let B be a fixed $n \times n$ matrix. V is the set of all $n \times n$ matrices A such that $AB = BA$.

5. Let S be a nonempty set. Let S' be a nonempty subset of S. Let V be the set of all functions f in $F(S)$ such that $f(x) = 0$ for all x in S'. Show that V is a subspace of $F(S)$.

6. Let V be the set of all differentiable functions from R to R. Show that V is a subspace of $F(R)$.

7. Let W be the subset of subspace V of Exercise 6 consisting of the functions f such that $f' = f$. Show that W is a subspace of V.

8. A function f in $F(R)$ is called an ***even function*** if $f(x) = f(-x)$ for all x in R. A function f is called an ***odd function*** if $f(-x) = -f(x)$ for all x in R.
   (a) Show that the subset of all even functions is a subspace of $F(R)$.
   (b) Show that the subset of all odd functions is a subspace of $F(R)$.

9. A ***magic square of order n*** is an $n \times n$ matrix each of whose rows and columns sum to the same value. This value is called the ***sum*** of the magic square. For example, the 2×2 matrix

$$\begin{bmatrix} 1 & 3 \\ 3 & 1 \end{bmatrix}$$

is a magic square of order 2 whose sum is 4. Let V_n be the set of all magic squares of order n.
   (a) Show that V_n is a subspace of $M_{n \times n}$.
   (b) Show that the subset W_n of magic squares of order n with sum 0 is a subspace of V_n.

10. Let V be the set of all continuous real-valued functions on the closed interval $[0, 1]$.
   (a) Show that V is a subspace of $F([0, 1])$.

(b) Let W be the subset of V defined by

$$W = \left\{ f \text{ in } V : \int_0^1 f(x)\, dx = 0 \right\}$$

Prove that W is a subspace of V.

11. For each of the following, determine whether or not the given matrix is a linear combination of the set.

$$\left\{ \begin{bmatrix} 1 & 2 & 1 \\ 0 & 0 & 0 \end{bmatrix}, \begin{bmatrix} 0 & 0 & 0 \\ 1 & 1 & 1 \end{bmatrix}, \begin{bmatrix} 1 & 0 & 1 \\ 1 & 2 & 3 \end{bmatrix} \right\}$$

(a) $\begin{bmatrix} 1 & 2 & 1 \\ 1 & 1 & 1 \end{bmatrix}$ (b) $\begin{bmatrix} 2 & 2 & 2 \\ 2 & 3 & 4 \end{bmatrix}$

(c) $\begin{bmatrix} 2 & 2 & 2 \\ 2 & 2 & 2 \end{bmatrix}$

12. Let N be the set of positive integers. Define functions f, g, and h in $F(N)$ by $f(n) = n + 1$, $g(n) = 1$, and $h(n) = 2n - 1$. Determine whether or not the set $\{f, g, h\}$ is linearly independent. Justify your answer.

13. Let N denote the set of positive integers. For each positive integer n let $f_n : N \to R$ be defined by

$$f_n(k) = \begin{cases} 0 & \text{for } k \neq n \\ 1 & \text{for } k = n \end{cases}$$

Let

$$S = \{f_1, f_2, \ldots, f_n, \ldots\}$$

(a) Show that S is a linearly independent subset of $F(N)$.

Let V be the subset of $F(N)$ of the functions which are zero except at finitely many elements of N.

(b) Show that V is a subspace of $F(N)$.
(c) Show that S is a basis for V.

14. For each of the following subsets of $F(R)$ determine whether or not the set is linearly independent.
(a) $\{\sin x, e^{-x}, e^x\}$
(b) $\{\sin x, \sin^2 x, \cos^2 x, 1\}$
(c) $\{e^x, e^{2x}, \ldots, e^{nx}, \ldots\}$

15. Let V and W be isomorphic finite-dimensional vector spaces. Let $T : V \to W$ be an isomorphism. Prove that if $\{\mathbf{x}_1, \mathbf{x}_2, \ldots, \mathbf{x}_n\}$ is a basis for V, then the set of images $\{T(\mathbf{x}_1), T(\mathbf{x}_2), \ldots, T(\mathbf{x}_n)\}$ is a basis for W.

16. Let V and W be isomorphic vector spaces.
(a) Prove that V is finite-dimensional if and only if W is finite-dimensional.
(b) Prove that if V is finite-dimensional, then V and W have the same dimension.

17. Use Exercise 15 and Examples 13 and 14 to find a basis for $L(R^n, R^m)$.

18. Let $S = \{s_1, s_2, \ldots, s_n\}$ be a finite set consisting of n elements. Define $T : F(S) \to R^n$ by

$$T(f) = (f(s_1), f(s_2), \ldots, f(s_n))$$

for any f in $F(S)$. Prove that T is an isomorphism.

19. Show that the vector space V of Exercise 13 is isomorphic to P, the vector space of all polynomial functions. [*Hint:* Choose $T : V \to P$ to be the transformation which takes a function f into the polynomial whose ith coefficient is $f(i)$.]

20. Let $T : V \to W$ be an isomorphism. Prove that the inverse, $T^{-1} : W \to V$ is also an isomorphism.

21. Suppose that we have a vector space V and a positive integer n such that any subset of V consisting of more than n vectors is linearly dependent. Suppose, furthermore, that V does have a linearly independent subset consisting of n vectors. Prove that any linearly independent subset of V consisting of n vectors is a basis for V. Prove that dim $V = n$.

22. Let V be an infinite-dimensional vector space. Prove that V contains an infinite linearly independent set. (*Hints:* Choose a nonzero vector $\mathbf{x}_1$ in V. Next, choose a vector $\mathbf{x}_2$ in V but not in the span of $\{\mathbf{x}_1\}$. Show that the process can be continued to obtain an infinite subset $\{\mathbf{x}_1, \mathbf{x}_2, \ldots, \mathbf{x}_k, \ldots\}$ of V such that for any k, $\mathbf{x}_{k+1}$ is not in the span of $\{\mathbf{x}_1, \mathbf{x}_2, \ldots, \mathbf{x}_k\}$. Now show that this infinite set is linearly independent.)

23. Prove the following variation of Corollary 8.2.4 for complex vector spaces: Any complex vector space with a finite basis consisting of n vectors is isomorphic to C^n.

24. Use Exercise 23 to prove that Theorem 8.2.5 is valid for complex vector spaces. Assume that the results of Section 3.2 are valid for C^n as well as for R^n.

8.3 LINEAR TRANSFORMATIONS AND MATRIX REPRESENTATIONS

In this section we study properties of linear transformations on abstract vector spaces. We shall see that for finite-dimensional spaces, linear transformations can be represented by matrices as was done in Chapter 4.

Isomorphisms are the only examples of linear transformations already encountered in Chapter 8. These were used to investigate the property of dimension. More generally, let V and W be vector spaces which are both real or both complex. Then a mapping $T: V \to W$ is a **linear transformation** if $T(\mathbf{x} + \mathbf{y}) = T(\mathbf{x}) + T(\mathbf{y})$ and $T(c\mathbf{x}) = cT(\mathbf{x})$ for any vectors $\mathbf{x}$ and $\mathbf{y}$ in V and any scalar c.

For a linear transformation $T: V \to W$, the **range**, $R(T)$, and the **null space**, $N(T)$, are defined as in Chapter 2, and the rank and nullity are defined as in Chapter 3. The range and null space are subspaces of W and V, respectively. However, the **rank** and **nullity** of T are defined only if the appropriate subspaces are finite-dimensional. It should be noted that if V is finite-dimensional, then so is $R(T)$ (see Exercise 10). Under this circumstance, both the rank and the nullity are defined, and the dimension theorem (Theorem 3.3.3) holds.

We begin with examples of linear transformations which arise quite naturally in the study of mathematics and related fields.

In the following example, use is made of the dimension theorem, which is valid for abstract vector spaces.

Example 1   Let $T: M_{n \times n} \to R$ be defined by $T(A) = \text{tr}(A)$, the trace of A. By Exercise 15 of Section 2.1, T is linear. In fact, the subspace W of Example 2 of the preceding section is precisely $N(T)$. Since T is onto, T has rank 1. Since $M_{n \times n}$ has dimension n^2, we have by the dimension theorem that $\dim N(T) = n^2 - 1$.

Example 2   (*Differential Operators*)   Let C^∞ denote the subset of $F(R)$ consisting of those functions which have derivatives of all orders. That is, a function $f: R \to R$ lies in C^∞ if the nth derivative of f exists for any positive integer n. Theorems from calculus imply that C^∞ is a subspace of $F(R)$. Consider the mapping $D: C^\infty \to C^\infty$ which takes a function into its derivative; that is, $D(f) = f'$ for all f in C^∞. Using the elementary properties of the derivative, we have that for any functions f and g in C^∞,

$$D(f + g) = (f + g)' = f' + g' = D(f) + D(f)$$

Similarly, $D(cf) = cD(f)$ for any scalar c. Thus, the mapping D is linear. By repeated applications of Theorem 2.4.3 and Corollary 2.4.7, we have that any mapping of the form

$$a_0 I + a_1 D + \cdots + a_n D^n$$

is also linear. Such a mapping is an example of a "differential operator." For example, the linear operator

$$2I + 3D - D^2$$

acts on the function $f(x) = x^3$ as

$$(2I + 3D - D^2)(x^3) = 2x^3 + 9x^2 - 6x$$

More generally, a **differential operator** is a mapping on C^∞ of the form above, where the a_i's are functions in C^∞. All differential operators are linear operators on C^∞. Many differential equations can be expressed in the context of differential operators just as systems of linear equations can be expressed in the context of matrices. For this reason, differential operators are important in the study of differential equations. For example, the differential equation

$$x^2 y'' - (\sin x)y' + 3y = 0$$

can be rewritten using the notation of differential operators as

$$(x^2 D^2 - (\sin x)D + 3I)y = 0$$

Thus, the solution set of the differential equation coincides with the null space of the differential operator

$$x^2 D^2 - (\sin x)D + 3I$$

Example 3   Let $C([a, b])$ be the set of all continuous real-valued functions defined on the closed interval $[a, b]$. It can be shown that $C([a, b])$ is a subspace of $F([a, b])$. Any continuous real-valued function f on $[a, b]$ is Riemann integrable. Let $T: C([a, b]) \to R$ be the mapping which takes a function f into its integral,

$$T(f) = \int_a^b f(x)\, dx$$

The linearity of T follows at once from the elementary properties of the Riemann integral. For example,

$$T(f + g) = \int_a^b [f(x) + g(x)]\, dx$$

$$= \int_a^b f(x)\, dx + \int_a^b g(x)\, dx$$

$$= T(f) + T(g)$$

Similarly, $T(cf) = cT(f)$ for any scalar c. Thus, T is linear.

Example 4   Let $T: M_{m \times n} \to M_{n \times m}$ be defined by $T(A) = A^t$, the transpose of A. The linearity of T is a consequence of Exercise 12 of Section 2.1. In fact, by Exercise 13 of Section 2.1, it follows that T is an isomorphism.

The definitions of *eigenvalue, eigenvector,* and *eigenspace,* as given in Chapter 6, apply to all vector spaces.

Example 5   Let $D: C^\infty \to C^\infty$ be the linear operator of Example 2. Let λ be any scalar and let f be the exponential function $f(x) = e^{\lambda x}$. Since $(e^{\lambda x})' = \lambda e^{\lambda x}$, we have that $D(f) = \lambda f$. Thus, f is an eigenvector of D, and λ is the eigenvalue corresponding to f. Since λ was chosen arbitrarily, we see that any scalar is an eigenvalue of D. Therefore, D has infinitely many eigenvalues, in contrast to linear operators on R^n.

Example 6   Show that the solution set to the differential equation

$$y'' + 4y = \mathbf{0}$$

coincides with the eigenspace of the differential operator D^2 corresponding to the eigenvalue $\lambda = -4$.

First observe that solutions to this differential equation lie in C^∞. For if y is a solution, then y must be at least twice differentiable. Since $y'' = -4y$, we have that y'' is twice differentiable. Therefore, y is 4 times differentiable. Repetitions of this argument lead to the conclusion that y has derivatives of any order. Since $y'' = D^2 y$, we can rewrite this differential equation as

$$D^2 y = -4y$$

But this last equation asserts that y is in the eigenspace of D^2 corresponding to the eigenvalue -4. Thus, the eigenspace and the solution set of the differential equation coincide. The functions $\sin 2x$ and $\cos 2x$ are examples of solutions to the differential equation. Notice that these functions are eigenvectors of D^2 corresponding to $\lambda = -4$.

Example 7   Let T be the linear operator of Example 4 which takes a matrix of order n into its transpose. Let A be any nonzero symmetric matrix. Since $A = A^t$, we have that $T(A) = A$. Therefore, A is an eigenvector of T, and 1 is the corresponding eigenvalue. We now show that -1 is also an eigenvalue of T. Recall that a square matrix B is *skew symmetric* if $B^t = -B$, that is, if $B_{ji} = -B_{ij}$ (see Exercise 10 of Section 5.3). For example, the matrix

$$\begin{bmatrix} 0 & 2 \\ -2 & 0 \end{bmatrix}$$

is skew symmetric. For a nonzero skew symmetric matrix B, $T(B) = B^t = -B$. Thus, B is an eigenvector of T, and -1 is the corresponding eigenvalue. It can be shown that 1 and -1 are the only eigenvalues of T (see Exercise 14).

Matrix Representations

We can use matrices to study linear transformations in abstract vector spaces provided that the spaces involved are finite-dimensional. The goal is to convert the process of finding the image of a vector under a linear transformation into the process of computing a matrix product. To do this, we must fix ordered bases for the vector spaces involved.

We first recall from Section 4.1 the definition of "coordinate vector relative to the chosen basis." Let V be an n-dimensional vector space with an ordered basis $S = \{x_1, x_2, \ldots, x_n\}$. Let x be a vector in V. Suppose that x is represented by the (unique) linear combination of the vectors of S

$$x = a_1 x_1 + a_2 x_2 + \cdots + a_n x_n$$

Then the n-tuple $[x]_S = (a_1, a_2, \ldots, a_n)$ is the *coordinate vector* of x *relative* to S.

Notice that the isomorphism $T_S: V \to R^n$ of Theorem 8.2.3 is associated with the coordinate vector in that $T(x) = [x]_S$ for any vector x in V.

For example, let P_n be the vector space of polynomials of degree at most n. We have seen that the set

$$S = \{1, x, x^2, \ldots, x^n\}$$

is an ordered basis for P_n. For the polynomial

$$f(x) = a_0 + a_1 x + a_2 x^2 + \cdots + a_n x^n$$

we have that

$$T_S(f) = [f]_S = (a_0, a_1, a_2, \ldots, a_n)$$

The coordinate vector is a concrete representation of an abstract vector. Of course, the representation depends on the choice of basis. We now consider a method of providing a concrete representation of a linear transformation. The result is a matrix which depends on the chosen bases.

Let $T: V \to W$ be a linear transformation from the finite-dimensional vector space V to the finite-dimensional vector space W. Let $S = \{x_1, x_2, \ldots, x_n\}$ and $S' = \{y_1, y_2, \ldots, y_m\}$ be ordered bases of V and W, respectively. For each vector x_j in S, $T(x_j)$ is a unique linear combination of the vectors of S'. Thus,

$$T(x_j) = a_{1j}y_1 + a_{2j}y_2 + \cdots + a_{mj}y_m$$

$$= \sum_{i=1}^{m} a_{ij}y_i$$

for unique scalars a_{ij}.

Definition   In the context of the discussion above, the $m \times n$ matrix A whose ijth entry is a_{ij} is called the ***matrix representation of T relative to the bases S and S'***. We denote this matrix symbolically by $[T]_S^{S'}$. If $V = W$ and $S = S'$, we call A the ***matrix representation of T relative to S*** and write $A = [T]_S$.

Compare this definition to the definition of "matrix representation" in Section 4.2. Suppose that $A = [T]_S^{S'}$, as above. Then, by definition the jth column of A is composed of the coefficients used in representing $T(x_j)$ as a linear combination of the vectors of S'. But these coefficients are the scalars used to represent $T(x_j)$ as the coordinate vector relative to S'. Thus, we have that the jth column of A is given by

$$^{j}A = [T(x_j)]_{S'}$$

Example 8   Let $D: P_3 \to P_2$ be the linear transformation which takes a polynomial in P_3 into its derivative. Let $S = \{1, x, x^2, x^3\}$, and let $S' = \{1, x, x^2\}$. Compute A, the matrix representation of T relative to S and S'.

As was observed above, the columns of A are the coordinate vectors relative to S'. Table 8.3.1 consists of the vectors of S, their images, and the coordinate vectors of these images relative to S'. These coordinate vectors are the columns of A. Since the coordinate vectors are the columns of A, we have that

$$A = \begin{bmatrix} 0 & 1 & 0 & 0 \\ 0 & 0 & 2 & 0 \\ 0 & 0 & 0 & 3 \end{bmatrix}$$

TABLE 8.3.1

$f(x)$ in S	$D(f(x))$	$[D(f(x))]_{S'}$
1	0	$\begin{bmatrix} 0 \\ 0 \\ 0 \end{bmatrix}$
x	1	$\begin{bmatrix} 1 \\ 0 \\ 0 \end{bmatrix}$
x^2	$2x$	$\begin{bmatrix} 0 \\ 2 \\ 0 \end{bmatrix}$
x^3	$3x^2$	$\begin{bmatrix} 0 \\ 0 \\ 3 \end{bmatrix}$

Example 9   Let $T: P_2 \to P_3$ be the transformation defined by

$$T(f(x)) = \int_0^x f(t)\, dt$$

for any polynomial $f(x)$ in P_2. For example, if $f(x) = 2 + 4x - 3x^2$, then

$$T(f(x)) = \int_0^x (2 + 4t - 3t^2)\, dt$$

$$= 2t + 2t^2 - t^3 \Big|_0^x$$

$$= 2x + 2x^2 - x^3$$

It can be shown that T is a linear transformation. Let $S = \{1, x, x^2\}$, $S' = \{1, x, x^2, x^3\}$, and $A = [T]_S^{S'}$. We compile a table as in the preceding example (Table 8.3.2). By the last column of Table 8.3.2, we have that

$$A = \begin{bmatrix} 0 & 0 & 0 \\ 1 & 0 & 0 \\ 0 & \frac{1}{2} & 0 \\ 0 & 0 & \frac{1}{3} \end{bmatrix}$$

The importance of matrix representations is that matrices are concrete numerical objects on which computational procedures can be applied. Now that we know

TABLE 8.3.2

$f(x)$ in S	$T(f(x))$	$[T(f(x))]_{S'}$
1	x	$\begin{bmatrix} 0 \\ 1 \\ 0 \\ 0 \end{bmatrix}$
x	$\dfrac{x^2}{2}$	$\begin{bmatrix} 0 \\ 0 \\ \frac{1}{2} \\ 0 \end{bmatrix}$
x^2	$\dfrac{x^3}{3}$	$\begin{bmatrix} 0 \\ 0 \\ 0 \\ \frac{1}{3} \end{bmatrix}$

how to represent abstract vectors with coordinate vectors and linear transformations with matrices, the next step is to show that these identifications preserve the essential relationship between a vector and linear transformation.

Let us make this idea precise. Suppose that $T: V \to W$ is linear, where V and W are n- and m-dimensional vector spaces, respectively. Let S and S' be ordered bases for V and W, respectively, and let A be the matrix representation of T relative to S and S'. We shall show that the action of A on coordinate vectors in R^n corresponds to the action of T on vectors in V. Let $\mathbf{x}$ be any vector in V. Then $[\mathbf{x}]_S$ is a vector in R^n, and $[T(\mathbf{x})]_{S'}$ is a vector in R^m. Also, L_A is a linear transformation from R^n to R^m. We shall show that L_A takes $[\mathbf{x}]_S$ into $[T(\mathbf{x})]_{S'}$. The significance of this result should be clear. If we identify vectors with their coordinate vectors and linear transformations with their matrix representations, then finding the image of a vector under the linear transformation amounts to multiplying two matrices.

Theorem 8.3.1 *Let V and W be finite-dimensional vector spaces, and let $T: V \to W$ be a linear transformation. Let S and S' be ordered bases for V and W, respectively, and let A be the matrix representation of T relative to S and S'. Then for any vector $\mathbf{x}$ in V, we have that*

$$[T(\mathbf{x})]_{S'} = A[\mathbf{x}]_S$$

It is useful to examine Theorem 8.3.1 from the functional point of view before proceeding with the proof. Since coordinate vectors relative to S and S' can be expressed as images of the isomorphisms T_S and $T_{S'}$, as defined in Theorem 8.2.3, we can rewrite the equation above as

$$T_{S'} T(\mathbf{x}) = L_A T_S(\mathbf{x})$$

Since $\mathbf{x}$ is an arbitrary vector in V, Theorem 8.3.1 is equivalent to the functional equation

$$T_{S'}\,T = L_A\,T_S$$

This equation tells us that the two ways of proceeding from V to R^m yield the same result, as follows

$$
\begin{array}{ccc}
V & \xrightarrow{\ \ T\ \ } & W \\
\Big\downarrow{\scriptstyle T_S} & & \Big\downarrow{\scriptstyle T_{S'}} \\
R^n & \xrightarrow{\ \ L_A\ \ } & R^m
\end{array}
$$

With these ideas in mind we attend to the proof of the theorem.

Proof

By the discussion above, the proof of the theorem is equivalent to showing that the transformations $T_{S'}\,T$ and $L_A\,T_S$ are identical. Since these transformations are compositions of linear transformations, they are linear. Furthermore, they have the same domain, namely, V. Thus, by Exercise 30 of Section 2.4, we need only show that $T_{S'}\,T$ and $L_A\,T_S$ agree on a basis for V. Consider the basis $S = \{\mathbf{x}_1, \mathbf{x}_2, \ldots, \mathbf{x}_n\}$. Suppose that $S' = \{\mathbf{y}_1, \mathbf{y}_2, \ldots, \mathbf{y}_m\}$. By Exercise 13, $T_S(\mathbf{x}_j)$ and $T_{S'}(\mathbf{y}_j)$ are the jth standard vectors of R^n and R^m, respectively. We shall make use of these facts below. By the definition of A, we have

$$T(\mathbf{x}_j) = \sum_{i=1}^{m} A_{ij}\mathbf{y}_i$$

Therefore,

$$T_{S'}\,T(\mathbf{x}_j) = T_{S'}\left(\sum_{i=1}^{m} A_{ij}\mathbf{y}_i\right)$$

$$= \sum_{i=1}^{m} A_{ij}\,T_{S'}(\mathbf{y}_i)$$

$$= \sum_{i=1}^{m} A_{ij}(\mathbf{e}_i)$$

$$= {}^{j}A$$

Furthermore,

$$L_A T_S(\mathbf{x}_j) = L_A(\mathbf{e}_j)$$

$$= A\mathbf{e}_j$$

$$= {}^j A \qquad \text{(by Theorem 2.1.2)}$$

We conclude that

$$T_{S'} T(\mathbf{x}_j) = L_A T_S (\mathbf{x}_j)$$

for all j. Therefore, $T_{S'} T = L_A T_S$. ∎

To illustrate how Theorem 8.3.1 works, we apply it to the linear transformations of Examples 8 and 9.

Example 10   Let D, S, S', and A be as in Example 8. Consider the polynomial $f(x) = 5 + 4x - x^2 + 2x^3$. Then $[f(x)]_S = (5, 4, -1, 2)$. Thus, by Theorem 8.3.1,

$$[D(f)]_{S'} = A[f]_S$$

$$= \begin{bmatrix} 0 & 1 & 0 & 0 \\ 0 & 0 & 2 & 0 \\ 0 & 0 & 0 & 3 \end{bmatrix} \begin{bmatrix} 5 \\ 4 \\ -1 \\ 2 \end{bmatrix}$$

$$= \begin{bmatrix} 4 \\ -2 \\ 6 \end{bmatrix}$$

This last vector is the coordinate vector for the polynomial $4 - 2x + 6x^2$, which is the derivative of $f(x)$.

Example 11   Let T, S, and A be as in Example 9. Consider the polynomial $f(x) = 3 - 2x + 6x^2$ in P_2. Then by Theorem 8.3.1,

$$[T(f(x))]_{S'} = A[f(x)]_S$$

$$= \begin{bmatrix} 0 & 0 & 0 \\ 1 & 0 & 0 \\ 0 & \frac{1}{2} & 0 \\ 0 & 0 & \frac{1}{3} \end{bmatrix} \begin{bmatrix} 3 \\ -2 \\ 6 \end{bmatrix}$$

$$= \begin{bmatrix} 0 \\ 3 \\ -1 \\ 2 \end{bmatrix}$$

This last vector is the coordinate vector for the polynomial

$$\int_0^x f(t)\, dt = 3x - x^2 + 2x^3$$

We can apply Theorem 8.3.1 to find the eigenvalues and eigenvectors of a linear operator. The next result tells us that a linear operator and its matrix representations have the same eigenvalues. Furthermore, the coordinate vectors of eigenvectors of the operator are eigenvectors of the matrix representation.

Theorem 8.3.2 *Let T be a linear operator on a finite-dimensional vector space V. Let S be an ordered basis for V, and let A be the matrix representation of T relative to S. Then a vector $\mathbf{x}$ in V is an eigenvector of T with eigenvalue λ if and only if $[\mathbf{x}]_S$ is an eigenvector of A with eigenvalue λ.*

Proof

Let $\mathbf{x}$ be an eigenvector of T corresponding to the eigenvalue λ. Applying Theorem 8.3.1 with $S = S'$, we have

$$A[\mathbf{x}]_S = [T(\mathbf{x})]_S = [\lambda\mathbf{x}]_S = \lambda[\mathbf{x}]_S$$

Therefore, $[\mathbf{x}]_S$ is an eigenvector of A corresponding to the eigenvalue λ.

Conversely, suppose that $A[\mathbf{x}]_S = \lambda[\mathbf{x}]_S$ for some nonzero vector $\mathbf{x}$ in V and some scalar λ. Then, by Theorem 8.3.1 we have

$$\begin{aligned}
T_S(T(\mathbf{x})) &= L_A T_S(\mathbf{x}) \\
&= A[\mathbf{x}]_S \\
&= \lambda[\mathbf{x}]_S \\
&= \lambda T_S(\mathbf{x}) \\
&= T_S(\lambda\mathbf{x})
\end{aligned}$$

Therefore, since T_S is an isomorphism, we have that $T(\mathbf{x}) = \lambda\mathbf{x}$.   ∎

A simple consequence of Theorem 8.3.2 is that T and A have the same eigenvalues.

Corollary 8.3.3 *Let T be a linear operator on a finite-dimensional vector space V, and let A be a matrix representation of T with respect to a fixed basis of V. Then A and T have the same eigenvalues.*

We omit the proof, leaving it as an exercise.

Example 12   We define the mapping $T: P_2 \to P_2$ by

$$T(f(x)) = f(0) + 3f(1)x + f(2)x^2$$

for any polynomial $f(x)$ in P_2. For example, if $f(x) = 2 + x - 2x^2$, then $f(0) = 2$, $f(1) = 1$, and $f(2) = -4$. Therefore, $T(f(x)) = 2 + 3x - 4x^2$. It can be shown that T is linear. We wish to find the eigenvalues and eigenvectors of T. To make things as simple as possible, we choose the set $S = \{1, x, x^2\}$ for our basis of P_2. Let A be the matrix representation of T relative to S. Since

$$T(1) = 1 + 3x + x^2$$

$$T(x) = 0 + 3x + 2x^2$$

$$T(x^2) = 0 + 3x + 4x^2$$

we have that

$$A = \begin{bmatrix} 1 & 0 & 0 \\ 3 & 3 & 3 \\ 1 & 2 & 4 \end{bmatrix}$$

By Corollary 8.3.3 the eigenvalues of T are the same as those of A. Using the methods of Chapter 6, we compute the characteristic polynomial of A to obtain

$$(x - 1)^2(x - 6)$$

We conclude that A, and therefore T, have the eigenvalues 1 and 6.

Eigenvectors corresponding to the eigenvalue 1: The eigenspace of A corresponding to $\lambda = 1$ is spanned by the vector $(0, -3, 2)$. Since $(0, -3, 2)$ is the coordinate vector of the polynomial $f(x) = -3x + 2x^2$, we may apply Theorem 8.3.2 to conclude that the nonzero multiples of $f(x)$ are the eigenvectors of T corresponding to 1.

Eigenvectors corresponding to the eigenvalue 6: The eigenspace of A corresponding to $\lambda = 6$ is spanned by $(0, 1, 1)$. Since $(0, 1, 1)$ is the coordinate vector for the polynomial $g(x) = x + x^2$, we have, as above, that the eigenvectors of T corresponding to 6 are the nonzero multiples of $g(x)$.

Exercises

In Exercises 1–7, determine whether or not the mapping T is a linear transformation. Justify your conclusions.

1. $T: M_{n \times n} \to R$ defined by $T(A) = \det A$.

2. $T: p \to p$ defined by $T(f)(x) = xf(x)$ for all x in R.

3. $T: P_2 \to R^3$ defined by $T(f) = (f(0), f(1), f(2))$.

4. $T: p \to p$ defined by $T(f)(x) = f^2(x)$ for all x in R.

5. $T: C([a, b]) \to C([a, b])$ defined by

$$T(f)(x) = \int_a^x f(t) \, dt \qquad \text{for } a \le x \le b$$

6. $T: M_{m \times n} \to M_{n \times m}$ defined by $T(A) = A^t$.

7. $T: C^\infty \to C^\infty$ defined by $T(f)(x) = e^x f''(x)$ for all x in R.

8. Show that the differential operator $T: C^\infty \to C^\infty$ defined by

$$T(f)(x) = (x^2 D^2 - (\sin x)D + 3I)f(x)$$

is linear.

9. Fix a nonzero vector $\mathbf{x}$ in R^n. Let $T: M_{m \times n} \to R^m$ be the mapping defined by $T(A) = A\mathbf{x}$.
   (a) Show that T is linear.
   (b) Show that T is onto. [*Hint:* Suppose that the ith component x_i of $\mathbf{x}$ is not zero. Let $\mathbf{y}$ be any vector in R^m. Let A be the $m \times n$ matrix whose ith column is $(1/x_i)\mathbf{y}$ and whose other columns are zero.]
   (c) Find the rank and the nullity of T.

10. Let $T: V \to W$ be a linear transformation. Prove that if V is finite-dimensional, then so is $R(T)$.

11. For the ordered basis

$$S = \left\{ \begin{bmatrix} 1 & 0 \\ 0 & 0 \end{bmatrix}, \begin{bmatrix} 0 & 0 \\ 1 & 0 \end{bmatrix}, \begin{bmatrix} 0 & 0 \\ 0 & 1 \end{bmatrix}, \begin{bmatrix} 0 & 1 \\ 0 & 0 \end{bmatrix} \right\}$$

of $M_{2 \times 2}$ and the matrix

$$A = \begin{bmatrix} 1 & 2 \\ 3 & 4 \end{bmatrix}$$

find the coordinate vector of A relative to S.

12. For the ordered basis $S = \{x^2, x, 1\}$ of P_2 find the coordinate vector of the polynomial $2 + x - 3x^2$ relative to S.

13. Let $S = \{\mathbf{x}_1, \mathbf{x}_2, \ldots, \mathbf{x}_n\}$ be an ordered basis for an n-dimensional vector space V. Show that for any j,

$$[\mathbf{x}_j]_S = \mathbf{e}_j$$

the jth standard vector of R^n.

14. Let $T: M_{2 \times 2} \to M_{2 \times 2}$ be defined by $T(A) = A^t$. Let

$$S = \left\{ \begin{bmatrix} 1 & 0 \\ 0 & 0 \end{bmatrix}, \begin{bmatrix} 0 & 1 \\ 0 & 0 \end{bmatrix}, \begin{bmatrix} 0 & 0 \\ 1 & 0 \end{bmatrix}, \begin{bmatrix} 0 & 0 \\ 0 & 1 \end{bmatrix} \right\}$$

   (a) Find the matrix representation of T relative to S.
   (b) In Example 7 it was shown that 1 and -1 are eigenvalues of T. Use part (a) to prove that these are the only eigenvalues of T.
   (c) Show that the nonzero symmetric and skew symmetric 2×2 matrices are the only eigenvectors of T.

15. For the differential operator $D^2 + D$:
   (a) Show that 1 and e^{-x} lie in the null space.
   (b) Show that for any real number a, e^{ax} is an eigenvector of $D^2 + D$ corresponding to eigenvalue $a^2 + a$.

16. Let $T: P_2 \to R$ be defined by

$$T(f) = \int_0^1 f(x) \, dx$$

Let $S = \{1, x, x^2\}$, $S' = \{1\}$, and let A be the matrix representation of T relative to S and S'.
   (a) Find A.
   (b) Show that for any polynomial $f(x) = c_0 + c_1 x + c_2 x^2$, we have

$$\int_0^1 f(x) \, dx = A \begin{bmatrix} c_0 \\ c_1 \\ c_2 \end{bmatrix}$$

17. Let $T: V \to W$ and $U: W \to Z$ be linear transformations, where V, W, and Z are finite-dimensional vector spaces. Let S, S', and S'' be ordered bases for V, W, and Z, respectively. Prove

$$[UT]_S^{S''} = [U]_{S'}^{S''}[T]_S^{S'}$$

[*Hint:* Apply Theorem 8.3.1 to $(UT)(\mathbf{x})$ and $U(T(\mathbf{x}))$.]

18. Let V be a finite-dimensional vector space, and let S and S' be ordered bases for V. Let $I: V \to V$ be the identity operator on V and let $P = [I]_S^{S'}$. Show that for any vector $\mathbf{x}$ in V,

$$[\mathbf{x}]_{S'} = P[\mathbf{x}]_S$$

P is called a *change of coordinate matrix*. Left multiplication by P converts a coordinate vector relative to S into a coordinate vector relative to S'.

19. In Exercise 9 of Section 8.2 we defined the vector space V_n of magic squares of order n and the subspace W_n of magic squares which sum to 0. Define $T: V_n \to R$ by $T(A) =$ the sum of A.
    (a) Prove that T is linear.
    (b) Prove that $N(T) = W_n$.

(c) Use the dimension theorem to prove that

$$\dim (V_n) = \dim (W_n) + 1$$

20. Prove Corollary 8.3.3.

21. Since a sequence is a mapping from N, the set of positive integers, to R, it follows that the set of sequences is equal to $F(N)$. Thus, we may view the set of sequences as a vector space. Let $E: F(N) \to F(N)$ be the mapping defined by $E(f)(n) = f(n + 1)$.
    (a) Prove that E is a linear operator. Recall the definition of a Fibonacci sequence from Section 6.3.
    (b) Prove that the set of Fibonacci sequences coincides with the null space of $E^2 - E - I$.
    (c) Use the results on Fibonacci sequences from Section 6.3 to find the nullity of $E^2 - E - I$.

8.4 INNER PRODUCT SPACES

The dot product was introduced in Section 1.6 to provide a stronger link between linear algebra and the geometry of R^n. The geometric properties of orthogonality were exploited in Chapter 7. These properties led to deep and elegant results about symmetric matrices.

Orthogonality arises naturally in certain examples of abstract vector spaces. For this reason there is a generalization of the dot product for abstract vector spaces. This abstract version is called an "inner product." Many of these examples involve integrals from calculus.

Definition   Let V be a real vector space. A *(real) inner product* on V is a real-valued function which assigns to any pair of vectors $\mathbf{x}$ and $\mathbf{y}$ in V a scalar, denoted $(\mathbf{x}, \mathbf{y})$, such that for any vectors $\mathbf{x}$, $\mathbf{y}$, and $\mathbf{z}$ in V and for any scalar a, the following axioms hold:
(a) $(\mathbf{x}, \mathbf{x}) > 0$ if $\mathbf{x} \neq \mathbf{0}$
(b) $(\mathbf{x}, \mathbf{y}) = (\mathbf{y}, \mathbf{x})$
(c) $(\mathbf{x} + \mathbf{y}, \mathbf{z}) = (\mathbf{x}, \mathbf{z}) + (\mathbf{y}, \mathbf{z})$
(d) $(a\mathbf{x}, \mathbf{y}) = a(\mathbf{x}, \mathbf{y})$
A real vector space with a fixed inner product is called a *(real) inner product space*.

The dot product for R^n is the classical example of an inner product. If we define $(\mathbf{x}, \mathbf{y}) = \mathbf{x} \cdot \mathbf{y}$ for $\mathbf{x}$ and $\mathbf{y}$ in R^n, it is a simple matter to verify that the dot product satisfies the properties of an inner product.

As was the case for abstract vector spaces, most of the results for dot products are valid for inner products. Usually, a proof of a result for inner product spaces requires little or no modification of the version for the dot product on R^n.

What follows are additional examples of inner products.

Example 1   Let $C([a, b])$ denote the vector space of continuous real-valued functions defined on $[a, b]$. For f and g in $C([a, b])$ we define

$$(f, g) = \int_a^b f(x)g(x) \, dx$$

We show that this definition determines an inner product on $C([a, b])$.

To verify axiom (a), let f be any nonzero function in $C([a, b])$. Then f^2 is continuous and nonnegative. Furthermore, for some x_0 in $[a, b]$, we have that $f^2(x_0) > 0$. It follows from the continuity of f^2 that

$$(f, f) = \int_a^b f^2(x) \, dx > 0$$

To verify axiom (b), let f and g be functions in $C([a, b])$. Then

$$(f, g) = \int_a^b f(x)g(x) \, dx = \int_a^b g(x)f(x) \, dx = (g, f)$$

We leave the verifications of axioms (c) and (d) as exercises.

Example 2   Recall the definition of the *trace* of a square matrix given in Exercise 15 of Section 2.1. Let $V = M_{n \times n}$, the vector space of $n \times n$ matrices. For A and B in V define

$$(A, B) = \text{tr}(AB^t)$$

To show that this determines an inner product on V, we must verify axioms (a), (b), (c), and (d) of the definition.

To verify axiom (a), let A be any nonzero matrix. Then

$$(A, A) = \text{tr}(AA^t)$$

$$= \sum_{i=1}^n (AA^t)_{ii}$$

$$= \sum_{i=1}^n \sum_{j=1}^n (A_{ij})^2$$

Since $A \neq O$, it follows that $(A_{ij})^2 > 0$ for some i and j and therefore, $(A, A) > 0$. This verifies axiom (a).

To verify axiom (b), let A and B be matrices in V. Then

$$(A, B) = \text{tr}(AB^t)$$

$$= \text{tr}(AB^t)^t \qquad \text{(Exercise 15 of Section 2.1)}$$

$$= \text{tr}(BA^t)$$

$$= (B, A)$$

We leave the verifications of axioms (c) and (d) as exercises.

As with the dot product, we can define the "length" of a vector in an inner product space. For a vector $\mathbf{x}$ in an inner product space V, the **norm** or **length** of $\mathbf{x}$, denoted $\|\mathbf{x}\|$, is defined by $\|\mathbf{x}\| = \sqrt{(\mathbf{x}, \mathbf{x})}$.

As stated earlier, the elementary properties of dot products are valid for all inner products. For example, parts (c), (d), and (e) of Theorem 1.6.1 are valid for arbitrary inner products. The proofs follow trivially from the definition of inner product. For example, to prove that $(\mathbf{x}, \mathbf{y} + \mathbf{z}) = (\mathbf{x}, \mathbf{y}) + (\mathbf{x}, \mathbf{z})$ for all $\mathbf{x}$, $\mathbf{y}$, and $\mathbf{z}$, we write

$$(\mathbf{x}, \mathbf{y} + \mathbf{z}) = (\mathbf{y} + \mathbf{z}, \mathbf{x}) = (\mathbf{y}, \mathbf{x}) + (\mathbf{z}, \mathbf{x}) = (\mathbf{x}, \mathbf{y}) + (\mathbf{x}, \mathbf{z})$$

appealing to axioms (b) and (c) of the definition of inner product. See the exercises for more of these kinds of results.

The Cauchy–Schwarz inequality (Theorem 1.6.3) and the triangle inequality (Corollary 1.6.4) as stated and proved for dot products are also valid for inner products. The proofs for the general setting are virtually identical to the proofs provided in Section 1.6. As we remarked earlier, the interesting examples of inner products involve integrals. For this reason, a typical application of the Cauchy–Schwarz inequality yields an inequality of integrals. For example, if f and g are continuous real-valued functions defined on $[a, b]$, we can apply the Cauchy–Schwarz inequality to the inner product of Example 1 to obtain (after squaring both sides)

$$\left[\int_a^b f(x)g(x)\, dx \right]^2 \leq \left[\int_a^b f^2(x)\, dx \right]\left[\int_a^b g^2(x)\, dx \right]$$

Orthogonality and the Gram–Schmidt Process

Let V be an inner product space. Two vectors $\mathbf{x}$ and $\mathbf{y}$ in V are **orthogonal** if $(\mathbf{x}, \mathbf{y}) = 0$. A subset S of V is **orthogonal** if any pair of vectors in S is orthogonal. The set S is **orthonormal** if S is orthogonal and each vector in S has norm 1 (is a **unit vector**).

These definitions are identical to the definitions introduced in Section 1.6 and Chapter 7. It was shown in Chapter 7 that any finite orthogonal set consisting of nonzero vectors is linearly independent (Theorem 7.1.1). This result is valid for arbitrary inner product spaces, and the proof is identical. Furthermore, if we employ the definition of an infinite linearly independent set given in Section 8.2, we can show that the result is valid if the orthogonal set is infinite (see the exercises).

Example 3   Let $f(x) = \sin x$ and $g(x) = \cos x$ be defined on the closed interval $[0, 2\pi]$. Verify that $S = \{f, g\}$ is an orthogonal subset of the space $C([0, 2\pi])$ with the inner product of Example 1.

Because S has only two vectors, we verify that they are orthogonal.

$$(f, g) = \int_0^{2\pi} \sin x \cos x \, dx$$

$$= \frac{1}{2} \sin^2 x \, \Big|_0^{2\pi}$$

$$= 0$$

Therefore, S is an orthogonal set.

A computational method for converting a linearly independent subset of R^n into an orthogonal set was introduced in Section 7.1. This method, called the Gram–Schmidt orthogonalization process, is valid for arbitrary inner product spaces. Its justification is identical to the proof of Theorem 7.1.5. The Gram–Schmidt process was used to replace an arbitrary basis of a subspace of R^n with an orthonormal basis. Because this can be done in any inner product space, we have the following result.

Theorem 8.4.1   *Any finite-dimensional subspace of an inner product space has an orthonormal basis.*

Example 4   We define an inner product on P_2 by

$$(f, g) = \int_{-1}^{1} f(x)g(x) \, dx$$

for any polynomials $f(x)$ and $g(x)$ in P_2. It is easy to verify that this does define an inner product for P_2. (See, for example, the arguments in Example 1.) We use the Gram–Schmidt process to convert the basis $S = \{1, x, x^2\}$ into an orthogonal basis for P_2. We then normalize the vectors of the orthogonal basis to obtain an orthonormal basis of P_2.

Using the notation of Theorem 7.1.5, let $\mathbf{y}_1 = \mathbf{x}_1 = 1$, $\mathbf{y}_2 = x$, and $\mathbf{y}_3 = x^2$. To compute $\mathbf{x}_2$ we first compute $(\mathbf{y}_2, \mathbf{x}_1)$.

$$(\mathbf{y}_2, \mathbf{x}_1) = \int_{-1}^{1} x \, dx = \left. \frac{x^2}{2} \right|_{-1}^{1} = 0$$

Therefore, $\mathbf{x}_2 = \mathbf{y}_2 = x$. We now compute $\mathbf{x}_3$.

$$(\mathbf{y}_3, \mathbf{x}_1) = \int_{-1}^{1} x^2 \, dx = \left. \frac{x^3}{3} \right|_{-1}^{1} = \frac{2}{3}$$

$$(\mathbf{y}_3, \mathbf{x}_2) = \int_{-1}^{1} x^2 x \, dx = 0$$

and

$$\|\mathbf{x}_1\|^2 = (\mathbf{x}_1, \mathbf{x}_1) = \int_{-1}^{1} 1 \, dx = 2$$

Therefore,

$$\mathbf{x}_3 = \mathbf{y}_3 - \frac{(\mathbf{y}_3, \mathbf{x}_1)}{\|\mathbf{x}_1\|^2} \mathbf{x}_1 - \frac{(\mathbf{y}_3, \mathbf{x}_2)}{\|\mathbf{x}_2\|^2} \mathbf{x}_2$$

$$= x^2 - \left(\frac{\frac{2}{3}}{2} \right) 1 - 0$$

$$= x^2 - \frac{1}{3}$$

Thus, the set $\{1, x, x^2 - \frac{1}{3}\}$ is an orthogonal basis for P_2. Next, we divide each basis vector by its norm.

We have seen that $(\mathbf{x}_1, \mathbf{x}_1) = 2$. Therefore, $\|\mathbf{x}_1\| = \sqrt{2}$. Furthermore,

$$(\mathbf{x}_2, \mathbf{x}_2) = \int_{-1}^{1} x^2 \, dx = \frac{2}{3}$$

and

$$(\mathbf{x}_3, \mathbf{x}_3) = \int_{-1}^{1} \left(x^2 - \frac{1}{3} \right)^2 dx = \frac{8}{45}$$

Therefore, $\|\mathbf{x}_2\| = \sqrt{2/3}$, and $\|\mathbf{x}_3\| = 2\sqrt{2/3}\sqrt{5}$. Dividing each $\mathbf{x}_i$ by its norm, we obtain the orthonormal basis

$$\left\{ \frac{1}{\sqrt{2}}, \frac{\sqrt{3}}{\sqrt{2}}x, \frac{3\sqrt{5}}{2\sqrt{2}}\left(x^2 - \frac{1}{3}\right) \right\}$$

for P_2.

The example above can be extended to P_n for any positive integer n by choosing $\{1, x, x^2, \ldots, x^n\}$ as the initial basis. As the Gram–Schmidt process is applied to polynomials of higher degree, the polynomials of lower degree are not changed. Thus, we obtain a sequence of polynomials, $p_0(x), p_1(x), \ldots, p_n(x), \ldots$ such that for any n, the first $n + 1$ polynomials in the sequence is an orthonormal basis for P_n. These polynomials are called the ***normalized Legendre polynomials***. We have computed the first three of these. Such polynomials are useful in representing solutions to certain differential equations. They are also used in the statistical design of experiments.

Complex Inner Product Spaces

The concept of "inner product" can be applied to complex vector spaces. In this case, the inner product is a complex-valued function. However, one of the axioms in the definition of inner product must be modified to avoid a mathematical contradiction (see Exercise 30).

Definition   A ***complex inner product*** on a complex vector space V is a complex-valued function which assign to any ordered pair of vectors $\mathbf{x}$ and $\mathbf{y}$ in V a complex scalar, denoted $(\mathbf{x}, \mathbf{y})$, such that the axioms (a), (c), and (d) for the definition of a real inner product are satisfied except that the scalar in axiom (d) is complex and such that axiom (b) is replaced by

$$(b')\quad \overline{(\mathbf{x}, \mathbf{y})} = (\mathbf{y}, \mathbf{x})$$

where the bar denotes the complex conjugate. A complex vector space with a complex inner product is called a ***complex inner product space***.

Example 5   We define a complex inner product on the complex vector space C^n as follows. For $\mathbf{x} = (x_1, x_2, \ldots, x_n)$ and $\mathbf{y} = (y_1, y_2, \ldots, y_n)$, let

$$(\mathbf{x}, \mathbf{y}) = \sum_{i=1}^{n} x_i \bar{y}_i$$

For example, if $\mathbf{x} = (1 + i, 2)$ and $\mathbf{y} = (3 + i, 1 - 2i)$, then

$$(\mathbf{x}, \mathbf{y}) = (1 + i)\overline{(3 + i)} + 2\overline{(1 - 2i)}$$

$$= (1 + i)(3 - i) + 2(1 + 2i)$$

$$= 6 + 6i$$

To show that this mapping is a complex inner product on C^n, we must verify axioms (a), (c), and (d) of the definition of inner product as well as axiom (b'). We shall verify axioms (a) and (b') and leave the rest as exercises. In what follows, let $\mathbf{x} = (x_1, x_2, \ldots, x_n)$ and $\mathbf{y} = (y_1, y_2, \ldots, y_n)$ be elements of C^n.

(a) Suppose that $\mathbf{x} \neq \mathbf{0}$. Then $x_j \neq 0$ for some j. So $x_j \bar{x}_j > 0$. Since $x_i \bar{x}_i \geq 0$ for all i, we have

$$(\mathbf{x}, \mathbf{x}) = \sum_{i=1}^{n} x_i \bar{x}_i > 0$$

(b')
$$(\overline{\mathbf{x}, \mathbf{y}}) = \overline{\sum_{i=1}^{n} x_i \bar{y}_i}$$

$$= \sum_{i=1}^{n} \overline{x_i \bar{y}_i}$$

$$= \sum_{i=1}^{n} \bar{x}_i y_i$$

$$= \sum_{i=1}^{n} y_i \bar{x}_i$$

$$= (\mathbf{y}, \mathbf{x})$$

Example 6   Let $[a, b]$ be a closed interval in R. Any complex function $f : [a, b] \to C$ can be expressed in the form $f(x) = f_1(x) + f_2(x)i$, where f_1 and f_2 are real-valued functions defined on $[a, b]$. The functions f_1 and f_2 are called the ***real*** and the ***imaginary parts*** of f, respectively. The function f is ***continuous*** if both f_1 and f_2 are continuous. For a continuous function f, we define the integral

$$\int_a^b f(x) \, dx = \int_a^b f_1(x) \, dx + \left[\int_a^b f_2(x) \, dx \right] i$$

Let V be the subset of $F([a, b], C)$ consisting of all the continuous functions from $[a, b]$ to C. It can be verified that V is a subspace of $F([a, b], C)$, and hence is a complex vector space. For any f and g in V, let

$$(f, g) = \int_a^b f(x)\overline{g(x)}\,dx$$

This defines an inner product on V. We give an example of how to compute (f, g) for specific functions f and g and then show that axiom (b′) is satisfied. The verifications of the other axioms are left as exercises.

Suppose that $f(x) = x + x^2 i$, $g(x) = 2 + 8xi$, $a = 0$, and $b = 1$. Then

$$(f, g) = \int_0^1 (x + x^2 i)\overline{(2 + 8xi)}\,dx$$

$$= \int_0^1 (x + x^2 i)(2 - 8xi)$$

$$= \int_0^1 (2x + 8x^3)\,dx - \left(\int_0^1 6x^2\,dx\right)i$$

$$= 3 - 2i$$

(b′) Let $f = f_1 + f_2 i$ and $g = g_1 + g_2 i$ be functions in V. Then

$$\overline{(f, g)} = \overline{\int_a^b f(x)\overline{g(x)}\,dx}$$

$$= \int_a^b \overline{f(x)\overline{g(x)}}\,dx$$

$$= \int_a^b \overline{f(x)}g(x)\,dx$$

$$= \int_a^b g(x)\overline{f(x)}\,dx$$

$$= (g, f)$$

Note that in going from the first to the second line of the proof above, we used the fact from calculus that the conjugate of an integral is the integral of the conjugate.

We conclude this section with several comments on complex inner product spaces. Complex inner products share most of the important properties of real inner products. For example, the Cauchy–Schwarz inequality and the Gram–Schmidt orthogonalization process are both valid.

One difference that does arise is due to axiom (b′):

If V is a complex inner product space, then for any vectors $\mathbf{x}$ and $\mathbf{y}$ in V, and for any scalar c,

$$(\mathbf{x}, c\mathbf{y}) = \bar{c}(\mathbf{x}, \mathbf{y})$$

(see Exercise 26). Compare this result with Exercise 14(d) for real inner product spaces.

Exercises

In Exercises 1–4, a real vector space V and a rule defining $(\mathbf{x}, \mathbf{y})$ are given. Determine if the rule defines an inner product on V. Justify your conclusion.

1. $V = R^n$, $(\mathbf{x}, \mathbf{y}) = 2(\mathbf{x} \cdot \mathbf{y})$
2. $V = R^n$, $(\mathbf{x}, \mathbf{y}) = |\mathbf{x} \cdot \mathbf{y}|$
3. Let V be any vector space on which are defined two inner products, $(\mathbf{x}, \mathbf{y})_1$ and $(\mathbf{x}, \mathbf{y})_2$, for $\mathbf{x}$ and $\mathbf{y}$ in V. Define $(\mathbf{x}, \mathbf{y})$ by

$$(\mathbf{x}, \mathbf{y}) = (\mathbf{x}, \mathbf{y})_1 + (\mathbf{x}, \mathbf{y})_2$$

4. $V = C([0, 2])$,

$$(f, g) = \int_0^1 f(x)g(x)\, dx$$

for f and g in V.

5. Let V be any real finite-dimensional vector space and let $S = \{\mathbf{x}_1, \mathbf{x}_2, \ldots, \mathbf{x}_n\}$ be an ordered basis for V. For $\mathbf{x}$ and $\mathbf{y}$ in V, define $(\mathbf{x}, \mathbf{y})$ by

$$(\mathbf{x}, \mathbf{y}) = [\mathbf{x}]_S \cdot [\mathbf{y}]_S$$

Prove that this defines an inner product on V.

6. Let A be an $n \times n$ positive definite matrix (see Section 7.2). For $\mathbf{x}$ and $\mathbf{y}$ in R^n define

$$(\mathbf{x}, \mathbf{y}) = (A\mathbf{x}) \cdot \mathbf{y}$$

Show that this defines an inner product on R^n.

7. Let A be an $n \times n$ invertible matrix. For $\mathbf{x}$ and $\mathbf{y}$ in R^n define

$$(\mathbf{x}, \mathbf{y}) = (A\mathbf{x}) \cdot (A\mathbf{y})$$

Show that this defines an inner product on R^n.

8. Verify axioms (c) and (d) of the definition of inner product for Example 1.
9. Verify axioms (c) and (d) of the definition of inner product for Example 2.
10. Find an orthogonal basis for the subspace of the inner product space of Example 1 for the interval $[0, 1]$ spanned by $\{1, e^x, e^{-x}\}$.
11. Use the inner product of Example 3 to show that $\sin mx$ and $\sin nx$ are orthogonal for any distinct integers m and n.
12. Use the methods of Example 4 to compute the fourth normalized Legendre polynomial.
13. Let V be a real inner product space. For any vector $\mathbf{y}$ in V we define the function $F_{\mathbf{y}}: V \to R$ by

$$F_{\mathbf{y}}(\mathbf{x}) = (\mathbf{x}, \mathbf{y}) \qquad \text{for all } \mathbf{x} \text{ in } V$$

Prove that $F_{\mathbf{y}}$ is a linear transformation.
14. Prove that for any inner product space V and any vectors $\mathbf{x}$, $\mathbf{y}$, and $\mathbf{z}$ in V, and any scalar c:
   (a) $(\mathbf{0}, \mathbf{x}) = (\mathbf{x}, \mathbf{0}) = 0$
   (b) $(\mathbf{x} - \mathbf{z}, \mathbf{y}) = (\mathbf{x}, \mathbf{y}) - (\mathbf{z}, \mathbf{y})$
   (c) $(\mathbf{y}, \mathbf{x} - \mathbf{z}) = (\mathbf{y}, \mathbf{x}) - (\mathbf{y}, \mathbf{z})$
   (d) $(\mathbf{x}, c\mathbf{y}) = c(\mathbf{x}, \mathbf{y})$
15. Prove that for any inner product space V, any vector $\mathbf{x}$ in V, and any scalar t:
   (a) $\|\mathbf{x}\| = 0$ if and only if $\mathbf{x} = \mathbf{0}$
   (b) $\|t\mathbf{x}\| = |t|\,\|\mathbf{x}\|$

16. Prove that for any real inner product space V, and vectors $\mathbf{y}$ and $\mathbf{z}$ in V:
    (a) If $(\mathbf{x}, \mathbf{z}) = 0$ for all $\mathbf{x}$ in V, then $\mathbf{z} = \mathbf{0}$.
    (b) If $(\mathbf{x}, \mathbf{y}) = (\mathbf{x}, \mathbf{z})$ for all $\mathbf{x}$ in V, then $\mathbf{y} = \mathbf{z}$.

17. Let V be an n-dimensional real inner product space with an orthonormal basis $S = \{\mathbf{x}_1, \mathbf{x}_2, \ldots, \mathbf{x}_n\}$. Prove that for any vectors $\mathbf{x}$ and $\mathbf{y}$ in V,

$$(\mathbf{x}, \mathbf{y}) = [\mathbf{x}]_S \cdot [\mathbf{y}]_S$$

The following definitions are used in Exercises 18, 19, and 29. Let V and W be inner product spaces. A linear transformation $T: V \to W$ is called a **linear isometry** if T is an isomorphism and $(T(\mathbf{x}), T(\mathbf{y})) = (\mathbf{x}, \mathbf{y})$ for any vectors $\mathbf{x}$ and $\mathbf{y}$ in V. The inner product spaces V and W are **isometric** if there exists a linear isometry from V to W.

18. Let V, W, and Z be inner product spaces. Prove:
    (a) V is isometric to itself.
    (b) If V is isometric to W, then W is isometric to V.
    (c) If V is isometric to W, and W is isometric to Z, then V is isometric to Z.

19. Prove that for any n-dimensional real inner product space V and any orthonormal basis S of V, the linear transformation $T_S: V \to R^n$ defined by $T_S(\mathbf{x}) = [\mathbf{x}]_S$ is a linear isometry. Prove that any n-dimensional real inner product space is isometric to R^n with the dot product.

20. Let V be a real inner product space. Prove that if $\{\mathbf{x}, \mathbf{y}\}$ is a linearly dependent subset of V, then

$$(\mathbf{x}, \mathbf{y})^2 = (\mathbf{x}, \mathbf{x})(\mathbf{y}, \mathbf{y})$$

21. Prove the converse of Exercise 20. For any real inner product space V, if $(\mathbf{x}, \mathbf{y})^2 = (\mathbf{x}, \mathbf{x})(\mathbf{y}, \mathbf{y})$, then $\{\mathbf{x}, \mathbf{y}\}$ is linearly dependent. *Hint:* If $\mathbf{x}$ or $\mathbf{y}$ is zero, then the result is trivial. Suppose that $\mathbf{x}$ and $\mathbf{y}$ are nonzero vectors. Show that

$$\left\| \mathbf{y} - \frac{(\mathbf{x}, \mathbf{y})}{(\mathbf{x}, \mathbf{x})} \mathbf{x} \right\| = 0$$

22. Let V be a finite-dimensional real inner product space. Prove the converse of Exercise 13. That is, prove that for any linear transformation $T: V \to R$, there exists a unique vector $\mathbf{y}$ in V such that $T = F_\mathbf{y}$. *Hint:* Let $\{\mathbf{x}_1, \mathbf{x}_2, \ldots, \mathbf{x}_n\}$ be an orthonormal basis for V. Define $\mathbf{y}$ by

$$\mathbf{y} = T(\mathbf{x}_1)\mathbf{x}_1 + T(\mathbf{x}_2)\mathbf{x}_2 + \cdots + T(\mathbf{x}_n)\mathbf{x}_n$$

23. (a) Prove that for any invertible matrix B, $B^t B$ is positive definite.
    (b) Use part (a) and Exercise 17 to prove the converse of Exercise 6; that is, show that for any inner product on R^n, there exists a positive definite matrix A such that

$$(\mathbf{x}, \mathbf{y}) = A\mathbf{x} \cdot \mathbf{y}$$

for any vectors $\mathbf{x}$ and $\mathbf{y}$ in R^n.

The following exercises are for complex inner product spaces.

24. Verify axioms (c) and (d) of the definition of the complex inner product for the space in Example 5.

25. Verify axioms (a), (c), and (d) of the definition of the complex inner product for the space in Example 6.

26. Suppose that V is a complex inner product space and $\mathbf{x}$ and $\mathbf{y}$ are vectors in V. Prove that

$$(\mathbf{x}, c\mathbf{y}) = \bar{c}(\mathbf{x}, \mathbf{y})$$

for any scalar c.

27. Show that the results of Exercise 15 hold for complex inner product spaces.

28. Show that the results of Exercise 16 hold for complex inner product spaces.

29. Prove the complex version of Exercise 19: Let V be an n-dimensional complex inner product space, and let C^n be the complex inner product space of Example 5. For any orthonormal basis S of V, the linear transformation $T_S: V \to C^n$ defined by $T_S(\mathbf{x}) = [\mathbf{x}]_S$ is a linear isometry.

30. Let V be a nonzero complex vector space. Show that there cannot be an inner product on V which satisfies axiom (b), rather than axiom (b'). [*Hint:* Suppose that one exists. Pick a nonzero vector $\mathbf{x}$ in V. Then $(\mathbf{x}, \mathbf{x}) > 0$ by axiom (a). Now compute $(i\mathbf{x}, i\mathbf{x})$.]

Key Words

9

Numerical Methods

In Section 1.5 we introduced Gaussian elimination to solve a system of linear equations. It was shown that it would take approximately $n^3/3$ multiplications to solve an $n \times n$ system. So for a system with $n > 10$, computation by hand becomes unrealistic. It is therefore important to consider what problems may arise when a computer is employed to solve such a system. Basically, there are three considerations.

First, the size of the *memory* of the computer will influence the size of the system that we can solve. Although there exist techniques for moving data from inside the computer to external storage devices and back, physical constraints still limit the amount of information that can be processed. It should be pointed out, however, that systems with $n = 80$ are manageable for a large class of microcomputers.

Second, the *speed* at which a computer can solve a system will basically be a function of the computer itself as well as the algorithm (procedure) that the computer employs. In this chapter we are concerned with the algorithms. Our measure of the relative speeds of the various algorithms will be the number of operations each algorithm requires. There are, of course, factors other than the number of operations which influence speed. However, the techniques for dealing with these factors are beyond the scope of this text.

Third, the *round-off errors* that computers invariably introduce into computations may greatly affect the final solution of the system. These errors are a function of the number of significant digits that the computer employs to represent a number. To be more precise, many computers represent a number x by the *floating-point form*

$$\hat{x} = \pm(.a_1 a_2 \cdots a_n) \times 10^k$$

371

where $a_1 \neq 0$, the a_i's, $i = 1, 2, \ldots, n$, are digits between 0 and 9, and $-N \leq k \leq N$. The number N depends on the size of the computer. For most computers $N \leq 77$; for many microcomputers $N \leq 37$. The number n is called the *number of significant digits* in the representation. For example,

$$x = \frac{8}{3} = 2.666\ldots$$

has either of the floating forms

$$0.266666666 \times 10^1 \quad \text{or} \quad 0.266666667 \times 10^1$$

in many microcomputers and calculators. In this case the number of significant digits is $n = 9$. The first form is the result of *truncating* or *chopping*, whereas the second form is the result of *rounding*. To see how these forms are produced, we first write the number x as

$$(.a_1 a_2 \cdots a_n a_{n+1} \cdots) \times 10^k$$

Under truncation, we simply use the first n digits and obtain

$$\hat{x} = \pm(.a_1 a_2 \cdots a_n) \times 10^k$$

Under the method of rounding, we truncate as before if $a_{n+1} < 5$. If $a_{n+1} \geq 5$, we add 1 to a_n and then truncate. Put another way, we add 5 to a_{n+1} and then truncate the resulting number.

Errors are usually produced in computations that use either of these representations. When $\hat{x}$ is used instead of x, two particular types of errors arise. First, there is the **absolute error** $x - \hat{x}$. Second, there is the **relative error** $(x - \hat{x})/x$. In general, the relative error is of more concern to us.

Additional errors arise even when the computer is working with numbers it can represent exactly. For example, suppose that

$$x = 0.212 \times 10^{10} \quad \text{and} \quad y = 0.1 \times 10^{-6}$$

On a calculator that uses nine significant digits and $N = 99$, we have $\hat{x} = x$ and $\hat{y} = y$. But the representation of $(x + y)$ is $0.212 \times 10^{10} = x = \hat{x}$. The reason for this is that the machine must align the decimal points of both numbers before it performs the addition. Because the calculator can remember only nine significant digits, the digits of the smaller number are lost. Symbolically, we have

$$(x + y)^\wedge \neq \hat{x} + \hat{y}$$

In addition to the machine errors that we have just discussed, the so-called *propagation of errors* due to repeated rounding will be greatly influenced by the choice of algorithm. In fact, such errors can be extreme for even very small systems.

All of these considerations also apply to computations other than those involved with the solution of systems. For example, the computation of eigenvalues and eigenvectors is also subject to these influences.

The methods that we shall consider in this chapter may be divided into two classes: *direct methods* and *iterative methods*. Microcomputer programs for some of these methods are listed in Appendix B. Numerical packages such as LINPACK or EISPACK which are designed for large computers are listed in the References.

9.1 DIRECT METHODS FOR SOLVING SYSTEMS OF LINEAR EQUATIONS

In this section we are concerned with the system $A\mathbf{x} = \mathbf{b}$, where A is an invertible matrix. We shall see that when Gaussian elimination can be applied without any row interchanges, it will be possible to "factor" A as a product of triangular matrices.

Pivoting

To see how machine errors affect Gaussian elimination, we consider the following system:

$$0.00300x + 59.18y = 59.21$$

$$5.282x - 6.230y = 46.59$$

The exact solution to this system is $x = 10$ and $y = 1$. Suppose that we have a computer which performs calculations in four significant digits with rounding. Recall that in Gaussian elimination, we must add $-(5.282/0.00300)$ times the first equation to the second equation to eliminate x from the second equation. This *multiplier* equals $-1760.\overline{6}$, which our computer represents as -1761. Using similar rounding, we obtain the new system

$$0.00300x + 59.18y = 59.21$$

$$-104200y = -104300$$

The solution to this system is represented by $y = 1.001$ and $x = -10$. So we see that although the error in y is only 0.001, the error in x is 20.

Now suppose that we interchange the two equations and apply Gaussian elimination to solve the system

$$5.282x - 6.230y = 46.59$$

$$0.00300x + 59.18y = 59.21$$

For this system we use the multiplier $(-0.00300/5.282)$, which the computer represents as -0.0005680. The new system is

$$5.282x - 6.230y = 46.59$$

$$59.18y = 59.18$$

The solution of this system is represented by $y = 1$ and $x = 10$. The improvement in accuracy due to one interchange is quite remarkable.

The difficulty that occurs in the first method is caused by the fact that the multiplier $-(5.282/0.00300)$ is relatively large compared to the coefficients in the second equation. We have seen that information is lost when large numbers are added to small numbers. The technique that helps to avoid this problem is called *pivoting*. Before we describe this technique, we introduce some terminology.

Recall that at each stage in Gaussian elimination, a multiple of one equation is used to eliminate an unknown from the remaining equations. The coefficient of this unknown in the given equation is called the *pivot*. For example, in the first system the pivot is 0.00300 and the associated multiplier used to eliminate x from the second equation is approximately -1761. In the second system, the pivot is 5.282 and the associated multiplier is approximately -0.0006. We have seen that if the multiplier is relatively large, significant rounding errors can occur. Thus, we should choose the pivot at each stage which has the smallest associated multipliers. This can be accomplished by choosing as our next pivot the coefficient of the unknown with the greatest magnitude (absolute value) from the remaining equations. We then perform an interchange and use this pivot to eliminate the unknown from the remaining equations. The technique is called *partial pivoting*. We illustrate the method in the next example.

Example 1   Solve the system associated with the augmented matrix given below using partial pivoting. Assume that you are using a computer that rounds to five digits.

$$\begin{bmatrix} 2 & 20 & 10 & 2 \\ 10 & 12 & 5 & 8 \\ 1 & 1 & -2 & -4 \end{bmatrix} \rightarrow \begin{bmatrix} 10 & 12 & 5 & 8 \\ 2 & 20 & 10 & 2 \\ 1 & 1 & -2 & -4 \end{bmatrix} \rightarrow \begin{bmatrix} 10 & 12 & 5 & 8 \\ 0 & 17.6 & 9 & 0.4 \\ 0 & -0.2 & -2.5 & -4.8 \end{bmatrix}$$

$$\rightarrow \begin{bmatrix} 10 & 12 & 5 & 8 \\ 0 & 17.6 & 9 & 0.4 \\ 0 & 0 & -2.3977 & -4.7955 \end{bmatrix}$$

Back-solving this system, we obtain $x = 1.0000$, $y = -1.0000$, and $z = 2.0000$. The exact solution is $x = 1$, $y = -1$, and $z = 2$.

We see that partial pivoting is easy to apply. It is also quite easy to program a computer to use this method. Another "enhancement" for Gaussian elimination is *total* or *complete pivoting*. For this method we begin by finding the coefficient A_{ij} with the largest magnitude. We then use the ith equation to eliminate x_j from all the remaining equations. We repeat this process until the final system can be backsolved as we did earlier for triangular systems. This method reduces round-off errors more than partial pivoting, but it is harder to program because it is necessary to keep track of the order in which the variables are eliminated. It is generally sufficient to use partial pivoting instead of total pivoting.

Another consideration in solving a system is *scaling*. Scaling is achieved by multiplying one or more of the equations by a nonzero scalar. We do this so that all or most of the coefficients in the system have the same relative size. This should be done *before* pivoting takes place as the next example illustrates.

Example 2   Consider the system

$$ax \quad\quad = a$$

$$x + \frac{1}{a} y = \frac{1}{a}$$

where a is a relatively small number. Clearly, the exact solution is $x = 1$ and $y = 1 - a$. To use our pivoting technique, we choose 1 as our pivot and perform an interchange, obtaining

$$x + \frac{1}{a} y = \frac{1}{a}$$

$$ax \quad\quad = a$$

After we eliminate x from the second equation, we obtain the system

$$x + \frac{1}{a} y = \frac{1}{a}$$

$$- \quad y = a - 1$$

As we begin to back-solve, we obtain $y = 1 - a$ which our computer will represent as 1 if a is relatively small. This yields $x = 0$ when substituted into the first equation. On the other hand, suppose that we *scale* the original system by multiplying the first equation by $1/a$. We then obtain the system

$$x \quad\quad = 1$$

$$x + \frac{1}{a} y = \frac{1}{a}$$

If we add -1 times the first equation to the second equation, we obtain the new second equation $(1/a)y = 1/a - 1$, the right side of which will be represented as $1/a$ because $1/a$ is relatively large. This yields the solution $y = 1$ and $x = 1$.

_____

As you can see, scaling resulted in a marked improvement. Unfortunately, no general procedure for scaling that works consistently is available.

Multipliers

We have mentioned that it takes approximately $n^3/3$ multiplications to solve the $n \times n$ system $A\mathbf{x} = \mathbf{b}$ by Gaussian elimination. In many applications the vector $\mathbf{b}$ will vary for a fixed $n \times n$ matrix A. For example, to find A^{-1} we can solve $A\mathbf{x} = \mathbf{e}_i$ for each i. The solutions of these systems yield the columns of A^{-1}. Because most of the work in solving $A\mathbf{x} = \mathbf{b}$ is done to reduce A to triangular form, it would be inefficient to repeat this for each vector $\mathbf{b}$. Thus, it would be desirable to develop a procedure for recording the row operations we use to reduce A to triangular form and then to apply these operations to the vector $\mathbf{b}$.

To see what is happening symbolically during this reduction, consider the augmented matrix $(A|\mathbf{b})$. To solve the system $A\mathbf{x} = \mathbf{b}$, we first reduce the augmented matrix to the triangular form $(A'|\mathbf{b}')$. Once we know A' and have a record of the row operations to produce $\mathbf{b}'$, we can quickly solve the system $A\mathbf{x} = \mathbf{c}$ for any column vector $\mathbf{c}$. To make things concrete, we consider the following system:

$$2x + y + 3z = 1$$
$$2x + 5y + 7z = 5$$
$$6x - 5y + 2z = -6$$

The augmented matrix for this system is

$$\begin{bmatrix} 2 & 1 & 3 & 1 \\ 2 & 5 & 7 & 5 \\ 6 & -5 & 2 & -6 \end{bmatrix}$$

To eliminate 2 from the second row, we multiply the first row by -1 and add the result to the second row. The multiplier -1 is denoted by m_{21}. We shall see that we must keep track of these multipliers to solve other systems in which A is the coefficient matrix. To eliminate 6 from the third row, we multiply the first row by -3 and add the result to the third row. In our new notation, this means that $m_{31} = -3$. After applying these operations, we obtain

$$\begin{bmatrix} 2 & 1 & 3 & 1 \\ 0 & 4 & 4 & 4 \\ 0 & -8 & -7 & -9 \end{bmatrix}$$

Similarly, we let $m_{32} = 2$ and obtain

$$\begin{bmatrix} 2 & 1 & 3 & 1 \\ 0 & 4 & 4 & 4 \\ 0 & 0 & 1 & -1 \end{bmatrix}$$

The solution for this vector $\mathbf{b}$ is $(1, 2, -1)$. To see how these multipliers may be used to obtain the constants in the triangular form above, we have

$$\begin{bmatrix} 1 \\ 5 \\ -6 \end{bmatrix} \rightarrow \begin{bmatrix} 1 \\ 5 + m_{21}(1) \\ -6 + m_{31}(1) \end{bmatrix} \rightarrow \begin{bmatrix} 1 \\ 5 + m_{21}(1) \\ -6 + m_{31}(1) + m_{32}[5 + m_{21}(1)] \end{bmatrix}$$

$$= \begin{bmatrix} 1 \\ 5 + (-1)(1) \\ -6 + (-3)(1) + (2)[5 + (-1)(1)] \end{bmatrix}$$

$$= \begin{bmatrix} 1 \\ 4 \\ -1 \end{bmatrix}$$

In terms of efficient storage in a computer, these multipliers may be overlaid in the zero entries they are used to produce. For our matrix A we have

$$\begin{bmatrix} 2 & 1 & 3 \\ 2 & 5 & 7 \\ 6 & -5 & 2 \end{bmatrix} \rightarrow \begin{bmatrix} 2 & 1 & 3 \\ -1 & 4 & 4 \\ -3 & -8 & -7 \end{bmatrix} \rightarrow \begin{bmatrix} 2 & 1 & 3 \\ -1 & 4 & 4 \\ -3 & 2 & 1 \end{bmatrix}$$

The positions of these multipliers serve as a reminder as to how they are used. For example, consider the system

$$2x + y + 3z = 8$$
$$2x + 5y + 7z = 16$$
$$6x - 5y + 2z = 9$$

Because we know the triangular form of the matrix A, we must find the associated constant vector $\mathbf{b}'$. We use the multipliers as follows:

$$\begin{bmatrix} 8 \\ 16 \\ 9 \end{bmatrix} \rightarrow \begin{bmatrix} 8 \\ 16 + (-1)8 \\ 9 + (-3)8 \end{bmatrix} = \begin{bmatrix} 8 \\ 8 \\ -15 \end{bmatrix} \rightarrow \begin{bmatrix} 8 \\ 8 \\ -15 + (2)(8) \end{bmatrix}$$

$$= \begin{bmatrix} 8 \\ 8 \\ 1 \end{bmatrix}$$

Thus, the triangular form of the augmented matrix is

$$\begin{bmatrix} 2 & 1 & 3 & 8 \\ 0 & 4 & 4 & 8 \\ 0 & 0 & 1 & 1 \end{bmatrix}$$

Back-solving yields the solution (2, 1, 1).

It should be pointed out that the alternative method of first finding A^{-1} and computing the solution $A^{-1}\mathbf{b}$ for each new vector $\mathbf{b}$ is less efficient.

LU-Decomposition

To obtain a more flexible approach to the reduction of the matrix A above to triangular form, we consider several approaches to writing A as a product of triangular matrices. First form the matrix L_1 equal to

$$\begin{bmatrix} 1 & 0 & 0 \\ m_{21} & 1 & 0 \\ m_{31} & 0 & 1 \end{bmatrix} = \begin{bmatrix} 1 & 0 & 0 \\ -1 & 1 & 0 \\ -3 & 0 & 1 \end{bmatrix}$$

Now form the product

$$L_1 A = \begin{bmatrix} 1 & 0 & 0 \\ -1 & 1 & 0 \\ -3 & 0 & 1 \end{bmatrix}\begin{bmatrix} 2 & 1 & 3 \\ 2 & 5 & 7 \\ 6 & -5 & 2 \end{bmatrix} = \begin{bmatrix} 2 & 1 & 3 \\ 0 & 4 & 4 \\ 0 & -8 & -7 \end{bmatrix}$$

Similarly, let L_2 be the matrix

$$\begin{bmatrix} 1 & 0 & 0 \\ 0 & 1 & 0 \\ 0 & m_{32} & 1 \end{bmatrix} = \begin{bmatrix} 1 & 0 & 0 \\ 0 & 1 & 0 \\ 0 & 2 & 1 \end{bmatrix}$$

We have

$$L_2 L_1 A = \begin{bmatrix} 1 & 0 & 0 \\ 0 & 1 & 0 \\ 0 & 2 & 1 \end{bmatrix}\begin{bmatrix} 2 & 1 & 3 \\ 0 & 4 & 4 \\ 0 & -8 & -7 \end{bmatrix} = \begin{bmatrix} 2 & 1 & 3 \\ 0 & 4 & 4 \\ 0 & 0 & 1 \end{bmatrix}$$

where we denote the last upper triangular matrix by U. The computation above is not surprising if we recall that the lower triangular matrices (a matrix is *lower triangular* if the entries above the diagonal are zero) which multiply A on the left act in the same way as the elementary matrices which are used to reduce A to triangular form. In fact, the lower triangular matrices are each products of these elementary matrices. We have just shown that $L_2 L_1 A = U$ or

$$A = (L_2 L_1)^{-1} U$$

which may be written as $A = LU$. We leave it as an exercise to show that the product of lower triangular matrices is lower triangular and that the inverse of a lower triangular matrix is lower triangular. Thus, L is lower triangular. Hence, we have factored A into what is called an **_LU-decomposition_**. Later, we shall discuss a better procedure for computing the matrices L and U.

Suppose that we have such a factorization. Consider the system

$$A\mathbf{x} = \mathbf{b}$$

or

$$LU\mathbf{x} = \mathbf{b}$$

If we let $\mathbf{y} = U\mathbf{x}$, then we have the system $L\mathbf{y} = \mathbf{b}$. Because this system is in triangular form, it is easy to solve for $\mathbf{y}$ by *forward substitution*. Once $\mathbf{y}$ is known, we find $\mathbf{x}$ by solving the triangular system $U\mathbf{x} = \mathbf{y}$. This method of factoring (approximately $n^3/3$ multiplication) and then solving (approximately n^2 operations) is more efficient than solving $A\mathbf{x} = \mathbf{b}$ for various column vectors $\mathbf{b}$ (approximately $n^3/3$ for each column vector).

To illustrate this technique, we shall solve the system $A\mathbf{x} = \mathbf{b}$ given below.

$$x + 2y + z = 1$$
$$x + y + 5z = 2$$
$$x + 2y + 2z = 3$$

This system was solved in Example 5 of Section 3.4 by computing the inverse of A. Suppose that we are given the LU-decomposition

$$\begin{bmatrix} 1 & 2 & 1 \\ 1 & 1 & 5 \\ 1 & 2 & 2 \end{bmatrix} = \begin{bmatrix} 1 & 0 & 0 \\ 1 & 1 & 0 \\ 1 & 0 & 1 \end{bmatrix}\begin{bmatrix} 1 & 2 & 1 \\ 0 & -1 & 4 \\ 0 & 0 & 1 \end{bmatrix}$$

We first apply forward substitution to solve the system $L\mathbf{y} = \mathbf{b}$, that is,

$$x' = 1$$
$$x' + y' = 2$$
$$x' + z' = 3$$

We obtain $x' = 1$, $y' = 1$, and $z' = 2$. Now we use backward substitution to solve the system $U\mathbf{x} = \mathbf{y}$, or

$$
\begin{aligned}
x + 2y + \ z &= 1 \\
- \ y + 4z &= 1 \\
z &= 2
\end{aligned}
$$

This system has the solution $x = -15$, $y = 7$, and $z = 2$, which agrees with the solution obtained earlier.

It should be noted that in the development of the LU-decomposition, we have assumed that no row interchanges are necessary. The elementary matrix corresponding to a row interchange is *not* lower triangular. However, if such an interchange is necessary, we may accomplish this by multiplying first by a ***permutation matrix*** P. In general, such a matrix P is formed by permuting (ordering) the rows of an identity matrix. The matrix PA is the matrix which results when the same permutation is applied to the rows of A. Actually in the case of one interchange, P is simply an elementary matrix. If an interchange is necessary at the ith step, then instead of the lower triangular matrix L_i, we must use matrix $L_i P_i$. Allowing for the possibility that P_i is the identity matrix, that is, no interchange is necessary, we now have

$$
U = (L_{n-1} P_{n-1}) \cdots (L_1 P_1) A
$$

In fact, it can be shown that if interchanges are necessary, then A can be factored as $A = PLU$, where P is a permutation matrix.

As we have seen, if an LU-decomposition exists for A, it may be possible to use Gaussian elimination without pivoting to find the matrices L and U. A more direct approach is to treat the entries of L and U as unknowns in the matrix equation $A = LU$. Because there are a total of $n(n + 1) = n^2 + n$ unknowns and n^2 equations, we are at liberty to specify n of these unknowns. As we have seen, Gaussian elimination produces a matrix L with 1's on the diagonal. For now, we assign 1's to the diagonal entries of L. We illustrate the method for the matrix

$$
A = \begin{bmatrix} 4 & 2 & 3 \\ -1 & 5 & 4 \\ 3 & 1 & 3 \end{bmatrix} = \begin{bmatrix} 1 & 0 & 0 \\ l_{21} & 1 & 0 \\ l_{31} & l_{32} & 1 \end{bmatrix} \begin{bmatrix} u_{11} & u_{12} & u_{13} \\ 0 & u_{22} & u_{23} \\ 0 & 0 & u_{33} \end{bmatrix}
$$

This matrix equation yields the equations

$$
\begin{aligned}
u_{11} &= 4 & u_{12} &= 2 \\
l_{21} u_{11} &= -1 & l_{21} u_{12} + u_{22} &= 5 \\
l_{31} u_{11} &= 3 & l_{31} u_{12} + l_{32} u_{22} &= 1
\end{aligned}
$$

and

$$u_{13} = 3$$

$$l_{21}u_{13} + u_{23} = 4$$

$$l_{31}u_{13} + l_{32}u_{23} + u_{33} = 3$$

Solving these equations successively yields

$$\begin{bmatrix} 4 & 2 & 3 \\ -1 & 5 & 4 \\ 3 & 1 & 3 \end{bmatrix} = \begin{bmatrix} 1 & 0 & 0 \\ -\frac{1}{4} & 1 & 0 \\ \frac{3}{4} & -\frac{1}{11} & 1 \end{bmatrix} \begin{bmatrix} 4 & 2 & 3 \\ 0 & \frac{11}{2} & \frac{19}{4} \\ 0 & 0 & \frac{13}{11} \end{bmatrix}$$

The program LU-DECOMP in Appendix B gives the matrices L and U for any matrix A possessing such a decomposition. Notice that the determinant of such a matrix A factored as above is now very easy to compute. In fact, $|A| = |L||U| = (1)(1)(1)(4)(\frac{11}{2})(\frac{13}{11}) = 26$.

A question that now arises concerns the uniqueness of the LU-decomposition, $A = LU$. If we let D be any invertible diagonal matrix, we may write

$$A = LU = (LD)(D^{-1}U)$$

Because the matrix LD is lower triangular and the matrix $D^{-1}U$ is upper triangular, we see that the factorization is not unique. This situation may be easily remedied. We call a matrix **unit upper triangular** if it is upper triangular and if all its diagonal entries are equal to 1. A similar definition holds for **unit lower triangular**. Clearly, if A is invertible and $A = LU$, we may choose a diagonal matrix D such that $A = LDU$, where L and U are unit lower and unit upper triangular matrices, respectively. We call this factorization of A an **LDU-decomposition**.

Theorem 9.1.1 *If A is an invertible matrix with an LU-decomposition $A = LU$, then A has a unique LDU-decomposition.*

Proof

We have already observed that if A has an LU-decomposition, then A has an LDU-decomposition. So suppose that

$$LDU = A = L'D'U'$$

are LDU-decompositions of A. Because A is invertible, it follows from Corollary 3.4.10 that all the matrices above are invertible. Therefore,

$$(L'D')^{-1}(LD) = U'U^{-1}$$

Since the left-hand side is a lower triangular matrix and the right-hand side is a unit upper triangular matrix, it follows that both sides equal the identity matrix. In particular,

$$U'U^{-1} = I \quad \text{or} \quad U' = U$$

A similar argument shows that $L' = L$, and so, $D' = D$. ∎

Other Decompositions (Optional)

Depending on how the diagonal matrix D above is treated, three interesting variants of the LU-decomposition arise. If we write

$$A = (LD)U$$

that is, U is chosen to be unit upper triangular, then the factorization is known as the **Crout decomposition**. If instead, we write

$$A = L(DU)$$

that is, L is chosen to be unit lower triangular, then the factorization is known as the **Doolittle decomposition**.

If A is a symmetric matrix, then we have

$$(LDU)^t = A^t = A = LDU$$

or

$$U^t D^t L^t = LDU$$

Because U^t is a unit lower triangular matrix and L^t is a unit upper triangular matrix, we may use the uniqueness of the LDU-decomposition to conclude that $L^t = U$. In this case, fewer equations have to be solved. We have to determine the LDU-decomposition

$$A = LDL^t$$

If the diagonal entries of D are positive, then D has a "square root" (see Exercise 20 of Section 6.2), denoted $D^{1/2} = \text{diag}(\sqrt{D_{11}}, \ldots, \sqrt{D_{nn}})$. In this case A can be shown to be a positive definite matrix. We leave it as an exercise to show that

$$A = L'L'^t$$

where $L' = LD^{1/2}$. This factorization is known as the **Cholesky decomposition**, and it can be shown to be less subject to rounding errors. Also, it can be proved that if A has order n, then the corresponding system $A\mathbf{x} = \mathbf{b}$ may be solved with only $n^3/6$ multiplications, that is, half as many as Gaussian elimination requires. The program CHOLESKY may be found in Appendix B.

For more information regarding these three methods of decomposition, the reader is advised to consult the texts listed under Numerical Analysis in the References.

Exercises

1. For each system:
   (i) Solve by Gaussian elimination with partial pivoting (if necessary).
   (ii) Find the LU-decomposition of the coefficient matrix A.
   (iii) Use your answer to (ii) to solve the system.
   (a) $12x + 5y + 8z = 9$
       $3x + 2y + z = 1$
       $x + 4y + 5z = 13$
   (b) $x + 4y + 3z = 15$
       $2x + 9y - 6z = 8$
       $3x + 9y + 10z = 41$
   (c) $2x + 3y + 4z = 13$
       $4x + 5y + 9z = 19$
       $-4x - 9y - 3z = 51$
   (d) $5x - 2y + 2z = 2$
       $-10x + 8y - 2z = 6$
       $20x - 20y + 5z = -25$

2. Find the multipliers for the coefficient matrix in Exercise 1(a) and use it to solve the new system with the constant vector $\mathbf{b}'$ given here.

   (a) $\mathbf{b}' = \begin{bmatrix} 17 \\ 1 \\ -5 \end{bmatrix}$ (b) $\mathbf{b}' = \begin{bmatrix} 2 \\ -3 \\ 29 \end{bmatrix}$

3. Find the multipliers for the coefficient matrix in Exercise 1(b) and use it to solve the new system with the constant vector $\mathbf{b}'$ given here.

   (a) $\mathbf{b}' = \begin{bmatrix} -2 \\ 47 \\ 19 \end{bmatrix}$ (b) $\mathbf{b}' = \begin{bmatrix} -17 \\ -39 \\ -36 \end{bmatrix}$

4. Let

   $$A = \begin{bmatrix} 1 & 4 \\ 3 & 13 \end{bmatrix} \quad \text{and} \quad B = \begin{bmatrix} 0 & 1 \\ 1 & 0 \end{bmatrix}$$

   (a) Show that the matrix A has an LU-decomposition, but that the matrix B does not.
   (b) Find the permutation matrix P such that PB has an LU-decomposition.

5. Let A be the matrix

   $$\begin{bmatrix} 1 & 2 & 1 \\ 1 & 1 & 5 \\ 1 & 2 & 2 \end{bmatrix}$$

   (a) Find the multipliers for the matrix A above.
   (b) Use the multipliers in part (a) to solve the systems $A\mathbf{x} = \mathbf{e}_i$, where $i = 1, 2, 3$.
   (c) Use the result of part (b) to determine the inverse of A.

*6. Use the program LU-DECOMP to find LU-decompositions of the coefficient matrices in Exercise 1. Then use your answer to find the determinants of these matrices.

*7. Use the program LU-DECOMP to find an LU-decomposition of the matrix

   $$\begin{bmatrix} 16 & 4 & 2 & 5 \\ 4 & 5 & 3 & -2 \\ 2 & 6 & 12 & 4 \\ 3 & -2 & 5 & 11 \end{bmatrix}$$

   Use your answer to find the determinant of the matrix.

8. Let A and B be lower triangular matrices of order n.
   (a) Prove that AB is lower triangular.
   (b) If A is invertible, then prove that A^{-1} is lower triangular.

* This problem should be solved by using one of the programs noted in Appendix B.

9. Suppose that P and Q are permutation matrices of order n.
   (a) Prove that PQ is a permutation matrix.
   (b) Prove that P is invertible and that P^{-1} is also a permutation matrix.
   (c) Suppose that $n = 3$. Prove that P is a product of elementary matrices each of which corresponds to a row interchange.
   (d) Use part (c) to prove that if $n = 3$, then $\det P = \pm 1$.

10. Suppose that A is a symmetric matrix and that the diagonal matrix D in the LDU-decomposition of A has positive diagonal entries. Prove that A may be decomposed as $A = L'L'^t$, where L' is a lower triangular matrix.

11. Use the code for the program LU-DECOMP to obtain a count of the number of multiplications and additions to determine the matrices L and U.

12. Use the code for the program CHOLESKY to obtain a count of the number of multiplications and additions to determine the matrix L'.

13. An $n \times n$ matrix A is said to be *tridiagonal* if $A_{ij} = 0$ for $j < i - 1$ or $j > i + 1$. For example, if $n = 5$, then A has the form

$$
\begin{bmatrix}
* & * & 0 & 0 & 0 \\
* & * & * & 0 & 0 \\
0 & * & * & * & 0 \\
0 & 0 & * & * & * \\
0 & 0 & 0 & * & *
\end{bmatrix}
$$

where $*$ represents any scalar. For a large class of tridiagonal matrices an LU-decomposition of the following form is possible.

$$
L = \begin{bmatrix}
1 & 0 & 0 & 0 & \cdots & 0 \\
* & 1 & 0 & 0 & \cdots & 0 \\
0 & * & 1 & 0 & \cdots & 0 \\
\vdots & \vdots & \vdots & \vdots & \cdots & \vdots \\
0 & 0 & 0 & 0 & \cdots & 1
\end{bmatrix}
$$

and

$$
U = \begin{bmatrix}
* & A_{12} & 0 & 0 & \cdots & 0 \\
0 & * & A_{23} & 0 & \cdots & 0 \\
0 & 0 & * & A_{34} & \cdots & 0 \\
0 & 0 & 0 & * & \cdots & 0 \\
\vdots & \vdots & \vdots & \vdots & & \vdots \\
0 & 0 & 0 & 0 & 0 & *
\end{bmatrix}
$$

For each of the following matrices A, set $A = LU$, where L and U have the forms given above. Then solve for the unknown scalars to determine L and U.

(a) $\begin{bmatrix} 2 & 1 & 0 & 0 \\ 3 & 4 & -2 & 0 \\ 0 & 5 & 5 & 1 \\ 0 & 0 & 1 & 6 \end{bmatrix}$

(b) $\begin{bmatrix} 4 & 2 & 0 & 0 \\ 1 & 1 & 2 & 0 \\ 0 & -4 & 6 & -2 \\ 0 & 0 & 1 & 4 \end{bmatrix}$

14. A well-known matrix for testing the accuracy of algorithms is the nth-order *Hilbert matrix* H, defined by $H_{ij} = 1/(i + j - 1)$, where $i, j = 1, \dots, n$. This matrix has the interesting property that it is invertible and its inverse has only integral entries. It is also positive definite.
   (a) Write down the nth-order Hilbert matrices for $n = 2, 3, 4, 5$.
      For the rest of the problem, assume that $n = 4, 5$.
   *(b) Use the program MATRIX to compute (approximately) H^{-1} and HH^{-1}.
   *(c) Use your answer to part (b) to guess the exact inverse of H. Now verify your guess by showing that $HH^{-1} = I$.
   *(d) Apply Gaussian elimination to the matrix $(H|I)$ via the program ROW REDUCTION to compute its inverse.
   *(e) Use the program CHOLESKY to determine the LU-decomposition of H. Use this decomposition and the program LINSYST to solve the systems $H\mathbf{x} = \mathbf{e}_i$ for $i = 1, \dots, n$. Now use your solutions to determine (immediately) H^{-1}.

9.2 ITERATIVE METHODS FOR SOLVING SYSTEMS OF LINEAR EQUATIONS

In Section 9.1 we developed several techniques for solving the system $Ax = b$, where A is an $n \times n$ matrix. Except for possible round-off errors, these methods produced exact solutions. In this section we develop iterative methods each of which produces a sequence of vectors that "converges" to the true solution of the system. We consider various conditions on the matrix A which guarantee such convergence, as well as properties of A which influence the "rate" of convergence.

In many of the applications that we have seen thus far, a sequence of vectors x_k was defined "iteratively" by equations of the form

$$x_{k+1} = Bx_k$$

for some matrix B and for $k = 0, 1, 2, \ldots$.

For example, in the context of Markov chains, the matrix B is a transition matrix and the vector x_0 represents the initial state of the process. Once x_0 is known, each of the vectors x_k is defined in terms of the preceding one. We say that the sequence $\{x_k\}$ is defined *iteratively*. The vectors x_k are called *iterates*. In terms of solving a system of equations, we are concerned with the question of whether or not a sequence converges to the true solution. The vectors x_k (or the sequence $\{x_k\}$) are said to *converge* to a vector x if the components of x_k converge to the corresponding components of x. If we denote the ith components of x_k and x by $x_k(i)$ and $x(i)$, respectively, then we may express convergence symbolically as

$$x_k(i) \to x(i) \qquad \text{for each } i \text{ as } k \to \infty$$

For example, $(2 + 1/k, 4, 7^{1/k}) \to (2, 4, 1)$ as $k \to \infty$.

To begin our study of iterative methods for finding a solution of the system $Ax = b$, we write $A = N - C$. There are a variety of ways of choosing the matrices N and C. For example, if

$$\mathbf{A} = \begin{bmatrix} 4 & 3 & 1 \\ -2 & 6 & 4 \\ 2 & 5 & 8 \end{bmatrix}$$

then we may choose

$$N = \begin{bmatrix} 4 & 0 & 0 \\ 0 & 6 & 0 \\ 0 & 0 & 8 \end{bmatrix} \quad \text{and} \quad C = \begin{bmatrix} 0 & -3 & -1 \\ 2 & 0 & -4 \\ -2 & -5 & 0 \end{bmatrix}$$

The system may be rewritten as

$$(N - C)x = b \qquad \text{or} \qquad Nx = Cx + b$$

Now suppose that we define the vectors $\mathbf{x}_k$ iteratively by

$$N\mathbf{x}_{k+1} = C\mathbf{x}_k + \mathbf{b}$$

where $\mathbf{x}_0$ is chosen arbitrarily. If the vectors $\mathbf{x}_k$ do converge to a vector $\mathbf{y}$, then it is easy to show that by letting k approach infinity in the equation above, we obtain

$$N\mathbf{y} = C\mathbf{y} + \mathbf{b} \quad \text{or} \quad (N - C)\mathbf{y} = \mathbf{b}$$

That is, $A\mathbf{y} = \mathbf{b}$. So $\mathbf{y}$ is indeed a solution. To solve for $\mathbf{x}_{k+1}$ easily in terms of $\mathbf{x}_k$, N must be chosen to be "nice." For example, if N is an invertible and either a diagonal or upper triangular matrix, we know that the system may be solved quickly. The goal is to choose matrices N and C so that the vectors $\mathbf{x}_k$ converge. There are a number of methods, each of which produce different choices of N and C. The two that we shall consider are the Jacobi method and the Gauss–Seidel method.

The Jacobi Method

Under the *Jacobi method* $N = D$ and $C = -(L + U)$, where D, L, and U are the diagonal, lower triangular, and upper triangular parts of A, respectively. For example, if

$$A = \begin{bmatrix} 4 & 3 & 1 \\ -2 & 6 & 4 \\ 2 & 5 & 8 \end{bmatrix}$$

then $D = \text{diag}\,(4, 6, 8)$,

$$L = \begin{bmatrix} 0 & 0 & 0 \\ -2 & 0 & 0 \\ 2 & 5 & 0 \end{bmatrix} \quad \text{and} \quad U = \begin{bmatrix} 0 & 3 & 1 \\ 0 & 0 & 4 \\ 0 & 0 & 0 \end{bmatrix}$$

It is easy to see that $A = L + D + U$. (These matrices should not be confused with those in the LDU-decomposition.) The vectors $\mathbf{x}_k$ are defined iteratively by

$$D\mathbf{x}_{k+1} = -(L + U)\mathbf{x}_k + \mathbf{b}$$

In terms of the components of the vector $\mathbf{x}_{k+1}$, we have

$$A_{ii}\mathbf{x}_{(k+1)}(i) = -\sum_{j \neq i} A_{ij}\mathbf{x}_k(j) + b_i \tag{1}$$

or

$$\mathbf{x}_{(k+1)}(i) = -\frac{1}{A_{ii}}\left(\sum_{j \neq i} A_{ij}\mathbf{x}_k(j) - b_i \right)$$

for $i = 1, \ldots, n$ (see Exercise 4). This equation illustrates the ease in which the vectors $\mathbf{x}_k$ can be computed. The program JACOBI that produces these vectors is listed in Appendix B. The initial vector $\mathbf{x}_0$ may be chosen to be the zero vector. To illustrate the Jacobi method, we consider the system given below.

$$5x + y + 2z = 13$$

$$3x + 8y - 3z = 10$$

$$-5x + 2y + 10z = 29$$

For this system, we have $N = \text{diag }(5, 8, 10)$ and

$$C = \begin{bmatrix} 0 & -1 & -2 \\ -3 & 0 & 3 \\ 5 & -2 & 0 \end{bmatrix}$$

In terms of the components of the iterates, if we let $\mathbf{x}_k = (x_k, y_k, z_k)$, equation (1) yields

$$5x_{k+1} = 13 - y_k - 2z_k$$

$$8y_{k+1} = 10 - 3x_k + 3z_k$$

$$10z_{k+1} = 29 + 5x_k - 2y_k$$

The program JACOBI produces the iterates shown in Table 9.2.1 (rounded to five places). The exact solution of the system is $x = 1$, $y = 2$, and $z = 3$. The fifteenth iteration seems to have produced a reasonable approximation. We shall give sufficient conditions for convergence after we discuss a variation of the Jacobi method, called the Gauss–Seidel method.

TABLE 9.2.1 Jacobi Iterations

k	$x_k(1)$	$x_k(2)$	$x_k(3)$
1	2.60000	1.25000	2.90000
2	1.19000	1.36250	3.95000
3	0.74750	2.28500	3.22250
4	0.85400	2.17813	2.81675
5	1.03768	1.98603	2.89138
6	1.04624	1.94514	3.02163
7	1.00232	1.99077	3.03409
8	0.98821	2.01192	3.00301
9	0.99641	2.00555	2.99172
10	1.00220	1.99824	2.99710
15	1.00010	1.99990	2.99994

Gauss–Seidel Method

Suppose that we rewrite the formula for computing the ith component of $\mathbf{x}_{k+1}$ under the Jacobi method in the form

$$A_{ii}\mathbf{x}_{(k+1)}(i) = -\sum_{j=1}^{i-1} A_{ij}\mathbf{x}_k(j) - \sum_{j=i+1}^{n} A_{ij}\mathbf{x}_k(j) + b_i$$

If it is the case that $\mathbf{x}_{k+1}(j)$ is closer to the true solution than $\mathbf{x}_k(j)$, it seems reasonable to use it in place of $\mathbf{x}_k(j)$ in the first sum in the formula above, that is,

$$A_{ii}\mathbf{x}_{(k+1)}(i) = -\sum_{j=1}^{i-1} A_{ij}\mathbf{x}_{k+1}(j) - \sum_{j=i+1}^{n} A_{ij}\mathbf{x}_k(j) + b_i \qquad (2)$$

The computation of the $\mathbf{x}_k$'s by this formula is known as the **Gauss–Seidel method**. The formula is equivalent to choosing $N = (D + L)$ and $C = -U$ (see Exercise 5). As a comparison to the Jacobi method, we write the iterations for the previous system of equations:

$$
\begin{aligned}
5x_{k+1} &= 13 & - \; y_k & \; - 2z_k \\
8y_{k+1} &= 10 - 3x_{k+1} & & + 3z_k \\
10z_{k+1} &= 29 + 5x_{k+1} - 2y_{k+1} &
\end{aligned}
$$

If we apply the GAUSS-SEIDEL program in Appendix B to this system, we obtain Table 9.2.2. By comparing the iterates in Tables 9.2.1 and 9.2.2, we see that the Gauss-Seidel seventh iterate is closer to the true solution than the Jacobi fifteenth iterate. In many applications the Gauss-Seidel method is superior. However, there do exist examples of systems where one of the methods leads to convergence but the other one does not.

TABLE 9.2.2 Gauss–Seidel Iterations

k	$\mathbf{x}_k(1)$	$\mathbf{x}_k(2)$	$\mathbf{x}_k(3)$
1	2.60000	0.27500	4.14500
2	0.88700	2.47175	2.84915
3	0.96599	1.95619	2.95176
4	1.01206	1.99239	3.00755
5	0.99850	2.00339	2.99857
6	0.99989	1.99950	3.00005
7	1.00008	1.99999	3.00004

Even when one of the methods does converge, the convergence may be painfully slow. There is a technique called the ***successive overrelaxation (SOR) method*** which can greatly improve the rate of convergence. To obtain an overview of this method, we shall return to the equation

$$N\mathbf{x} = C\mathbf{x} + \mathbf{b}$$

In the Jacobi method $N = D$ and $C = -(L + U)$, whereas in the Gauss–Seidel method, $N = (D + L)$ and $C = -U$. Now consider the iterative relation

$$(D + wL)\mathbf{x}_{k+1} = [(1 - w)D - wU]\mathbf{x}_k + w\mathbf{b}$$

where w is a parameter called the ***relaxation parameter***, introduced to accelerate the rate of convergence. Notice that when $w = 1$, the method coincides with the Gauss–Seidel method. The parameter w is usually chosen to be a number between 1 and 2. However, it can be shown that if $0 < w < 2$, the SOR iterates converge to the solution of $A\mathbf{x} = \mathbf{b}$ if A is positive definite. The interested reader may obtain more information about this method from the texts listed under Numerical Analysis in the References.

Convergence of the Jacobi and Gauss–Seidel Methods _____

Before we consider the convergence of these two methods, we need to introduce a property possessed by an important class of matrices.

Definition   Let A be an $n \times n$ matrix. A is said to be ***(strictly) diagonally dominant*** if

$$|A_{ii}| > \sum_{j \neq i} |A_{ij}|$$

for $i = 1, 2, \ldots, n$.

_____

So a matrix is diagonally dominant if the magnitude of each diagonal entry is larger than the sum of the magnitudes of the other entries in the corresponding row.

Example 1   Consider the matrix

$$A = \begin{bmatrix} 8 & 3 & 1 \\ -1 & 6 & 4 \\ 2 & 5 & -9 \end{bmatrix}$$

Because $8 > 3 + 1 = 4$, $6 > 1 + 4 = 5$, and $9 > 2 + 5 = 7$, we have that A is diagonally dominant.

_____

Theorem 9.2.1   *If the matrix A is diagonally dominant, then both the Jacobi and Gauss–Seidel iterates converge to the true solution of $A\mathbf{x} = \mathbf{b}$ for any initial vector $\mathbf{x}_0$.*

An additional result holds only for the Gauss–Seidel method.

Theorem 9.2.2   *If the matrix A is positive definite, then the Gauss–Seidel iterates converge to the true solution of $A\mathbf{x} = \mathbf{b}$ for any initial vector $\mathbf{x}_0$.*

We shall omit the proof of Theorem 9.2.2. Before we discuss the proof of Theorem 9.2.1, in terms of the Jacobi iterates, we shall need the following definition.

Definition   Let M be an $n \times n$ matrix. We define the **spectral radius** of M to be the largest of the magnitudes of the eigenvalues of M.

Although most of the matrices that we will encounter here only have real eigenvalues, the definition includes the case where some of the eigenvalues may not be real. We will show in Section 9.3 that the spectral radius ρ of M satisfies

$$\rho \leq \max_i \sum_{j=1}^{n} |M_{ij}|$$

For the matrix A of Example 1, the sums in the inequality are 12, 11, and 16. Therefore, we can conclude that all of the real eigenvalues lie in the interval $[-16, 16]$.

We shall see that the spectral radius plays an important role in the rate of convergence of the iterates.

The Convergence Rate of the Jacobi Iterates (Optional)

We begin by letting $\mathbf{x}^*$ denote the true solution of $A\mathbf{x} = \mathbf{b}$. Then, because $A = N - C$, we have

$$N\mathbf{x}^* = C\mathbf{x}^* + \mathbf{b}$$

Subtracting this equation from

$$N\mathbf{x}_{k+1} = C\mathbf{x}_k + \mathbf{b}$$

we obtain

$$N(\mathbf{x}_{k+1} - \mathbf{x}^*) = C(\mathbf{x}_k - \mathbf{x}^*)$$

or

$$N\mathbf{j}_{k+1} = C\mathbf{j}_k$$

where $\mathbf{j}_k = \mathbf{x}_k - \mathbf{x}^*$ represents the ***kth error vector***, that is, the difference between the actual solution and the kth approximation. For the vectors $\mathbf{x}_k$ to converge to the solution $\mathbf{x}^*$, we need that the error vectors converge to the zero vector. To understand what is involved in determining such convergence, we write

$$\mathbf{j}_{k+1} = N^{-1}C\mathbf{j}_k$$

or

$$\mathbf{j}_{k+1} = M\mathbf{j}_k$$

where $M = N^{-1}C$. As in previous iterative relationships, the equation above yields

$$\mathbf{j}_k = M^k\mathbf{j}_0$$

If M is diagonalizable with eigenvalues $\lambda_1, \ldots, \lambda_n$, then we can write $M = PDP^{-1}$, where $D = \text{diag}(\lambda_1, \ldots, \lambda_n)$. Using this expression, we have

$$\mathbf{j}_k = PD^kP^{-1}\mathbf{j}_0$$

If we carry out the multiplication of the right side, we have that the ith coordinate of $\mathbf{j}_k$ may be written in the form

$$\mathbf{j}_k(i) = \sum_{j=1}^{n} e_{ij}\lambda_j^k \tag{3}$$

for scalars e_{ij}, where $i = 1, \ldots, n$ (see Exercise 6). If we let ρ denote the spectral radius of M, then we have

$$|\mathbf{j}_k(i)| \leq \sum_{j=1}^{n} |e_{ij}\lambda_j^k|$$

$$\leq \rho^k \sum_{j=1}^{n} |e_{ij}|$$

$$\leq \rho^k m \tag{4}$$

where m is defined as the sum in the preceding inequality. Because the speed of convergence is determined by the rate at which the components of the error vectors approach zero, we see that the spectral radius of M is of great importance. For example, if $\rho = \frac{1}{2}$, then

$$|\mathbf{j}_k(i)| \leq \frac{m}{2^k}$$

for each i. That is, the upper bound for the error is reduced by half after each iteration. A similar computation may be employed even if we do not use the assumption that the matrix M is diagonalizable. However, that approach requires a knowledge of the "Jordan canonical form" of a matrix.

It is not difficult to show that if the coefficient matrix A of the system $A\mathbf{x} = \mathbf{b}$ is diagonally dominant, then the matrix $M = N^{-1}C$ in the Jacobi method has spectral radius less than 1 (see Exercise 7). So (4) allows us to conclude that the Jacobi iterates converge. The same conclusion holds for the Gauss–Seidel iterates, but it is harder to prove.

Comments

When should iterative methods be used to solve the system $A\mathbf{x} = \mathbf{b}$? In practice, iterative methods are seldom used for solving systems of small size. The main reason for this is that the time required to obtain sufficient accuracy in the iterates exceeds that of the direct methods, such as Gaussian elimination. On the other hand, in problems such as determining an approximate solution to a partial differential equation, the matrix A may be quite large, say, 1000×1000, but has many zero entries (such a matrix is called *sparse*). If the pattern of these zeros is easy to program, then less computer storage is necessary. For example, if the order of A is 1000 and there are only five nonzero entries in each row, then the storage requirement for A is only 5000 locations as opposed to 1 million locations. If Gaussian elimination is applied, then many of these zero entries are replaced by nonzero entries and more memory is required.

In addition, our iterative procedure, written in the form

$$N\mathbf{x}_{k+1} = C\mathbf{x}_k + \mathbf{b}$$

requires approximately n^2 multiplications per iteration, whereas Gaussian elimination requires $n^3/3$ multiplications. It may be that the iterate $\mathbf{x}_k$ is a suitable approximation to the true solution long before $kn^2 = n^3/3$ or $k = n/3$, in which case the indirect method is faster.

Exercises

1. For each system:
   (i) Find the exact solution by Gaussian elimination.
   (ii) Write the first two iterates using the Jacobi method with $\mathbf{x}_0 = \mathbf{0}$.
   (iii) Write the first two iterates using the Gauss–Seidel method with $\mathbf{x}_0 = \mathbf{0}$.
   (a) $2x - 3y = 3$
        $x + 4y = 7$
   (b) $-5x + 3y = 22$
        $-2x + 4y = 20$
   (c) $4x - 2y = -20$
        $3x + 6y = 15$
   (d) $5x + 2y = 5$
        $x + 4y = -17$

2. Repeat (ii) and (iii) of Exercise 1 for each of the following systems, but now use the initial vectors given.
   (a) For the system in Exercise 1(a), use $\mathbf{x}_0 = (3.2, 1.5)$.
   (b) For the system in Exercise 1(b), use $\mathbf{x}_0 = (-2.4, 3)$.
   (c) For the system in Exercise 1(c), use $\mathbf{x}_0 = (-20, 15)$.

3. For each matrix:
   (i) Compute the matrices N and C used in the Jacobi method.
   (ii) Compute the matrices N and C used in the Gauss–Seidel method.

   (a) $\begin{bmatrix} 6 & 3 & -4 \\ -2 & 1 & 0 \\ 2 & 9 & 8 \end{bmatrix}$ (b) $\begin{bmatrix} 3 & -2 & 5 \\ 4 & 7 & 8 \\ 1 & -3 & 4 \end{bmatrix}$

4. Verify that the vector equation

$$D\mathbf{x}_{k+1} = -(L + U)\mathbf{x}_k + \mathbf{b}$$

   used in the Jacobi method yields equation (1).

5. Verify that the choices $N = D + L$ and $C = -U$ used in the Gauss–Seidel method yield equation (2).

6. Verify equation (3).

7. Suppose that A is a 3×3 matrix which is diagonally dominant. Prove that the matrix $M = N^{-1}C$ in the Jacobi method has spectral radius less than 1 by showing that the sums of the absolute values of the entries in each row are less than 1.

8. For the system $A\mathbf{x} = \mathbf{b}$ let $A = N - C$, where N is invertible. Define vectors $\mathbf{x}_k$ iteratively by

$$N\mathbf{x}_{k+1} = C\mathbf{x}_k + \mathbf{b}$$

   Prove that if the initial vector $\mathbf{x}_0$ is equal to the true solution of the system, then all of the $\mathbf{x}_k$'s are equal to $\mathbf{x}_0$.

*9. For each system below, the exact solution is given.

   (i) Use the program JACOBI to count the number of iterates that must be produced to be within 10^{-4} of the exact solution.
   (ii) Use the program GAUSS–SEIDEL to count the number of iterates that must be produced to be within 10^{-4} of the exact solution.

   (a) $8x + 4y - 2z = 10$
       $4x + 7y + z = -26$
       $6x + 2y + 10z = -22;$ $\mathbf{x} = (3, -5, -3)$
   (b) $6x - 2y + 3z = -16$
       $5x + 7y - 2z = 12$
       $4x + y + 3z = 5;$ $\mathbf{x} = (-3, 5, 4)$
   (c) $8x + 3y - 2z + w = 44$
       $4x + 6y + 3z + 2w = 42$
       $2x - y + 7z + 3w = 18$
       $4x + 2y - 3z + 8w = 12;$ $\mathbf{x} = (5, 3, 2, -1)$
   (d) $-6x + 3y + z + w = -4$
       $4x + 8y - 2z = 62$
       $3x + 7y + z - w = 35$
       $4x + 4y + z + 8w = 59;$
       $\mathbf{x} = (3, 5, -5, 4)$
   (e) $-6x + 3y + z + w = -4$
       $4x + 8y - 2z = 62$
       $3x + y + 7z - w = -25$
       $4x + 4y + z + 8w = 59;$
       $\mathbf{x} = (3, 5, -5, 4)$

10. Although the coefficient matrices of the systems in parts (d) and (e) of Exercise 9 differ only in two entries, convergence occurred in part (e) but not in part (d). Speculate why this happens.

* This problem should be solved by using one of the programs noted in Appendix B.

9.3 GERSCHGORIN'S THEOREM

In many applications that involve eigenvalues, it is not necessary to know all the eigenvalues. In fact, in applications involving vibrations, statistical analyses of variance, differential equations, or extrema of special functions, it is often only necessary to know either the largest or smallest eigenvalue. In some mathematical applications, it may only be important to know if all the eigenvalues are nonnegative. Gerschgorin's theorem will provide a simple method of determining the approximate location of the eigenvalues of a given matrix. The proof of this interesting result is surprisingly simple. We will assume throughout this section that the entries of all matrices are real; however, we will not assume that the eigenvalues are real.

Theorem 9.3.1 (***Gerschgorin's Disk Theorem***)   *Let A be an n × n matrix and for i = 1, ..., n denote by r_i the following sum*:

$$r_i = \sum_{j \neq i} |A_{ij}|$$

Let C_i denote the interval (disk in the complex case) about A_{ii} of radius r_i. Then every eigenvalue is in some C_i.

Proof

Let λ be an eigenvalue of A with corresponding eigenvector $\mathbf{x} = (x_1, \ldots, x_n)$. Let x_i be the coordinate of $\mathbf{x}$ with the largest absolute value. Because $\mathbf{x}$ is a nonzero vector, it follows that $x_i \neq 0$. Comparing the ith components of both sides of $\lambda \mathbf{x} = A\mathbf{x}$, we have

$$\lambda x_i = \sum_{j=1}^{n} A_{ij} x_j \tag{1}$$

By using equation (1) and elementary properties of absolute values, we have

$$|\lambda x_i - A_{ii} x_i| \leq \sum_{j \neq i} |A_{ij}||x_j|$$

So

$$|x_i||\lambda - A_{ii}| \leq |x_i| \sum_{j \neq i} |A_{ij}|$$

$$= |x_i| r_i$$

Thus,

$$|\lambda - A_{ii}| \leq r_i \tag{2}$$

This inequality is equivalent to the statement that the eigenvalue λ is in the interval C_i.   ■

We call r_i the **radius** of the ith row, and we call C_i a **Gerschgorin interval** (**disk**).

Comments

1. Because A and A^t have the same eigenvalues (see Exercise 25 of Section 6.1) as well as the same diagonal entries, a similar result holds if the radius is

defined in terms of the entries of the ith column. Thus, to obtain more precision, choose either the ith column or ith row depending on which possesses the smaller radius.

2. Recall from Section 7.3 that symmetric matrices have only real eigenvalues. We will use this result in the examples below. So the application of Gerschgorin's theorem will involve only intervals rather than disks when the given matrix is symmetric. We will deal with disks in the latter part of this section.

Example 1   Exercise 20 of Section 6.2 states that diagonalizable matrices have "square roots" if all the eigenvalues are nonnegative. Because the matrix below is symmetric, it is diagonalizable and it has only real eigenvalues.

$$\begin{bmatrix} 5 & 3 & 1 \\ 3 & 4 & 1 \\ 1 & 1 & 4 \end{bmatrix}$$

For this matrix, the intervals of Gerschgorin's theorem have centers 5, 4, and 4 with respective radii 4, 4, and 2. Thus, all of the eigenvalues are included within the intervals

$$[1, 9], \quad [0, 8], \quad \text{and} \quad [2, 6]$$

So all the eigenvalues are nonnegative and therefore the matrix has a square root.

Consider the matrix

$$A = \begin{bmatrix} 3 & 0.01 & -0.01 \\ 0.01 & 4 & 0.01 \\ -0.01 & 0.01 & 5 \end{bmatrix}$$

Because A is symmetric, the eigenvalues lie within one or more of the intervals

$$[2.98, 3.02], \quad [3.98, 4.02], \quad \text{and} \quad [4.98, 5.02]$$

That is, the eigenvalues are all close to one or more of the numbers 3, 4, or 5. The matrix A has the property that each of its diagonal entries is much larger in absolute value than the other entries of the corresponding row. In other words A is diagonally dominant (see Section 9.2).

Diagonally dominant matrices are *stable* relative to their eigenvalues in the following sense: If the entries of the matrix are changed by a relatively "small amount," say, due to round-off error, the eigenvalues do not change very much. This is true because the intervals C_i do not change significantly. These ideas involve the notion of "conditioning," which is an important topic of numerical analysis.

**Corollary
9.3.2**

If A is diagonally dominant, then A is invertible.

Proof
Let λ be an eigenvalue of A. Then λ is in some interval C_i. By inequality (2)

$$|A_{ii}| - |\lambda| \le |A_{ii} - \lambda| \le r_i$$

Therefore,

$$0 < |A_{ii}| - r_i \le |\lambda|$$

By hypothesis, the left side of this inequality is positive, so $\lambda \ne 0$. Thus, A has no zero eigenvalues; that is, the solution space of $A\mathbf{x} = \mathbf{0}$ consists only of the zero vector, so A is invertible. ∎

Example 2   Let

$$A = \begin{bmatrix} 5 & 3 & 1 \\ 2 & -4 & 0 \\ 3 & -2 & 6 \end{bmatrix}$$

It is easy to verify by inspection that A is diagonally dominant. Hence, A is invertible.

One of the most useful consequences of Gerschgorin's theorem is that it gives an upper bound for the magnitude of the largest eigenvalue, that is, the spectral radius of the matrix (see Section 9.2).

**Corollary
9.3.3**

Let A be an $n \times n$ matrix and let ρ denote the spectral radius of A. Then

$$\rho \le \max_i \sum_{j=1}^{n} |A_{ij}|$$

Proof
Let λ be the eigenvalue of A with the largest magnitude. By Gerschgorin's theorem, λ is in some interval C_i. So, by equation (2),

$$|\lambda - A_{ii}| \le r_i$$

or

$$|\lambda| \le r_i + |A_{ii}| = \sum_{j=1}^{n} |A_{ij}|$$ ∎

There are many extensions of Gerschgorin's theorem which allow for better estimates of the eigenvalues. Although the proofs of many of these results are inappropriate for this text, we will include the statement of one of the more useful extensions.

Theorem 9.3.4   *Let A be an n × n matrix. Let C_i denote the interval (disk in the complex case) about A_{ii} of radius r_i. If the union S of k of the intervals is disjoint from the other n − k intervals, then S contains k eigenvalues (including multiplicities).*

Example 3   How many distinct eigenvalues does the matrix A given below possess?

$$A = \begin{bmatrix} 2 & 0.6 & 0.04 \\ 0.01 & 3 & 0.04 \\ 0.20 & 0.1 & 4 \end{bmatrix}$$

It is easy to see (Exercise 6) that all three of the intervals (disks) are disjoint from each other. Hence, each of the intervals must contain one eigenvalue. From this result, we may in fact conclude that A is diagonalizable because each eigenvalue must have multiplicity 1.

Complex Case

As we indicated parenthetically in the statement of Gerschgorin's theorem, the C_i's may be considered to be disks in the complex plane. For elementary facts about complex numbers, the reader should consult Appendix A. The results that we have stated thus far remain valid in the complex case. In fact, the proof of Gerschgorin's theorem for matrices with complex entries is identical to the one given.

Consider the matrix

$$A = \begin{bmatrix} 0 & -1 \\ 1 & 0 \end{bmatrix}$$

A represents rotation by 90°. Because we were only concerned with *real* eigenvalues, we noted earlier that such a rotation has no eigenvalues. If we reconsider the characteristic equation

$$x^2 + 1 = 0$$

then we see that in fact A has two complex eigenvalues, namely, i and $-i$. This agrees with the complex version of Gerschgorin's theorem (see Figure 9.3.1).

Care must be taken when we are interested in the *sign* of the eigenvalues. Complex numbers have considerably more possibilities than being positive, negative, or zero. If we consider the matrix A of Example 1, we cannot conclude from

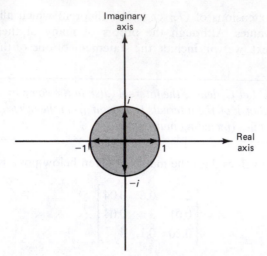

Figure 9.3.1

Gerschgorin's theorem alone that all the eigenvalues are nonnegative. In that example, we only have to deal with intervals because the matrix is symmetric. Consider the matrix

$$A = \begin{bmatrix} 2 & -0.5 & 0.5 \\ 0 & 5 & 1 \\ 1 & -2 & 8 \end{bmatrix}$$

The disks are given in Figure 9.3.2.

From Theorem 9.3.4 we may conclude that one eigenvalue is in C_1 and the other two eigenvalues are in the union of C_2 and C_3. However, because the eigenvalues are not necessarily real numbers, this information does not tell us anything about the signs of the eigenvalues.

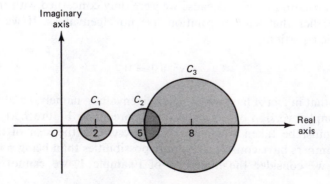

Figure 9.3.2

Exercises

1. For each of the following matrices, first determine the Gerschgorin intervals using either the columns or rows to compute the radii. Second, compute the actual eigenvalues and compare your results.

(a) $\begin{bmatrix} 3 & -2 & 0 \\ -2 & 3 & 0 \\ 0 & 0 & 5 \end{bmatrix}$ (b) $\begin{bmatrix} 1 & 2 & -1 \\ 1 & 0 & 1 \\ 4 & -4 & 5 \end{bmatrix}$

2. Use Corollary 9.3.2 to prove that the following matrices are invertible.

(a) $\begin{bmatrix} 3 & 2 & 0 \\ 2 & 5 & -2 \\ 4 & 1 & 8 \end{bmatrix}$ (b) $\begin{bmatrix} 3 & 1 & 1 \\ -4 & 9 & 2 \\ 2 & -1 & 6 \end{bmatrix}$

3. Use Gerschgorin's theorem or Theorem 9.3.4 to estimate the eigenvalues of the following matrix to within two decimal places.

$$\begin{bmatrix} 1.000 & 0.002 & 0.003 \\ 0.002 & -3.000 & -0.001 \\ 0.003 & -0.001 & -6.000 \end{bmatrix}$$

4. Use Theorem 9.3.4 to prove that the following matrix is diagonalizable.

$$\begin{bmatrix} -7 & 2 & 2 \\ 1 & 6 & 2 \\ 1 & -1 & 0 \end{bmatrix}$$

5. Use Theorem 9.3.4 to determine how many eigenvalues (including multiplicities) of the following symmetric matrix are positive.

$$\begin{bmatrix} 6 & 1 & 1 & 2 \\ 1 & -4 & 2 & 0 \\ 1 & 2 & 5 & -1 \\ 2 & 0 & -1 & 7 \end{bmatrix}$$

6. Verify that the Gerschgorin intervals in Example 3 are disjoint.

7. Let A be a transition matrix. Use Corollary 9.3.3 (applied to the columns) to prove that the eigenvalue of A of largest magnitude is 1. [*Hint:* First prove that $(1, 1, \ldots, 1)$ is an eigenvector of A^t.]

8. Suppose that A is an invertible matrix. Use Corollary 9.3.3 and Exercise 21 of Section 6.1 to establish an inequality for the magnitude of the smallest eigenvalue of A.

9.4 NUMERICAL METHODS FOR DETERMINING EIGENVALUES AND EIGENVECTORS

We have seen the vast importance of eigenvalues and eigenvectors in many of the applications that we have studied so far. In most of these applications we only had to contend with matrices of order at most 3. In these cases we computed the characteristic polynomial and then found its zeros. The computation of the characteristic polynomial involved the calculation of a determinant by cofactor expansion. We showed in Section 5.2 that the number of operations needed in this calculation alone is astronomical even for relatively small matrices. Also, after the characteristic polynomial is known, it may be quite difficult (or impossible) to obtain its exact zeros. Therefore, we need to consider methods that will allow us to at least approximate the eigenvalues and eigenvectors of relatively large matrices. As in Section 9.2, the methods we examine here are iterative. Many of the computations used here also hold for complex eigenvalues.

The Power Method

Perhaps the most common method to estimate eigenvalues and eigenvectors is the *power method*. It is also one of the easiest methods to understand. This method allows us to estimate the eigenvalue of largest magnitude as well as a corresponding eigenvector. In many applications it is only this eigenvalue and eigenvector that are of interest. To make things concrete, we shall consider the case of a 2×2 diagonalizable matrix A. In this case we have that A possesses linearly independent eigenvectors $\mathbf{y}_1$ and $\mathbf{y}_2$ with corresponding eigenvalues λ_1 and λ_2. The discussion that follows generalizes immediately to the case where A is an $n \times n$ matrix. We use the result that if $A\mathbf{x} = \lambda\mathbf{x}$, then $A^k\mathbf{x} = \lambda^k\mathbf{x}$. Now let $\mathbf{x}_0$ be any vector in R^2. Because $\mathbf{y}_1$ and $\mathbf{y}_2$ form a basis of R^2, we have

$$\mathbf{x}_0 = a\mathbf{y}_1 + b\mathbf{y}_2$$

for some scalars a and b. From our comment above, we have

$$
\begin{aligned}
A^k\mathbf{x}_0 &= aA^k\mathbf{y}_1 + bA^k\mathbf{y}_2 \\
&= a\lambda_1^k\mathbf{y}_1 + b\lambda_2^k\mathbf{y}_2 \\
&= \lambda_1^k\left(a\mathbf{y}_1 + b\left(\frac{\lambda_2}{\lambda_1}\right)^k\mathbf{y}_2\right)
\end{aligned}
$$

Now we make the critical assumption that $|\lambda_1| > |\lambda_2|$. It follows from this assumption that $(\lambda_2/\lambda_1)^k$ approaches 0 as k approaches infinity. This observation allows us to conclude that

$$A^k\mathbf{x}_0 \doteq \lambda_1^k a\mathbf{y}_1$$

for large values of k, where $\doteq$ means "is approximately equal to." Thus, if $a \neq 0$, we obtain an approximation to a multiple of the eigenvector $\mathbf{y}_1$ corresponding to the eigenvalue of largest magnitude. We call the eigenvalue of largest magnitude the *dominant eigenvalue*. Any eigenvector associated with this eigenvalue is called a *dominant eigenvector*.

It would be quite inefficient to compute A^k and then multiply the result by $\mathbf{x}_0$ (see Exercise 4). Instead, we define the vectors $\mathbf{x}_k$ iteratively by

$$\mathbf{x}_{k+1} = A\mathbf{x}_k$$

It then follows that $\mathbf{x}_k = A^k\mathbf{x}_0$ (see Exercise 4), so

$$\mathbf{x}_k \doteq \lambda_1^k a\mathbf{y}_1$$

Notice that if $|\lambda_1| > 1$, then the components of the approximation $\mathbf{x}_k$ become arbitrarily large in magnitude. On the other hand, if $|\lambda_1| < 1$, then the components of $\mathbf{x}_k$ approach 0. To avoid either of these two situations, we *scale* each iterate before the next one is computed. There are many ways of accomplishing this scaling. The one that we shall consider here requires that each iterate is divided by its component of largest magnitude. It can be shown (see Exercise 5) that this scaling does not affect the convergence of the sequence. To illustrate these ideas, consider the matrix

$$A = \begin{bmatrix} 1 & 1 \\ 6 & 2 \end{bmatrix}$$

Suppose that we let $\mathbf{x}_0 = (1, 1)$. Then

$$\mathbf{x}_1 = A\mathbf{x}_0 = \begin{bmatrix} 2 \\ 8 \end{bmatrix}$$

The component of largest magnitude is 8, so we form the scaled iterate

$$\mathbf{x}_1' = \begin{bmatrix} \frac{2}{8} \\ \frac{8}{8} \end{bmatrix} = \begin{bmatrix} 0.25 \\ 1.00 \end{bmatrix}$$

Now form

$$\mathbf{x}_2 = A\mathbf{x}_1' = \begin{bmatrix} 1.25 \\ 3.50 \end{bmatrix}$$

Notice that if $A\mathbf{x}_1' \doteq \lambda_1 \mathbf{x}_1'$, then λ_1 is approximately equal to 3.50 because the second component of $\mathbf{x}_1'$ is equal to 1. Similar reasoning holds at the kth step. That is, the approximation to λ_1 is the component of $\mathbf{x}_k$ which corresponds to the component of $\mathbf{x}_{k-1}'$ that equals 1. In Table 9.4.1, we apply the power method with initial vector $\mathbf{x}_0 = (1, 1)$ to the matrix A and list the first six iterates $\mathbf{x}_i' = (\mathbf{x}_i'(1), \mathbf{x}_i'(2))$ and the accompanying approximations m_i to λ_1. The results are rounded to four places. From this table we obtain an estimate of 3.9978 for the dominant eigenvalue and $(0.3334, 1)$

TABLE 9.4.1 Power Method Iterations

i	m_i	$\mathbf{x}_i'(1)$	$\mathbf{x}_i'(2)$
1	8.0000	0.2500	1
2	3.5000	0.3571	1
3	4.1429	0.3276	1
4	3.9655	0.3348	1
5	4.0087	0.3330	1
6	3.9978	0.3334	1

for a dominant eigenvector. The true eigenvalues of A are 4 and -1 with corresponding eigenvectors $(\frac{1}{3}, 1)$ and $(1, -2)$.

Comments

1. The rate at which the process converges depends on the size of the quotient $|\lambda_2/\lambda_1|$. If the quotient is very close to 1, the convergence may be quite slow.

2. The closer the initial vector $\mathbf{x}_0$ is to the true dominant eigenvector, the better the approximations.

3. If the scalar a in $\mathbf{x}_0 = a\mathbf{y}_1 + b\mathbf{y}_2$ is 0, then, in theory, the iterates will not converge to $\mathbf{y}_1$. On the other hand, because of round-off errors, some iterate when expressed as a linear combination of $\mathbf{y}_1$ and $\mathbf{y}_2$ will probably possess a nonzero coefficient of $\mathbf{y}_1$. In this situation the iterates will at first appear to converge to $\mathbf{y}_2$, but then will converge to $\mathbf{y}_1$.

4. The general case for an $n \times n$ matrix A proceeds in exactly the same way. We assume that the eigenvalues of A satisfy

$$|\lambda_1| > |\lambda_2| \geq |\lambda_3| \cdots \geq |\lambda_n|$$

5. Even if A is not diagonalizable but does have a unique dominant eigenvalue, the iterates will again converge to $\mathbf{y}_1$. (This result can be proven from the theory of Jordan canonical forms.)

6. The case when $|\lambda_1| = |\lambda_2|$ or when λ_1 is not real requires modifications of the method that go beyond the scope of this text.

7. Another common method of scaling uses the ***Rayleigh quotient***

$$\frac{A\mathbf{x}_k \cdot \mathbf{x}_k}{\mathbf{x}_k \cdot \mathbf{x}_k} \doteq \frac{\lambda_1 \mathbf{x}_k \cdot \mathbf{x}_k}{\mathbf{x}_k \cdot \mathbf{x}_k}$$

$$= \lambda_1$$

It can be shown that if the matrix A is symmetric, the Rayleigh quotients converge to the dominant eigenvalue approximately twice as fast as in the general case.

The (Shifted) Inverse Power Method

We have just seen that the power method produces approximations to the dominant eigenvalue (and eigenvector). Now we will describe a method that will allow us to compute approximations to the other eigenvalues and their eigenvectors.

We begin with an $n \times n$ matrix A which possesses n linearly independent eigenvectors $\mathbf{y}_1, \ldots, \mathbf{y}_n$ and corresponding eigenvalues $\lambda_1, \ldots, \lambda_n$, respectively. It is easy to see that

$$(A - \alpha I)\mathbf{y}_i = (\lambda_i - \alpha)\mathbf{y}_i \tag{1}$$

for any scalar α. So if the matrix $(A - \alpha I)$ is invertible, we obtain from equation (1) that

$$(A - \alpha I)^{-1}\mathbf{y}_i = \frac{1}{(\lambda_i - \alpha)}\,\mathbf{y}_i$$

Thus, the matrix $(A - \alpha I)^{-1}$ has the eigenvalue $1/(\lambda_i - \alpha)$ with the corresponding eigenvector $\mathbf{y}_i$. It is easy to see that if $\alpha = 0$, then the eigenvalues of A^{-1} are the reciprocals of the eigenvalues of A. Therefore, if we are only interested in finding the eigenvalue of A of smallest magnitude and a corresponding eigenvector, we need only apply the power method to A^{-1}.

In the general case, we define the iterates $\mathbf{x}_k$ by

$$\mathbf{x}_{k+1} = (A - \alpha I)^{-1}\mathbf{x}_k$$

By the power method (applied to $(A - \alpha I)^{-1}$), these iterates will produce an approximation to the dominant eigenvalue (and corresponding eigenvector) of $(A - \alpha I)^{-1}$, that is, the value $1/(\lambda_i - \alpha)$, where λ_i is the eigenvalue of A which is the closest to the scalar α. For example, suppose that we choose $\alpha = 5$ and that the eigenvalues of A are 9, 6, and 2. Then the eigenvalues of $(A - 5I)^{-1}$ are $1/(9-5) = \frac{1}{4}$, $1/(6-5) = 1$, and $1/(2-5) = -\frac{1}{3}$. In this case, if the power method is applied to $(A - 5I)^{-1}$, then we will obtain an approximation to the eigenvalue 1 of $(A - 5I)^{-1}$ and a corresponding eigenvector, which is also an eigenvector of A. Because the value of α is known, we can obtain an approximation to the eigenvalue 6 of A. This method, which applies the power method to the matrix $(A - \alpha I)^{-1}$, is called the (**shifted**) **inverse power method**. The procedure may be summarized as follows.

INVERSE POWER METHOD FOR A MATRIX A

1. Choose a scalar α.
2. Apply the power method to the matrix $(A - \alpha I)^{-1}$.
3. The iterations will produce approximations to the eigenvalue of A which is closest to α. The approximations will also produce a corresponding eigenvector of A.

At this stage the reader may wonder how to choose the scalar α. Thanks to Gerschgorin's theorem, we know the intervals on the real line (or disks in the complex plane) which contain the eigenvalues of the matrix A. If α is chosen as the center of one of these intervals, the inverse power method will produce approximations to one of the eigenvalues (and a corresponding eigenvector) which lie in this interval.

Comments

1. As a practical procedure, we obtain the iterates by solving the systems

$$(A - \alpha I)\mathbf{x}_{k+1} = \mathbf{x}_k$$

This may be accomplished by any of the methods we have developed for solving systems thus far. For example, once we obtain the LU-decomposition of $(A - \alpha I)$, we may use it repeatedly to solve for $\mathbf{x}_{k+1}$ in the systems above.

2. Recall that the rate at which the power method converges depends on the magnitude of the ratios of the nondominant eigenvalues to the dominant one. For the inverse power method, we must consider the ratios

$$\frac{1/(\lambda_i - \alpha)}{1/(\lambda_j - \alpha)} = \frac{(\lambda_j - \alpha)}{(\lambda_i - \alpha)}$$

When this ratio is the largest, say q, it can be shown that the $(k + 1)$st error is approximately q times the kth error. Thus, if the eigenvalues are relatively far apart, and if α is chosen to be close to one of the eigenvalues, the convergence to that eigenvalue is quite rapid.

Deflation Methods

As before, we begin with an $n \times n$ matrix A which possesses n linearly independent eigenvectors $\mathbf{y}_1, \ldots, \mathbf{y}_n$ and corresponding eigenvalues $\lambda_1, \ldots, \lambda_n$. Suppose that we know one of the eigenvalues, say, λ_1 and a corresponding eigenvector. The idea behind a *deflation method* is to construct an $n \times n$ matrix A' that has eigenvalues $0, \lambda_2, \ldots, \lambda_n$ and whose eigenvectors have a simple relationship to the eigenvectors of A. If λ_1 is the dominant eigenvalue of A, the power method may be applied to A' to find the next largest eigenvalue of A. In any case, the method may be repeated on each new ("deflated") matrix followed by the power or inverse power method to find the remaining eigenvalues and eigenvectors of A.

There are a number of deflation procedures which may be found in the sources listed in the References. We will restrict ourselves to the special case of a symmetric matrix.

Theorem 9.4.1 *Let A be an n × n symmetric matrix with (not necessarily distinct) eigenvalues $\lambda_1, \ldots, \lambda_n$ and corresponding orthonormal eigenvectors $\mathbf{y}_1, \ldots, \mathbf{y}_n$. Then the eigenvalues of the matrix*

$$A' = A - \lambda_1 \mathbf{y}_1 \mathbf{y}_1^t$$

are $0, \lambda_2, \ldots, \lambda_n$, with corresponding eigenvectors $\mathbf{y}_1, \ldots, \mathbf{y}_n$.

Notice that because $\mathbf{y}_1$ is an $n \times 1$ column vector, it follows that $\mathbf{y}_1\mathbf{y}_1^t$ is an $n \times n$ matrix. In addition, because of Corollary 7.2.4, it is always possible to construct an orthonormal basis of eigenvectors of A.

Proof

Recall that for any column vectors $\mathbf{x}$ and $\mathbf{y}$, we may write $\mathbf{x} \cdot \mathbf{y} = \mathbf{x}^t\mathbf{y}$. Because the $\mathbf{y}_i$'s are orthonormal, we have

$$
\begin{aligned}
A'\mathbf{y}_i &= (A - \lambda_1\mathbf{y}_1\mathbf{y}_1^t)\mathbf{y}_i \\
&= A\mathbf{y}_i - \lambda_1\mathbf{y}_1(\mathbf{y}_1^t\mathbf{y}_i) \\
&= \lambda_i\mathbf{y}_i - \lambda_1\mathbf{y}_1(\mathbf{y}_1 \cdot \mathbf{y}_i) \\
&= \begin{cases} \lambda_1\mathbf{y}_1 - \lambda_1\mathbf{y}_1(1) & \text{if } i = 1 \\ \lambda_i\mathbf{y}_i \ - \lambda_1\mathbf{y}_1(0) & \text{if } i > 1 \end{cases} \\
&= \begin{cases} 0 & \text{if } i = 1 \\ \lambda_i\mathbf{y}_i & \text{if } i > 1 \end{cases}
\end{aligned}
$$

This proves the theorem. ∎

We shall illustrate the theorem with an example. Let

$$
A = \begin{bmatrix} 2 & 0 & 36 \\ 0 & 3 & 0 \\ 36 & 0 & 23 \end{bmatrix}
$$

The eigenvalues of A are -25, 3, and 50 with corresponding eigenvectors

$$
\begin{bmatrix} -4 \\ 0 \\ 3 \end{bmatrix}, \quad \begin{bmatrix} 0 \\ 1 \\ 0 \end{bmatrix}, \quad \text{and} \quad \begin{bmatrix} 3 \\ 0 \\ 4 \end{bmatrix}
$$

respectively. Suppose that we let $\lambda_1 = 50$. Then we take

$$
\mathbf{y}_1 = \frac{1}{5}\begin{bmatrix} 3 \\ 0 \\ 4 \end{bmatrix} = \begin{bmatrix} 0.6 \\ 0 \\ 0.8 \end{bmatrix}
$$

So

$$
\mathbf{y}_1\mathbf{y}_1^t = \begin{bmatrix} 0.36 & 0 & 0.48 \\ 0 & 0 & 0 \\ 0.48 & 0 & 0.64 \end{bmatrix}
$$

Therefore,

$$A' = A - 50\mathbf{y}_1\mathbf{y}_1^t = \begin{bmatrix} -16 & 0 & 12 \\ 0 & 3 & 0 \\ 12 & 0 & -9 \end{bmatrix}$$

As a check of Theorem 9.4.1, we note that the characteristic polynomial of A' is

$$f(x) = x^3 + 22x^2 - 75x$$
$$= x(x + 25)(x - 3)$$

Thus, the eigenvalues of A' are 0, -25, and 3. We leave it as an exercise to verify the remainder of Theorem 9.4.1.

Comments

1. Deflation methods are generally subject to rounding errors and should not be used to find *all* of the eigenvalues. The reason for this is that after each eigenvalue is found, a new deflated matrix is produced. Any errors introduced in the previously deflated matrix propagate, causing the next deflated matrix to be subject to additional errors, and so on.

2. To avoid the situation described in comment 1, the estimated eigenvalues obtained in the deflation method should serve as initial approximations in the inverse power method applied to the original matrix A.

3. For symmetric matrices, there is a well-known technique for finding all the eigenvalues. It is called the *QL algorithm*.

We conclude this chapter by stating that we have only provided a glimpse of the available numerical techniques in dealing with the solution of systems of linear equations and the determination of eigenvalues and eigenvectors. The References contain several sources which themselves contain many additional references. Due to space limitations, we have omitted several basic topics that the reader may wish to pursue. Among them are *ill-conditioned* matrices, reduction to *Hessenberg form*, the *Householder transformation*, the *singular value decomposition*, and *QR factorization*.

Exercises

1. For each symmetric matrix shown, its dominant eigenvalue λ and dominant eigenvector $\mathbf{y}$ are given.
   (i) Find the deflated matrix A'.
   (ii) Find the nonzero eigenvalues of A' and their corresponding eigenvectors.

(iii) Use your results to verify Theorem 9.4.1.

(a) $\begin{bmatrix} 3 & 1 \\ 1 & 3 \end{bmatrix}$, $\lambda = 4$, $\mathbf{y} = \begin{bmatrix} 1 \\ 1 \end{bmatrix}$

(b) $\begin{bmatrix} 5 & 2 \\ 2 & 2 \end{bmatrix}$, $\lambda = 6$, $\mathbf{y} = \begin{bmatrix} 2 \\ 1 \end{bmatrix}$

(c) $\begin{bmatrix} 1 & 12 & 0 \\ 12 & -6 & 0 \\ 0 & 0 & -5 \end{bmatrix}, \lambda = -15, \mathbf{y} = \begin{bmatrix} 3 \\ -4 \\ 0 \end{bmatrix}$

(d) $\begin{bmatrix} 2 & -2 & 0 \\ -2 & 1 & -2 \\ 0 & -2 & 0 \end{bmatrix}, \lambda = 4, \mathbf{y} = \begin{bmatrix} 2 \\ -2 \\ 1 \end{bmatrix}$

2. Verify Theorem 9.4.1 using the matrix

$$\begin{bmatrix} 2 & 0 & 36 \\ 0 & 3 & 0 \\ 36 & 0 & 23 \end{bmatrix}$$

given as an example in the section.

3. Assume the notation of Theorem 9.4.1.
   (a) Prove that the deflated matrix A' is symmetric.
   (b) Prove that

   $$A = \sum_{i=1}^{n} \lambda_i \mathbf{y}_i \mathbf{y}_i^t$$

   (c) Prove that

   $$A' = \sum_{i=2}^{n} \lambda_i \mathbf{y}_i \mathbf{y}_i^t$$

4. The vectors $\mathbf{x}_k$, $k = 1, 2, \ldots$, used in the power method are defined by the relationship

   $$\mathbf{x}_k = A\mathbf{x}_{k-1} \qquad (1)$$

   where A is an $n \times n$ matrix.
   (a) Prove that equation (1) is equivalent to

   $$\mathbf{x}_k = A^k \mathbf{x}_0 \qquad (2)$$

   for $k = 1, 2, \ldots$.
   (b) Prove that it requires $(k - 1)n^3 + n^2$ multiplications to compute $\mathbf{x}_k$ by equation (2).
   (c) Prove that it requires only kn^2 multiplications to compute $\mathbf{x}_k$ by equation (1).
   (d) Use parts (b) and (c) to compare the number of multiplications that are

required to compute $\mathbf{x}_k$ by equations (1) and (2) for $n = 4$ and $k = 5, 10, 100$.

5. Suppose that the nonzero vectors $\mathbf{x}_k$, $k = 1, 2, \ldots$, converge to the nonzero vector $\mathbf{x}$. Let λ_k be the component of $\mathbf{x}_k$ of largest magnitude and let λ be the component of $\mathbf{x}$ of largest magnitude.
   (a) Prove that λ_k converges to λ.
   (b) Prove that the vectors $(1/\lambda_k)\mathbf{x}_k$ converge to the vector $(1/\lambda)\mathbf{x}$.
   (c) How do parts (a) and (b) apply to the power method?

*6. Use the program POWER METHOD to verify the values of λ and $\mathbf{y}$ for the matrices in Exercise 1.

*7. Use the program POWER METHOD to approximate the dominant eigenvalue and eigenvector for the following matrices. (Terminate the iterations when two consecutive approximations of the dominant eigenvalue are within 10^{-4} of each other.)

   (a) $\begin{bmatrix} 5 & 2 & 1 \\ -1 & 4 & 3 \\ 2 & 4 & 5 \end{bmatrix}$ (b) $\begin{bmatrix} 10 & 2 & 3 \\ 0 & 20 & 4 \\ 3 & 4 & 18 \end{bmatrix}$

   (c) $\begin{bmatrix} 10 & 2 & 3 \\ 2 & 20 & 4 \\ 3 & 4 & 18 \end{bmatrix}$ (d) $\begin{bmatrix} 2 & -1 & 1 \\ -2 & 3 & 1 \\ 1 & -1 & 2 \end{bmatrix}$

8. Suppose that the eigenvalues of A satisfy

   $$|\lambda_1| > |\lambda_2| > |\lambda_3| > \cdots > |\lambda_n|$$

   with corresponding eigenvectors $\mathbf{y}_1, \ldots, \mathbf{y}_n$, respectively.
   (a) If the initial vector $\mathbf{x}_0$ in the power method is a linear combination of the $\mathbf{y}_i$'s for $i \geq 2$ with no zero coefficients, then prove that the iterates converge to a multiple of $\mathbf{y}_2$.
   (b) If the initial vector $\mathbf{x}_0$ in the power method is a linear combination of the $\mathbf{y}_i$'s for $i \geq 1$ with no zero coefficients, and if A is now replaced by $(A - \lambda_1 I)$, then

* This problem should be solved by using one of the programs noted in Appendix B.

prove that the iterates converge to a multiple of $\mathbf{y}_i$, where i is chosen so that $(\lambda_i - \lambda_1)$ is the largest in magnitude among $(\lambda_k - \lambda_1)$, $k = 1, \ldots, n$.

9. Suppose that λ is an eigenvalue of multiplicity 2 of a matrix A. Suppose also that A is diagonalizable and that λ is a dominant eigenvalue. Prove that for a typical initial vector, the iterates of the power method will converge to λ and to an eigenvector corresponding to λ.

10. Suppose that the eigenvalues of a 4×4 matrix A are 1, 3, 9, and 14, and that the inverse power method with $\alpha = 7$ is applied.

To what eigenvalue of A will the iterates of the power method converge?

11. Suppose that A is a symmetric matrix and that for some nonzero vector $\mathbf{x}$, q denotes the Rayleigh quotient, that is,

$$q = \frac{A\mathbf{x} \cdot \mathbf{x}}{\mathbf{x} \cdot \mathbf{x}}$$

Let c and d denote the smallest and largest eigenvalues of A. Prove that $c \le q \le d$. (*Hint:* Express $\mathbf{x}$ as a linear combination of the vectors in an orthonormal basis of eigenvectors.)

Key Words

An Introduction to Linear Programming

What follows is a brief introduction to linear programming and the simplex method. For greater depth, the reader is advised to consult the References.

10.1 LINEAR PROGRAMMING PROBLEMS

Systems of linear inequalities have attained a high degree of importance since World War II. A large number of problems can be solved by choosing a solution to such a system which "optimizes" a particular linear expression. To illustrate this type of problem, we consider the following two examples.

Example 1 An independent truck driver can deliver two kinds of desks: a small one which weighs 100 pounds, and a large one which weighs 200 pounds. The driver can deliver up to 30 desks of both types at one time. Furthermore, he can carry up to 5000 pounds of freight at one time. Suppose that he receives $15 for each small desk and $20 for each large desk that he delivers. How many of each desk should he carry to receive the largest income?

Let x be the number of small desks, and let y be the number of large desks that he should deliver in order to maximize his income. We can express the *constraints* given in this example by means of certain linear inequalities involving x and y. For example, since the driver does not deliver a negative number of desks, it follows that $x \geq 0$ and $y \geq 0$. Furthermore, since the total number of desks that can be delivered at one time is at most 30, we have that

$$x + y \leq 30$$

Because each small desk weighs 100 pounds, and each large desk weighs 200 pounds, the total weight of the desks is given by $100x + 200y$. This total cannot exceed the limit of 5000 pounds. Therefore,

$$100x + 200y \leq 5000$$

This inequality can be simplified to

$$x + 2y \leq 50$$

If we let p be the total income due to the driver for delivering the desks, then $p = 15x + 20y$. We wish to maximize p *subject to the constraints* above. We summarize this problem as follows: Maximize the function $p = 15x + 20y$ subject to the constraints

$$x \geq 0, \qquad y \geq 0$$

and

$$x + y \leq 30$$
$$x + 2y \leq 50$$

Example 2   Let us take a second look at Example 2 of Section 1.4. In that example we used linear equations to solve the problem of deciding how to combine eggs, milk, and orange juice in order to obtain a drink with 540 calories of energy and 25 grams of protein. Suppose that we are given the energy and protein content and the costs of each of the foods according to Table 10.1.1. We reformulate the problem: How much of each ingredient should be blended to produce a drink with *at least* 540 calories of energy and at least 25 grams of protein so that the cost is a minimum?

Let x, y, and z denote the quantities of eggs, cups of milk, and cups of orange juice, respectively. As in the first example, x, y, and z are nonnegative. Of the following system of linear inequalities, the first expresses the requirement that the energy content of the drink is at least 540 calories, and the second expresses the requirement that the protein content is at least 25 grams.

$$80x + 160y + 110z \geq 540$$
$$6x + 9y + 2z \geq 25$$

Let p denote the cost of a drink consisting of x eggs, y cups of milk, and z cups of orange juice. Then $p = 9x + 18y + 12z$. We wish to find nonnegative values of x, y, and z which satisfy the linear inequalities above so that p is minimized.

TABLE 10.1.1

	1 Egg	1 Cup Milk	1 Cup Orange Juice
food energy (calories)	80	160	110
protein (grams)	6	9	2
cost (cents)	9	18	12

In each of the preceding examples, the solution to the problem requires finding the maximum or minimum value of a real-valued linear function on the solution set of a system of linear inequalities.

Definition   A *linear programming problem* is a system of m linear inequalities in n nonnegative unknowns $x_1, x_2, \ldots, x_n$ of the form

$$a_{11}x_1 + a_{12}x_2 + \cdots + a_{1n}x_n \le (\ge) b_1$$
$$a_{21}x_1 + a_{22}x_2 + \cdots + a_{2n}x_n \le (\ge) b_2$$
$$\vdots \qquad \vdots \qquad \qquad \vdots \qquad \qquad \vdots$$
$$a_{m1}x_1 + a_{m2}x_2 + \cdots + a_{mn}x_n \le (\ge) b_m$$

and a linear function of the form

$$p = c_1x_1 + c_2x_2 + \cdots + c_nx_n$$

with the requirement to find a vector $\mathbf{x} = (x_1, x_2, \ldots, x_n)$ which maximizes or minimizes the value of p subject to the given inequalities.

The function p is called the *objective function* of the problem and the inequalities are called the *constraints* of the problem. Any vector $\mathbf{x}$ that satisfies the constraints is called a *feasible solution*. The set of all feasible solutions is called the *feasible region*. A feasible solution that maximizes or minimizes p is called an *optimal solution*.

In general, a linear programming problem can have one, many, or no optimal solutions.

A Graphical Approach

The feasible region of a linear programming problem in n unknowns is a subset of R^n. As such, it can be viewed as a geometrical object. For example, in the cases $n = 2$ and $n = 3$, the region is often a polygon or a polyhedron. Polygons and polyhedra contain certain points called *vertices*. The vertex and its analog in higher dimensions ($n > 3$), the *extreme point*, are used in solving linear programming problems. Since we shall make no formal use of it, we omit the precise definition of "extreme point" [see (10) in the References]. Furthermore, we shall adhere to the more familiar term of "vertex." Polygons and polyhedra are examples of *bounded* sets. A subset S of R^n is **bounded** if there is a positive number r such that $\|\mathbf{x}\| \leq r$ for all $\mathbf{x}$ in S. For example, a subset of R^2 is bounded if it is contained inside a circle.

The following theorem gives a sufficient condition for the existence of an optimal solution to a linear programming problem and describes the location of at least one such solution.

Theorem 10.1.1

Let S be the feasible region of a linear programming problem.

(a) *If S is bounded, then the problem has an optimal solution.*

(b) *In any case, if the linear programming problem has an optimal solution, then at least one such solution occurs at a vertex of S.*

If the feasible region of a linear programming problem lies in the xy-plane, it can be sketched. With the help of Theorem 10.1.1 the sketch can be used to find the optimal solution.

In general, this approach is not very practical because the feasible region of a linear programming problem in n unknowns is a subset of R^n. Such a set can be visualized for $n = 2$, and with some difficulty, for $n = 3$. With more than three unknowns the approach is useless. However, the method does serve as an aid in understanding linear programming problems.

In the next example we apply Theorem 10.1.1 to solve the linear programming problem given in Example 1.

Example 3

Use Theorem 10.1.1 to solve the problem of Example 1.

The feasible region of the linear programming problem of Example 1 is shaded in Figure 10.1.1. This region lies in the first quadrant because x and y are nonnegative.

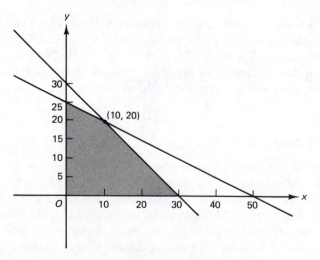

Figure 10.1.1

TABLE 10.1.2

x	0	0	10	30
y	0	25	20	0
p	0	500	550	450

Furthermore, because of the linear inequalities, we have that the feasible region lies below the lines with equations

$$x + y = 30 \quad \text{and} \quad x + 2y = 50$$

including those segments of the lines which act as boundaries. Referring to Figure 10.1.1, we see that the feasible region is a quadrilateral with vertices $(0, 0)$, $(0, 25)$, $(10, 20)$, and $(30, 0)$. Hence it is bounded. By Theorem 10.1.1, an optimal solution exists and occurs at one or more of the vertices of this quadrilateral. We evaluate $p = 15x + 20y$ at these vertices and list the results in Table 10.1.2. Of the four vertices, p has the greatest value at $(10, 20)$. This is the optimal solution for the problem, and the maximum value of p is 550. In the context of Example 1, the truck driver should haul 10 small desks and 20 large desks to receive the maximum income of $550.

Example 4

A dietician wishes to provide a mixture of egg and milk which contains at least 480 calories of food energy, 30 grams of protein, and 1600 international units (IU) of vitamin A. The nutritional contents of eggs and milk are given in Table 10.1.3. Suppose, furthermore, that large eggs cost 7 cents each and milk costs 13 cents per cup. How should the eggs and milk be combined to produce a mixture at a minimum cost?

Let x and y designate the number of large eggs and cups of milk required for the mixture, respectively. Then x and y are nonnegative. Of the following system of linear inequalities, the first one expresses the requirement that the energy content of the mixture is at least 480 calories. The second inequality expresses the requirement that the protein content is at least 30 grams. The third inequality expresses the requirement that the vitamin A content of the mixture is at least 1600 international units.

$$80x + 160y \geq 480$$

$$6x + 9y \geq 30$$

$$590x + 350y \geq 1600$$

TABLE 10.1.3

	Food Energy (Calories)	Protein (Grams)	Vitamin A (IU)
egg (1 large)	80	6	590
milk (1 cup)	160	9	350

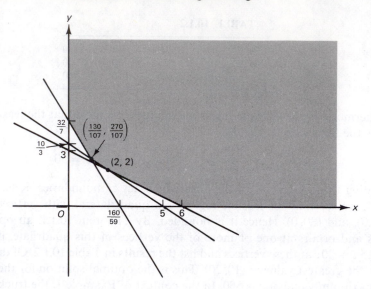

Figure 10.1.2

The inequalities of this system can be simplified to

$$x + 2y \geq 6$$

$$2x + 3y \geq 10$$

$$59x + 35y \geq 160$$

Let p denote the cost of a mixture of x large eggs and y cups of milk. Then

$$p = 7x + 13y$$

We wish to find nonnegative values of x and y that satisfy the linear inequalities so that p is minimized. The feasible region of this problem is the subset of the first quadrant of the xy-plane consisting of all points above the graphs of the three linear equations

$$x + 2y = 6 \qquad 2x + 3y = 10 \qquad 59x + 35y = 160$$

including those segments of the lines which act as boundaries. The region is shaded in Figure 10.1.2. The goal is to find the point in the feasible region (Figure 10.1.2) for which p has the minimum value. Since the feasible region is not bounded, Theorem 10.1.1(a) does not apply. One way to remedy this situation is to choose a large value for p, for example, $p = 100$, and then draw the line $p = 100 = 7x + 13y$ in the first

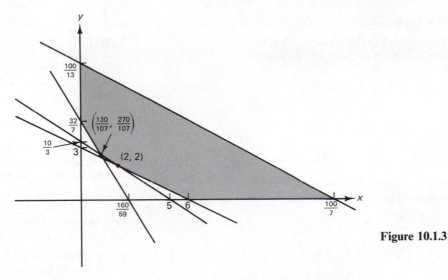

Figure 10.1.3

quadrant of the xy-plane. Observe that every feasible point below this line yields a value of p that is less than 100. Thus, if we adjoin the inequality

$$7x + 13y \leq 100$$

to the previous system of inequalities, it follows that the feasible region of the expanded system is a polygon, and hence bounded (see Figure 10.1.3). Furthermore, the minimum value of p on this polygon coincides with the minimum value of p on the original region, as shown in Figure 10.1.2. Thus the original problem has an optimal solution. By Theorem 10.1.1(b) an optimal solution occurs at a vertex of the original feasible region (see Figure 10.1.2). The vertices of the feasible region as shown in Figure 10.1.2 and the values of p rounded to the nearest tenth at each vertex are given in Table 10.1.4. Clearly, p has a minimum value at $(130/107, 270/107)$, and this value is (rounded to the nearest tenth) $p = 33.7$. Thus, for each portion of the mixture, the dietician should use 130/107 of a large egg and 270/107 of a cup of milk.

TABLE 10.1.4

x	0	6	2	$\frac{130}{107}$
y	$\frac{32}{7}$	0	2	$\frac{270}{107}$
p	59.4	42	40	33.7

Exercises

1. Maximize each of the following objective functions subject to the linear inequalities.

$$2x + y \leq 10$$

$$x + y \leq 8$$

$$x, y \geq 0$$

   (a) $x + 2y$ (b) $3x + 2y$

2. Maximize each of the following objective functions subject to the linear inequalities

$$x + 3y \leq 6$$

$$-x + y \leq 1$$

$$x + y \leq 3$$

$$x, y \geq 0$$

   (a) $4x + y$ (b) $x + 2y$
   (c) $-2x + y$ (d) $x + 4y$

3. Minimize each of the following objective functions subject to the linear inequalities

$$2x + 3y \geq 6$$

$$x + 5y \geq 5$$

$$x, y \geq 0$$

   (a) $x + 3y$ (b) $2x + y$

4. A manufacturer can produce two kinds of birdhouses: the chunky style which requires more wood but less labor, and the slim style which requires less wood but more labor. The costs of wood and labor for each kind of birdhouse are given in Table 10.1.5.

TABLE 10.1.5

	Wood	Labor
chunky	$4	$2
slim	$2	$6

There are $300 worth of wood and $600 worth of labor now available for a production run of birdhouses. Each birdhouse produces a profit of $2.
(a) How many of each kind of birdhouse should be made to maximize the profit?
(b) What is the maximum profit?

5. A distributor of adventure games gives away two kinds of dice to his customers: a transparent one and an opaque one, both of which he makes in his office. The cutting time, inking time, and cost for each kind of die are given in Table 10.1.6.

TABLE 10.1.6

	Cutting Time in Minutes	Inking Time in Minutes	Cost in Cents
transparent die	5	3	20
opaque die	2	6	10

For a certain production run the distributor must hire a cutter and an inker for at least 6 hours and 8 hours, respectively.
(a) How many of each kind of die should be made to minimize his cost?
(b) What is the minimum cost?

10.2 THE SIMPLEX ALGORITHM

The typical linear programming problem consists of many linear inequalities and unknowns. Therefore, it is impossible to approach the problem using the methods employed in Section 10.1. Although Theorem 10.1.1 is still valid, it may be impossible to visualize the graph of a feasible set. Furthermore, because of the immense number of unknowns and inequalities involved, it is desirable to employ a method of solving the problem which can be implemented on a high-speed computer.

There is a computational method for proceeding from one vertex to a better one until an optimal solution is found. Developed in the 1940s by George B. Danzig, a professor at Stanford University, the **simplex algorithm** relies on the use of elementary row operations of a matrix. In its simplest form, which we now describe, the algorithm is applied to certain linear programming problems.

Definition   A linear programming problem is called *standard* if its constraints are given by a system of linear inequalities of the form

$$a_{11}x_1 + a_{12}x_2 + \cdots + a_{1n}x_n \le b_1$$
$$a_{21}x_1 + a_{22}x_2 + \cdots + a_{2n}x_n \le b_2$$
$$\vdots \qquad \vdots \qquad \qquad \vdots \qquad \vdots$$
$$a_{m1}x_1 + a_{m2}x_2 + \cdots + a_{mn}x_n \le b_m$$

with $b_i \ge 0$ for all i and $x_j \ge 0$ for all j, and such that the objective function

$$p = c_1x_1 + c_2x_2 + \cdots + c_nx_n$$

is to be *maximized* subject to these constraints.

The linear programming problems of Examples 1 and 3 of Section 10.1 are standard. However, the linear programming problems of Examples 2 and 4 of Section 10.1 are not standard, for two reasons: all the inequalities are in the wrong direction, and the objective functions are minimized rather than maximized.

Since we are familiar with the methods of solving systems of linear equations, we convert the inequalities above into linear equations. This is done by introducing a new set of nonnegative unknowns, one for each of the m inequalities. Since the left-hand side of each inequality is less than or equal to the right-hand side, for each $i = 1, 2, \ldots, m$, there is a nonnegative number u_i such that

$$a_{i1}x_1 + a_{i2}x_2 + \cdots + a_{in}x_n + u_i = b_i$$

The unknowns $u_1, u_2, \ldots, u_m$ are called *slack variables*. Their introduction provides us with m linear equations in $n + m$ unknowns. Furthermore, all of these unknowns are nonnegative. By adjoining the additional equation

$$-c_1x_1 - c_2x_2 - \cdots - c_nx_n + p = 0$$

we can regard p, the value of the objective function, as an unknown. This gives us a total of $m + 1$ equations in $n + m + 1$ unknowns:

$$a_{11}x_1 + a_{12}x_2 + \cdots + a_{1n}x_n + u_1 \qquad\qquad = b_1$$

$$a_{21}x_1 + a_{22}x_2 + \cdots + a_{2n}x_n \qquad + u_2 \qquad = b_2$$

$$\vdots \qquad \vdots \qquad\qquad \vdots \quad \vdots$$

$$a_{m1}x_1 + a_{m2}x_2 + \cdots + a_{mn}x_n \qquad\qquad + u_m \quad = b_m$$

$$-c_1x_1 - c_2x_2 - \cdots - c_nx_n \qquad\qquad\qquad + p = 0$$

As an example of such a system, consider the linear programming problem of Example 1 of Section 10.1. Since there are two inequalities, there are two slack variables. In place of subscripted notation, we shall call our slack variables u and v. Then the system of linear equations for Example 1 of Section 10.1 is

$$x + \quad y + u \qquad\qquad = 30$$

$$x + 2y + \qquad v \qquad = 50$$

$$-15x - 20y \qquad\qquad + p = \quad 0$$

Given a standard linear programming problem involving m inequalities and n unknowns, we wish to choose a nonnegative solution to this system which maximizes the objective function p. We begin with the entries of the augmented matrix of this system displayed in tabular form with the variables of the system placed above the appropriate columns (see Table 10.2.1). Table 10.2.1 is called the ***initial tableau*** of the linear programming problem. The bottom row of the tableau is called the ***objective row*** since it is associated with the objective function. For example, the initial tableau of the problem of Example 1 of Section 10.1 is represented in Table 10.2.2.

We shall perform elementary row operations on the initial tableau of a standard linear programming problem in order to transform it into other tableaux according to certain rules embodied in the simplex algorithm. Each such tableau will be associated with a "vertex" of the feasible region of the linear programming problem. For example, we shall see that the initial tableau is associated with the zero vector. The simplex algorithm describes a method of going from one tableau to another so that the value of the objective function at the associated vertices increases. Ideally, the process leads to an optimal solution.

We begin our discussion of the algorithm by describing how a vertex is associated with a tableau. Suppose that the linear programming problem has n unknowns $x_1, x_2, \ldots, x_n$ and m inequalities. Then there are m slack variables $u_1, u_2, \ldots, u_m$. In the initial tableau (Table 10.2.1) the columns corresponding to the slack variables are the first m standard vectors of R^{m+1}. As we shall see, for each subsequent tableau, these standard vectors are the columns of m of the $m + n$ variables $x_1, x_2, \ldots,$

TABLE 10.2.1 The Initial Tableau

x_1	x_2	$\cdots$	x_n	u_1	u_2	$\cdots$	u_m	p	
a_{11}	a_{12}	$\cdots$	a_{1n}	1	0	$\cdots$	0	0	b_1
a_{21}	a_{22}	$\cdots$	a_{2n}	0	1	$\cdots$	0	0	b_2
$\vdots$	$\vdots$		$\vdots$	$\vdots$	$\vdots$		$\vdots$	$\vdots$	$\vdots$
a_{m1}	a_{m2}	$\cdots$	a_{mn}	0	0	$\cdots$	1	0	b_m
$-c_1$	$-c_2$	$\cdots$	$-c_n$	0	0	$\cdots$	0	1	0

$x_n, u_1, u_2, \ldots, u_m$. For any tableau, a variable with such a corresponding column is called a **basic** variable. There are m basic variables associated with each tableau. The other n variables are called **nonbasic** variables. Since a tableau is the augmented matrix of a system of linear equations, we can obtain a solution to the system by setting the nonbasic variables equal to zero. The basic variables assume the values of the constants of the system. The resulting $(n + m)$-tuple is called a **basic solution** to the linear programming problem. For example, for the initial tableau of Table 10.2.2, the basic solution is $(0, 0, 30, 50)$. A basic solution is called **feasible** if all of its components are nonnegative. As long as the constants of the system are nonnegative, the basic solution is feasible. In this case, the basic solution can be associated with a vertex of the feasible region of the linear programming problem. This vertex is obtained by forming the vector in R^n consisting of the first n components of the basic solution. In the case of the initial tableau given in Table 10.2.2, this vertex is the origin of R^2. For this reason, row operations performed on a tableau are chosen to avoid the introduction of negative numbers in the rightmost column (the constants of the system) of the tableau.

The other consideration in choosing row operations is to shift to a basic solution which increases the value of the objective function. This is done by eliminating negative entries in the objective (bottom) row of the tableau. Suppose that by a sequence of elementary row operations we obtain the objective row

$$[g_1 \quad g_2 \quad \cdots \quad g_n \quad h_1 \quad \cdots \quad h_m \quad 1 \quad k]$$

TABLE 10.2.2

x	y	u	v	p	
1	1	1	0	0	30
1	2	0	1	0	50
-15	-20	0	0	1	0

which contains no negative entries. Then this row contains the coefficients of the linear equation

$$g_1 x_1 + g_2 x_2 + \cdots + g_n x_n + h_1 u_1 + \cdots + h_m u_m + p = k$$

If this equation is solved for p, we obtain

$$p = k - g_1 x_1 - g_2 x_2 - \cdots - g_n x_n - h_1 u_1 - \cdots - h_m u_m$$

Since the g_i's, h_i's, x_i's, and u_i's are all nonnegative, the maximum value of p is k. The coefficients of the basic variables in this row are zeros and the nonbasic variables are set equal to zero. The result is that the equation above reduces to $p = k$ for the basic solution, and this solution is optimal.

With these considerations in mind, we describe the rules necessary to transform one tableau into the next tableau. As we do this, we shall apply these rules to the tableaux for the linear programming problem given in Example 1 of Section 10.1. The initial tableau for this problem is given in Table 10.2.2.

The first step is to choose the entry in the objective row which contains the negative number of greatest absolute value. In the case of Table 10.2.2 this number is -20, which appears in the second column and is associated with the variable y. We call this column the ***pivot column***. Next, choose one of the first m rows, to be called the ***pivot row***, to eliminate all nonzero entries of the pivot column (except in the pivot row) by means of elementary row operations. The pivot row is chosen so that the resulting row operations guarantee nonnegative entries in the last column. This can be achieved by choosing the row with a positive entry in the pivot column so that the ratio of the entry of the last column to the entry in the pivot column is smallest. In our case the entries of the pivot column are 1 and 2. So both entries are positive. The ratio for the first row is $30/1 = 30$, and the ratio for the second row is $50/2 = 25$. Therefore, we choose the second row as our pivot row. The entry in the pivot column which is also in the pivot row is called the ***pivot entry***. Next, divide the second row by the pivot entry, in this case 2, to obtain

x	y	u	v	p	
1	1	1	0	0	30
$\frac{1}{2}$	1	0	$\frac{1}{2}$	0	25
-15	-20	0	0	1	0

We now perform the elementary row operations necessary to convert the remaining entries of the pivot column to zeros, thus obtaining the second tableau (see Table 10.2.3). Notice that the columns of the second tableau associated with the variables y and u now contain the first two standard vectors of R^3. These variables are the new basic variables. As a consequence of the pivoting process, v loses and y gains the status

TABLE 10.2.3 The Second Tableau

x	y	u	v	p	
$\frac{1}{2}$	0	1	$-\frac{1}{2}$	0	5
$\frac{1}{2}$	1	0	$\frac{1}{2}$	0	25
-5	0	0	10	1	500

of a basic variable. It follows that $(0, 25, 5, 0)$ is the new basic solution, and $(0, 25)$ is the corresponding vertex (see Figure 10.1.1).

Since the objective row contains a negative entry, we must proceed to a third tableau. In this case it is clear that the first column is the pivot column because it contains the only negative entry. Dividing the last entry of each of the top two rows by the corresponding entry of the pivot column, we obtain the ratios 10 and 50. Since the smaller ratio is determined by the first row, this row becomes the pivot row. If we multiply the entries of the pivot row by 2 and use the result to eliminate the nonzero entries of the first column below the first row, then we obtain the third tableau (see Table 10.2.4). As can be seen, x and y are the basic variables and u and v are the nonbasic variables of the third tableau. For this tableau, $(10, 20, 0, 0)$ is the basic solution, and $(10, 20)$ is the associated vertex (see Figure 10.1.1). Since there are no negative entries in the last row, we conclude that the basic solution is optimal. At this vertex, $p = 550$. Notice that 550 is the lower right entry of this tableau. This is the same result as that obtained in Example 3 of Section 10.1.

We now summarize the simplex method for a standard linear programming problem of m inequalities in n unknowns, and objective function p.

1. *Form the initial tableau* as in Table 10.2.1.

2. *Determine the pivot column.* Choose the column with the negative entry in the objective row of greatest absolute value. If there is more than one such entry then choose any one of these. If there are no negative entries, then proceed to step 7. Otherwise, proceed to step 3.

3. *Determine the pivot row.* For each row with a positive entry in the pivot column, divide the last entry of the row by the entry of the pivot column. The row with the smallest ratio is the pivot row. If there is more than one such row, then choose any of these.

TABLE 10.2.4 The Third Tableau

x	y	u	v	p	
1	0	2	-1	0	10
0	1	-1	1	0	20
0	0	10	5	1	550

4. *Divide the pivot row by the pivot entry.* The result is a 1 in the pivot entry.

5. *Convert the other entries of the pivot column to zeros.* Use elementary row operations of type 3 to accomplish this. The result is the next tableau.

6. *Test for the optimal solution.* If there are any negative entries in the objective (last) row, then the optimal solution has not been reached. Return to step 2 using the new tableau. Otherwise, the optimal solution has been reached. Continue to the next step.

7. *Form the basic solution* $(x_1, x_2, \ldots, x_n, u_1, u_2, \ldots, u_m)$ *associated with this tableau.* Set each nonbasic variable equal to zero. Then it follows that each basic variable is equal to the entry in the last column of the row containing the 1 in the column for the variable.

8. *Form the optimal solution.* The vector $\mathbf{x} = (x_1, x_2, \ldots, x_n)$ is an optimal solution of the linear programming problem. The optimal value of the objective function is contained in the lower right corner of the tableau.

We illustrate the procedure with another example.

Example 1   A business school employs both computer and finance instructors. On a weekly basis, the computer instructors are available for a total of 250 hours of instruction and the finance instructors are available for a total of 400 hours of instruction. There are three possible courses that can be offered by the school: Junior Business, Intermediate Business, and Senior Business. Table 10.2.5 lists the number of hours of weekly instruction required for each section of the three courses and the profit that the school makes from each section of the course that it offers. Furthermore, due to space limitations, the school can accommodate a maximum of 30 sections of all courses. How many sections of each course should be offered for the greatest profit?

Let x, y, and z be the number of sections of the Junior, Intermediate, and Senior courses, respectively. By Table 10.2.5, the number of hours required on a weekly basis for computer instruction is given by $5x + 10y + 10z$. But this amount of time is limited to 500. Therefore,

$$5x + 10y + 10z \leq 250$$

TABLE 10.2.5

	Junior Business	Intermediate Business	Senior Business
computer instruction	5	10	10
finance instruction	10	10	30
profit	$200	$300	$100

TABLE 10.2.6 Initial Tableau

x	y	z	u	v	w	p	
1	2	2	1	0	0	0	50
1	1	3	0	1	0	0	40
1	1	1	0	0	1	0	30
−200	−300	−100	0	0	0	1	0

Similarly, since there are at most 400 hours available weekly for finance instruction, we have

$$10x + 10y + 30z \le 400$$

Finally, because the school can accommodate a maximum of 30 sections of all courses, we have

$$x + y + z \le 30$$

We can simplify the first two inequalities to obtain the system

$$x + 2y + 2z \le 50$$
$$x + y + 3z \le 40$$
$$x + y + z \le 30$$

By Table 10.2.5, the profit p is given by

$$p = 200x + 300y + 100z$$

Let u, v, and w be the three slack variables for this linear programming problem. Then with p as the objective function, the initial tableau of the problem is as given in Table 10.2.6. Since -300 is the negative entry of the last row of greatest absolute value, the second column of the tableau is the pivot column. For each row above the objective

TABLE 10.2.7 Tableau 2

x	y	z	u	v	w	p	
$\frac{1}{2}$	1	1	$\frac{1}{2}$	0	0	0	25
$\frac{1}{2}$	0	2	$-\frac{1}{2}$	1	0	0	15
$\frac{1}{2}$	0	0	$-\frac{1}{2}$	0	1	0	5
−50	0	200	150	0	0	1	7500

TABLE 10.2.8 Tableau 3

x	y	z	u	v	w	p	
0	1	1	1	0	−1	0	20
0	0	2	0	1	−1	0	10
1	0	0	−1	0	2	0	10
0	0	200	100	0	100	1	8000

row, we divide the last entry of the row by the entry in the pivot column. The resulting ratios are 25, 40, and 30. Since the first ratio is the smallest, the first row is the pivot row. We divide this row by the pivot entry 2 and eliminate the other nonzero entries in the pivot column to obtain the second tableau, given in Table 10.2.7. Since the only negative entry of the objective row of the new tableau occurs in the first column, this column is the pivot column of tableau 2. If we divide each last entry of the top three rows of tableau 2 by the corresponding entry in the first column, then we obtain the ratios 50, 30, and 10. Since the third ratio is the smallest, the third row of tableau 2 is the pivot row. We carry out steps 4 and 5 of the procedure to obtain the third tableau, given in Table 10.2.8. Because none of the entries of the last row are negative, we can determine the optimal solution from tableau 3. The first three standard vectors of R^4 lie in the columns corresponding to the variables x, y, and v. These are the basic variables. The nonbasic variables z, u, and w are set equal to zero, and we have that $x = 10$, $y = 20$, and $v = 10$. The optimal solution is obtained by ignoring the slack variables. Thus the optimal solution is $(10, 20, 0)$. Furthermore, the maximum value of p is 8000. We conclude that the business school will make the maximum profit of \$8000 by offering 10 sections of the Junior Business course, 20 sections of the Intermediate Business course, and no sections of the Senior Business course.

Exercises

1. Use the simplex method to work Exercise 1 of Section 10.1.

2. Use the simplex method to work Exercise 2 of Section 10.1.

3. Maximize each of the following objective functions subject to the linear inequalities

$$5x + 3y + 2z \le 20$$
$$x + y + z \le 6$$
$$x, y, z \ge 0$$

(a) $x - y + z$ (b) $13x + 8y + z$
(c) $x + 2y + 3z$

4. Maximize each of the following objective functions subject to the linear inequalities

$$2x + y + 2z \le 24$$
$$2x + 3y + z \le 24$$
$$2x + 3y + 3z \le 30$$
$$x, y, z, \ge 0$$

(a) $x + y + z$ (b) $3x + y + z$

5. Maximize each of the following objective functions subject to the linear inequalities

$$x + 2y + 3z + w \le 19$$
$$2x + y + z + w \le 15$$
$$x + y + z + w \le 11$$
$$x, y, z, w \ge 0$$

(a) $x + 3y + z + 2w$ (b) $4x + 4z + 3w$
(c) $3x + y + 2z + w$ (d) $x + 2y + z + 3w$

6. A health food store sells three kinds of packaged dried fruit mixtures: the regular, the iron-rich, and the super. The amounts (in pounds) of each dried fruit in each package of mixture are given in Table 10.2.9.

TABLE 10.2.9 Packaged Dried Fruit Mixtures

Mixture	Prunes	Apples	Peaches
regular	1	1	1
iron-rich	2	0.5	1
super	1	2	2

The profit on each package of regular, iron-rich, and super mix is $1.00, $1.00, and $1.50, respectively. The store currently has a stock of 700 pounds of prunes and 500 pounds each of apples and peaches.
(a) How many packages of each type of mixture should be prepared from the current stock for a maximum profit?
(b) What is the maximum profit?

7. A small furniture company can produce three kinds of bookcases: the compact, the regular, and the deluxe. The two employees available for the construction of these bookcases are a carpenter and a finisher. The carpenter is available for 30 hours per week, and the finisher is available for 40 hours per week. The amounts of time (in hours) required by the carpenter and the finisher for each book case are given in Table 10.2.10.

TABLE 10.2.10

Bookcase	Hours of Carpentry	Hours of Finishing
compact	1	2
regular	2	1
deluxe	2	2

The company makes a profit of $20 each for a compact or a regular bookcase and $30 each for a deluxe bookcase. Assume that the company can sell all of the bookcases that it produces.
(a) How many of each bookcase should be produced each week for a maximum profit?
(b) What is the maximum profit?

Key Words

References

Linear Algebra (Advanced)

1. Friedberg, S., A. Insel, and L. Spence, *Linear Algebra*. Englewood Cliffs, N.J.: Prentice-Hall, Inc., 1979.
2. Hoffman K., and R. Kunze, *Linear Algebra* (2nd ed.). Englewood Cliffs, N.J.: Prentice-Hall, Inc., 1971.
3. Noble B., and J. Daniel, *Applied Linear Algebra* (2nd ed.). Englewood Cliffs, N.J.: Prentice-Hall, Inc., 1977.

Numerical Linear Algebra

4. Burden, R., J. Faires, and A. Reynolds, *Numerical Analysis* (2nd ed.). Boston, Mass. Prindle, Weber, and Schmidt, 1981.
5. Johnson, L., and R. Riess, *Numerical Analysis* (2nd ed.). Reading, Mass.: Addison-Wesley, 1982.
6. McWorter, W., Jr., An Algorithm for the Characteristic Polynomial. *Mathematics Magazine, 56* (1983) 168–175.
7. G. Stewart, *Introduction to Matrix Computation*. New York: Academic Press, 1973.
8. Strang, G., *Linear Algebra and Its Applications* (2nd ed.). New York: Academic Press, 1980.

Linear Programming

9. Adams, W., A. Gewirtz, and L. Quintas, *Elements of Linear Programming*. New York: Van Nostrand Reinhold Co., 1969.
10. Smythe, W. Jr., and L. Johnson, *Introduction to Linear Programming with Applications*. Englewood Cliffs, N.J.: Prentice-Hall, Inc., 1966.

Computer Packages

11. Garbow, B., et al., *EISPACK, Lecture Notes in Computer Science*, Vol. 51. New York: Springer Verlag, 1971.

12. Dongarra, J. Bunch, and C. Stewart, *LINPACK Users' Guide*. Philadelphia: SIAM, 1979.

Appendix A

COMPLEX NUMBERS

Throughout this text the word *scalar* has been used almost interchangeably with *real number*. However, many of the results that we have established may be reformulated to hold when scalars are allowed to be *complex numbers*.

Definition A *complex number* z is an expression of the form

$$z = a + bi$$

where a and b are real numbers. The real number a is called the ***real part*** of z, and the real number b is called the ***imaginary part*** of z. Two complex numbers are ***equal*** if their real and imaginary parts are equal. The set of all complex numbers is denoted by C.

For example, $z = 3 + (-2)i = 3 - 2i$ is a complex number. The real part of z is 3 and the imaginary part of z is -2. If the imaginary part of a complex number z is zero, we identify z with the real part; that is, $z = a + 0i = a$. In this way R may be considered as a subset of C.

The arithmetic operations on R can be extended to C. We define the ***sum*** and ***product*** of $z = a + bi$ and $w = c + di$ as follows:

$$z + w = (a + c) + (b + d)i$$

and

$$zw = (ac - bd) + (bc + ad)i$$

From this definition, we see that

$$i^2 = (i)(i) = (0 + 1i)(0 + 1i) = -1$$

These rules are easy to remember because they are carried out just as if $a + bi$ were an algebraic expression with $i^2 = -1$.

Example 1   The product of $2 + 3i$ and $4 - 5i$ is

$$(2 + 3i)(4 - 5i) = 8 + 2i - 15i^2$$
$$= 8 + 2i - 15(-1)$$
$$= 23 + 2i$$

It is very convenient to attach a geometric interpretation to complex numbers. We consider z as a vector in a plane with two axes: the ***real axis*** and the ***imaginary axis***. With this interpretation, a sum of complex numbers corresponds to the sum of vectors (see Figure A.1). The ***absolute value*** (or ***modulus***) of z, denoted $|z|$, corresponds to the length of a vector, and it is defined as the nonnegative real number

$$|z| = \sqrt{a^2 + b^2}$$

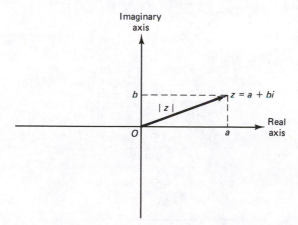

Figure A.1

The following properties of absolute value for complex numbers z and w are easy to verify:

1. $|z| \geq 0$
2. $|zw| = |z||w|$
3. $|z + w| \leq |z| + |w|$

Definition   The (**complex**) **conjugate** of the complex number $z = a + bi$ is denoted $\bar{z}$ and it is defined by

$$\bar{z} = a - bi$$

Following are some useful properties of conjugates for complex numbers z and w:

1. $\overline{(z + w)} = \bar{z} + \bar{w}$.
2. $z + \bar{z}$ is real.
3. $z\bar{z} = |z|^2$
4. $\overline{zw} = \bar{z}\bar{w}$.

It is now easy to compute the *quotient* of two complex numbers. For example, suppose that $z = a + bi$ and $w = c + di \neq 0$. Then

$$\frac{z}{w} = \frac{a + bi}{c + di} \cdot \frac{c - di}{c - di}$$

$$= \frac{(ac + bd) + (bc - ad)i}{c^2 + d^2}$$

$$= \frac{ac + bd}{c^2 + d^2} + \frac{bc - ad}{c^2 + d^2} i$$

Example 2   Compute the quotient $(1 + 4i)/(3 - 2i)$.

$$\frac{1 + 4i}{3 - 2i} = \frac{1 + 4i}{3 - 2i} \cdot \frac{3 + 2i}{3 + 2i}$$

$$= \frac{-5 + 14i}{9 + 4}$$

$$= \frac{-5}{13} + \frac{14}{13} i$$

Our last definition is for the expression e^z, where e is the number whose natural logarithm is 1, and $z = a + bi$. We define

$$e^z = e^{a+bi} = e^a(\cos b + i \sin b)$$

Although this definition may seem strange at first, it can be shown that all the laws regarding exponents are preserved. For example,

$$e^{z+w} = e^z e^w$$

Notice that if z is real, the definition of e^z reduces to the familiar exponential. Our final result is the ***fundamental theorem of algebra***.

Theorem   *Any polynomial may be factored into a product of linear factors.*

For example,

$$f(x) = x^3 - 2x^2 + x - 2 = (x - 2)(x + i)(x - i)$$

The theorem also holds in the case where the coefficients are complex; however, we will not have need for this more general version.

Appendix B

COMPUTER PROGRAMS

Below is a description of 12 computer programs which are useful for illustrating many of the concepts developed in this text as well as for providing computational tools for working a large number of the exercises. The list is divided into two parts. The first part contains those programs which are developed in Chapters 9 and 10 and the program ROW REDUCTION. The latter program is useful in a variety of ways, for example, in the solution of systems of linear equations, in the determination of linear independence, and in the computation of the determinant of a matrix. The code for these programs is listed at the end of this appendix.

The second part of the list contains those programs which, in addition to the earlier ones, are included on a diskette available from Prentice-Hall, Inc.[1] Although all of the programs are self-explanatory, we provide short descriptions of their functions.

It should be noted that the purpose of these programs is instructive. Professional *canned software* is listed in the References.

Programs Whose Codes Are Listed in This Appendix _____

    1. ROW REDUCTION: The program allows the user to perform the elementary row operations on a matrix.

[1] *Solutions Diskette to Introduction to Linear Algebra with Applications*, S. Friedberg and A. Insel, Prentice-Hall, Englewood Cliffs, N.J., © 1986.

432

2. *LU*-DECOMP: For a given matrix *A*, the program produces a unit lower triangular matrix *L* and an upper triangular matrix *U* such that $A = LU$.

3. CHOLESKY: For a given positive definite matrix *A*, the program produces the Cholesky decomposition of *A*, that is, a lower triangular matrix *L* such that $A = LL^t$.

4. JACOBI: The program produces the Jacobi approximations to the solution of a square system of linear equations.

5. GAUSS–SEIDEL: The program produces the Gauss–Seidel approximations to the solution of a square system of linear equations. For many systems, the program requires fewer iterations for a given accuracy than the program JACOBI.

6. POWER METHOD: The program applies the power method to produce approximations to the dominant eigenvalue and a dominant eigenvector of a matrix.

7. SIMPLEX: The program employs the simplex method to solve a standard linear programming problem.

Additional Programs That Are Given on the Diskette

8. MATRIX: The program accepts and stores matrices and then computes requested arithmetic expressions including determinants. For example, if the matrices *A*, *B*, and *C* are entered, then the program can compute an expression such as

$$D = (2A + B^t)^{-1}C^5 \text{ and } \det (A + BC)$$

9. LINSYST: The program solves a system of linear equations and presents the solution in vector form:

$$\mathbf{x} = \mathbf{x}_0 + t_1\mathbf{x}_1 + \cdots + t_k\mathbf{x}_k$$

10. CHAR POLYN: The program computes the characteristic polynomial of a square matrix. In most cases the characteristic polynomial is presented in partially factored form. The program is based on an algorithm developed in (6) in the References.

11. GRAM–SCHMIDT: The program applies the Gram–Schmidt process to a set of vectors to produce an orthogonal set of vectors.

12. ROTATION: The program will accept a (three-dimensional) polyhedral figure and will rotate it about any specified coordinate axes by any specified angles.

Row Reduction _____

```
100   REM     **** ROW REDUCTION
105   REM     THIS PROGRAM ALLOWS THE USER TO PERFORM THE ELEMENTARY ROW
110   REM     OPERATIONS ON A GIVEN MATRIX.
115   ONERR  GOTO 365
120   HOME : CLEAR
125   REM   S IS THE NO. OF PLACES AFTER THE DECIMAL
130   S = 3
135   S1 = 10 ^ S
140   VTAB 3: HTAB 15: PRINT "ROW REDUCTION"
145   VTAB 6: PRINT "ENTER THE DIMENSIONS OF THE MATRIX"
150   PRINT "IN THE FORM M,N."
155   PRINT : INPUT M,N
160   DIM A(M,N): DIM R(M,N)
165   PRINT : PRINT "TYPE THE REQUESTED ENTRIES."
170   PRINT
175   FOR I = 1 TO M
180   FOR J = 1 TO N
185   PRINT "A(";I",",";J;") = ";: INPUT A(I,J)
190   NEXT J: PRINT : NEXT I
195   GOTO 365
200   PRINT : PRINT "IS THIS CORRECT? (Y/N)"
205   PRINT : INPUT Z$
210   IF Z$ = "N" GOTO 120
215   GOTO 365
220   PRINT : PRINT "WHICH OPERATION IS DESIRED?"
225   PRINT : PRINT "  1 : INTERCHANGE ROWS I AND J"
230   PRINT "  2 : MULTIPLY ROW I BY C"
235   PRINT "  3 : MULTIPLY ROW I BY 1/C"
240   PRINT "  4 : ADD A/B TIMES ROW I TO ROW J"
245   PRINT "  5 : QUIT"
250   PRINT : INPUT X
255   ON X GOTO 265,295,295,330,430
260   GOTO 365
265   PRINT : PRINT "ENTER I,J."
270   PRINT : INPUT I,J
275   FOR K = 1 TO N
280   C = A(I,K):A(I,K) = A(J,K):A(J,K) = C
285   NEXT K
290   GOTO 365
295   PRINT : PRINT "ENTER I,C."
300   PRINT : INPUT I,C
305   IF X = 3 THEN C = 1 / C
310   FOR K = 1 TO N
315   A(I,K) = C * A(I,K)
320   NEXT K
325   GOTO 365
330   PRINT : PRINT "ENTER A,B,I,J."
335   PRINT : INPUT A,B,I,J
340   C = A / B
345   FOR K = 1 TO N
350   A(J,K) = A(J,K) + C * A(I,K)
355   NEXT K
360   REM    PRINT ROUTINE
365   HOME :L = 0
370   FOR I = 1 TO M
375   FOR J = 1 TO N
```

```
380 R(I,J) =   INT (S1 * A(I,J) + .5) / S1
385 LS =  LEN ( STR$ (R(I,J)))
390  IF L < LS THEN L = LS
395  NEXT J: NEXT I:L = L + 2:T = 21 - N * L / 2
400  FOR I = 1 TO M
405  FOR J = 1 TO N
410  PRINT  TAB( T + L * (J - 1))R(I,J);
415  NEXT J: PRINT : PRINT : NEXT I
420  IF Z$ = "" GOTO 200
425  GOTO 220
430  END
```

LU-Decomp

```
100  REM    **** LU-DECOMPOSITION
105  REM    THIS PROGRAM COMPUTES THE UNIT LOWER TRIANGULAR MATRIX
110  REM     AND THE UPPER TRIANGULAR MATRIX IN THE LU-DECOMPOSITION
115  REM  OF AN INVERIBLE MATRIX A.
120  HOME : CLEAR
125  REM     IT IS ASSUMED THAT A MAY BE REDUCED TO TRIANGULAR FORM
130  REM  WITHOUT ANY ROW INTERCHANGES.
135  REM  S IS THE NO. OF PLACES AFTER THE DECIMAL
140  S = 3
145  S1 = 10 ^ S
150  PRINT : PRINT : PRINT "LU-DECOMPOSITION"
155  PRINT : PRINT "WHAT IS THE ORDER OF THE MATRIX A?"
160  PRINT : INPUT N
165  DIM A(N,N,1),R(N,N)
170  PRINT : PRINT "TYPE THE REQUESTED ENTRIES"
175  PRINT
180  FOR I = 1 TO N
185  FOR J = 1 TO N
190  PRINT "A(";I;",";J;") = ";: INPUT A(I,J,0)
195  NEXT J: PRINT : NEXT I
200  HOME :Z = 0: GOSUB 320
205  PRINT : PRINT "IS THIS CORRECT? (Y/N)"
210  PRINT : INPUT Q$
215  IF Q$ = "N" GOTO 120
220  FOR J = 1 TO N
225  FOR I = 1 TO J
230 A(I,J,1) = A(I,J,0)
235  IF I = J THEN  GOTO 255
240  FOR K = I TO N
245 A(K,J,0) = A(K,J,0) - A(I,J,1) * A(K,I,0)
250  NEXT K
255  NEXT I
260  IF A(J,J,1) = 0 THEN  GOTO 380
265  FOR I = J TO N
270 A(I,J,0) = A(I,J,0) / A(J,J,1)
275  NEXT I: NEXT J
280  PRINT : PRINT "PRESS ANY KEY TO SEE L.": GET C$
285  HOME : PRINT : PRINT
290  PRINT "    THE MATRIX L IS:"
295  GOSUB 320
300  PRINT : PRINT "PRESS ANY KEY TO SEE U.":Z = 1: GET C$
305  PRINT : PRINT : PRINT "THE MATRIX U IS:"
310  GOSUB 320: END
```

```
315   REM     PRINT ROUTINE
320   PRINT : PRINT :L = 0
325   FOR I = 1 TO N
330   FOR J = 1 TO N
335   R(I,J) =   INT (S1 * A(I,J,Z) + .5) / S1
340   LS =  LEN ( STR$ (R(I,J)))
345   IF L < LS THEN L = LS
350   NEXT J: NEXT I:L = L + 2
355   T = 21 - N * L / 2
360   FOR I = 1 TO N: FOR J = 1 TO N
365   PRINT  TAB( T + L * (J - 1))R(I,J);
370   NEXT J: PRINT : PRINT : NEXT I
375   RETURN
380   HOME
385   PRINT : PRINT "FACTORIZATION IS IMPOSSIBLE."
```

Cholesky

```
100   REM     **** CHOLESKY DECOMPOSITION
105   REM    FOR A POSITIVE DEFINITE MATRIX A, A LOWER TRIANGULAR
110   REM   MATRIX L IS COMPUTED SUCH THAT A = LL-TRANSPOSE.
115   HOME : CLEAR
120   REM  S IS THE NO. OF PLACES AFTER THE DECIMAL
125   S = 3
130   S1 = 10 ^ S
135   PRINT : PRINT : PRINT "CHOLESKY DECOMPOSITION"
140   PRINT : PRINT "WHAT IS THE ORDER OF THE MATRIX A ?"
145   PRINT : INPUT N
150   DIM  A(N,N),R(N,N)
155   PRINT : PRINT "TYPE THE REQUESTED ENTRIES": PRINT
160   FOR I = 1 TO N
165   FOR J = 1 TO N
170   PRINT "A(";I;",";J;") = ";: INPUT A(I,J)
175   NEXT J: PRINT : NEXT I
180   HOME : GOSUB 280
185   PRINT : PRINT "IS THIS CORRECT? (Y/N)"
190   PRINT : INPUT Q$
195   IF Q$ = "N" GOTO 115
200   PRINT : PRINT "PRESS ANY KEY TO SEE THE MATRIX L"
205   PRINT "SUCH THAT A = L'L": GET X$
210   FOR J = 1 TO N
215   IF J = 1 THEN   GOTO 240
220   FOR I = 1 TO J - 1
225   FOR K = I TO N
230   A(K,J) = A(K,J) - A(J,I) * A(K,I)
235   NEXT K: NEXT I
240   IF A(J,J) <  = 0 THEN   GOTO 340
245   S =  SQR (A(J,J))
250   FOR I = J TO N
255   A(I,J) = A(I,J) / S
260   NEXT I: NEXT J: HOME
265   PRINT "    THE MATRIX L IS:"
270   PRINT : GOSUB 280: END
275   REM     PRINT ROUTINE
280   L = 0
285   FOR I = 1 TO N
290   FOR J = 1 TO N
```

```
295 R(I,J) =  INT ((S1 * A(I,J)) + .5) / S1
300 LS =  LEN ( STR$ (R(I,J)))
305  IF L < LS THEN L = LS
310  NEXT J: NEXT I:L = L + 2
315 T = 21 - N * L / 2
320  FOR I = 1 TO N: FOR J = 1 TO N
325  PRINT  TAB( T + L * (J - 1))R(I,J);
330  NEXT J: PRINT : PRINT : NEXT I
335  RETURN
340  HOME
345  PRINT : PRINT "FACTORIZATION IS IMPOSSIBLE."
350  END
```

Jacobi

```
100  REM    **** JACOBI METHOD
105  REM    THE METHOD PRODUCES ITERATIONS TO SOLVE THE SYSTEM AX=B,
110  REM    WHERE A IS A MATRIX OF ORDER N WITH NONZERO DIAGONAL ENTRIES
115 HOME : CLEAR
120  PRINT "JACOBI METHOD FOR SOLVING AX=B"
125  PRINT "BY ITERATIONS."
130  PRINT : PRINT "WHAT IS THE ORDER OF A?"
135  PRINT : INPUT N
140  DIM A(N,N),B(N),X(N,1)
145  PRINT : PRINT "TYPE THE REQUESTED ENTRIES."
150  PRINT :L = 0
155  FOR I = 1 TO N: FOR J = 1 TO N
160  PRINT "A(";I;",";J;") = ";: INPUT A(I,J)
165  IF I = J AND A(I,J) = 0 THEN  GOTO 345
170 LS =  LEN ( STR$ (A(I,J)))
175  IF L < LS THEN L = LS
180  NEXT J: PRINT : NEXT I
185  HOME :L = L + 2
190 T = 21 - N * L / 2
195  FOR I = 1 TO N: FOR J = 1 TO N
200  PRINT  TAB( T + L * (J - 1))A(I,J);
205  NEXT J
210  PRINT : PRINT : NEXT I
215  PRINT : PRINT "IS THIS CORRECT? (Y/N)"
220  PRINT : INPUT Q$
225  IF Q$ = "N" GOTO 115
230  HOME : PRINT "TYPE THE ENTRIES OF B"
235  PRINT
240  FOR I = 1 TO N
245  PRINT "B(";I;") = ";: INPUT B(I)
250  NEXT I
255 L = 0:P = 0
260 L = L + 1:P = 1 - P
265  FOR I = 1 TO N
270 S = B(I)
275  FOR J = 1 TO N
280  IF I < > J THEN S = S - A(I,J) * X(J,1 - P)
285  NEXT J
290 X(I,P) = S / A(I,I)
295  NEXT I
300  HOME : PRINT
305  PRINT "ITERATION ";L: PRINT
```

```
310   FOR J = 1 TO N
315   PRINT "X(";J;") = ";X(J,P)
320   NEXT J
325   PRINT : PRINT "PRESS Q TO QUIT AND ANY OTHER KEY TO SEE THE NEXT ITERATION"
330   GET Z$
335   IF Z$ = "Q" THEN  END
340   GOTO 260
345   HOME
350   PRINT "THE JACOBI METHOD DOES NOT ALLOW A"
355   PRINT "ZERO DIAGONAL ENTRY."
```

Gauss-Seidel

```
100   REM      **** GAUSS-SEIDEL METHOD
105   REM   THE METHOD PRODUCES ITERATIONS TO SOLVE THE SYSTEM AX=B,
110   REM     WHERE A IS A MATRIX OF ORDER N WITH NONZERO DIAGONAL ENTRIES
115   HOME : CLEAR
120   PRINT "GAUSS-SEIDEL METHOD FOR SOLVING AX=B"
125   PRINT "BY ITERATIONS."
130   PRINT : PRINT "WHAT IS THE ORDER OF A?"
135   PRINT : INPUT N
140   DIM A(N,N),B(N),X(N)
145   PRINT : PRINT "TYPE THE REQUESTED ENTRIES."
150   PRINT :L = 0
155   FOR I = 1 TO N: FOR J = 1 TO N
160   PRINT "A(";I;",";J;") = ";: INPUT A(I,J)
165   IF I = J AND A(I,J) = 0 THEN  GOTO 345
170 LS =  LEN ( STR$ (A(I,J)))
175   IF L < LS THEN L = LS
180   NEXT J: PRINT : NEXT I
185   HOME :L = L + 2
190 T = 21 - N * L / 2
195   FOR I = 1 TO N: FOR J = 1 TO N
200   PRINT  TAB( T + L * (J - 1))A(I,J);
205   NEXT J
210   PRINT : PRINT : NEXT I
215   PRINT : PRINT "IS THIS CORRECT? (Y/N)"
220   PRINT : INPUT Q$
225   IF Q$ = "N" GOTO 115
230   HOME : PRINT "TYPE THE ENTRIES OF B"
235   PRINT
240   FOR I = 1 TO N
245   PRINT "B(";I;") = ";: INPUT B(I)
250   NEXT I
255 L = 0
260 L = L + 1
265   FOR I = 1 TO N
270 S = B(I)
275   FOR J = 1 TO N
280   IF I <  > J THEN S = S - A(I,J) * X(J)
285   NEXT J
290 X(I) = S / A(I,I)
295   NEXT I
300   HOME : PRINT
305   PRINT "ITERATION ";L: PRINT
310   FOR J = 1 TO N
315   PRINT "X(";J;") = ";X(J)
320   NEXT J
```

```
325   PRINT : PRINT "PRESS Q TO QUIT AND ANY OTHER KEY TO SEE THE NEXT ITERATION"
330   GET Z$
335   IF Z$ = "Q" THEN   END
340   GOTO 260
345   HOME
350   PRINT "THE GAUSS-SEIDEL DOES NOT ALLOW A"
355   PRINT "ZERO DIAGONAL ENTRY."
```

Power Method

```
100   REM    **** POWER METHOD
105   REM    THE METHOD PRODUCES APPROXIMATIONS TO THE
110   REM    DOMINANT EIGENVALUE AND DOMINANT EIGENVECTOR.
115   HOME : CLEAR
120   PRINT : PRINT "THE POWER METHOD"
125   PRINT : PRINT "M(K) REPRESENTS THE KTH APPROXIMATION"
130   PRINT "FOR THE EIGENVALUE OF LARGEST MAGNITUDE"
135   PRINT : PRINT "X(I) REPRESENTS THE ITH COORDINATE OF"
140   PRINT "THE VECTOR WHICH APPROXIMATES THE"
145   PRINT "CORRESPONDING EIGENVECTOR"
150   PRINT
155   PRINT : PRINT "WHAT IS THE ORDER OF THE MATRIX?"
160   PRINT : INPUT N
165   HOME
170   PRINT : PRINT "TYPE THE REQUESTED ENTRIES."
175   PRINT
180   DIM A(N,N),X(N),Y(N)
185 L = 0
190   FOR I = 1 TO N
195   FOR J = 1 TO N
200   PRINT "A(";I;",";J;") = ";: INPUT A(I,J)
205 LS =  LEN ( STR$ (A(I,J)))
210   IF L < LS THEN L = LS
215   NEXT J: PRINT : NEXT I
220   HOME :L = L + 2
225 T = 21 - N * L / 2
230   FOR I = 1 TO N
235   FOR J = 1 TO N
240   PRINT  TAB( T + L * (J - 1))A(I,J);
245   NEXT J: PRINT : PRINT : NEXT I
250   PRINT : PRINT "IS THIS CORRECT? (Y/N)"
255   PRINT : INPUT Q$
260   HOME
265   IF Q$ = "N" GOTO 115
270   REM   CHOOSE THE INITIAL VECTOR
275 E = 1
280   FOR I = 1 TO N
285 X(I) = 0
290   NEXT I
295 X(E) = 1
300   REM   BEGIN JTH APPROXIMATION
305 J = 1
310   FOR I = 1 TO N
315 Y(I) = 0
320   FOR K = 1 TO N
325 Y(I) = Y(I) + A(I,K) * X(K)
330   NEXT K
335   NEXT I
```

```
340    REM    SELECT THE COMPONENT OF Y OF LARGEST MAGNITUDE
345  H = 0:P = 0
350    FOR I = 1 TO N
355    IF  ABS (Y(I)) > P THEN P =  ABS (Y(I)):H = I
360    NEXT I
365    IF P = 0 THEN  GOTO 490
370    REM  SCALE THE VECTORS
375    FOR I = 1 TO N
380  X(I) = Y(I) / Y(H)
385    NEXT I
390    REM    PRINT THE ITERATE AND THE APPROXIMATION
395    REM  TO THE LARGEST EIGENVALUE
400    HOME
405    PRINT "STEP ";J;" WITH INITIAL VECTOR E(";E;")"
410    PRINT : PRINT "M(";J;") = ";Y(H)
415    PRINT
420    FOR I = 1 TO N
425    PRINT "X(";I;") = ";X(I)
430    NEXT I
435    PRINT
440    IF E = N THEN  GOTO 450
445    PRINT "PRESS I TO CHANGE THE INITIAL VECTOR TO E(";E + 1;")"
450    PRINT
455    PRINT "PRESS Q TO QUIT AND ANY OTHER KEY TO"
460    PRINT "SEE THE NEXT STEP"
465    GET Z$
470    IF Z$ = "Q" THEN  END
475    IF Z$ = "I" GOTO 535
480    PRINT :J = J + 1
485    GOTO 310
490    HOME
495    PRINT "THE CHOSEN INITIAL VECTOR, E(";E;"),
500    PRINT "DOES NOT YIELD A SOLUTION."
505    PRINT : PRINT
510    IF E = N GOTO 550
515    PRINT "PRESS ANY KEY TO RERUN THE PROGRAM"
520    PRINT "WITH THE SAME MATRIX AND WITH THE"
525    PRINT "INITIAL VECTOR E(";E + 1;")"
530    GET Q$
535    HOME :E = E + 1
540    IF E > N THEN  GOTO 550
545    GOTO 280
550    PRINT "THERE ARE NO MORE STANDARD VECTORS."
```

Simplex

```
100    REM    **** SIMPLEX METHOD
105    REM    FOR SOLVING A STANDARD LINEAR PROGRAMMING PROBLEM
110    HOME : CLEAR
115    PRINT "SIMPLEX ALGORITHM FOR SOLVING A STANDARD"
120    PRINT "LINEAR PROGRAMMING PROBLEM"
125    PRINT : PRINT
130    INPUT "INPUT NO OF UNKNOWNS N = ";N
135    PRINT
140    INPUT "INPUT NO OF INEQUALITIES M = ";M
145  P = M + N + 1
150    DIM X(N),A(M + 1,P),L(M)
155    REM  TABLEAU SETUP
```

```
160   HOME
165   PRINT "N = ";N; SPC( 5)"M = ";M
170   PRINT
175   PRINT "INPUT THE FIRST N COLUMNS OF THE INITIAL TABLEAU"
180   PRINT : PRINT :L = 0
185   FOR I = 1 TO M: FOR J = 1 TO N
190   PRINT "A(";I;",";J;: INPUT ") = ";A(I,J)
195   LS = LEN ( STR$ (A(I,J)))
200   IF L < LS THEN L = LS
205   NEXT J: PRINT : NEXT I:L = L + 2
210   T = 21 - N * L / 2
215   HOME : FOR I = 1 TO M: FOR J = 1 TO N
220   PRINT TAB( T + L * (J - 1))A(I,J);: NEXT J: PRINT : PRINT : NEXT I
225   PRINT : PRINT : PRINT "IS THIS CORRECT? (Y/N)"
230   INPUT G$
235   IF G$ = "N" THEN  GOTO 100
240   HOME : PRINT "INPUT THE LAST COLUMN OF"
245   PRINT "THE INITIAL TABLEAU": PRINT : PRINT
250   FOR I = 1 TO M
255   PRINT "B";I;: INPUT " = ";A(I,P)
260   A(I,N + I) = 1: PRINT
265   NEXT I
270   HOME : FOR I = 1 TO M
275   PRINT "B";I;" = ";A(I,P)
280   NEXT I: PRINT : PRINT : PRINT "IS THIS CORRECT? (Y/N)": INPUT G$
285   IF G$ = "N" THEN  GOTO 240
290   HOME : PRINT "INPUT THE ENTRIES OF THE OBJECTIVE ROW"
295   PRINT "OF THE INITIAL TABLEAU": PRINT : PRINT
300   FOR J = 1 TO N
305   PRINT "C";J;: INPUT " = ";A(M + 1,J)
310   PRINT
315   NEXT J
320   HOME : FOR J = 1 TO N: PRINT "C";J;" = ";A(M + 1,J): NEXT J
325   PRINT : PRINT : PRINT "IS THIS CORRECT? (Y/N)": INPUT G$
330   IF G$ = "N" THEN  GOTO 290
335   FOR J = 1 TO N
340   A(M + 1,J) =  - A(M + 1,J)
345   NEXT J
350   REM  FIND PIVOT COLUMN
355   HOME : FOR J = 1 TO N
360   C = 0:CJ = 0
365   FOR J = 1 TO P - 1
370   IF A(M + 1,J) < C THEN C = A(M + 1,J):CJ = J
375   NEXT J
380   IF CJ = 0 THEN  GOTO 505
385   REM  FIND PIVOT ROW
390   RI = 0:RQ =  - 1
395   FOR I = 1 TO M
400   IF A(I,CJ) < = 0 THEN  GOTO 420
405   T = A(I,P) / A(I,CJ)
410   IF RQ =  - 1 THEN RQ = T:RI = I
415   IF T < RQ THEN RQ = T:RI = I
420   NEXT I
425   IF RI = 0 THEN  PRINT "THERE IS NO MAXIMUM VALUE": END
430   REM  DIVIDE ROW BY PIVOT ENTRY
435   D = A(RI,CJ)
440   FOR J = 1 TO P
445   A(RI,J) = A(RI,J) / D
```

```
450   NEXT J
455   REM   DO ROW OPERATION
460   FOR I = 1 TO M + 1
465   IF I = RI THEN  GOTO 495
470 E = A(I,CJ)
475   FOR J = 1 TO P
480   IF A(RI,J) = 0 THEN  GOTO 490
485 A(I,J) = A(I,J) - A(RI,J) * E
490   NEXT J
495   NEXT I
500   GOTO 350
505   REM   OUPUT OPTIMAL SOLUTION
510   HOME
515   PRINT "OPTIMAL SOLUTION": PRINT
520   FOR J = 1 TO N
525 Z = 0
530   FOR I = 1 TO M
535   IF A(I,J) = 0 THEN  GOTO 555
540   IF A(I,J) <  > 1 THEN X(J) = 0:I = M:Z = 0: GOTO 555
545   IF Z <  > 0 THEN X(J) = 0:I = M:Z = 0: GOTO 555
550 Z = I:X(J) = A(I,P)
555   NEXT I
560   IF Z > 0 AND L(Z) = 1 THEN X(J) = 0
565 L(Z) = 1
570   NEXT J
575   FOR I = 1 TO N
580   PRINT "X(";I;") = ";X(I)
585   NEXT I: PRINT
590   PRINT "P = ";A(M + 1,P)
```

Selected Answers for the Exercises

Chapter 1

Section 1.1

1. (a) Yes;   (c) yes;   (e) no
2. (a) $\mathbf{x} = (-25\sqrt{2}, 25\sqrt{2})$;
   (c) $(3, -4)$;   (e) $(-3, 2)$
4. (a) $(50 + 150\sqrt{2}, 150\sqrt{2})$;   (b) 337.21 mph
6. $(4, 3)$
7. $(7, 8)$
12. $Q = (3, 2)$

Section 1.2

1. (a) Yes;   (c) no
2. (a) $(3, 5, 4)$
3. (a) $x = 2 + t, y = -1 + 3t, z = 3 + t$
4. (a) $x = 2 + 2t, y = 3 + 2t, z = 5 - 4t$
5. $(0, 8, 9)$
6. (a) $(100, 150, 300, 100)$
8. $\mathbf{v} = (150, 150\sqrt{3}, 10)$

Section 1.3

1. For the following, the sets are closed under:

	(a)	(c)	(e)
Scalar multiplication	Yes	No	Yes
Vector addition	No	Yes	Yes

2. (a) no, not closed under either operation;
   (c) no, not closed under either operation;
   (e) no, not closed under scalar multiplication;
   (f) yes;
   (h) no, not closed under scalar multiplication
4. (a) Yes;   (c) no;   (e) yes;   (g) yes
5. (a) Yes;   (c) no;   (e) no

Section 1.4

2. (a) $(-1, 2, 3)$;
   (c) $(-2, 2, 0) + s(-8, 3, 1)$;
   (e) $s(0, -\frac{1}{2}, 1, 0) + t(1, -1, 0, 1)$;
   (h) $(-5, 4, 0, 0) + s(-7, 3, 1, 0)$
      $+ t(5, -2, 0, 1)$
4. The numbers are 2, 2, and 6.
5. 2 packs from A, 3 packs from B, and 1 pack from C

Section 1.5

1. (a) Yes; (c) no; (e) no; (g) yes; (i) yes

3. (a) $(2, 3, -1)$; (c) inconsistent;
(e) $(4, 2, 0, 0) + s(-1, 0, 1, 0)$
$+ t(-2, -1, 0, 1)$

5. (a) $c \neq \frac{3}{2}$, none

6. (a) Consistent; (c) inconsistent;
(e) inconsistent; (g) consistent

8. (a) Yes; (c) no; (e) yes

Section 1.6

1. (a) $\sqrt{29}$; (c) 2

2. (a) $\sqrt{5}$; (c) $\sqrt{11}$; (e) $\sqrt{6}$

3. (a) $\mathbf{x} \cdot \mathbf{y} = 1$, $\cos \theta = 1/\sqrt{55}$
(c) $\mathbf{x} \cdot \mathbf{y} = -33$, $\cos \theta = -33/65$
(e) $\mathbf{x} \cdot \mathbf{y} = 6$, $\cos \theta = 3\sqrt{2}/7$

4. (a) $\|\mathbf{x}\| = \sqrt{10}$, $\|\mathbf{y}\| = \sqrt{10}$, $\|\mathbf{x} + \mathbf{y}\| = 2\sqrt{5}$,
$|\mathbf{x} \cdot \mathbf{y}| = 0$
(c) $\|\mathbf{x}\| = \sqrt{6}$, $\|\mathbf{y}\| = 2\sqrt{6}$, $\|\mathbf{x} + \mathbf{y}\| = \sqrt{6}$

10. $S = \{(3, 1, 0), (-2, 0, 1)\}$

15. $x - 3y - z = -4$

20. (a) $(-7, 1, 4)$; (c) $(-7, 1, 1, 8)$

Chapter 2

Section 2.1

1. (a) $\begin{bmatrix} 8 & 2 \\ 4 & 6 \end{bmatrix}$; (b) $\begin{bmatrix} -13 & -10 \\ 1 & 0 \end{bmatrix}$;

(d) $\begin{bmatrix} 26 \\ 8 \end{bmatrix}$; (e) not defined; (h) $\begin{bmatrix} 9 \\ 17 \end{bmatrix}$

2. (b) $\begin{bmatrix} 5 & 0 \\ 25 & 20 \end{bmatrix}$; (c) $\begin{bmatrix} 19 & 2 \\ 7 & 6 \end{bmatrix}$;

(d) $\begin{bmatrix} 29 & 56 & 23 \\ 7 & 8 & 9 \end{bmatrix}$;

(e) not defined; (h) $\begin{bmatrix} -35 & -30 \\ 45 & 10 \end{bmatrix}$

7. (a) 34,900 people in the city and 35,100 people in the suburbs;
(b) 30,718 people in the city and 39,282 people in the suburbs

8. (b) $A_\theta = 1/\sqrt{2} \begin{bmatrix} 1 & -1 \\ 1 & 1 \end{bmatrix}$, $1/\sqrt{2} \begin{bmatrix} 3 \\ -1 \end{bmatrix}$

20. (a) 1 and 3, 1 and 4, 2 and 4, 2 and 5, 3 and 5; (b) 2

22. (a) 1 and 5; (b) 1 and 2, 2 and 4;
(c) 5, 3, 5, 3, and 7

24. (a) $\begin{bmatrix} 72 & -245 \\ 53 & 243 \end{bmatrix}$; (c) $\begin{bmatrix} 17 & -5 \\ 22 & 0 \\ -2 & 21 \\ 38 & -14 \end{bmatrix}$;

(e) $\begin{bmatrix} 33 & 36 & -11 & 64 \\ -2 & 14 & 34 & 3 \end{bmatrix}$

Section 2.2

3. (a) $\left[\begin{array}{cc|cc} 5 & 4 & 6 & -1 \\ \hline 9 & 4 & 6 & -1 \\ -9 & 1 & -13 & 0 \end{array}\right]$;

(b) $\left[\begin{array}{cccc} -2 & 4 & 6 & 0 \\ -1 & 6 & 8 & 2 \\ 11 & -2 & -8 & 10 \\ \hline 3 & 2 & 1 & 4 \end{array}\right]$; (c) $(-4 | 2)$

13. (a) 36,900; (b) 0.626; (c) 0.524, 0.719, 0.818

Section 2.3

1. (a) $\begin{bmatrix} 3 & 2 & 0 \\ 0.6 & 0 & 0 \\ 0 & 0.2 & 0 \end{bmatrix}$; (b) $\begin{bmatrix} 420 \\ 60 \\ 12 \end{bmatrix}$,

$\begin{bmatrix} 15588 \\ 2786 \\ 166 \end{bmatrix}$

2. (a) All;
(b) none from the first; one each from the second

3. $\begin{bmatrix} 0.8 & 0 & 0 \\ 0.2 & 1 & 0.3 \\ 0 & 0 & 0.7 \end{bmatrix} \begin{bmatrix} 0.6 & 0 \\ 0.4 & 0.5 \\ 0 & 0.5 \end{bmatrix} \begin{bmatrix} p \\ q \end{bmatrix}$

4. $\begin{bmatrix} 0.644 & 0.628 \\ 0.356 & 0.372 \end{bmatrix}$

7. (a)

k	Sun	Noble	Honored	Stinkard
1	100	300	500	7700
2	100	400	800	7300
3	100	500	1200	6800

   (b) 10

Section 2.4

1. 3, 2

3. 2, 3

4. (a) $(4, 4)$ and $(2, 1)$;
   (c) $L_{A+B}(1, 0, 1) = (6, 5)$

5. (b) $(1, 5)$ and $(1, -3, 2)$

6. (a) No; (c) yes; (e) yes; (g) yes; (i) no

7. (a) Yes; (c) no; (e) yes

8. (a) $(3, 6)$ and $(20, 12)$

9. (c) $(UT)(x, y) = (x + y, 2x - 2y, 6x)$

10. (a) For 8(a) the standard matrix of T is

$$\begin{bmatrix} 1 & -1 \\ 1 & 0 \end{bmatrix}.$$

11. (a) For 9(c) the standard matrix of T is

$$\begin{bmatrix} 1 & 1 \\ 1 & -1 \\ 2 & 0 \end{bmatrix}.$$

15. (a) $\begin{bmatrix} 0 & 1 \\ 0 & 0 \end{bmatrix}$; (c) $\begin{bmatrix} 1 & 1 & 1 \\ 2 & 0 & 0 \end{bmatrix}$;

   (d) $\begin{bmatrix} 0 & 3 \\ 2 & -1 \\ 1 & 1 \end{bmatrix}$; (g) $\begin{bmatrix} 4 & 0 & 0 \\ 0 & 4 & 0 \\ 0 & 0 & 4 \end{bmatrix}$;

   (h) $(2 \quad -1 \quad 1)$

17. $T(3, 2) = (7, 3, 17)$

32. (a) $\begin{bmatrix} 29 \\ 32 \\ -2 \\ 40 \end{bmatrix}$; (b) $\begin{bmatrix} 67 \\ 88 \\ 9 \\ 121 \end{bmatrix}$

Section 2.5

1.

	(i)	(ii)	(iii)	(iv)
(a)	y-axis	x-axis	No	No
(b)	**0**	R^2	Yes	Yes
(c)	**0**	R^2	Yes	Yes

2. (a) (i) $\{\mathbf{0}\}$, (ii) $\{(2, 1), (-1, 1)\}$;
   (c) (i) $\{(1, 1, -2)\}$, (ii) $\{(1, 0, 2), (1, 0, 0),$ $(1, 0, 1)\}$
   (f) (i) $\{(0, 1, 0), (0, 0, 1)\}$, (ii) $\{1\}$;
   (h) (i) $\{(0, 0, 1)\}$, (ii) $\{(1, 0, 0), (0, 1, 0)\}$

3. (b) $\mathbf{x} = (1, 1, 1)$

Chapter 3

Section 3.1

1. (a) Linearly independent;
   (c) linearly independent;
   (e) linearly dependent;
   (g) linearly dependent

3. (a) $\{(1, 2)\}$; (c) $\{(1, 1, 1), (0, 0, -2)\}$;
   (f) $\{(1, -1, 1), (0, 1, 2), (0, 0, 7)\}$;
   (g) $\{(1, 2, 3, 4), (0, -1, -2, -3), (0, 0, 1, 2)\}$

4. (a) Yes; (c) no; (e) yes; (h) yes

21. (c) $\{(1, 1, 4, 5, 6), (0, 1, -5, -6, -11),$ $(0, 0, -17, -15, -27)\}$

Section 3.2

2. (a) Add $(0, 0, 1)$.
   (b) Add $(1, 0, -1)$.
   (c) Don't add anything.
   (d) Add the standard basis for R^5.
   (e) Add the vectors $(0, 0, 1, 0)$ and $(0, 0, 0, 1)$.

3. (a) (i) 2, (ii) 2; (c) (i) 2, (ii) 3;
   (d) (i) 1, (ii) 2; (g) (i) 2, (ii) 3

4. (a) $\{(3, 0, 1), (0, 3, 1)\}$, 2;
   (c) $\{(-2, 1, 0, 0), (1, 0, 1, 0), (-3, 0, 0, 1)\}$, 3;
   (e) $\{(0, 1, 0, 0, 0), (0, 0, 1, 1, 0), (1, 0, 0, 0, 1)\}$, 3

10. (c) Add the vectors $\mathbf{e}_4$ and $\mathbf{e}_5$.

Section 3.3

1.

	(a)	(c)	(e)
Rank	3	2	2
Nullity	0	2	2

2.

	(a)	(c)	(e)
Rank	1	3	2
Nullity	2	0	1

3. (a) Neither one-to-one nor onto;
(c) one-to-one and onto;
(e) neither one-to-one nor onto

4.

	(a)	(c)	(e)	(g)
Rank	m	n	1	1
Nullity	$n-m$	0	$n-1$	$n-1$

20. (a) Rank $= 2$ and nullity $= 3$.

Section 3.4

1. (a) Yes;   (c) yes;   (e) no;   (g) no
2. (a) $(1, 1)$
3. (a) Yes
5. (a) Yes;   (c) no;   (e) yes
6. (a) $T^{-1}(x, y) = (-4x + 3y, 3x - 2y)$
7. (a) Yes;   (c) no
23. (a) Yes
24. (a) Not invertible

Section 3.5

1. (a) $\begin{bmatrix} 0 & 1 \\ 1 & 0 \end{bmatrix}$;   (b) $\begin{bmatrix} 1 & 0 & 0 \\ 0 & 1 & 0 \\ 3 & 0 & 1 \end{bmatrix}$

2. (a) $\begin{bmatrix} 1 & 2 \\ 0 & 1 \end{bmatrix}$;   (b) $\begin{bmatrix} 0 & 1 & 0 \\ 1 & 0 & 0 \\ 0 & 0 & 1 \end{bmatrix}$

4. (b) $\begin{bmatrix} 4 & -1 & -1 \\ -2 & 1 & 0 \\ 1 & -1 & 1 \end{bmatrix}$;

(f) $\begin{bmatrix} 2 & -1 & 0 \\ -5 & -1 & 2 \\ 2 & 1 & -1 \end{bmatrix}$;

(g) $\begin{bmatrix} 0 & -2 & 1 & 1 \\ 6 & 1 & -3 & -1 \\ 3 & 0 & -1 & -1 \\ -5 & 0 & 2 & 1 \end{bmatrix}$

6. (b) $T^{-1}(x, y, z) = (x, -x + y, x - y + z)$
(d) $T^{-1}(x, y, z, w) = (6y + 3z - 5w,$
$-2x + y, x - 3y - z + 2w,$
$x - y - z + w)$

14. $4 : 3 : 4$

16. 37.25, 62.25, and 22.5 units

19. (a) $T^{-1}(x, y, z) = (1.5x - y - 0.5z,$
$-3.5x + y + 0.5z, 5x - y - z)$

Section 3.6

1. (a) 1;   (b) 2;   (d) 3;   (f) 0
2. (a) Inconsistent;   (c) consistent
7. Basis $= \{(-1.2, 3.4, 1, 0), (-1.2, -0.6, 0, 1)\}$

Chapter 4

Section 4.1

1. (a) $(11, 1)$;   (b) $(0, 5, 6)$
2. (a) $(8, -2)$;   (c) $(19, -4, -3)$

3. $\dfrac{(\sqrt{3}x + y)^2}{36} + \dfrac{(-x + \sqrt{3}y)^2}{16} = 1$

5. (a) $\begin{bmatrix} 2 & -15 \\ -1 & 11 \end{bmatrix}$;   (b) $\begin{bmatrix} -16 & 0 & -21 \\ 3 & 0 & 4 \\ 12 & 1 & 16 \end{bmatrix}$

6. (b) $(-90, 17, 67)$

12. (a) $\begin{bmatrix} 13 \\ -2 \\ -5 \end{bmatrix}$;   (c) $\begin{bmatrix} -46 \\ 12 \\ 18 \end{bmatrix}$;

(e) $\begin{bmatrix} 21 \\ -7 \\ -6 \\ 0 \end{bmatrix}$

Section 4.2

1. (a) $\begin{bmatrix} 1 & 1 \\ 3 & 0 \end{bmatrix}$; (d) $\begin{bmatrix} 10 & 19 & 16 \\ -5 & -8 & -8 \\ 2 & 2 & 3 \end{bmatrix}$

2. (a) (i) $\begin{bmatrix} 0 & 1 \\ 1 & 0 \end{bmatrix}$; (ii) $\begin{bmatrix} 1 & 0 \\ 3 & -1 \end{bmatrix}$

(c) (i) $\begin{bmatrix} 0 & 0 & 1 \\ -1 & 0 & 1 \\ 0 & 1 & 0 \end{bmatrix}$; (ii) $\begin{bmatrix} 0 & 1 & 0 \\ 0 & 1 & -1 \\ 1 & 0 & -1 \end{bmatrix}$

3. (a) $\begin{bmatrix} 1 & -1 \\ 2 & -1 \end{bmatrix}$

4. (a) $\{(-1, 1), (1, 1)\}$; (b) $\begin{bmatrix} 0 & -1 \\ -1 & 0 \end{bmatrix}$;

(c) $T(x, y) = (-y, -x)$

6. (a) $\begin{bmatrix} 1 & 0 \\ 0 & 0 \end{bmatrix}$; (b) $\begin{bmatrix} 0.2 & 0.4 \\ 0.4 & 0.8 \end{bmatrix}$

15. (b) $\begin{bmatrix} 1 & 0 & 0 \\ 0 & 1 & 0 \\ 0 & 0 & 0 \end{bmatrix}$

17. (a) $\begin{bmatrix} -93.5 & -125 & -19.5 & -361 \\ 39.5 & 51 & 4.5 & 150 \\ -30.5 & -41 & -7.5 & -120 \\ 12.5 & 18 & 5.5 & 51 \end{bmatrix}$

18. For 15(c) the standard matrix is

$$\begin{bmatrix} 0.93 & -0.14 & 0.21 \\ -0.14 & 0.71 & 0.43 \\ 0.21 & 0.43 & 0.36 \end{bmatrix}$$

Chapter 5

Section 5.1

1. 18

3. 14

4. (a) 21; (c) -22; (e) -9

8. (a) -4

9. (a) 6; (c) 3

10. (a) 25; (b) 250

13. (a) $-\frac{3}{2}$

15. (a) $c = 2$, $\mathbf{x} = (2, 1)$

16. (a) $x = \frac{14}{13}$, $y = -\frac{8}{13}$

Section 5.2

1. 54

2. -40

3. 160

6. 654

8. 154

10. -170

11. $-\frac{1}{4}$

13. -24

15. -24

17. 18

23. (a) 8; (c) -60

26. (a) $x = -4$, $y = 6$, $z = 2$;
(c) $x = 0$, $y = -1$, $z = 2$

29. (a) $(-4, 5, 6)$

32. (b) 4

Section 5.3

1. (a) 0; (c) 0

4. (a) 9; (c) 3

13. (a) -180; (c) -51

Chapter 6

Section 6.1

1. 0, $\{\mathbf{e}_1, \ldots, \mathbf{e}_n\}$

3. 1, $\{\mathbf{e}_1, \mathbf{e}_2\}$; 0, $\{\mathbf{e}_3\}$

5. 1, $\{(1, 1)\}$; 0, $\{(1, -1)\}$

7. No eigenvalues

8. 0, $\{\mathbf{y}\}$

9. $f(x) = x^2 - 2x - 3$; 3, $\{(1, 2)\}$; -1, $\{(-1, 2)\}$

12. $f(x) = x^2 - 2x + 1$; 1, $\{(1, 1)\}$

16. $f(x) = (x - 2)^2(x - 1)$; 2, $\{(-1, 3, 0)$,
$(1, 0, 3)\}$; 1, $\{(2, 1, 8)\}$

30. (d) (i) $f(x) = (x - 3)^2(x - 2)^2$;
(ii) 3, $\{(1.2857, 1.5714, 1, 0)$,
$(-2.1429, -2.2857, 0, 1)\}$; 2,
$\{(0, 1, 1, 0), (1, -1, 0, 1)\}$

Section 6.2

1. Yes

3. No

5. No

7. $P = \begin{bmatrix} 1 & 1 & 1 \\ 0 & 1 & 2 \\ 0 & 0 & 1 \end{bmatrix}$ and $D = \begin{bmatrix} 1 & 0 & 0 \\ 0 & 2 & 0 \\ 0 & 0 & 3 \end{bmatrix}$

9. $\begin{bmatrix} -12010 & 28394 \\ -14197 & 30581 \end{bmatrix}$

11. $\begin{bmatrix} 13731 & 5306 \\ 7959 & 466 \end{bmatrix}$

13. $\begin{bmatrix} -1931 & 254 & -1093 \\ 2059 & -126 & 1093 \\ 8236 & -508 & 4373 \end{bmatrix}$

15. $A = \begin{bmatrix} 2 & 2 & 0 \\ 0 & 3 & 0 \\ 0 & 1 & 2 \end{bmatrix}$

26. (a) $A = \begin{bmatrix} -1.8 & -2.6 & 5.8 \\ -12 & -4 & 15 \\ -4.8 & -3.6 & 9.8 \end{bmatrix}$

Section 6.3

1. $r(n) = 3(2^n) + 4(-1)^n$; $r(6) = 196$

3. (a) $r(n) = (\frac{2}{5})4^n + (\frac{3}{5})(-1)^n$; $r(6) = 1639$

5. (a) $r(1) = 2$, $r(2) = 7$, $r(3) = 20$;
   (b) $r(n) = 2r(n-1) + 3r(n-2)$;
   (c) $r(n) = (\frac{3}{4})(3^n) + (\frac{1}{4})(-1)^n$

7. $\begin{bmatrix} r(n+2) \\ r(n+1) \\ r(n) \end{bmatrix} = \begin{bmatrix} 4 & -2 & 5 \\ 1 & 0 & 0 \\ 0 & 1 & 0 \end{bmatrix} \begin{bmatrix} r(n+1) \\ r(n) \\ r(n-1) \end{bmatrix}$

9. (b) Not regular; (d) regular

12. (b) $\begin{bmatrix} 0.75 & 0.35 \\ 0.25 & 0.65 \end{bmatrix}$

   (c) 56.52% will be college educated in three generations.

   (d) $\frac{7}{12}$

14. (a) $\begin{bmatrix} 0.40 & 0.20 & 0.20 \\ 0.10 & 0.70 & 0.20 \\ 0.50 & 0.10 & 0.60 \end{bmatrix}$

(b) 25.2%, 33.4%, and 41.4% will be living in the city, suburbs, and country, respectively, after 3 years.

(c) 25%, 34.95%, and 40.05% will be living in the city, suburbs, and country, respectively, after 8 years.

18. $u = 5e^{3t} + 10e^{-t}$, $v = 10e^{3t} - 20e^{-t}$

20. $y = -ae^t - 2be^{2t} - ce^{3t}$, $u = ae^t + be^{2t} + ce^{3t}$, $v = 2ae^t + 4be^{2t} + 4ce^{3t}$

21. (a) $u = 100e^{-2t} + 800e^t$, $v = 100e^{-2t} + 200e^t$
   (b) There will be 4017 foxes and 16,069 rabbits after 3 years. The ratio is approximately 0.24998 after 3 years.
   (c) $\frac{1}{4}$, no

22. $y = a + be^{4t} + ce^{-2t}$

Chapter 7

Section 7.1

1. (a) The coefficients are $5/\sqrt{2}$, $3/\sqrt{2}$.
   (c) The coefficients are $5/\sqrt{2}$, $-3/\sqrt{6}$, $3/\sqrt{3}$.

3. (a) $\{(1, -1), (\frac{5}{2}, \frac{5}{2})\}$;
   (c) $\{(1, 2, 3), (\frac{19}{14}, \frac{10}{14}, -\frac{13}{14})\}$;
   (g) $\{(1, 1, -1, 0), (\frac{1}{3}, \frac{1}{3}, \frac{2}{3}, 1), (-\frac{2}{5}, \frac{3}{5}, \frac{1}{5}, -\frac{3}{5})\}$

6. (a) $(\frac{11}{5})(1, 2)$; (c) $(0, 5, 8)$; (d) $(\frac{1}{3})(1, -1, 1)$

21. $y = \frac{1}{3} + 2x$

23. $y = -6.35 + 2.1x$

25. (a) $y = 50 + 40t - 16t^2$;
   (b) $y = 48.93 + 43.14t - 17.15t^2$

26. $y = 1.42 + 0.49x + 0.38x^2 + 0.73x^3$

Section 7.2

1. $P = \frac{1}{\sqrt{2}} \begin{bmatrix} 1 & -1 \\ 1 & 1 \end{bmatrix}$; $D = \text{diag}(4, 2)$

2. $P = \frac{1}{\sqrt{5}} \begin{bmatrix} 2 & 1 \\ 1 & -2 \end{bmatrix}$; $D = \text{diag}(5, -5)$

5. $P = \frac{1}{3} \begin{bmatrix} -2 & 1 & 2 \\ 2 & 2 & 1 \\ 1 & -2 & 2 \end{bmatrix}$; $D = \text{diag}(0, 3, 6)$

8. $P = \frac{1}{5} \begin{bmatrix} -4 & 0 & 3 \\ 0 & 5 & 0 \\ 3 & 0 & 4 \end{bmatrix}$; $D = \text{diag}(25, -3, -50)$

9. (a) $\dfrac{1}{\sqrt{5}}\begin{bmatrix} 1 & 2 \\ -2 & 1 \end{bmatrix}$

10. (a) $\dfrac{1}{5}\begin{bmatrix} -4 & 0 & 3 \\ 0 & 5 & 0 \\ 3 & 0 & 4 \end{bmatrix}$

11. (a), (b), (d), and (f) are orthogonal transformations; the rest are not.

23. (a) $A = \begin{bmatrix} 2 & 1 \\ 1 & 2 \end{bmatrix}$; (b) $P = \dfrac{1}{\sqrt{2}}\begin{bmatrix} 1 & 1 \\ -1 & 1 \end{bmatrix}$;

   (c) $x'^2 + 3y'^2 = 1$; ellipse

25. (a) No; (c) yes; (e) no

Section 7.3

1. $\begin{bmatrix} \dfrac{\sqrt{3}}{2\sqrt{2}} & -\dfrac{1}{2} & \dfrac{\sqrt{3}}{2\sqrt{2}} \\[2mm] \dfrac{1}{2\sqrt{2}} & \dfrac{\sqrt{3}}{2} & \dfrac{1}{2\sqrt{2}} \\[2mm] -\dfrac{1}{\sqrt{2}} & 0 & \dfrac{1}{\sqrt{2}} \end{bmatrix}$

6. (a) $\begin{bmatrix} 0.8660 & 0 & 0.5000 \\ 0 & 1 & 0 \\ -0.5000 & 0 & 0.8660 \end{bmatrix}$.

   $\begin{bmatrix} 1 & 0 & 0 \\ 0 & 0.7071 & -0.7071 \\ 0 & 0.7071 & 0.7071 \end{bmatrix}$

   $= \begin{bmatrix} 0.8660 & 0.3536 & 0.3536 \\ 0 & 0.7071 & -0.7071 \\ -0.5000 & 0.6124 & 0.6124 \end{bmatrix}$

   (b) $\begin{bmatrix} 1.5731 \\ 0 \\ 0.7247 \end{bmatrix}$

Chapter 8

Section 8.2

1. No
2. Yes

3. No
4. Yes
11. (a) Yes: $x_1 + x_2$; (b) yes: $x_1 + x_2 + x_3$;
    (c) no
12. No: $h = 2f - 3g$
14. (a) Yes; (b) no; (c) yes
17. Let $\{e_1, e_2, \ldots, e_n\}$ and $\{e'_1, e'_2, \ldots, e'_m\}$ be the standard ordered bases for R^n and R^m, respectively. For each i and j, let T_{ij} be the linear transformation from R^n to R^m defined by

$$T_{ij}(e_k) = \begin{cases} \mathbf{0} & k \neq j \\ e'_i & k = j \end{cases}$$

Then $\{T_{ij}\}$ is the required basis.

Section 8.3

1. No
2. Yes
3. Yes
4. No
5. Yes
6. Yes
7. Yes
11. $[A]_S = (1 \quad 3 \quad 4 \quad 2)$
12. $[2 + x - 3x^2]_S = (-3 \quad 1 \quad 2)$

14. (a) $[T]_S = \begin{bmatrix} 1 & 0 & 0 & 0 \\ 0 & 0 & 1 & 0 \\ 0 & 1 & 0 & 0 \\ 0 & 0 & 0 & 1 \end{bmatrix}$

   (b) $[T]_S$ has characteristic polynomial
   $(1 - t)^2(t^2 - 1)$.

15. (a) $(D^2 + D)(1) = D^2(1) + D(1) = 0 + 0 = 0$;
    $(D^2 + D)(e^{-x}) = (e^{-x})'' + (e^{-x})'$
    $= e^{-x} - e^{-x} = 0$
    (b) $(D^2 + D)(e^{ax}) = a^2 e^{ax} + a e^{ax} = (a^2 + a)e^{ax}$

16. (a) $(1 \quad \frac{1}{2} \quad \frac{1}{3})$

Section 8.4

1. Yes
2. No
3. Yes
4. No

10. $\left\{ 1, e^x - e + 1, e^{-x} + e^{-1} - 1 \right.$

$\left. + \dfrac{(e + e^{-1} - 3)(e^x - e + 1)}{(-\frac{1}{2})e^2 + 2e - \frac{3}{2}} \right\}$

12. $(5\sqrt{7}/2\sqrt{2})(x^3 - \frac{3}{5}x)$

Chapter 9

Section 9.1

1. (a) (i) $x = -1, y = 1, z = 2$

(ii) $A = \begin{bmatrix} 1 & 0 & 0 \\ \frac{1}{4} & 1 & 0 \\ \frac{1}{12} & \frac{43}{9} & 1 \end{bmatrix} \begin{bmatrix} 12 & 5 & 8 \\ 0 & \frac{3}{4} & -1 \\ 0 & 0 & \frac{82}{9} \end{bmatrix}$

(d) (i) $x = 2, y = 3, z = -1$

(ii) $A = \begin{bmatrix} 1 & 0 & 0 \\ -2 & 1 & 0 \\ 4 & -3 & 1 \end{bmatrix} \begin{bmatrix} 5 & -2 & 2 \\ 0 & 4 & 2 \\ 0 & 0 & 3 \end{bmatrix}$

2. (a) $x = 2, y = -3, z = 1$;

(b) $x = -4, y = 2, z = 5$

4. (a) $A = \begin{bmatrix} 1 & 0 \\ 3 & 1 \end{bmatrix} \begin{bmatrix} 1 & 4 \\ 0 & 1 \end{bmatrix}$;

(b) $P = \begin{bmatrix} 1 & 0 \\ 0 & 1 \end{bmatrix}$

7. $\begin{bmatrix} 1 & 0 & 0 & 0 \\ 0.25 & 1 & 0 & 0 \\ 0.125 & 1.375 & 1 & 0 \\ 0.188 & -0.688 & 0.763 & 1 \end{bmatrix}$.

$\begin{bmatrix} 16 & 4 & 2 & 5 \\ 0 & 4 & 2.5 & -3.25 \\ 0 & 0 & 8.313 & 7.844 \\ 0 & 0 & 0 & 1.842 \end{bmatrix}$

13. (a) $A = \begin{bmatrix} 1 & 0 & 0 & 0 \\ \frac{3}{2} & 1 & 0 & 0 \\ 0 & 2 & 1 & 0 \\ 0 & 0 & \frac{1}{9} & 1 \end{bmatrix} \begin{bmatrix} 2 & 1 & 0 & 0 \\ 0 & \frac{5}{2} & -2 & 0 \\ 0 & 0 & 9 & 1 \\ 0 & 0 & 0 & \frac{53}{9} \end{bmatrix}$

14. (c) $H^{-1} = \begin{bmatrix} 16 & -120 & 240 & -140 \\ -120 & 1200 & -2700 & 1680 \\ 240 & -2700 & 6480 & -4200 \\ -140 & 1680 & -4200 & 2800 \end{bmatrix}$

Section 9.2

1. (a) (i) $x = (3, 1)$;

(ii) $x_1 = (1.5, 1.75), x_2 = (4.125, 1.375)$;

(iii) $x_1 = (1.5, 1.375)$,

$x_2 = (3.5625, 0.859375)$

(c) (i) $x = (-3, 4)$;

(ii) $x_1 = (-5, 2.5), x_2 = (-3.75, 5)$;

(iii) $x_1 = (-5, 5), x_2 = (-2.5, 3.75)$

3. (a) (i) $N = \begin{bmatrix} 6 & 0 & 0 \\ 0 & 1 & 0 \\ 0 & 0 & 8 \end{bmatrix}$,

$C = \begin{bmatrix} 0 & -3 & 4 \\ 2 & 0 & 0 \\ -2 & -9 & 0 \end{bmatrix}$

(ii) $N = \begin{bmatrix} 6 & 0 & 0 \\ -2 & 1 & 0 \\ 2 & 9 & 8 \end{bmatrix}$,

$C = \begin{bmatrix} 0 & -3 & 4 \\ 0 & 0 & 0 \\ 0 & 0 & 0 \end{bmatrix}$

9. (a) (i) 12,   (ii) 8;   (c) (i) 30,   (ii) 11;

(d) (i) diverges,   (ii) diverges

10. The matrix in 9(e) is "closer" to being diagonally dominant than the matrix in 9(d).

Section 9.3

1. (a) $[1, 5]$ and $[5, 5]$; eigenvalues are 1 and 5.

3. $1, -3,$ and -6.

5. Three are positive.

Section 9.4

1. (a) $A' = \begin{bmatrix} 1 & -1 \\ -1 & 1 \end{bmatrix}, \lambda_2 = 2, y_2 = \begin{bmatrix} -1 \\ 1 \end{bmatrix}$

(c) $A' = \dfrac{1}{5} \begin{bmatrix} 32 & 24 & 0 \\ 24 & 18 & 0 \\ 0 & 0 & -25 \end{bmatrix}$

7. (a) $\lambda = 8.3976, x(18) = (0.6137, 0.5426, 1)$;

(c) $\lambda = 23.97, x(24) = (0.3218, 1, 0.8317)$

Chapter 10

Section 10.1

1. (a) $x = 0$, $y = 8$, $p = 16$
2. (a) $x = 3$, $y = 0$, $p = 12$;   (c) $x = 0$, $y = 1$, $p = 1$
3. (a) $x = \frac{15}{7}$, $y = \frac{4}{7}$, $p = \frac{27}{7}$
5. (a) 50 transparent dice and 55 opaque dice; (b) \$15.50

Section 10.2

3. (a) $x = \frac{8}{3}$, $y = 0$, $z = \frac{10}{3}$, $p = 6$;
   (c) $x = 0$, $y = 0$, $z = 6$, $p = 18$
5. (a) $x = 0$, $y = 8$, $z = 0$, $w = 3$, $p = 30$;
   (c) $x = 5.2$, $y = 0$, $z = 4.6$, $w = 0$, $p = 24.8$
7. (a) 10 compact bookcases, no regular bookcases, and 10 deluxe bookcases.
   (b) \$500.00

Index

LIST OF PROCEDURES